International
REVIEW OF
Neurobiology
Volume 71

International

REVIEW OF

Neurobiology

Volume 71

GABA in Autism and Related Disorders

EDITED BY

DIRK M. DHOSSCHE

Department of Psychiatry and Human Behavior
University of Mississippi Medical Center
Jackson, Mississippi 39216, USA

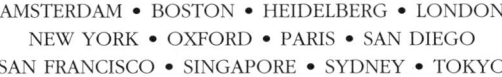

AMSTERDAM • BOSTON • HEIDELBERG • LONDON
NEW YORK • OXFORD • PARIS • SAN DIEGO
SAN FRANCISCO • SINGAPORE • SYDNEY • TOKYO
Academic Press is an imprint of Elsevier

ELSEVIER

Elsevier Academic Press
525 B Street, Suite 1900, San Diego, California 92101-4495, USA
84 Theobald's Road, London WC1X 8RR, UK

This book is printed on acid-free paper.

For all information on all Elsevier Academic Press publications
visit our Web site at www.books.elsevier.com

ISBN-13: 978-0-12-366872-7
ISBN-10: 0-12-366872-7

PRINTED IN THE UNITED STATES OF AMERICA
05 06 07 08 09 9 8 7 6 5 4 3 2 1

Working together to grow
libraries in developing countries

www.elsevier.com | www.bookaid.org | www.sabre.org

ELSEVIER   BOOK AID
            International   Sabre Foundation

CONTENTS

Autism: Neuropathology, Alterations of the GABAergic System, and Animal Models

CHRISTOPH SCHMITZ, IMKE A. J. VAN KOOTEN, PATRICK R. HOF, HERMAN VAN ENGELAND, PAUL H. PATTERSON, HARRY W. M. STEINBUSCH

The Role of GABA in the Early Neuronal Development

MARTA JELITAI AND EMÍLIA MADARASZ

GABAergic Signaling in the Developing Cerebellum

CHITOSHI TAKAYAMA

Insights into GABA Functions in the Developing Cerebellum

MÓNICA L. FISZMAN

Role of GABA in the Mechanism of the Onset of Puberty in Non-Human Primates

EI TERASAWA

Rett Syndrome: A Rosetta Stone for Understanding the Molecular Pathogenesis of Autism

JANINE M. LASALLE, AMBER HOGART, AND KAREN N. THATCHER

GABAergic Cerebellar System in Autism: A Neuropathological and Developmental Perspective

GENE J. BLATT

Reelin Glycoprotein in Autism and Schizophrenia

S. HOSSEIN FATEMI

Is There A Connection Between Autism, Prader-Willi Syndrome, Catatonia, and GABA?

DIRK M. DHOSSCHE, YARU SONG, AND YIMING LIU

Alcohol, GABA Receptors, and Neurodevelopmental Disorders

UJJWAL K. ROUT

Effects of Secretin on Extracellular GABA and Other Amino Acid Concentrations in the Rat Hippocampus

HANS-WILLI CLEMENT, ALEXANDER PSCHIBUL, AND EBERHARD SCHULZ

Predicted Role of Secretin and Oxytocin in the Treatment of Behavioral and Developmental Disorders: Implications for Autism

MARTHA G. WELCH AND DAVID A. RUGGIERO

Immunological Findings in Autism

HARI HAR PARSHAD COHLY AND ASIT PANJA

Correlates of Psychomotor Symptoms in Autism

LAURA STOPPELBEIN, SARA SYTSMA-JORDAN, AND LEILANI GREENING

GABRB3 Gene Deficient Mice: A Potential Model of Autism Spectrum Disorder

TIMOTHY M. DELOREY

The Reeler Mouse: Anatomy of a Mutant

GABRIELLA D'ARCANGELO

Shared Chromosomal Susceptibility Regions Between Autism and Other Mental Disorders

Yvon C. Chagnon

CONTRIBUTORS

Numbers in parentheses indicate the pages on which the authors' contributions begin.

Gene J. Blatt (167), Department of Anatomy and Neurobiology, Boston University School of Medicine, Boston, Massachusetts 02118, USA

Yvon C. Chagnon (419), Genetic and Molecular Psychiatry Unit, Robert-Giffard Research Center, Laval University, Beauport, Québec G1J 2G3, Canada

Hans-Willi Clement (239), Department of Child and Adolescent Psychiatry, University of Freiburg, Freiburg D-79104, Germany

Hari Har Parshad Cohly (317), Department of Biology, Jackson State University, Jackson, Mississippi 39217, USA

Gabriella D'Arcangelo (383), The Cain Foundation Laboratories, Texas Children's Hospital, Departments of Pediatrics, Neuroscience, Programs in Developmental Biology, Translational Biology, and Molecular Medicine, Baylor College of Medicine, Houston, Texas 77030, USA

Timothy M. DeLorey (359), Molecular Research Institute, Mountain View, California 94043, USA

Dirk M. Dhossche (189), Department of Psychiatry & Human Behavior, University of Mississippi Medical Center, Jackson, Mississippi 39216, USA

S. Hossein Fatemi (179), Department of Psychiatry, University of Minnesota Medical School, Minneapolis, Minnesota 55455, USA

Mónica L. Fiszman (95), Instituto de Investigaciones Farmacologicas-CONICET, Ciudad de Buenos Aires 1113, Argentina

Leilani Greening (343), Department of Psychiatry, University of Mississippi Medical Center, Jackson, Mississippi 39216, USA

Patrick R. Hof (1), Department of Neuroscience, Mount Sinai School of Medicine, New York, New York 10029, USA

Amber Hogart (131), Medical Microbiology and Immunology and Rowe Program in Human Genetics, School of Medicine, University of California, Davis, California 95616, USA

Marta Jelitai (27), Laboratory of Neural Cell and Developmental Biology, Institute of Experimental Medicine, Hungarian Academy of Sciences, Budapest 1083, Hungary

Janine M. LaSalle (131), Medical Microbiology and Immunology and Rowe Program in Human Genetics, School of Medicine, University of California, Davis, California 95616, USA

Yiming Liu (189), Department of Chemistry, Jackson State University, Jackson, Mississippi 39217, USA

Emília Madarasz (27), Laboratory of Neural Cell and Developmental Biology, Institute of Experimental Medicine, Hungarian Academy of Sciences, Budapest 1083, Hungary

Asit Panja (317), Department of Medicine, Division of Gastroenterology, University of Medicine and Dentistry of New Jersey-Robert Wood Johnson Medical School, New Brunswick, New Jersey 08903, USA

Paul H. Patterson (1), California Institute of Technology, Division of Biology, Pasadena, California 91125, USA

Alexander Pschibul (239), Department of Child and Adolescent Psychiatry, University of Freiburg, Freiburg D-79104, Germany

Ujjwal K. Rout (217), Department of Surgery, Division of Pediatric Surgery Research Laboratories, University of Mississippi Medical Center, Jackson, Mississippi 39216, USA

David A. Ruggiero (273), Department of Psychiatry, Division of Neuroscience, Columbia University College of Physicians & Surgeons, New York, New York 10032, USA; Department of Anatomy & Cell Biology, Columbia University College of Physicians & Surgeons, New York, New York 10032, USA

Christoph Schmitz (1), Department of Psychiatry and Neuropsychology, Division of Cellular Neuroscience, Maastricht University, Maastricht 6100 MD, The Netherlands; European Graduate School of Neuroscience (EURON), Maastricht, Limburg 6200 MD, The Netherlands; California Institute of Technology, Division of Biology, Pasadena, California 91125, USA

Eberhard Schulz (239), Department of Child and Adolescent Psychiatry, University of Freiburg, Freiburg D-79104, Germany

Yaru Song (189), Department of Chemistry, Jackson State University, Jackson, Mississippi 39217, USA

Harry W. M. Steinbusch (1), Department of Psychiatry and Neuropsychology, Division of Cellular Neuroscience, Maastricht University, Maastricht 6100 MD, The Netherlands; European Graduate School of Neuroscience (EURON), Maastricht, Limburg 6200 MD, The Netherlands

Laura Stoppelbein (343), Department of Psychiatry, University of Mississippi Medical Center, Jackson, Mississippi 39216, USA

Sara Sytsma-Jordan (343), Department of Psychology, University of Southern Mississippi, Hattiesburg, Mississippi 39406, USA

Chitoshi Takayama (63), Department of Molecular Neuroanatomy, Hokkaido University School of Medicine, Sapporo 060-8638, Japan

Ei Terasawa (113), Department of Pediatrics and Wisconsin National Primate Research Center, University of Wisconsin, Madison, Wisconsin 53715, USA

Karen N. Thatcher (131), Medical Microbiology and Immunology and Rowe Program in Human Genetics, School of Medicine, University of California, Davis, California 95616, USA

Herman van Engeland (1), Department of Child and Adolescent Psychiatry, Rudolf Magnus Institute of Neuroscience, University Medical Center Utrecht, Utrecht 3508 GA, The Netherlands

Imke A. J. van Kooten (1), Department of Child and Adolescent Psychiatry, Rudolf Magnus Institute of Neuroscience, University Medical Center Utrecht, Utrecht 3508 GA, The Netherlands; Department of Psychiatry and Neuropsychology, Division of Cellular Neuroscience, Maastricht University, Maastricht 6100 MD, The Netherlands; European Graduate School of Neuroscience (EURON), Maastricht, Limburg 6200 MD, The Netherlands

Martha G. Welch (273), Department of Psychiatry, Division of Neuroscience, Columbia University College of Physicians & Surgeons, New York, New York 10032, USA

ACKNOWLEDGMENTS

I would like to thank the series editors, particularly Ron Bradley, and the editorial board of International Review of Neurobiology, for giving me the opportunity to edit this new volume. Cindy Minor, Senior Developmental Editor, Elsevier Academic Press, San Diego, was an attentive guide during this trip.

Research support from the Thrasher Research Fund, Salt Lake City in Utah, is acknowledged.

My special thanks go to Eugene Roberts for writing the foreword. Our discussions on ways to improve the integration of basic and clinical research were stimulating and refreshing.

Dirk M. Dhossche

PREFACE

"The most valuable lesson that knowledge can teach us is that its creation depends upon a continuous line of human relationships and traditions that go far back into the past. That continuity is an unbroken thread. It links cultures and peoples; it brings tolerance and understanding; it delivers hope and compassion"

Richard Horton, BSc, MB, FRCP, FMedScie
Editor-in-Chief, The Lancet
The 2004 Elsevier Library Connect Medical Library Lecture

Autism refers to a group of disorders with prominent autistic symptoms, i.e., Autistic Disorder, Asperger Disorder, Childhood Disintegrative Disorder, Rett's Disorder, and Pervasive Developmental Disorder not otherwise specified. These disorders are early-onset behavioral syndromes with a broad range of severity, characterized by lifelong impaired communication, impaired social interactions, and repetitive interests and behavior.

Autism has increasingly come into the limelight. Autism afflicts our young at a much higher rate than the early prevalence studies suggested. The prevalence of the whole spectrum is now considered about 6 per 1000, but some studies suggest it may be as high as 9 or 10 per 1000. Some concern has been raised about a possible increase in prevalence, but changes in diagnostic methodology and ascertainment strategy complicate comparisons across time.

Autism is diagnosed on clinical grounds. There are no diagnostic biomarkers for autistic disorders, except for Rett's Disorder where a causal genetic defect on the X chromosome has been identified. Autism occurs about four times more frequent in males than in females, with an even higher ratio in milder forms. Only in a small proportion of cases can a medical or neurological disorder be found. In the majority of cases, a strong genetic component is suspected, but the pattern of inheritance is complex. Twin studies show a 60% to 91% concordance rate in monozygotic twins, for narrowly and broadly defined phenotypes respectively. In contrast, there are no observations of concordance in dizygotic twins under narrow phenotypic definition; there is 10% concordance under broader phenotypic definition. Sibling recurrence rate is about 4.5%. This pattern of sharply increasing risk for first-degree relatives and monozygotic twins suggests the involvement of multiple genes interacting with one another to lead to disease susceptibility. No definite genetic or environmental causes for autism have been found. Some behavioral and pharmacological treatments have shown limited benefits in some individuals. No cure for autism has been found yet.

Leo Kanner was the first to describe the core features of autism, in 1943. He reports that the syndrome "differed markedly and uniquely from anything reported so far." Subsequent studies have confirmed the phenotypic validity of autism and have highlighted differences with childhood psychoses and other developmental disorders in cardinal symptoms (e.g., absence of hallucinations in autistic children), course of illness, intellectual functioning, sex distribution, social class, brain abnormalities, age of onset, and family history of schizophrenia. However, boundaries among the different types of autism, and between autism and other disorders, particularly early-onset psychotic disorders and some developmental disorders, are not always clear.

Since Kanner's seminal observations, valuable autism research has been done albeit at the expense of exploring links between autism and other early-onset disorders. It has become increasingly clear that some autistic symptoms may also be present in other disorders that are not in the group of autistic disorders (e.g., childhood schizophrenia), and that some of the molecular events leading to autistic development may not be unique to autism. Research advances in autism and other disorders reveal phenomenological, biochemical, and genetic areas of overlap. The decision to include chapters on schizophrenia, Prader-Willi syndrome, catatonia, and Fetal Alcohol syndrome, collectively referred to as "related disorders" in the title of this volume, acknowledges the importance of exploring areas of overlap between autism and certain non-autistic developmental disorders.

γ-aminobutyric acid (GABA) was discovered in the brain in 1950 by Eugene Roberts. Since then, GABA has risen from obscure brain compound status to stardom. GABA is now considered one of the most important neurotransmitters and developmental signals, with involvement in the pathophysiology of epilepsy, depression, anxiety, alcoholism, and psychosis. Knowledge on the complexity of GABA function has increased exponentially. Undoubtedly, GABA has not revealed all its secrets yet. The actions of GABA are multifaceted and Janus-faced, showing excitation and inhibition at different stages of development. It seems only natural to explore the role of GABA in autism and other developmental disorders.

This volume covers some clinical, but mostly basic research on GABA in the developing brain as it may relate to onset of autism and related developmental disorders. Although the evidence that GABA dysfunction is associated with autism is still limited, findings seem to converge. The chapters, grouped in three sections, are preceded by a comprehensive overview of GABAergic abnormalities in autism by van Kooten et al. (Chapter 1)

In the first section "GABA in Early Development" (Chapters 2–5), morphological and functional aspects of GABA signaling in cerebral and cerebellar development are reviewed by Takayama, Jelitai & Madarasz, and Fiszman in their respective chapters. Terasawa summarizes findings on the importance of GABA function in onset of puberty, a transitional period associated with the emergence and exacerbation of various neuropsychiatric disturbances.

In the second section, "Neurobiology of Autism and Related Disorders" (Chapters 6–14), new findings on common genetic mechanisms in Rett syndrome, Angelman syndrome, and autism are presented by Hogart, Thatcher & LaSalle. Blatt describes abnormalities of the GABAergic cerebellar system in autism. Dhossche, Song & Liu propose GABA hypotheses of autism, Prader-Willi syndrome, and catatonia. Preliminary evidence of plasma GABA as familial marker of autism is presented. Abnormal reelin signaling is reported by Fatemi to be present in both autism and schizophrenia. Rout highlights the role of GABA in prenatal alcohol exposure. The immunological abnormalities in autism are described by Cohly and Panja. Stoppelbein, Sytsma-Jordan & Greening review the psychological and biochemical correlates of abnormal movements in autism. Two chapters by Clement, Pschibul & Schulz, and Welch & Ruggiero examine the role of hypothalamic neuropeptides, including secretin, in autism and related disorders.

In the final section "Genetics of Autism and Related Disorders" (Chapters 15–17), Chagnon examines shared genetic risk factors between autism and major mental disorders. New models of autism, i.e., the GABRB3 and reeler mouse, are presented by D'Arcangelo and DeLorey.

Autism lacks truly effective treatments. The Interagency Autism Coordinating Committee Autism Research Matrix (April 2004, http://www.nimh.nih.gov/autismiacc/congapprcommrep.pdf) that is endorsed by the U.S. National Institutes of Health lists the following goals for achievement within 7–10 years: 1/evidence that 25% of cases of autism can be prevented from symptom expression through early identification and early treatment; 2/methods that allow 90% of individuals with autism to develop speech; 3/the identification of genetic and non-genetic causes of autism (and possible interactions between them), and; 4/the development of efficacious drug treatments that target core symptoms of autism. Three of these four goals concern existing or new treatments.

Some leaders in the field informally predict that new treatments for autism will be found by serendipity, similarly to the discovery of antipsychotic and antidepressant compounds in the 1950s and 1960s. The secretin saga of the last few years offers a salient example. Despite the initial enthusiasm based on a single, but widely publicized, case of dramatic improvement of autism after secretin infusion and a small case-series, large controlled trials have now shown that in most children secretin is not more effective than placebo. Is secretin a dead end? Is secretin only effective in a yet unspecified group of autistic children? Is secretin only effective when administered during a critical phase of autistic development? If so, how much secretin needs to be administered and how often? These questions largely remain unanswered. Arguably, I believe it is still (and possibly even more) relevant now to find out more about the biochemical and behavioral properties of secretin (and other neuropeptides) as described in Chapters 11 and 12.

Bredero, a famous 16th century Dutch poet, writer, and playwright, signed off on all his work with the phrase "'t kan verkeren" (*things may change*). His motto may also apply to the discovery of new treatments for autism. The history of psychiatry shows us that the discovery of antipsychotics, antidepressants, and other highly effective treatments for severe neuropsychiatric conditions such as electroconvulsive therapy has not led to major discoveries of etiological factors involved in psychotic and affective disorders. Maybe we should be prepared to accept that effective behavioral techniques, drug treatments, or other somatic therapies, might be discovered unexpectedly even if the causes of autism remain as elusive as before. Based on the contributions in this book, I predict that any such treatments will have important effects on GABA. As stated in the foreword by Eugene Robert, the stage may be set for great progress, and Bredero's motto can be honored again: "'t kan verkeren".

Dirk M. Dhossche

FOREWORD

The excellent papers in this volume juxtapose current knowledge of GABAergic function with various areas of knowledge about autism spectrum disorders (ASD) and add much potential building material for the construction of bridges of understanding that eventually may lead to early diagnosis, prevention, and treatments of this dread complex group of diseases. To help round out the picture, one hopes that another volume soon will follow dealing with the glutamatergic excitatory system and the maintenance of the delicate balance between the GABAergic and glutamatergic systems.

Under normal conditions, inhibitory projection and local-circuit neurons play crucial roles in information processing in nervous systems. They are the elements of the nervous system that prevent its lapse into a tyrannical synchronic state of paroxysmal discharge, total inactivity, or chaos by adjusting the timing, sensitivity, and versatility of the processes by which information is received, interpreted, and acted upon. They determine in neuronal assemblies at any particular moment which neurons shall act as groups and which shall act alone and the frequencies and sequences of these activities. Adequate function of the GABA-releasing neurons, the major inhibitory system in the vertebrate nervous system, is essential for maintenance of appropriate information-processing states in various functional modes. The GABA system of inhibitory neurons is a pervasive one and appears to control activities in most areas of the CNS.

The nervous system is highly restrained, inhibitory neurons acting like reins that serve to keep the neuronal ''horses'' from running wild. In coherent behavioral sequences, innate or learned, preprogrammed circuits are *released* to function at various rates and in various combinations. This is accomplished largely by *disinhibition* of pacemaker neurons whose activities are under the dual tonic inhibitory controls of local circuit GABAergic neurons and of GABAergic projection neurons coming from neural command centers. According to this view, disinhibition is permissive, and excitatory input to pacemaker neurons has mainly a modulatory role. Disinhibition, acting in conjunction with intrinsic pacemaker and modulatory glutamatergic inputs is the major organizing principle in nervous system function.

Thus, all activities in nervous systems are critically modulated and shaped by ever-shifting local and global balances between inhibition and excitation, which are largely mediated by the GABAergic and glutmatergic transmitter systems. The spectrum of abnormalities in ASD is remindful of abnormalities that result from gross imbalances between these systems. For example, the deficit in glutamatergic excitatory sensory input produced in normal adults during periods of sensory isolation results in occurrence of hallucinations that are rapidly reversed by return of the individuals to a normal environment. Convulsive seizures in otherwise normal infants with a simple dietary deficiency of pyridoxine (vitamin B_6) was induced by feeding a commercial infant formula from which the vitamin was inadvertently omitted. The seizures were completely eliminated almost immediately after intramuscular injection of pyridoxine because of an extremely rapid conversion of the injected pyridoxine to the co-enzymatically active pyridoxal phosphate, the association of the latter with a suboptimally functional glutamic acid decarboxylase in nerve terminals of inhibitory nerves, accelerated synthesis of GABA therein, and the release of GABA from the terminals onto postsynaptic receptor sites. The consequent desynchronization resulting from the reinstitution of normal neural inhibition prevented the pyroxysmal discharges of groups of neurons whose firing was causing the seizures.

There are many substances that can upregulate or downregulate either GABAergic or glutamatergic transmitter systems. Although a number have been tested in autistic children, none has yet attained the legitimacy of "standard" treatment. However, the possibility is not precluded that judicious use of some of these substances administered alone or in combinations together with skillful clinical observation will yield useful treatments in the future.

At this time, I am skeptical about our ability to develop effective, relatively non-toxic "designer" drugs for ASD spectrum conditions based on logical extensions of current knowledge of brain neurotransmitter and neuromodulator systems. Rather, I believe that ingenious manipulation of structures and uses of substances arrived at unexpectedly or during empirical screening may lead to therapeutic progress which, in turn, will give rise to new tools for study of basic biological mechanisms.

At one time, it was anticipated that new knowledge of amino acid and vitamin metabolism would lead quickly to the synthesis of analogues of these substances with antibacterial potency. Much effort and money were expended in such endeavors with no success. Instead, the "antibiotic revolution" stemmed from the concerted exploitation of an observation that a mold contaminating a bacterial culture caused bacteria in its vicinity to undergo lysis and that the broth in which the fungus was grown was bacteriocidal for many common pathogenic organisms. The mechanism of action of penicillin was elucidated only many years after millions of lives had been saved by its use.

Most drugs effective in treatment of nervous system disorders arose from the folk medicine of ancient civilizations, from empirical searches among many substances for specific desired effects, or from unexpected observations. For example, the anticonvulsant properties of bromide and phenobarbital were discovered by chance. The development of diphenylhydantoin (Dilantin) for treatment of seizures resulted from a planned search for compounds capable of suppressing electroshock convulsions in laboratory animals while not showing the degree of sedation or impairment of consciousness seen with barbiturates. Meprobamate (Miltown) was selected as an anti-anxiety drug from tests of over 1200 compounds. The first benzodiazepine to be used clinically, chlordiazepoxide (Librium), was noted to have a "taming" effect in several species of animals. This led to the trial of the drug for anti-anxiety effects and to the synthesis of diazepam (Valium) and a number of other benzodiazepines that currently are used in treatment of anxiety and tension, sleep disorders, epilepsy, muscle spasms and a variety of so-called psychosomatic disorders. It was approximately 20 years after initiation of clinical use of benzodiazepines that one of the mechanisms of their action was linked to function of the receptor-chloride-ionophore system regulated by GABA, thereby sparking great progress in our understanding of that system. Barbiturates, benzodiazepines, bromide, and ethanol were found to converge in their actions at the level of the GABA receptor complex, leading to further understanding of its components and their relationships.

Although for some time there has been much information about the glutamatergic and GABAergic transmitter systems, respectively, few widely clinically useful drugs based on this knowledge have been devised to date. On the other hand, curiosity about the analgesic and anti-spastic properties of Baclofen, a failed GABA-mimetic in classical Cl^- channel conductance paradigms, led to the discovery of $GABA_B$ receptors, which are coupled to K^+ channels and to many new therapeutic possibilities.

Over the years I have engaged in studies related to problems of epilepsy, schizophrenia, Huntington's disease, senile dementia of the Alzheimer's type, hepatic coma, multiple sclerosis, and spinal cord injury. After perusing the papers which will appear in this volume and surveying the current literature, I am adding ASD to the list of my concerns and hope that among the other readers of this material there will be some who do likewise. The stage may be set for great progress.

Eugene Roberts, Ph.D., FAIC
Department of Neurobiochemistry

AUTISM: NEUROPATHOLOGY, ALTERATIONS OF THE GABAERGIC SYSTEM, AND ANIMAL MODELS

Christoph Schmitz,[†,‡,1] Imke A. J. van Kooten,[*,†,‡] Patrick R. Hof,[§]
Herman van Engeland,[*] Paul H. Patterson,[¶] and Harry W. M. Steinbusch[†,‡]

*Department of Child and Adolescent Psychiatry, Rudolf Magnus Institute of Neuroscience
University Medical Center Utrecht, Utrecht 3508 GA, The Netherlands
†Department of Psychiatry and Neuropsychology, Division of Cellular Neuroscience
Maastricht University, Maastricht 6100 MD, The Netherlands
‡European Graduate School of Neuroscience (EURON)
Maastricht, Limburg 6200 MD, The Netherlands
§Department of Neuroscience, Mount Sinai School of Medicine
New York, New York 10029, USA
¶California Institute of Technology, Division of Biology, Pasadena, California 91125, USA

Autism is a neurodevelopmental disorder with a strong genetic component and several known environmental risk factors. Neuropathological studies have shown consistent abnormalities in the limbic system, cerebellum, and cerebral cortex. Several findings suggest a role for the GABAergic system in autism

[1]Corresponding author. Department of Psychiatry and Neuropsychology, Division of Cellular Neuroscience, Maastricht University, P.O. Box 616, 6100 MD Maastricht, The Netherlands.

INTERNATIONAL REVIEW OF
NEUROBIOLOGY, VOL. 71
DOI: 10.1016/S0074-7742(05)71001-1

1

neuropathology. There are reports of elevated plasma GABA levels, reduction of the GABAergic system enzymes, and decreased availability of GABA in autistic patients. Autism has a reported heritability of 60–90%. Abnormalities in the 15q11–13 region have been found in autistic people, and the GABA$_A$ receptor genes are located in this region. In addition, GABA dysfunction may occur in conjunction with Reelin. Abnormalities in the gene encoding for Reelin have been implicated in autism, and Reelin and GABA play an important role in the development of minicolumns. Compared to controls, minicolumns are more numerous, smaller, and less compact in autistic patients. Several studies provided evidence for the role of GABA receptors in tangential migration of neurons. Furthermore, GABA regulates cell proliferation in some brain regions. Because the underlying causes of the reduced GABA system function in autism are not well understood, it is important to develop animal models of autism, which can give more insights into the neuropathology and behavioral aspects of the disease. Animal models of autism include misregulation of genes implicated in the disorder, as well as the use of known environmental risk factors. In the future, investigation of human autism tissues and animal models, in combination with implementation of new techniques such as design-based stereology and gene expression, may result in the elucidation of the etiology of autism.

I. Introduction

Autism is currently viewed as a genetically determined neurodevelopmental disorder (Bailey *et al.*, 1996), defined by the presence of marked social deficits, specific language abnormalities, and stereotyped, repetitive behaviors (American Psychiatric Association, 1994). Approximately 20% of the autistic subjects show macroencephaly, defined as head circumference above the 97th percentile (Aylward *et al.*, 2002; Bailey *et al.*, 1993; Courchesne *et al.*, 2003; Davidovitch *et al.*, 1996; van Karnebeek *et al.*, 2002). However, this macroencephaly is not present until after the first year of life (Courchesne *et al.*, 2003). Although evidence of increased head circumference (Aylward *et al.*, 2002; Bailey *et al.*, 1993; Courchesne *et al.*, 2003; Davidovitch *et al.*, 1996; Fombonne, 2000; van Karnebeek *et al.*, 2002), brain weight (Bailey *et al.*, 1998; Casanova *et al.*, 2002b; Courchesne *et al.*, 1999; Kemper and Bauman, 1998), and brain volume (Aylward *et al.*, 2002; Courchesne *et al.*, 2001; Sparks *et al.*, 2002) has been described in autism, the underlying biological mechanisms remain to be determined. These observations could be due to increased neurogenesis, gliogenesis, or synaptogenesis, disturbed neuroblast migration, decreased apoptosis or synaptic pruning, or combinations of these effects (Palmen *et al.*, 2004). In addition, the detection of abnormalities in neurotransmitter systems, such as γ-aminobutyric acid (GABA) in autistic patients, suggests that it could be worthwhile to concentrate research on these systems. A good approach is the use of animal models that mimic features of autism.

In this chapter we examine the literature on autism neuropathology, the role of the GABAergic system in this disorder, and the relevance of rodent models with autistic features.

II. Neuropathologic Alterations in Specific Brain Regions in Autism

A. LIMBIC SYSTEM

Bauman and Kemper were the first to report neuropathological findings in autism. They demonstrated increased cell packing density and smaller neurons in several regions of the limbic system, including the hippocampus, subiculum, amygdala, entorhinal cortex, mammillary bodies, and septal nuclei (Bauman, 1991; Bauman and Kemper, 1985, 1987, 1990; Bailey et al., 1998; Kemper and Bauman, 1993; Raymond et al., 1996). This pattern of small, closely packed neurons with reduced dendritic arbors could reflect features of an immature brain (Jacobsen, 1991). Bailey et al. (1998) demonstrated increased cell packing density in all CA regions of the hippocampus in one out of five autistic cases. Raymond et al. (1996) investigated the dendritic morphology of hippocampal neurons in two well-documented autistic patients and one age-matched control. Using the Golgi stain, this group found smaller neurons and less dendritic branching of both CA1 and CA4 hippocampal neurons in autistic subjects compared to controls. These findings further suggest a curtailment of maturation, a feature previously highlighted by Kemper and Bauman (1993).

Neuropathologic findings are rather consistent with respect to the limbic system, whereas MRI studies are mostly conflicting. For example, compared to controls, volumes of the limbic system of autistic subjects are reported to be increased (Howard et al., 2000; Sparks et al., 2002), decreased (Aylward et al., 1999), or unchanged (Howard et al., 2000; Piven et al., 1998).

B. CEREBELLUM AND BRAINSTEM

Williams et al. (1980) were the first to examine neuropathological alterations in the cerebellum in autism. In one out of four brains from autistic patients they found reduced Purkinje cell density. In a study by Ritvo et al. (1986), all autistic cases showed a decreased number of Purkinje cells in the cerebellar vermis and hemisphere. Kemper and Bauman (1993) replicated this finding. In another study five adult autistic cases had a low number of cerebellar Purkinje cells, although this feature was not found in a 4-year-old autistic male (Bailey *et al.*, 1998). In contrast, no abnormalities were found in the cerebellum of a 16-year-old female with autism and severe mental retardation (Guerin *et al.*, 1996). Because of the consistent findings of decreased numbers of Purkinje cells in the cerebellum, Fatemi *et al.* (2002b) examined the size of these cells. They showed a 24% decrease in mean Purkinje cell size in autistic brains. It should be mentioned that cerebellar Purkinje cells are the final targets of projections from the inferior olivary nucleus. A decrease in Purkinje cells may result in an abnormal development of the olivary projections to the cerebellar nuclei (Palmen *et al.*, 2004). Many research groups have found a decrease in the number of cerebellar Purkinje cells without significant gliosis in the cerebellum (Bailey *et al.*, 1998; Fatemi *et al.*, 2002b; Kemper and Bauman, 1993). Most MRI studies have shown smaller cerebellar hemispheres (Courchesne, 2004; Murakami *et al.*, 1989) or vermis (Courchesne, 2004; Hashimoto *et al.*, 1995; but see Nowell *et al.*, 1990; Piven *et al.*, 1997).

Furthermore, two out of six brains from autistic patients in the study of Kemper and Bauman (1993) demonstrated enlarged neurons in deep cerebellar nuclei and inferior olivary nucleus, whereas in the older subjects (>22 years) these neurons were small and pale, with normal numbers. In all brains, the inferior olivary nucleus did not show retrograde loss of neurons. Moreover, age-related abnormalities in the cerebellar nuclei and the inferior olive have been reported (Bauman, 1991). Apart from reports on limbic alterations, Bailey *et al.* (1998) have demonstrated olivary dysplasia in three of five autistic cases. In another two autistic subjects they found ectopic neurons related to the olivary complex (Bailey *et al.*, 1998). Finally, Rodier *et al.* (1996) reported that the brain of a 21-year-old autistic woman with mental retardation and comorbid epilepsy, exhibited near-complete absence of the facial nucleus and superior olive along with shortening of the brainstem between the trapezoid body and the inferior olive.

C. Migration Abnormalities and Cortical Dysgenesis

Reelin is a signaling molecule that plays a role in migration and lamination of neurons during embryogenesis. In adult life Reelin is involved in synaptic plasticity (Fatemi, 2000b, 2002, 2004). In the cerebellar cortex, Fatemi *et al.* (2001a,b) found a >40% reduction in Reelin and a 34–51% reduction in Bcl-2 levels in autistic subjects. Very recently, Fatemi *et al.* (2005) observed reduced Reelin signaling in the frontal cortex in autism. There is an association of decreased Reelin levels with disturbed neuronal migration and lamination of the cerebral and cerebellar cortex in mice (Gonzalez *et al.*, 1997) and humans

(Persico *et al.*, 2001; Piven *et al.*, 1990). Furthermore, decreased Reelin levels in the blood have been related to severe mental retardation and hypoplastic cerebellum, both of which have been reported in autism. The decreased Bcl-2 levels might inhibit apoptosis. Both decreased Bcl-2 and increased P53 levels were correlated to mental retardation and have been suggested to result in a greater propensity for cell death (Fatemi and Halt, 2001a). Cortical dysgenesis was identified in four of six (all mentally handicapped) autistic cases (Bailey *et al.*, 1998). These brains showed thickened cortices, high neuronal density, presence of neurons in the molecular layer, irregular laminar patterns, and poor gray-white matter boundaries.

Given the importance of cortical cellular organization during development, Casanova *et al.* (2002a) investigated the morphology of cell minicolumns. Compared to controls, there were more numerous, smaller, and less compact minicolumns in autistic subjects (Casanova *et al.*, 2002a,b). However, the functional significance of these changes in minicolumns is still unclear (Hutsler *et al.*, 2003).

D. CHOLINERGIC SYSTEM

Hohmann *et al.* (1998) emphasized the importance of the cholinergic system during brain development. They reported a delay in cortical development, permanent changes in cortical architecture, and cognitive function when cholinergic innervation is disrupted during early postnatal development. Furthermore, Bauman and Kemper (1994) demonstrated larger cholinergic neurons in young autistic cases compared to smaller cholinergic neurons at an older age. Perry *et al.* (2001) investigated levels of cholinergic enzyme and receptor activities in the frontal and parietal cerebral cortex. Compared to non-autistic mentally retarded cases, the autistic cases display 30% lower muscarinic M1 receptor binding in the parietal cortex. In addition, a reduction in $\alpha 4$ nicotinic receptor binding was found in all groups as compared to controls. In the cerebellum, the nicotinic receptor ($\alpha 3$ and $\alpha 4$ subunit) was reduced by 40–50%, whereas the $\alpha 7$ subunit was increased in autism (Lee *et al.*, 2002).

III. GABAergic Abnormalities in Autism

The GABAergic system has also been suggested to be involved in autism (Blatt *et al.*, 2001; Cook *et al.*, 1998; Schroer *et al.*, 1998). GABA is implicated in various psychiatric disorders, including schizophrenia (Caruncho *et al.*, 2004; Guidotti *et al.*, 2000), mood disorders (Sanacora *et al.*, 1999), anxiety disorders

(Goddard *et al.*, 2001), and autism (Cook *et al.*, 1998; Schroer *et al.*, 1998). Changes in GABA levels have been found in platelets of 18 of 18 autistic children (Rolf *et al.*, 1993) and in the plasma and urine of one male (case study) autistic child (Cohen, 1999, 2000). Dhossche *et al.* (2002) hypothesized that the elevated plasma levels of GABA could reflect a compensatory increase in presynaptic GABA release in response to hyposensitivity of a subset of GABA receptors. In turn, this could produce increased postsynaptic activation of other, normal GABA receptors subtypes, resulting in complex alterations of GABAergic function throughout the brain in autistic people. By introducing a GABA-transaminase agonist, the GABA plasma levels can be lowered (Cohen, 2002). During normal GABA catabolism, GABA-transaminase is responsible for the conversion of GABA into succinic semialdehyde (SSA) (Tillakaratne *et al.*, 1995). Such a GABA-transaminase agonist can activate GABA-transaminase enzyme activity by causing a reduction of plasma GABA levels in the brain. They proposed that this could result in an elevated signaling between axons and oligodendrocytes in the corpus callosum, which might result in a decrease of autistic features due to abnormal development of axons in the corpus callosum (Cohen, 2002). Cohen (2000) also reported increased levels of ammonia in the liver of the same autistic child, illustrating a possible link between the liver and infantile autism.

Blatt *et al.* (2001) investigated the hippocampal density and distribution of neurotransmitter receptors from the GABAergic, serotonergic (5-HT), cholinergic, and glutamatergic systems in autistic patients and controls (age range 16–24 years). They reported that only the GABAergic receptor system was significantly reduced in autism. It has been suggested that some autistic patients respond abnormally to benzodiazepines, possibly due to pre-existing GABAergic dysfunction or abnormalities in the GABA benzodiazepine receptor complex (Garreau *et al.*, 1993). Furthermore, [3][H]-flunitrazepam-labeled benzodiazepine binding sites and [3][H]-muscimol labeled $GABA_A$ receptors were reduced in the hippocampus of autistic people (Blatt *et al.*, 2001), which provides evidence for abnormal benzodiazepine receptor complexes in the hippocampus of these people. Fatemi *et al.* (2002c) found reduced levels of both glutamic acid decarboxylases (GAD)65 and GAD67 in autistic parietal and cerebellar cortex, implying a deficit in GABA. Lower GABA levels could reduce the threshold for developing seizures, which are often associated with autism (Bailey *et al.*, 1998; Blatt *et al.*, 2001).

The heritability of autism is 60–90% (Andres, 1998; Bespalova and Buxbaum, 2003). Abnormalities on the long arm (15q11–13) of the maternally derived chromosome 15 have been found in a small proportion of autistic people (Dhossche *et al.*, 2002; Schroer *et al.*, 1998). These aberrations include duplications and deletions involving the proximal long arm of this chromosome (Schroer *et al.*, 1998). Three $GABA_A$ receptor subunit genes, GABRB3, GABRA5, and GABRG3, are located in the proximal arm of the 15q region (Schroer *et al.*, 1998). In addition, using fluorescence *in situ* hybridization, Silva *et al.* (2002)

showed tetrasomy in the 15q11–13 chromosomal region in a female child with autism. This tetrasomy could result in an excess of GABA receptors, leading to behavioral problems, including hyperactivity and epilepsy, a frequent comorbidity of autism. Linkage studies for this chromosomal region have produced contradictory results (Bass et al., 2000; Buxbaum et al., 2002; Cook et al., 1998; Maestrini et al., 1999; Martin et al., 2000; Menold et al., 2001; Muhle et al., 2004; Schroer et al., 1998). As previously mentioned, Casanova et al. (2002a, 2003) found narrower cell minicolumns in brains from autistics. In this regard it might be of interest that abnormalities in GABAergic interneurons have been suggested to be associated with narrowing of cell minicolumns (Nishikawa et al., 2002). Furthermore, Reelin appears to play a role in the development of minicolumns (Nishikawa et al., 2002), which may imply a relationship between the GABAergic system, minicolumns, Reelin, and autism.

IV. GABA and Brain Development

A. EARLY DEVELOPMENT

Several studies have pointed out that GABAergic synapses and receptors are generated and active before glutamatergic synapses in all brain regions (Ben-Ari et al., 2004; Chen et al., 1995; Koller et al., 1990; Walton et al., 1993). However, GABAergic interneurons have a longer migration journey before they reach their final destination (Ben-Ari et al., 2004). Furthermore, Tyzio et al. (1999) have suggested that the first GABAergic synapses are probably located on the apical dendrites of pyramidal neurons and not on the soma.

Many different GABAergic receptor subunits are expressed in the embryonic and/or adult brain. The change in subunit composition is essential for normal development in specific brain regions (Lujan et al., 2005). Each subunit of the $GABA_A$ receptor exhibits a unique regional and temporal expression profile during brain development (Laurie et al., 1992). The $\alpha2$, $\alpha3$, and $\alpha5$ subunits are dominantly expressed during embryonic development whereas others dominate postnatal or adult brain (Fritschy et al., 1994; Laurie et al., 1992). For example, $\alpha1$ subunit expression is low at birth and increases during the first postnatal week, while the $\alpha2$ subunit decreases progressively (Fritschy et al., 1994).

During early development, GABA causes depolarization (i.e., excitation) due to a relatively high concentration of intracellular Cl^- (Miles, 1999; Owens et al., 1996). Depolarization occurs via the opening of GABA-gated Cl^- channels, which also increases the intracellular Ca^+ concentration (Jelitai et al., 2004; Miles, 1999; Owens et al., 1996). Both GABA excitation and Ca^+ influx may be important for plasticity, synaptic connections, and for establishing neural networks (Kriegstein *et al.*, 2001). Consistent with this observation, in primary cultures of several embryonic and neonatal brain tissues, GABA exerts a variety of neurotrophic actions such as promotion of neurite extension, synaptogenesis, and the synthesis of its own receptors (Hansen *et al.*, 1987). Moreover, excitatory GABAergic interneurons can generate giant depolarizing potentials (GDPs), which in turn cause primitive network-driven patterns of electrical activity in all developing circuits (Ben-Ari, 2002). GDPs are hallmarks of developing neural networks and constitute the first synaptic pattern observed in the developing rat hippocampus (between P0 and P10) (Ben-Ari, 2001; Ben-Ari *et al.*, 1989).

During maturation, GABA induces the opposite effect; Cl^- ions are pumped out and the cell becomes hyperpolarized. Thus, GABA switches from an excitatory to an inhibitory neurotransmitter during development (Miles, 1999). In the rat this switch occurs around birth and is due to the expression of the K^+/Cl^- cotransporter 2 (KCC2) (Herlenius *et al.*, 2004; Miles, 1999; Rivera *et al.*, 1999). KCC2 expression increases progressively from embryonic stages to postnatal day 15 (P15) (Stein *et al.*, 2004). This elevation occurs simultaneously with the GABA switch from excitatory to inhibitory (Stein *et al.*, 2004). When the GABA switch occurs in humans is not known (Herlenius *et al.*, 2001).

B. Migration

Migration is a process of neuronal movement essential for the establishment of normal brain architecture. Most cortical neurons migrate from their original site to their final destination in the cerebral cortex using the radial pathway from cortical ventricular zones (VZs) into the neuronal layers. In contrast, a subpopulation of cells including GABAergic neurons move tangentially within the intermediate zone (Hatten, 1999; O'Rourke *et al.*, 1997; Rakic, 1995; Rapp and Bachevalier, 1993). GABAergic neurons arise in the medial and lateral ganglionic eminences and migrate dorsally into the developing cortex (Rapp and Bachevalier, 1993). As early as embryonic day 10 (E10), GABA is located near the target destinations of migratory neurons and in the migrating neurons themselves in the developing mouse brain *in vivo* (Del Rio *et al.*, 2000). Different GABA concentrations promote migration of GABAergic and non-GABAergic neuronal cortical subpopulations (Behar *et al.*, 1996). For example, in cortical regions, femtomolar concentrations of GABA stimulate directed migration (chemotaxis), whereas micromolar levels stimulate chemokinesis (random motility) of more mature neurons (Behar *et al.*, 1996, 1998). Lopez-Bendito *et al.* (2003) showed a modified distribution of tangential migration neurons within the cortex upon blockade of $GABA_B$ receptors with a specific antagonist (CGP52432) in embryonic rat organic cultures. Furthermore, blocking of GABA receptors with saclofen or picrotoxin resulted only in a delay of cell movements, but not a complete arrest of migration. These findings could indicate that GABA receptor activation in the developing cortex modulates the rate of cell migration rather than initiating it (Behar et al., 2000). Furthermore, it has been suggested that migration, mediated by the GABAergic system, can act through Ca^{2+} ions, which alter cell movements by changing the dynamics of the cytoskeletal remodeling (Gomez et al., 1999)

Other factors regulate tangential migration as well (Marin et al., 2001). Mitogenic factors stimulate the movement of cells, such as hepatocyte growth factor/scattered factor (HGF/SF) and neurotrophic factors. In slice cultures, for example, exogenous HGF/SF increases the number of cells that migrate away from the subpallial telencephalon, while anti-HGF/SF antibodies inhibit cell movement (Powell et al., 2001). Little is known about the substrates that are used by migrating neurons. As described by Parnavelas (2000), it is possible that migrating interneurons use axons as a substrate in the cortex. Nevertheless, it is still unclear whether migrating interneurons interact with fiber tracts in vivo (Wichterle et al., 2001). Finally, the factors that guide different migratory streams through appropriate pathways towards their targets (Marin and Rubenstein, 2001), including the netrin/Dcc (Livesey et al., 1997; Parnavelas, 2000), Slit/Robo (Parnavelas, 2000; Yuan et al., 1999), and semaphorin/neuropilin systems (Skaliora et al., 1998).

C. PROLIFERATION

The process of proliferation is responsible for generating the correct number of specific cell types in the correct sequence in the brain (Lujan et al., 2005). Growth factors, neurotransmitters, and their receptors have been implicated in the extrinsic regulation of cell proliferation in the developing telencephalon (Nguyen et al., 2001). In contrast to microglia and astrocytes, neurons exhibit very low rates of proliferation in culture (Eliason et al., 2002). Several studies have demonstrated a positive effect of GABA on cell proliferation. Ben-Yaakov et al. (2003) demonstrated GABA-dependent cell proliferation in the hippocampus. GABA also promotes cell proliferation in cultures of cerebellar progenitors, with no effect on cell survival (Fiszman et al., 1999). Furthermore, Haydar et al. (2000) showed that exogenous GABA increases proliferation by shortening the cell cycle in the neocortical ventricular zone (VZ) of the embryonic cerebrum in organotypic cultures (Haydar et al., 2000). The reverse effect was found in the subventricular zone (SVZ) (Haydar et al., 2000). The same group demonstrated that the effect seen in the VZ was mediated by $GABA_A$ receptors (Haydar et al., 2000). Activation of $GABA_A$ receptors also influences DNA synthesis (Haydar et al., 2000). However, an in situ study demonstrated that $GABA_A$ receptors, triggered by GABA or muscimol (a $GABA_A$ receptor agonist) negatively regulate

DNA synthesis in neural progenitors in the rat embryonic neocortical VZ between E16 and E19 (LoTurco *et al.*, 1995).

In contrast, GABA (and glutamate) has been implicated in the reduction of the number of proliferating cells in dissociated or organotypic cultures of the neocortex (LoTurco *et al.*, 1995). Moreover, Luk and Sadikot (2001) found no GABAergic effect on cell proliferation in the rodent neostriatum. In their study of parvalbumin-immunoreactive progenitors, they only revealed an effect of GABA on cell survival. However, Fiszman *et al.* (1999) demonstrated that GABA and GABA$_A$ receptor agonists do not influence survival in cerebellar granule cells *in vitro*. This finding could be due to a calcium influx via voltage gated calcium channel activation, which subsequently activates the MAPK cascade (Fiszman *et al.*, 1999).

D. DIFFERENTIATION

Neuronal differentiation is another step in brain development that seems to be regulated by early glutamate- and GABA-mediated signaling. During early cortical neuronal differentiation, Cajal-Retzius (CR) cells play a key role in regulating cortical lamination (Lujan *et al.*, 2005), and they produce Reelin (Nishikawa *et al.*, 2002). Reeler (Reelin mutant) mice display severe cortical laminar disruption (Nishikawa *et al.*, 2002). In addition, CR cells express diverse neurotransmitters and GABA$_A$ receptors, indicating that they can respond to GABA produced by nearby neurons (Mienville *et al.*, 1999). In cultured embryonic hippocampal neurons, GABA$_A$ receptor activation increases neurite outgrowth and maturation of GABAergic interneurons (Barbin *et al.*, 1993). Such activation also plays a role in the morphological development of cortical neurons via membrane depolarization (Maric *et al.*, 2001). In addition, switches in GABA$_A$ receptor unit composition (Owens *et al.*, 1999) and changes in expression of components involved in GABA synthesis, storage, and release may mediate the transition from embryonic to adult GABAergic signaling (Somogyi *et al.*, 1995).

Several factors regulate the differentiation of tangentially migrating neurons, such as Dlx1, Dlx2, and Mash1. These genes may control the timing of neuronal differentiation and induce the GABAergic phenotype. Dlx1 and Dlx2 are linked homeobox genes, whereas Mash1 is a basic helix-loop-helix gene that can induce Dlx1 (Fode *et al.*, 2000). Normally, these three transcription factors are expressed in the VZ and SVZ progenitor cells in the anterior peduncular area (AEP), lateral (LGE), and medial ganglionic eminence (MGE) (Bulfone *et al.*, 1993; Eisenstat *et al.*, 1999; Guillemot *et al.*, 1993; Porteus *et al.*, 1994). Studies with mice lacking Dlx1, Dlx2, or Mash1 have revealed insights in the role of these transcription factors during differentiation. Anderson *et al.* (1997) have shown that the functional loss of Dlx1 and Dlx2 blocks the differentiation of late-born subpallial telencephalic neurons. Interestingly, double mutations of Dlx1/Dlx2 have a four-fold reduction in GABAergic interneurons (Anderson *et al.*, 2001). Functional loss of Mash1 leads to premature differentiation of several early-born cell populations (Casarosa *et al.*, 1999). In Mash1 mutants, more GABAergic interneurons are lost in the marginal zone of the cortex as compared to the intermediate zone (Marin *et al.*, 2000). Additionally, gain-of-function studies revealed that Dlx and Mash genes both could induce aspects of the GABAergic phenotype (Fode *et al.*, 2000).

V. Animal Models of Autism

Animal models are very useful for determining the role of genes and environment, understanding the pathogenesis, and for testing potential therapeutic approaches. Moreover, animal models need not be perfect mimics of human diseases in order to be valuable. Autism is a heterogeneous disorder in which most of the susceptibility genes have not yet been identified. Nonetheless, there are several genetic changes that do entail a high risk for autism, and mouse models of these changes share some features of the human disorder. Furthermore, models based on environmental risk factors are valuable. Finally, brain lesion models are of interest. The following section describes a variety of animal models that show features of autism.

A. GENETIC MANIPULATION

1. *X-chromosome Loci*

Four loci on the X-chromosome have been identified in autism thus far. These genes are the Fragile X mental retardation protein (Fmr1), methyl-CpG-binding protein type 2 (MECP2), and neuroligin (NLGN) 3 and 4. Fmr1 is silenced in Fragile X syndrome (FXS), a condition that often includes autism symptoms (Wassink *et al.*, 2001). The Fmr1 knockout (KO) mouse displays increased dendritic spine density in the visual and somatosensory cortices, with a greater number of spines with an immature appearance (Comery *et al.*, 1997; Galvez *et al.*, 2003; Nimchinsky *et al.*, 2001). These features have also been found in several cortical areas in FXS humans (Irwin *et al.*, 2001). However, there appears to be a decrease in dendritic branching in the human hippocampus (Raymond *et al.*, 1996). Clearly, more neuropathologic studies are needed to relate this animal model to autism. The NLGN1, 3, and 4 genes map to three loci associated with predisposition to autism, 3q26, Xp22.3, and Xq13, respectively.

Mutations in NLGN3 and 4 are associated with autism and, in some cases, with mental retardation, a feature often associated with autism (Jamain *et al.*, 2003; Laumonnier *et al.*, 2004). Chih *et al.* (2004) and Comoletti *et al.* (2004) found that these NLGN3 and 4 mutations lead to loss of protein processing and loss of the capacity for stimulation of synapse formation. Detailed neuropathology and behavioral analysis of the MECP2 KO remains to be reported.

2. *15q11-q13 Locus*

As previously mentioned, the 15q11-q13 locus is relatively small and has been linked to autism in several studies (Buxbaum *et al.*, 2002; Cook *et al.*, 1998; Dhossche *et al.*, 2002; Menold *et al.*, 2001; Muhle *et al.*, 2004; Schroer *et al.*, 1998). Jiang *et al.* (2004) have identified Ube3a, a genetic locus for Angelman syndrome (AS), which is located in this region and shares some clinical features with autism. The same group investigated a mouse model with a maternal null mutation, which shows a lack of Ube3a expression in cerebellar Purkinje cells and in the hippocampus. Both features have been implicated in autism (Jiang *et al.*, 1998). The relationship of the 15q11-q13 locus and the GABAergic system was previously described. However, more neuropathology is required to establish the relationship between $GABA_A$ receptor KO mice and autism.

3. *Serotonin*

It is generally agreed that there are serotonin (5-HT) abnormalities in autism. 5-HT levels in platelets are increased (Anderson *et al.*, 2002; Cook *et al.*, 1996); however, the underlying mechanism of this elevation is unclear. Therefore, the relevance of these changes for the brain is not well understood. Some studies have emphasized the importance of 5-HT during fetal brain development (Gaspar *et al.*, 2003; Whitaker-Azmitia, 2001), where this transmitter plays a role in neurogenesis, neuronal differentiation, neuropil formation, axon myelination, and synaptogenesis *in vivo* (Whitaker-Azmitia, 2001). Thus, altering 5-HT levels during development could lead to relevant models for autism. Whitaker-Azmitia and colleagues (2001) have investigated 5-HT depleted neonatal rat pups and found decreased dendritic length and spine density in the hippocampus, the same features Raymond *et al.* (1996) observed in autism. Furthermore, neonatal disruption of 5-HT tracts causes alterations in cortical morphogenesis in rodents (Connell *et al.*, 2004).

Studies on KO mice have revealed insights into the role of 5-HT in both early embryonic and postnatal development (Gaspar *et al.*, 2003; Gingrich *et al.*, 2003). For example, KO experiments showed that the $5-HT_{2B}$ receptor is involved in the regulation of neurogenesis, cell specification, and cell survival during early development. At later developmental stages, depending on the brain region, such control can be mediated by the 5-HT$_{1A}$ (regulation of dendrite growth) and 5-HT$_{2A/2C}$ (regulation of axonal growth) receptors. Unfortunately, detailed neuropathology in these mice, particularly with relevance to known changes in the autistic brain, is missing.

4. DLX

DLX genes regulate the development of a subset of cortical and striatal neurons. Two of the linkage loci for autism, 2q31.1 and 7q21.3, contain the DLX1/2 and DLX 5/6 complexes, respectively. Stuhmer et al. (2002) reported that mutations in DLX2 and 5 genes alter the development of GABA neurons in the forebrain. As stated earlier, the GABAergic system is involved in the pathology of autism. Thus, DLX mutant mice will be of interest.

5. Engrailed

The gene *Engrailed 2* (En2) is located on chromosome 7, which has been linked to autism (Gharani *et al.*, 2004). Mouse mutants of En2 and autistic individuals display similar cerebellar morphological abnormalities. En2 KO mice show a loss of Purkinje cells (Gharani *et al.*, 2004; Liu *et al.*, 2001), as do autistic brains, where the loss appears in stripes or patches (Bailey *et al.*, 1998). Other features shared between the En2 KO mice and human autistic cases are deficiencies in the number of deep nuclear, granule, and inferior olive neurons. Recently, increased neuronal packing, a smaller hippocampus, and ectopic location of neuronal subgroups in the amygdala in En2 KO mice have been linked to autism (Kuemerle *et al.*, 1997). There is now a need for detailed behavioral studies of these mice.

6. Reelin

The *Reelin* gene, located on chromosome 7, has been linked to autism. Conflicting reports exist about the association of its polymorphisms with autism (Bonora *et al.*, 2003; Krebs *et al.*, 2002; Persico *et al.*, 2001). In the adult brain, Reelin is normally expressed by GABAergic neurons. In patients, Guidotti *et al.* (2000) suggested that there might be a correlation between Reelin and GAD67, one of the two molecular forms of GAD. This interaction is clearly expressed in the cerebellum, where Reelin regulates dendritic sprouting in GABAergic Purkinje cells (Curran *et al.*, 1998). These features were also reported in the heterozygous Reeler mouse (Tueting *et al.*, 1999). Interestingly, in the male heterozygous Reeler mouse (rl+/−), loss of Purkinje cells has been reported between 3 and 16 months of age. Furthermore, mutant Reeler mice, in which Reelin is absent, display severe cortical laminar disruption (Nishikawa *et al.*, 2002).

B. Environmental Factors

1. *Thalidomide and Valproic Acid*

Thalidomide has been associated with a marked increase in the incidence of autism (Stromland *et al.*, 1994). In rats, thalidomide exposure on E9 causes increased plasma, hippocampal, and frontal cortex 5-HT as well as an altered distribution of 5-HT in neurons in the raphe nuclei (Narita *et al.*, 2002). The offspring of women taking valproic acid (VPA) during early pregnancy have an increased risk for autism (Costa *et al.*, 2004). The offspring of pregnant rats given VPA show a reduced number of Purkinje cells, decreased cerebellar volume, and a decreased cell number in the cranial nerve motor nuclei (Ingram *et al.*, 2000). In addition, neurons in the inferior olive that innervate Purkinje cells are also reduced in number, as are those in the deep nuclei targets of Purkinje cells in the nucleus interpositus (Rodier *et al.*, 1996). As with thalidomide, VPA exposure on E9 causes hyperserotonemia in the mouse hippocampus, frontal cortex, and cerebellum (Narita *et al.*, 2002). These observations all parallel human autistic pathologic findings. The use of thalidomide and VPA has provided important insights into autism and has led to useful animal models as well.

2. *Maternal Infection*

Maternal infection increases the risk of autism in the offspring. For example, prenatal exposure to rubella virus increases the incidence of autism (Chess, 1977). Furthermore, pregnant mice infected with influenza virus at E9.5 yield adult offspring that display histological and behavioral abnormalities found in autism and schizophrenia (Shi *et al.*, 2003). These mice have smaller brain sizes at birth, but macroencephaly in adulthood (Fatemi *et al.*, 2002a; Shi *et al.*, 2003). This neonatal undergrowth followed by overgrowth mirrors the pattern seen in autism. With respect to brain pathology, the offspring of maternally infected mice display a selective loss of Purkinje cells in lobule VII, as well as thinning of the neocortex and hippocampus, pyramidal cell atrophy, reduced levels of Reelin immunoreactivity, and changes in neuronal nitric oxide synthase expression and synaptosome associated protein of 25kDa (SNAP-25) (Fatemi *et al.*, 1998, 2000a, 2002a; Shi *et al.*, 2003). Some of these features mimic autism pathology.

3. *Postnatal Viral Infection*

The most extensively studied rodent model involving postnatal viral infection utilizes intracerebral injection of Borna disease virus (BDV) within 12 hrs of birth (Hornig *et al.*, 2003; Pletnikov *et al.*, 2002). Several aspects of this model are relevant to autism, including alterations in 5-HT levels in various brain regions, loss of Purkinje cells in the cerebellum, and granule cells in the dentate gyrus (Dietz *et al.*, 2004). Postnatal infection with lymphocytic choriomeningitis virus

(LCMV) also leads to acute loss of neurons in the cerebellum and delayed loss of neurons in the hippocampus (Pearce, 2003). There is also an association between cytomegalovirus (CMV) infection and autism, which could be followed up in animals (Yamashita *et al.*, 2003).

C. LESIONS

1. *Cerebellum*

Neuropathologic observations in autism are rather consistent in respect to the cerebellum. Therefore, it is of particular interest to study genetic, surgical, and toxin lesions of the cerebellum. Recently, numerous mutations and toxic insults were associated with diverse patterns of Purkinje cell loss (Sarna and Hawkes, 2003). Subpopulations of Purkinje cells are less vulnerable to death, probably because they express the neuroprotective protein HSP25/27. The same feature is seen in mouse models of Niemann-Pick disease type A/B and C (Sarna and Hawkes, 2003). Accordingly, the pattern of cell loss in each case is specific to the type of insult. Thus, the selective loss of Purkinje cells in lobule VII in the influenza model is an important parallel with autism.

The various models discussed here are the primary ones showing neuro-pathologies that mimic some of the features found in autism. However, striking behavioral features of autism can be assayed in animals, such as stereotypic and repetitive behaviors, enhanced anxiety, abnormal pain sensitivity, disturbed sleep patterns, deficient maternal bonding/affiliation, and deficits in sensorimotor gating (prepulse inhibition, PPI). While the maternal infection, thalidomide, and VPA models have been studied behaviorally, much remains to be done in this respect with the other models.

VI. Discussion

In this chapter we have examined the literature on autism neuropathology, the role of the GABAergic system in this disorder, and the relevance of mouse models showing features of autism. With respect to the neuropathology of autism, consistent findings have emerged for the limbic system, cerebellum, and cerebral cortex. Neuropathologic data of the limbic system show increased cell packing density and smaller neurons. These observations might be explained by an arrest of normal development (Kemper and Bauman, 1998). A decreased number of cerebellar Purkinje cells without significant gliosis and features of cortical dys-genesis have been reported by several different research groups. These findings suggest a largely prenatal origin of autism, which is supported by the epidemiological findings with thalidomide, VPA, and maternal infection. Furthermore, age-related abnormalities in the inferior olive and cerebellar nuclei have been found in autism. As the inferior olive project to the cerebellar Purkinje cells, a decrease in Purkinje cell number can be caused by a very early, abnormal development of these projections (Palmen *et al.*, 2004). Cortical dysgenesis might affect correct lamination of the cortex and can cause abnormalities in the process of cell death. In line with this are findings of reduced Reelin levels and Bcl-2. Reelin also plays a role in the development of minicolumns, a feature that has been reported in autism. Furthermore, abnormalities in GABAergic neurons are associated with narrowing of the cell minicolumns. However, those studies involve small sample sizes, the use of quantification techniques not free from bias, and high percentages of autistic subjects with comorbid mental retardation or epilepsy.

As with the cholinergic system, several studies have reported a reduction in GABA function, availability, and activity in autism. Furthermore, a decrease in GABA receptor binding has been shown in autism. An imbalance in the availability of GABA receptors subunits may alter receptor activity and hence change the activity of the brain's major inhibitory neurotransmitter (Schroer *et al.*, 1998). As a consequence, the threshold for developing seizures, a frequent comorbidity of autism, might be reduced. From a genetic point of view, three $GABA_A$ receptor subunit genes have been located in the proximal arm of chromosome 15. Abnormalities in this chromosomal region have been found in a small proportion of autistic people.

As autism is a neurodevelopmental disorder, it is important to elucidate the role of GABA during development. Evidence indicates that GABA plays a role in several developmental processes including cell migration, proliferation, and differentiation. As these processes might be affected in autism, GABA dysfunction could account for some neuropathology seen in autism. For example, GABA dysfunction may occur in conjunction with Reelin and therefore affect neuronal migration and the arrangement of cortical minicolumns in autism. The results reported on the influence of GABA on proliferation are somewhat controversial. A positive influence on GABA cell survival could reflect a GABA mediated decrease in apoptosis during development and in turn might explain the increased head circumference, brain weight, and brain volume reported in autism. However, when abundant cells are not functioning properly, they would die, resulting in normalization of, or a decrease in, brain volume later in childhood. However, in those subgroups of autistic subjects in which this brain enlargement is still present in adulthood, this compensatory cell death may not take place (Courchesne, 2004). Instead, these autistic subjects might be able to recruit these 'extra' neurons, possibly resulting in increased dendritic growth and thus in increased brain volume, still present in adolescence.

Although the neuropathologic results in autistic subjects are revealing, animal models are essential for better understanding of the pathophysiology, cause, and treatment of autism. Although no animal model has been created that perfectly mimics all aspects of a human disease, striking parallels between autism and animal models have already been reported. For example, in genetic models, mouse mutants of En2 display decreased Purkinje cell number (Gharani et al., 2004). In NLGN mouse mutants, a loss of protein processing and a loss of the capacity for synapse stimulation are found (Chih et al., 2004). DLX mutations are associated with abnormalities in the development of GABAergic neurons (Stuhmer et al., 2002). The recent linkages of NLGN, DLX, and En2 to autism offer possibilities for animal models, particularly if introducing the relevant, specific mutations (as opposed to simple KOs) can cause interesting pathology and behavior. In addition, infection with Borna disease virus shows behavioral disturbances in sensorimotor, emotional, and social activity, together with a decrease in the number of Purkinje cells (Hornig et al., 2002). Maternal infection models display selective Purkinje cell loss (Patterson, 2002; Shi et al., 2003). This model also shows macroencephaly in the offspring, atrophy of pyramidal cells, and reduced Reelin immunoreactivity. Furthermore, these mice display deficits in social behavior, social interaction, and PPI.

Acknowledgments

Our research on autism was supported by the Korczak foundation (to H.v.E.), the US National Alliance for Autism Research (to C.S. and P.R.H.), the CAN and McKnight Foundations (to P.H.P.), and by NIH grants MH66392 (to P.R.H.) and MH067978 (to P.H.P.).

References

American Psychiatric Association (1994). "Diagnostic and Statistical Manual of Mental Disorders (DSMIV)." 4th ed. American Psychiatric Association, Washington, DC.

Anderson, G. M., Gutknecht, L., Cohen, D. J., Brailly-Tabard, S., Cohen, J. H., Ferrari, P., Roubertoux, P. L., and Tordjman, S. (2002). Serotonin transporter promoter variants in autism: Functional effects and relationship to platelet hyperserotonemia. *Mol. Psychiatry* **7**, 831–836.

Anderson, S. A., Qiu, M., Bulfone, A., Eisenstat, D. D., Meneses, J., Pedersen, R., and Rubenstein, J. L. (1997). Mutations of the homeobox genes Dlx-1 and Dlx-2 disrupt the striatal subventricular zone and differentiation of late born striatal neurons. *Neuron* **19**, 27–37.

Anderson, S. A., Marin, O., Horn, C., Jennings, K., and Rubenstein, J. L. (2001). Distinct cortical migrations from the medial and lateral ganglionic eminences. *Development* **128**, 353–363.

Andres, C. (1998). Molecular genetics and animal models in autistic disorder. *Brain Res. Bull.* **57**, 109–119.

Aylward, E. H., Minshew, N. J., Goldstein, G., Honeycutt, N. A., Augustine, A. M., Yates, K. O., Barta, P. E., and Pearlson, G. D. (1999). MRI volumes of amygdala and hippocampus in non-mentally retarded autistic adolescents and adults. *Neurology* **53,** 2145–2150.

Aylward, E. H., Minshew, N. J., Field, K., Sparks, B. F., and Singh, N. (2002). Effects of age on brain volume and head circumference in autism. *Neurology* **59,** 175–183.

Bailey, A., Luthert, P., Bolton, P., Le Couteur, A., Rutter, M., and Harding, B. (1993). Autism and megalencephaly. *Lancet* **341,** 1225–1226.

Bailey, A., Phillips, W., and Rutter, M. (1996). Autism: Towards an integration of clinical, genetic, neuropsychological, and neurobiological perspectives. *J. Child Psychol. Psychiatry* **37,** 89–126.

Bailey, A., Luthert, P., Dean, A., Harding, B., Janota, I., Montgomery, M., Rutter, M., and Lantos, P. (1998). A clinicopathological study of autism. *Brain* **121,** 889–905.

Barbin, G., Pollard, H., Gaiarsa, J. L., and Ben-Ari, Y. (1993). Involvement of GABAA receptors in the outgrowth of cultured hippocampal neurons. *Neurosci. Lett.* **152,** 150–154.

Bass, M. P., Menold, M. M., Wolpert, C. M., Donnelly, S. L., Ravan, S. A., Hauser, E. R., Maddox, L. O., Vance, J. M., Abramson, R. K., Wright, H. H., Gilbert, J. R., Cuccaro, M. L., De Long, G. R., and Pericak-Vance, M. A. (2000). Genetic studies in autistic disorder and chromosome 15. *Neurogenetics* **2,** 219–226.

Bauman, M. L. (1991). Microscopic neuroanatomic abnormalities in autism. *Pediatrics* **87,** 791–796.

Bauman, M., and Kemper, T. L. (1985). Histoanatomic observations of the brain in early infantile autism. *Neurology* **35,** 866–874.

Bauman, M. L., and Kemper, T. L. (1987). Limbic involvement in a second case of early infantile autism. *Neurology* **37,** 147.

Bauman, M. L., and Kemper, T. L. (1990). Limbic and cerebellar abnormalities are also present in an autistic child of normal intelligence. *Neurology* **40,** 359.

Bauman, M. L., and Kemper, T. L. (1994). "The Neurobiology of Autism." Johns Hopkins University, Baltimore.

Behar, T. N., Li, Y. X., Tran, H. T., Ma, W., Dunlap, V., Scott, C., and Barker, J. L. (1996). GABA stimulates chemotaxis and chemokinesis of embryonic cortical neurons via calcium-dependent mechanisms. *J. Neurosci.* **16,** 1808–1818.

Behar, T. N., Schaffner, A. E., Scott, C. A., O'Connell, C., and Barker, J. L. (1998). Differential response of cortical plate and ventricular zone cells to GABA as a migration stimulus. *J. Neurosci.* **18,** 6378–6387.

Behar, T. N., Schaffner, A. E., Scott, C. A., Greene, C. L., and Barker, J. L. (2000). GABA receptor antagonists modulate postmitotic cell migration in slice cultures of embryonic rat cortex. *Cereb. Cortex* **10,** 899–909.

Ben-Ari, Y. (2001). Developing networks play a similar melody. *Trends Neurosci.* **24,** 353–360.

Ben-Ari, Y. (2002). Excitatory actions of GABA during development: The nature of the nurture. *Nat. Rev. Neurosci.* **3,** 728–739.

Ben-Ari, Y., Cherubini, E., Corradetti, R., and Gaiarsa, J. L. (1989). Giant synaptic potentials in immature rat CA3 hippocampal neurones. *J. Physiol.* **416,** 303–325.

Ben-Ari, Y., Khalilov, I., Represa, A., and Gozlan, H. (2004). Interneurons set the tune of developing networks. *Trends Neurosci.* **27,** 422–427.

Ben-Yaakov, G., and Golan, H. (2003). Cell proliferation in response to GABA in postnatal hippocampal slice culture. *Int. J. Dev. Neurosci.* **21,** 153–157.

Bespalova, I. N., and Buxbaum, J. D. (2003). Disease susceptibility genes for autism. *Ann. Med.* **35,** 274–281.

Blatt, G. J., Fitzgerald, C. M., Guptill, J. T., Booker, A. B., Kemper, T. L., and Bauman, M. L. (2001). Density and distribution of hippocampal neurotransmitter receptors in autism: An autoradiographic study. *J. Autism. Dev. Disord.* **31,** 537–543.

Bonora, E., Beyer, K. S., Lamb, J. A., Parr, J. R., Klauck, S. M., Benner, A., Paolucci, M., Abbott, A., Ragoussis, I., Poustka, A., Bailey, A. J., and Monaco, A. P. (2003). Analysis of reelin as a candidate gene for autism. *Mol. Psychiatry* **8,** 885–892.

Bulfone, A., Puelles, L., Porteus, M. H., Frohman, M. A., Martin, G. R., and Rubenstein, J. L. (1993). Spatially restricted expression of Dlx-1, Dlx-2 (Tes-1), Gbx-2, and Wnt-3 in the embryonic day 12.5 mouse forebrain defines potential transverse and longitudinal segmental boundaries. *J. Neurosci.* **13,** 3155–3172.

Buxbaum, J. D., Silverman, J. M., Smith, C. J., Greenberg, D. A., Kilifarski, M., Reichert, J., Cook, E. H., Jr., Fang, Y., Song, C. Y., and Vitale, R. (2002). Association between a GABRB3 polymorphism and autism. *Mol. Psychiatry* **7,** 311–316.

Caruncho, H. J., Dopeso-Reyes, I. G., Loza, M. I., and Rodriguez, M. A. (2004). A GABA, reelin, and the neurodevelopmental hypothesis of schizophrenia. *Crit. Rev. Neurobiol.* **16,** 25–32.

Casanova, M. F., Buxhoeveden, D. P., Switala, A. E., and Roy, E. (2002a). Minicolumnar pathology in autism. *Neurology* **58,** 428–432.

Casanova, M. F., Buxhoeveden, D. P., Switala, A. E., and Roy, E. (2002b). Neuronal density and architecture (Gray Level Index) in the brains of autistic patients. *J. Child Neurol.* **17,** 515–521.

Casanova, M. F., Buxhoeveden, D., and Gomez, J. (2003). Disruption in the inhibitory architecture of the cell minicolumn: Implications for autisim. *Neuroscientist* **9,** 496–507.

Casarosa, S., Fode, C., and Guillemot, F. (1999). Mash1 regulates neurogenesis in the ventral telencephalon. *Development* **126,** 525–534.

Chen, G., Trombley, P. Q., and van den Pol, A. N. (1995). GABA receptors precede glutamate receptors in hypothalamic development; differential regulation by astrocytes. *J. Neurophysiol.* **74,** 1473–1484.

Chess, S. (1977). Follow-up report on autism in congenital rubella. *J. Autism Child Schizophr.* **7,** 69–81.

Chih, B., Afridi, S. K., Clark, L., and Scheiffele, P. (2004). Disorder-associated mutations lead to functional inactivation of neuroligins. *Hum. Mol. Genet.* **13,** 1471–1477.

Cohen, B. I. (1999). Elevated levels of plasma and urine gamma-aminobutyric acid – a case study for an autistic child. *Autism* **3,** 437–440.

Cohen, B. I. (2000). Infantile autism and the liver – a possible connection. *Autism* **4,** 441–442.

Cohen, B. I. (2002). Use of a GABA-transaminase agonist for treatment of infantile autism. *Med. Hypotheses* **59,** 115–116.

Comery, T. A., Harris, J. B., Willems, P. J., Oostra, B. A., Irwin, S. A., Weiler, I. J., and Greenough, W. T. (1997). Abnormal dendritic spines in fragile X knockout mice: Maturation and pruning deficits. *Proc. Natl. Acad. Sci. USA* **94,** 5401–5404.

Comoletti, D., De Jaco, A., Jennings, L. L., Flynn, R. E., Gaietta, G., Tsigelny, I., Ellisman, M. H., and Taylor, P. (2004). The Arg451Cys-neuroligin-3 mutation associated with autism reveals a defect in protein processing. *J. Neurosci.* **24,** 4889–4893.

Connell, S., Karikari, C., and Hohmann, C. F. (2004). Sex-specific development of cortical monoamine levels in mouse. *Brain Res. Dev. Brain Res.* **151,** 187–191.

Cook, E. H., and Leventhal, B. L. (1996). The serotonin system in autism. *Curr. Opin. Pediatr.* **8,** 348–354.

Cook, E. H., Jr., Courchesne, R. Y., Cox, N. J., Lord, C., Gonen, D., Guter, S. J., Lincoln, A., Nix, K., Haas, R., Leventhal, B. L., and Courchesne, E. (1998). Linkage-disequilibrium mapping of autistic disorder, with 15q11–13 markers. *Am. J. Hum. Genet.* **62,** 1077–1083.

Costa, L. G., Aschner, M., Vitalone, A., Syversen, T., and Soldin, O. P. (2004). Developmental neuropathology of environmental agents. *Annu. Rev. Pharmacol. Toxicol.* **44,** 87–110.

Courchesne, E., Muller, R. A., and Saitoh, O. (1999). Brain weight in autism: Normal in the majority of cases, megalencephalic in rare cases. *Neurology* **52,** 1057–1059.

Courchesne, E., Karns, C. M., Davis, H. R., Ziccardi, R., Carper, R. A., Tigue, Z. D., Chisum, H. J., Moses, P., Pierce, K., Lord, C., Lincoln, A. J., Pizzo, S., Schreibman, L., Haas, R. H.,

Akshoomoff, N. A., and Courchesne, R. Y. (2001). Unusual brain growth patterns in early life in patients with autistic disorder: An MRI study. *Neurology* **57,** 245–254.

Courchesne, E., Carper, R., and Akshoomoff, N. (2003). Evidence of brain overgrowth in the first year of life in autism. *JAMA* **290,** 337–344.

Courchesne, E. (2004). Brain development in autism: Early overgrowth followed by premature arrest of growth. *Ment. Ret. Dev. Dis.* **10,** 106–111.

Curran, T., and D'Arcangelo, G. (1998). Role of reelin in the control of brain development. *Brain Res. Brain Res. Rev.* **26,** 285–294.

Davidovitch, M., Patterson, B., and Gartside, P. (1996). Head circumference measurements in children with autism. *J. Child Neurol.* **11,** 389–393.

Del Rio, J. A., Martinez, A., Auladell, C., and Soriano, E. (2000). Developmental history of the subplate and developing white matter in the murine neocortex. Neuronal organization and relationship with the main afferent systems at embryonic and perinatal stages. *Cereb. Cortex* **10,** 784–801.

Dhossche, D., Applegate, H., Abraham, A., Maertens, P., Bland, L., Bencsath, A., and Martinez, J. (2002). Elevated plasma gamma-aminobutyric acid (GABA) levels in autistic youngsters: Stimulus for a GABA hypothesis of autism. *Med. Sci. Monit.* **8,** 1–6.

Dietz, D., Vogel, M., Rubin, S., Moran, T., Carbone, K., and Pletnikov, M. (2004). Developmental alterations in serotoninergic neurotransmission in Borna disease virus (BDV)-infected rats: A multidisciplinary analysis. *J. Neurovirol.* **10,** 267–277.

Eisenstat, D. D., Liu, J. K., Mione, M., Zhong, W., Yu, G., Anderson, S. A., Ghattas, I., Puelles, L., and Rubenstein, J. L. (1999). DLX-1, DLX-2, and DLX-5 expression define distinct stages of basal forebrain differentiation. *J. Comp. Neurol.* **414,** 217–237.

Eliason, D. A., Cohen, S. A., Baratta, J., Yu, J., and Robertson, R. T. (2002). Local proliferation of microglia cells in response to neocortical injury *in vitro*. *Brain Res. Dev. Brain Res.* **137,** 75–79.

Fatemi, S. H. (2002). The role of Reelin in pathology of autism. *Mol. Psychiatry* **7,** 919–920.

Fatemi, S. H. (2004). Reelin glycoprotein: Structure, biology and roles in health and disease. *Mol. Psychiatry* **10,** 251–257.

Fatemi, S. H., Sidwell, R., Kist, D., Akhter, P., Meltzer, H. Y., Bailey, K., Thuras, P., and Sedgwick, J. (1998). Differential expression of synaptosome-associated protein 25 kDa [SNAP-25] in hippocampi of neonatal mice following exposure to human influenza virus *in utero*. *Brain Res.* **800,** 1–9.

Fatemi, S. H., Cuadra, A. E., El-Fakahany, E. E., Sidwell, R. W., and Thuras, P. (2000a). Prenatal viral infection causes alterations in nNOS expression in developing mouse brains. *Neuroreport* **11,** 1493–1496.

Fatemi, S. H., Earle, J. A., and McMenomy, T. (2000b). Reduction in Reelin immunoreactivity in hippocampus of subjects with schizophrenia, bipolar disorder and major depression. *Mol. Psychiatry* **5,** 654–663, 571.

Fatemi, S. H., and Halt, A. R. (2001a). Altered levels of Bcl2 and p53 proteins in parietal cortex reflect deranged apoptotic regulation in autism. *Synapse* **42,** 281–284.

Fatemi, S. H., Halt, A. R., Stary, J. M., Realmuto, G. M., and Jalali-Mousavi, M. (2001b). Reduction in anti-apoptotic protein Bcl-2 in autistic cerebellum. *Neuroreport* **12,** 929–933.

Fatemi, S. H., Earle, J., Kanodia, R., Kist, D., Emamian, E. S., Patterson, P. H., Shi, L., and Sidwell, R. (2002a). Prenatal viral infection leads to pyramidal cell atrophy and macrocephaly in adulthood: Implications for genesis of autism and schizophrenia. *Cell. Mol. Neurobiol.* **22,** 25–33.

Fatemi, S. H., Halt, A. R., Realmuto, G., Earle, J., Kist, D. A., Thuras, P., and Merz, A. (2002b). Purkinje cell size is reduced in cerebellum of patients with autism. *Cell. Mol. Neurobiol.* **22,** 171–175.

Fatemi, S. H., Halt, A. R., Stary, J. M., Kanodia, R., Schulz, S. C., and Realmuto, G. R. (2002c). Glutamic acid decarboxylase 65 and 67 kDa proteins are reduced in autistic parietal and cerebellar cortices. *Biol. Psychiatry* **52,** 805–810.

Fatemi, S. H., Snow, A. V., Stary, J. M., Araghi-Niknam, M., Reutiman, T. J., Lee, S., Brooks, A. I., and Pearce, D. A. (2005). Reelin signalling is impaired in autism. *Biol. Psychiatry* **57,** 777–787.

Fiszman, M. L., Borodinsky, L. N., and Neale, J. H. (1999). GABA induces proliferation of immature cerebellar granule cells grown *in vitro. Brain Res. Dev. Brain Res.* **115,** 1–8.

Fode, C., Ma, Q., Casarosa, S., Ang, S. L., Anderson, D. J., and Guillemot, F. (2000). A role for neural determination genes in specifying the dorsoventral identity of telencephalic neurons. *Genes Dev.* **14,** 67–80.

Fombonne, E. (2000). Is a large head circumference a sign of autism? *J. Autism Dev. Disord.* **30,** 365.

Fritschy, J. M., Paysan, J., Enna, A., and Mohler, H. (1994). Switch in the expression of rat GABAA-receptor subtypes during postnatal development: An immunohistochemical study. *J. Neurosci.* **14,** 5302–5324.

Galvez, R., Gopal, A. R., and Greenough, W. T. (2003). Somatosensory cortical barrel dendritic abnormalities in a mouse model of the fragile X mental retardation syndrome. *Brain Res.* **971,** 83–89.

Garreau, B., Herry, D., Zilbovicius, M., Samson, Y., Guerin, P., and Lelord, G. (1993). Theoretical aspects of the study of benzodiazepine receptors in infantile autism. *Acta Paedopsychiatr.* **56,** 133–138.

Gaspar, P., Cases, O., and Maroteaux, L. (2003). The developmental role of serotonin: News from mouse molecular genetics. *Nat. Rev. Neurosci.* **4,** 1002–1012.

Gharani, N., Benayed, R., Mancuso, V., Brzustowicz, L. M., and Millonig, J. H. (2004). Association of the homeobox transcription factor, ENGRAILED 2, 3, with autism spectrum disorder. *Mol. Psychiatry* **9,** 474–484.

Gingrich, J. A., Ansorge, M. S., Merker, R., Weisstaub, N., and Zhou, M. (2003). New lessons from knockout mice: The role of serotonin during development and its possible contribution to the origins of neuropsychiatric disorders. *CNS Spectr.* **8,** 572–577.

Goddard, A. W., Mason, G. F., Almai, A., Rothman, D. L., Behar, K. L., Petroff, O. A., Charney, D. S., and Krystal, J. H. (2001). Reductions in occipital cortex GABA levels in panic disorder detected with 1h-magnetic resonance spectroscopy. *Arch. Gen. Psychiatry* **58,** 556–561.

Gomez, T. M., and Spitzer, N. C. (1999). *In vivo* regulation of axon extension and pathfinding by growth-cone calcium transients. *Nature* **397,** 350–355.

Gonzalez, J. L., Russo, C. J., Goldowitz, D., Sweet, H. O., Davisson, M. T., and Walsh, C. A. (1997). Birthdate and cell marker analysis of scrambler: A novel mutation affecting cortical development with a reeler-like phenotype. *J. Neurosci.* **17,** 9204–9211.

Guerin, P., Lyon, G., Barthelemy, C., Sostak, E., Chevrollier, V., Garreau, B., and Lelord, G. (1996). Neuropathological study of a case of autistic syndrome with severe mental retardation. *Dev. Med. Child. Neurol.* **38,** 203–211.

Guidotti, A., Auta, J., Davis, J. M., Di-Giorgi-Gerevini, V., Dwivedi, Y., Grayson, D. R., Impagnatiello, F., Pandey, G., Pesold, C., Sharma, R., Uzunov, D., and Costa, E. (2000). Decrease in reelin and glutamic acid decarboxylase67 (GAD67) expression in schizophrenia and bipolar disorder: A postmortem brain study. *Arch. Gen. Psychiatry* **57,** 1061–1069.

Guillemot, F., and Joyner, A. L. (1993). Dynamic expression of the murine Achaete-Scute homologue Mash-1 in the developing nervous system. *Mech. Dev.* **42,** 171–185.

Hansen, G. H., Meier, E., Abraham, J., and Schousboe, A. (1987). Trophic effects of GABA on cerebellar granule cells in culture. *In* "Neurology and Neurobiology; Neurotrophic Activity of GABA During Development" (D. A. Redbum, Ed.), Vol. 32, pp. 109–138. Liss, New York.

Hashimoto, T., Tayama, M., Murakawa, K., Yoshimoto, T., Miyazaki, M., and Harada, M. (1995). Development of the brainstem and cerebellum in autistic patients. *J. Autism. Dev. Disord.* **25,** 1–18.

Hatten, M. E. (1999). Central nervous system neuronal migration. *Annu. Rev. Neurosci.* **22,** 511–539.

Haydar, T. F., Wang, F., Schwartz, M. L., and Rakic, P. (2000). Differential modulation of proliferation in the neocortical ventricular and subventricular zones. *J. Neurosci.* **20,** 5764–5774.

Herlenius, E., and Lagercrantz, H. (2001). Neurotransmitters and neuromodulators during early human development. *Early Hum. Dev.* **65,** 21–37.

Herlenius, E., and Lagercrantz, H. (2004). Development of neurotransmitter systems during critical periods. *Exp. Neurol.* **190**(Suppl. 1), S8–S21.

Hohmann, C. F., and Berger-Sweeney, J. (1998). Cholinergic regulation of cortical development and plasticity. New twists to an old story. *Perspect. Dev. Neurobiol.* **5,** 401–425.

Hornig, M., Mervis, R., Hoffman, K., and Lipkin, W. I. (2002). Infectious and immune factors in neurodevelopmental damage. *Mol. Psychiatry* **7,** S34–S35.

Hornig, M., Briese, T., and Lipkin, W. I. (2003). Borna disease virus. *J. Neurovirol.* **9,** 259–273.

Howard, M. A., Cowell, P. E., Boucher, J., Broks, P., Mayes, A., Farrant, A., and Roberts, N. (2000). Convergent neuroanatomical and behavioural evidence of an amygdala hypothesis of autism. *Neuroreport.* **11,** 2931–2935.

Hutsler, J., and Galuske, R. A. (2003). Hemispheric asymmetries in cerebral cortical networks. *Trends Neurosci.* **26,** 429–435.

Ingram, J. L., Peckham, S. M., Tisdale, B., and Rodier, P. M. (2000). Prenatal exposure of rats to valproic acid reproduces the cerebellar anomalies associated with autism. *Neurotoxicol. Teratol.* **22,** 319–324.

Irwin, S. A., Patel, B., Idupulapati, M., Harris, J. B., Crisostomo, R. A., Larsen, B. P., Kooy, F., Willems, P. J., Cras, P., Kozlowski, P. B., Swain, R. A., Weiler, I. J., and Greenough, W. T. (2001). Abnormal dendritic spine characteristics in the temporal and visual cortices of patients with fragile-X syndrome: A quantitative examination. *Am. J. Med. Genet.* **98,** 161–167.

Jacobsen, M. (1991). "Developmental Neurobiology." Plenum Press, New York.

Jamain, S., Quach, H., Betancur, C., Rastam, M., Colineaux, C., Gillberg, I. C., Soderstrom, H., Giros, B., Leboyer, M., Gillberg, C., and Bourgeron, T. (2003). Mutations of the X-linked genes encoding neuroligins NLGN3 and NLGN4 are associated with autism. *Nat. Genet.* **34,** 27–29.

Jelitai, M., Anderova, M., Marko, K., Kekesi, K., Koncz, P., Sykova, E., and Madarasz, E. (2004). Role of gamma-aminobutyric acid in early neuronal development: Studies with an embryonic neuroectodermal stem cell clone. *J. Neurosci. Res.* **76,** 801–811.

Jiang, Y. H., Armstrong, D., Albrecht, U., Atkins, C. M., Noebels, J. L., Eichele, G., Sweatt, J. D., and Beaudet, A. L. (1998). Mutation of the Angelman ubiquitin ligase in mice causes increased cytoplasmic p53 and deficits of contextual learning and long-term potentiation. *Neuron* **21,** 799–811.

Jiang, Y. H., and Beaudet, A. L. (2004). Human disorders of ubiquitination and proteasomal degradation. *Curr. Opin. Pediatr.* **16,** 419–426.

Kemper, T. L., and Bauman, M. L. (1993). The contribution of neuropathologic studies to the understanding of autism. *Neurol. Clin.* **11,** 175–187.

Kemper, T. L., and Bauman, M. (1998). Neuropathology of infantile autism. *J. Neuropathol. Exp. Neurol.* **57,** 645–652.

Koller, H., Siebler, M., Schmalenbach, C., and Muller, H. W. (1990). GABA and glutamate receptor development of cultured neurons from rat hippocampus, septal region, and neocortex. *Synapse* **5,** 59–64.

Krebs, M. O., Betancur, C., Leroy, S., Bourdel, M. C., Gillberg, C., and Leboyer, M. (2002). Absence of association between a polymorphic GGC repeat in the $5'$ untranslated region of the reelin gene and autism. *Mol. Psychiatry* **7,** 801–804.

Kriegstein, A. R., and Owens, D. F. (2001). GABA may act as a self-limiting trophic factor at developing synapses. *Sci. STKE. 2001,* PE1.

Kuemerle, B., Zanjani, H., Joyner, A., and Herrup, K. (1997). Pattern deformities and cell loss in Engrailed-2 mutant mice suggest two separate patterning events during cerebellar development. *J. Neurosci.* **17,** 7881–7889.

Laumonnier, F., Bonnet-Brilhault, F., Gomot, M., Blanc, R., David, A., Moizard, M. P., Raynaud, M., Ronce, N., Lemonnier, E., Calvas, P., Laudier, B., Chelly, J., Fryns, J. P., Ropers, H. H., Hamel, B. C., Andres, C., Barthelemy, C., Moraine, C., and Briault, S. (2004). X-linked mental retardation and autism are associated with a mutation in the NLGN4 gene, a member of the neuroligin family. *Am. J. Hum. Genet.* **74,** 552–557.

Laurie, D. J., Wisden, W., and Seeburg, P. H. (1992). The distribution of thirteen GABAA receptor subunit mRNAs in the rat brain. III. Embryonic and postnatal development. *J. Neurosci.* **12,** 4151–4172.

Lee, M., Martin-Ruiz, C., Graham, A., Court, J., Jaros, E., Perry, R., Iversen, P., Bauman, M., and Perry, E. (2002). Nicotinic receptor abnormalities in the cerebellar cortex in autism. *Brain* **125,** 1483–1495.

Liu, A., and Joyner, A. L. (2001). Early anterior/posterior patterning of the midbrain and cerebellum. *Annu. Rev. Neurosci.* **24,** 869–896.

Livesey, F. J., and Hunt, S. P. (1997). Netrin and netrin receptor expression in the embryonic mammalian nervous system suggests roles in retinal, striatal, nigral, and cerebellar development. *Mol. Cell. Neurosci.* **8,** 417–429.

Lopez-Bendito, G., Lujan, R., Shigemoto, R., Ganter, P., Paulsen, O., and Molnar, Z. (2003). Blockade of GABA(B) receptors alters the tangential migration of cortical neurons. *Cereb. Cortex* **13,** 932–942.

LoTurco, J. J., Owens, D. F., Heath, M. J., Davis, M. B., and Kriegstein, A. R. (1995). GABA and glutamate depolarize cortical progenitor cells and inhibit DNA synthesis. *Neuron* **15,** 1287–1298.

Lujan, R., Shigemoto, R., and Lopez-Bendito, G. (2005). Glutamate and GABA receptor signalling in the developing brain. *Neuroscience* **130,** 567–580.

Luk, K. C., and Sadikot, A. F. (2001). GABA promotes survival but not proliferation of parvalbumin-immunoreactive interneurons in rodent neostriatum: An *in vivo* study with stereology. *Neuroscience* **104,** 93–103.

Maestrini, E., Lai, C., Marlow, A., Matthews, N., Wallace, S., Bailey, A., Cook, E. H., Weeks, D. E., and Monaco, A. P. (1999). Serotonin transporter (5-HTT) and gamma-aminobutyric acid receptor subunit beta3 (GABRB3) gene polymorphisms are not associated with autism in the IMGSA families. The International Molecular Genetic Study of Autism Consortium. *Am. J. Med. Genet.* **88,** 492–496.

Maric, D., Liu, Q. Y., Maric, I., Chaudry, S., Chang, Y. H., Smith, S. V., Sieghart, W., Fritschy, J. M., and Barker, J. L. (2001). GABA expression dominates neuronal lineage progression in the embryonic rat neocortex and facilitates neurite outgrowth via GABA(A) autoreceptor/Cl- channels. *J. Neurosci.* **21,** 2343–2360.

Marin, O., Anderson, S. A., and Rubenstein, J. L. (2000). Origin and molecular specification of striatal interneurons. *J. Neurosci.* **20,** 6063–6076.

Marin, O., and Rubenstein, J. L. (2001). A long, remarkable journey: Tangential migration in the telencephalon. *Nat. Rev. Neurosci.* **2,** 780–790.

Martin, E. R., Menold, M. M., Wolpert, C. M., Bass, M. P., Donnelly, S. L., Ravan, S. A., Zimmerman, A., Gilbert, J. R., Vance, J. M., Maddox, L. O., Wright, H. H., Abramson, R. K., De Long, G. R., Cuccaro, M. L., and Pericak-Vance, M. A. (2000). Analysis of linkage disequilibrium in gamma-aminobutyric acid receptor subunit genes in autistic disorder. *Am. J. Med. Genet.* **96,** 43–48.

Menold, M. M., Shao, Y., Wolpert, C. M., Donnelly, S. L., Raiford, K. L., Martin, E. R., Ravan, S. A., Abramson, R. K., Wright, H. H., Delong, G. R., Cuccaro, M. L., Pericak-Vance, M. A., and Gilbert, J. R. (2001). Association analysis of chromosome 15 gabaa receptor subunit genes in autistic disorder. *J. Neurogenet.* **15**, 245–259.

Mienville, J. M., and Pesold, C. (1999). Low resting potential and postnatal upregulation of NMDA receptors may cause Cajal-Retzius cell death. *J. Neurosci.* **19**, 1636–1646.

Miles, R. (1999). Neurobiology. A homeostatic switch. *Nature* **397**, 215–216.

Muhle, R., Trentacoste, S. V., and Rapin, I. (2004). The genetics of autism. *Pediatrics* **113**, 472–486.

Murakami, J. W., Courchesne, E., Press, G. A., Yeung-Courchesne, R., and Hesselink, J. R. (1989). Reduced cerebellar hemisphere size and its relationship to vermal hypoplasia in autism. *Arch. Neurol.* **46**, 689–694.

Narita, N., Kato, M., Tazoe, M., Miyazaki, K., Narita, M., and Okado, N. (2002). Increased monoamine concentration in the brain and blood of fetal thalidomide- and valproic acid-exposed rat: Putative animal models for autism. *Pediatr. Res.* **52**, 576–579.

Nguyen, L., Rigo, J. M., Rocher, V., Belachew, S., Malgrange, B., Rogister, B., Leprince, P., and Moonen, G. (2001). Neurotransmitters as early signals for central nervous system development. *Cell Tissue Res.* **305**, 187–202.

Nimchinsky, E. A., Oberlander, A. M., and Svoboda, K. (2001). Abnormal development of dendritic spines in FMR1 knock-out mice. *J. Neurosci.* **21**, 5139–5146.

Nishikawa, S., Goto, S., Hamasaki, T., Yamada, K., and Ushio, Y. (2002). Involvement of reelin and Cajal-Retzius cells in the developmental formation of vertical columnar structures in the cerebral cortex: Evidence from the study of mouse presubicular cortex. *Cereb. Cortex* **12**, 1024–1030.

Nowell, M. A., Hackney, D. B., Muraki, A. S., and Coleman, M. (1990). Varied MR appearance of autism: Fifty-three pediatric patients having the full autistic syndrome. *Magn. Reson. Imaging* **8**, 811–816.

O'Rourke, N. A., Chenn, A., and McConnell, S. K. (1997). Postmitotic neurons migrate tangentially in the cortical ventricular zone. *Development* **124**, 997–1005.

Owens, D. F., Boyce, L. H., Davis, M. B., and Kriegstein, A. R. (1996). Excitatory GABA responses in embryonic and neonatal cortical slices demonstrated by gramicidin perforated-patch recordings and calcium imaging. *J. Neurosci.* **16**, 6414–6423.

Owens, D. F., Liu, X., and Kriegstein, A. R. (1999). Changing properties of GABA(A) receptor-mediated signalling during early neocortical development. *J. Neurophysiol.* **82**, 570–583.

Palmen, S. J. M. C., Van Engeland, H., Hof, P. R., and Schmitz, C. (2004). Neuropathological findings in autism. *Brain* **127**, 1–12.

Parnavelas, J. G. (2000). The origin and migration of cortical neurones: New vistas. *Trends Neurosci.* **23**, 126–131.

Patterson, P. H. (2002). Maternal infection: Window on neuroimmune interactions in fetal brain development and mental illness. *Curr. Opin. Neurobiol.* **12**, 115–118.

Pearce, B. D. (2003). Modeling the role of infections in the etiology of mental illness. *Clin. Neurosci. Res.* **3**, 271.

Perry, E. K., Lee, M. L., Martin-Ruiz, C. M., Court, J. A., Volsen, S. G., Merrit, J., Folly, E., Iversen, P. E., Bauman, M. L., Perry, R. H., and Wenk, G. L. (2001). Cholinergic activity in autism: Abnormalities in the cerebral cortex and basal forebrain. *Am. J. Psychiatry* **158**, 1058–1066.

Persico, A. M., D'Agruma, L., Maiorano, N., Totaro, A., Militerni, R., Bravaccio, C., Wassink, T. H., Schneider, C., Melmed, R., Trillo, S., Montecchi, F., Palermo, M., Pascucci, T., Puglisi-Allegra, S., Reichelt, K. L., Conciatori, M., Marino, R., Quattrocchi, C. C., Baldi, A., Zelante, L., Gasparini, P., and Keller, F. (2001). Reelin gene alleles and haplotypes as a factor predisposing to autistic disorder. *Mol. Psychiatry* **6**, 150–159.

Piven, J., Berthier, M. L., Starkstein, S. E., Nehme, E., Pearlson, G., and Folstein, S. (1990). Magnetic resonance imaging evidence for a defect of cerebral cortical development in autism. *Am. J. Psychiatry* **147**, 734–739.

Piven, J., Saliba, K., Bailey, J., and Arndt, S. (1997). An MRI study of autism: The cerebellum revisited. *Neurology* **49,** 546–551.

Piven, J., Bailey, J., Ranson, B. J., and Arndt, S. (1998). No difference in hippocampus volume detected on magnetic resonance imaging in autistic individuals. *J. Autism. Dev. Disord.* **28,** 105–110.

Pletnikov, M. V., Rubin, S. A., Vogel, M. W., Moran, T. H., and Carbone, K. M. (2002). Effects of genetic background on neonatal Borna disease virus infection-induced neurodevelopmental damage. II. Neurochemical alterations and responses to pharmacological treatments. *Brain Res.* **944,** 108–123.

Porteus, M. H., Bulfone, A., Liu, J. K., Puelles, L., Lo, L. C., and Rubenstein, J. L. (1994). DLX-2, MASH-1, and MAP-2 expression and bromodeoxyuridine incorporation define molecularly distinct cell populations in the embryonic mouse forebrain. *J. Neurosci.* **14,** 6370–6383.

Powell, E. M., Mars, W. M., and Levitt, P. (2001). Hepatocyte growth factor/scatter factor is a motogen for interneurons migrating from the ventral to dorsal telencephalon. *Neuron* **30,** 79–89.

Rakic, P. (1995). Radial versus tangential migration of neuronal clones in the developing cerebral cortex. *Proc. Natl. Acad. Sci. USA* **92,** 11323–11327.

Rapp, P. R., and Bachevalier, J. (1993). "Fundamental Neuroscience." Academic Press, San Diego, California.

Raymond, G. V., Bauman, M. L., and Kemper, T. L. (1996). Hippocampus in autism: A Golgi analysis. *Acta Neuropathol. (Berl.)* **91,** 117–119.

Ritvo, E. R., Freeman, B. J., Scheibel, A. B., Duong, T., Robinson, H., Guthrie, D., and Ritvo, A. (1986). Lower Purkinje cell counts in the cerebella of four autistic subjects: Initial findings of the UCLA-NSAC Autopsy Research Report. *Am. J. Psychiatry* **143,** 862–866.

Rivera, C., Voipio, J., Payne, J. A., Ruusuvuori, E., Lahtinen, H., Lamsa, K., Pirvola, U., Saarma, M., and Kaila, K. (1999). The K+/Cl− co-transporter KCC2 renders GABA hyperpolarizing during neuronal maturation. *Nature* **397,** 251–255.

Rodier, P. M., Ingram, J. L., Tisdale, B., Nelson, S., and Romano, J. (1996). Embryological origin for autism: Developmental anomalies of the cranial nerve motor nuclei. *J. Comp. Neurol.* **370,** 247–261.

Rolf, L. H., Haarmann, F. Y., Grotemeyer, K. H., and Kehrer, H. (1993). Serotonin and amino acid content in platelets of autistic children. *Acta. Psychiatr. Scand.* **87,** 312–316.

Sanacora, G., Mason, G. F., Rothman, D. L., Behar, K. L., Hyder, F., Petroff, O. A., Berman, R. M., Charney, D. S., and Krystal, J. H. (1999). Reduced cortical gamma-aminobutyric acid levels in depressed patients determined by proton magnetic resonance spectroscopy. *Arch. Gen. Psychiatry* **56,** 1043–1047.

Sarna, J. R., and Hawkes, R. (2003). Patterned Purkinje cell death in the cerebellum. *Prog. Neurobiol.* **70,** 473–507.

Schroer, R. J., Phelan, M. C., Michaelis, R. C., Crawford, E. C., Skinner, S. A., Cuccaro, M., Simensen, R. J., Bishop, J., Skinner, C., Fender, D., and Stevenson, R. E. (1998). Autism and maternally derived aberrations of chromosome 15q. *Am. J. Med. Genet.* **76,** 327–336.

Shi, L., Fatemi, S. H., Sidwell, R. W., and Patterson, P. H. (2003). Maternal influenza infection causes marked behavioral and pharmacological changes in the offspring. *J. Neurosci.* **23,** 297–302.

Silva, A. E., Vayego-Lourenco, S. A., Fett-Conte, A. C., Goloni-Bertollo, E. M., and Varella-Garcia, M. (2002). Tetrasomy 15q11-q13 identified by fluorescence in situ hybridization in a patient with autistic disorder. *Arq. Neuropsiquiatr.* **60,** 290–294.

Skaliora, I., Singer, W., Betz, H., and Puschel, A. W. (1998). Differential patterns of semaphorin expression in the developing rat brain. *Eur. J. Neurosci.* **10,** 1215–1229.

Somogyi, R., Wen, X., Ma, W., and Barker, J. L. (1995). Developmental kinetics of GAD family mRNAs parallel neurogenesis in the rat spinal cord. *J. Neurosci.* **15,** 2575–2591.

Sparks, B. F., Friedman, S. D., Shaw, D. W., Aylward, E. H., Echelard, D., Artru, A. A., Maravilla, K. R., Giedd, J. N., Munson, J., Dawson, G., and Dager, S. R. (2002). Brain structural abnormalities in young children with autism spectrum disorder. *Neurology* **59,** 184–192.

Stein, V., Hermans-Borgmeyer, I., Jentsch, T. J., and Hubner, C. A. (2004). Expression of the KCl cotransporter KCC2 parallels neuronal maturation and the emergence of low intracellular chloride. *J. Comp. Neurol.* **468,** 57–64.

Stromland, K., Nordin, V., Miller, M., Akerstrom, B., and Gillberg, C. (1994). Autism in thalidomide embryopathy: A population study. *Dev. Med. Child. Neurol.* **36,** 351–356.

Stuhmer, T., Anderson, S. A., Ekker, M., and Rubenstein, J. L. (2002). Ectopic expression of the Dlx genes induces glutamic acid decarboxylase and Dlx expression. *Development* **129,** 245–252.

Tillakaratne, N. J., Medina-Kauwe, L., and Gibson, K. M. (1995). gamma-Aminobutyric acid (GABA) metabolism in mammalian neural and nonneural tissues. *Comp. Biochem. Physiol. A. Physiol.* **112,** 247–263.

Tueting, P., Costa, E., Dwivedi, Y., Guidotti, A., Impagnatiello, F., Manev, R., and Pesold, C. (1999). The phenotypic characteristics of heterozygous reeler mouse. *Neuroreport* **10,** 1329–1334.

Tyzio, R., Represa, A., Jorquera, I., Ben-Ari, Y., Gozlan, H., and Aniksztejn, L. (1999). The establishment of GABAergic and glutamatergic synapses on CA1 pyramidal neurons is sequential and correlates with the development of the apical dendrite. *J. Neurosci.* **19,** 10372–10382.

van Karnebeek, C. D., van Gelderen, I., Nijhof, G. J., Abeling, N. G., Vreken, P., Redeker, E. J., van Eeghen, A. M., Hoovers, J. M., and Hennekam, R. C. (2002). An aetiological study of 25 mentally retarded adults with autism. *J. Med. Genet.* **39,** 205–213.

Walton, M. K., Schaffner, A. E., and Barker, J. L. (1993). Sodium channels, GABAA receptors, and glutamate receptors develop sequentially on embryonic rat spinal cord cells. *J. Neurosci.* **13,** 2068–2084.

Wassink, T. H., Piven, J., and Patil, S. R. (2001). Chromosomal abnormalities in a clinic sample of individuals with autistic disorder. *Psychiatr. Genet.* **11,** 57–63.

Whitaker-Azmitia, P. M. (2001). Serotonin and brain development: Role in human developmental diseases. *Brain Res. Bull.* **56,** 479–485.

Wichterle, H., Turnbull, D. H., Nery, S., Fishell, G., and Alvarez-Buylla, A. (2001). *In utero* fate mapping reveals distinct migratory pathways and fates of neurons born in the mammalian basal forebrain. *Development* **128,** 3759–3771.

Williams, R. S., Hauser, S. L., Purpura, D. P., De Long, G. R., and Swisher, C. N. (1980). Autism and mental retardation: Neuropathologic studies performed in four retarded persons with autistic behavior. *Arch. Neurol.* **37,** 749–753.

Yamashita, Y., Fujimoto, C., Nakajima, E., Isagai, T., and Matsuishi, T. (2003). Possible association between congenital cytomegalovirus infection and autistic disorder. *J. Autism Dev. Disord.* **33,** 455–459.

Yuan, W., Zhou, L., Chen, J. H., Wu, J. Y., Rao, Y., and Ornitz, D. M. (1999). The mouse SLIT family: Secreted ligands for ROBO expressed in patterns that suggest a role in morphogenesis and axon guidance. *Dev. Biol.* **212,** 290–306.

THE ROLE OF GABA IN THE EARLY NEURONAL DEVELOPMENT

Marta Jelitai and Emília Madarasz

Laboratory of Neural Cell and Developmental Biology
Institute of Experimental Medicine, Hungarian Academy of Sciences
Budapest 1083, Hungary

A large body of experimental evidences demonstrates that GABA signaling possess an inherent capability for potent regulation of almost all steps of neuronal differentiation and neural tissue formation. The transient GABA production is an inherent feature of many differentiating neuronal populations, regardless of the future neurotransmitter phenotype.

While in the mature CNS GABA acts mainly as a synaptic neurotransmitter and elicits phasic responses, in the developing nervous tissue GABA acts as an autocrine/paracrine signal molecule and its main roles are executed through tonic signaling. In young differentiating neurons, most important effects of GABA are mediated through $GABA_A$ receptors and result in membrane depolarization, and in an increased $[Ca^{2+}]_I$, which in turn can stimulate multiple cellular processes. $GABA_B$ receptor-mediated effects, which can cause hyperpolarization and reduce $[Ca^{2+}]_I$, appear in later phases of tissue genesis, and seem to balance $GABA_A$ signaling.

The effects of GABA had been demonstrated in the entire period of neural tissue genesis, from proliferation, through migration and differentiation of

INTERNATIONAL REVIEW OF
NEUROBIOLOGY, VOL. 71
DOI: 10.1016/S0074-7742(05)71002-3

27

neuronal precursors, up to the synapse formation and circuit refinement by maturing neurons. The most intriguing developmental role that has been attributed to GABA is the generation and maintenance of activity waves in the period of functional network formation.

I. Introduction

Nearly all organisms contain and synthesize γ-aminobutyric acid (GABA), from bacteria (Ackermann, 1910) through plants (Steward et al., 1949) up to humans (Elliott and Jasper, 1959). In plants, GABA serves as an important metabolic compound, but its role as a signal molecule was also demonstrated (Bouche et al., 2003). The presence of GABA and the similarity of GABA signaling in a broad range of tissues and organisms argue for the importance of phylogenetically conserved GABA-functions.

In the mammalian brain, GABA was recognized more than 50 years ago (Awapara et al., 1950). Soon after, strong evidence indicated that GABA acts as an inhibitory neurotransmitter in both the vertebrate and invertebrate nervous system. Beside the accumulation at presynapses and in the postsynaptic densities, different types of GABA receptors were shown in extrasynaptic locations on various types of neurons (Farrant and Nusser, 2005) and also on non-neuronal cells (Watanabe et al., 2002). During the last 20 years, it was revealed that most neuronal precursors respond to GABA by depolarization in defined phases of differentiation. GABA-induced anion fluxes result in hyperpolarization in later stages of development, in conjunction with the maturation of neuronal ion-homeostasis (Owens and Kriegstein, 2002). In the developing mammalian central nervous system, many neurons and glial cells produce and release GABA. Besides its excitatory or inhibitory neurotransmitter functions, GABA serves also as an autocrine/paracrine signal molecule, and plays important regulatory roles in the entire period of neural development.

In this chapter we give a summary of the recent knowledge of the development of GABA signaling and of the GABA actions, which contribute to the formation of the neural tissue. The first part describes the developmental changes in the composition and distribution of the GABA signaling system. In the second part, we intend to summarize the recent understanding of GABA effects on distinct steps of neuronal differentiation. The chapter does not focus on glial development, and the important modulatory roles of glial cells in almost all aspects of GABA signaling are only occasionally mentioned.

II. The GABA Signaling System in the Developing CNS

Components of the GABA signaling system, GABA and GABA synthesizing enzymes, GABA receptors, GABA transporters, and GABA metabolizing mechanisms appear early in development (Barker *et al.*, 1998). In many areas of the developing nervous system, the components of GABA signaling machinery show transient overexpression.

A. THE PRESENCE OF GABA IN THE DEVELOPING BRAIN TISSUE

In the extracellular environment of the adult brain, GABA is present in concentrations between 0.5–1 μM (Lerma *et al.*, 1986). In the surroundings of adult progenitors in the subventricular zone, however, elevated (2–3 μM) GABA concentrations were reported (Bolteus *et al.*, 2005). In immature human brain, the GABA content in the cerebro-spinal fluid exceeds the adult level (Hedner *et al.*, 1982). The environment of developing neural cells is enriched in GABA (Benitez-Diaz *et al.*, 2003).

The elevated GABA-level in the developing neural tissue is the result, in part, of immature functioning of the blood brain barrier (Engelhardt, 2003). In neonates, enhanced transport-rate of GABA was demonstrated from the choroid plexus to the CSF (Al-Sarraf, 2002).

An important shift in the balance between GABA release and uptake has been shown as another source to enhance the extracellular GABA concentration in the developing neural tissue. Many populations of developing neuronal precursors produce and release GABA in defined stages of differentiation. *In vitro* studies demonstrated a continuous GABA-release from developing hippocampal cells (Valeyev *et al.*, 1993) and revealed a continuously elevated (3–10 μM) GABA concentration in the environment of neural stem cells (Jelitai *et al.*, 2004). Immunocytochemical studies on young embryos showed that GABA-containing cells and fibers are present, well before the onset of synaptogenesis (see Box 1). In the period of the formation of the cortical plate, the early developing Cajal-Retzius cells (Marin-Padilla, 1998) and cortico-petal fibre tracts contain GABA (Lauder *et al.*, 1986). Sparse GABA-containing cells were also detected in the developing rat striatum as early as E13 (Fiszman *et al.*, 1993). The number of GABA-positive cells increased, in parallel with the immigration of future interneurons derived from the subventricular zone (Lavdas *et al.*, 1999). By E16, many layers of the developing rat forebrain cortex contain GABA-positive cells (Lauder *et al.*, 1986). GABA-containing cells were also demonstrated in the human neocortex in early periods of its organization (Zecevic and Milosevic, 1997). Besides some transient "organizer" and future GABAergic neurons (Box 1), many precursors also seem to produce and release GABA, transiently (Verney, 2003).

Box 1

The origin and development of GABAergic neurons is best understood in the development of the cerebral cortex. As a first sign of cortical development, Cajal-Retzius cells are migrating from the ventricular zone to the marginal zone along radial glial cells (Rakic, 1972). These cells produce reelin (D'Arcangelo *et al.*, 1995) and GABA and play an important role in directing the arrangement of later arriving neuronal precursors.

The majority of cortical GABAergic interneuron precursors are generated in the basal ganglia anlage (Anderson *et al.*, 1999), in the medial (MGE) lateral (LGE), caudal (CGE), and retro-bulbar ganglionic eminences (Xu *et al.*, 2004), and migrate to their cortical destination by taking tangential and radial routes (Kriegstein and Noctor, 2004). The precise origin and migratory paths of GABAergic interneuron populations may differ in different species. In humans, more than half of the cortical interneurons were reported to derive from the cortical subventricular zone (Letinic *et al.*, 2002).

GABA is distributed diffusely throughout the cytoplasm of immature neuronal precursors, as vesicular GABA storage mechanisms develop relatively late (Minelli *et al.*, 2003a). Spontaneous GABA release was shown from cell bodies, growth cones, and along the neurites of developing neurons (Balcar *et al.*, 1983). The mechanism of spontaneous GABA release, however, is not properly understood. *In vitro* (Taylor *et al.*, 1990) and *in vivo* (Demarque *et al.*, 2002) evidences indicate that there is an early, Ca-independent GABA release, which can be provoked by K^+-influx, and so, presumably, by depolarization.

Later, in the periods of neuronal network formation, GABA is secreted—at least in part—through reverse action of transporters (Attwell *et al.*, 1993), which can be stimulated by mild depolarization (Belhage *et al.*, 1993).

Specific GABA transporters (GATs) (see Box 2) have a pivotal role in the maintenance of extracellular GABA concentration. In adults, GATs can terminate GABA signals by reuptake of GABA into nerve terminals and into the surrounding glial cells. In early phases of the neural tissue development, however, GABA-uptake seems to underfunction (Demarque *et al.*, 2002).

B. GABA TRANSPORTERS

GABA transporters display characteristic cellular distribution and regional expression patterns (Borden, 1996). To date, four GABA transporters, GAT1-3 and BGT-1 (in nomenclature used for mouse GAT1-4) (Liu *et al.*, 1993) have been cloned (Guastella *et al.*, 1990).

Box 2

The cerebral cortex contains GAT-1 at the highest level, and GAT-3, GAT-2, and BGT-1 in decreasing amount, respectively. GAT-1 and GAT-3 are expressed exclusively in the brain. The various GATs display different ion-dependency and pharmacological properties (Gadea and Lopez-Colome, 2001). The transporters can undergo rapid redistribution between surface and intracellular compartments, and their function can be regulated by phosphorylation (Hansra *et al.*, 2004) and by intermolecular interactions (Quick *et al.*, 2004).

In mature GABAergic neurons, a distinct carrier, the vesicular GABA transporter (VGAT), sorts GABA in at least two separate pools: the cytoplasmic and the vesicular pool. For synaptic release, GABA is loaded into synaptic vesicles by VGAT and then liberated from nerve terminals by calcium dependent exocytosis.

GAT-1 was demonstrated both in neurons and astrocytes. During synaptogenesis, it is transiently overexpressed in both neuronal and astrocytic cell bodies (Xia *et al.*, 1993). In the marginal zone of developing cerebral cortex, GAT-1 mRNA appears at late embryonic stages in both humans (Hachiya and Takashima, 2001) and rats (Jursky and Nelson, 1996). The expression is weak in the neonatal cortex, and increases gradually to the mature pattern reached after one month of birth in rodents (Jursky and Nelson, 1996). Efficient GABA uptake through functional GAT-1 was demonstrated only in postnatal periods (Sabau *et al.*, 1999). In the developing cortex, however, postnatal GABA uptake transiently exceeds the adult level (Coyle and Yamamura, 1976).

The expression of GAT-2 is more pronounced in the developing than in the mature CNS (Liu *et al.*, 1993). GAT-2 containing astrocytic processes is more abundant around blood vessels in the neonatal than in adult cortex (Ikegaki *et al.*, 1994). In adults, GAT-2 was detected in epithelial and glial cells and also in neurons (Conti *et al.*, 1999). GAT-2 is expressed in the pial meninges and along the arachnoid trabeculae. GAT-2 seems to regulate the GABA level in the CSF and the GABA transport across the blood-brain barrier in the developing CNS.

GAT-3 is expressed mainly by glial cells, but it was demonstrated in perinatal cortical GABAergic neurons, as well (Durkin *et al.*, 1995). In late embryonic stages, GAT-3 expression has been demonstrated in the marginal and intermediate zones of the cerebral cortex (Evans *et al.*, 1996). In neonates, GAT-3 is present throughout the cortical wall (Jursky and Nelson, 1999). In contrast to GAT-1, GABA-uptake through GAT-3 could be demonstrated in the periods of fetal cortico-genesis (Conti *et al.*, 2004). GAT-3, together with GAT-2, was detected also around the cortical blood vessels (Minelli *et al.*, 2003b) in neonates. The presence of functional GAT-3 in regions of migrating neuronal precursors indicates that this GABA-transporter plays a role in the regulation of the tonic GABA-level in the early periods of neural tissue genesis. The functional importance of GATs and the onset of GABA uptake in various regions of developing CNS, however, have not been fully elucidated, yet.

C. Synthesis of GABA: GADs

GABA is synthesized primarily from glutamate in a reaction that is catalysed by two glutamic acid decarboxylase (GAD) enzymes: GAD65 and GAD67 coded by distinct genes (Erlander *et al.*, 1991). The two forms seem to synthesize GABA for specific pools, which serve distinct functions (Soghomonian and Martin, 1998). The GAD67 isoform is diffusely distributed in the cytoplasm and produces GABA mainly for the metabolic pool and for extrasynaptic release (Kanaani *et al.*, 1999). The majority of the GAD65 enzyme isoforms are associated with membranes and are enriched in the neuronal terminals (Martin, 2004). The association with synaptic vesicles indicates that GAD65 is involved in the synthesis of vesicular GABA, which mediates fast synaptic communication (Hsu *et al.*, 2000).

GAD enzymes were detected as early as the ninth day after conception in non-neural regions of mouse embryos (Maddox and Condie, 2001). During embryonic development, different GAD67 transcripts, truncated the "embryonic" (25-kDa and 44-kDa) GAD proteins are produced (Bond *et al.*, 1990). The enzymatically inactive GAD25 is widespread at earlier developmental stages (from E10.5-E12.5 in mouse) (Szabo *et al.*, 1994), in regions characterized by proliferation and initial differentiation of future neurons. There are post-mitotic neuronal precursors in the embryonic (E13) rat spinal cord, which are GAD+ but GABA− (Ma *et al.*, 1992) and express enzymatically inactive GAD25 (Behar *et al.*, 1993). GAD44, the enzymatically active embryonic form, was detected in mouse CNS in a period from E11 to P21 (Szabo *et al.*, 1994). Embryonic forms of GAD 67 are transiently expressed also in the developing telencephalon, mainly at sites with active neurogenesis and containing migrating or postmigratory neuronal precursors (Behar *et al.*, 1994). The adult form of GAD67 appears around mid-gestation, rises rapidly until birth, and reaches the adult level in mouse 4 weeks after birth (Szabo *et al.*, 1994). With the expression of the full-length form of GAD67, the percentage of GABA+ cells increases (Behar *et al.*, 1993). The sequential appearance of the individual GAD forms clearly indicates that they perform different developmental functions. The widespread though transient appearance of GAD44 suggests that this embryonic GAD protein might contribute to transient GABA production, while for the 25 kDa form, some distinct regulatory functions can be postulated (Varju *et al.*, 2002).

D. METABOLIC REMOVAL OF GABA

The fraction of GABA taken up by neurons is mainly recycled into synaptic vesicles, and used repeatedly for synaptic communication. In astrocytes, a large part of GABA is metabolized via the citrate cycle. Through this pathway, GABA can serve as an intermediary metabolic compound (Waagepetersen et al., 1999). GABA is transformed to succinic semialdehyde by transamination reaction catalysed by the GABA transaminase (GABA-T) enzyme (Balazs et al., 1970). GABA-T is associated with the inner mitochondrial membrane (Salganicoff and Derobertis, 1965) and its activity plays an important role in setting the GABA-level of the local tissue-environment. Therapeutical modification of GABA-T activity provides a possible tool for setting extracellular GABA-concentration in various psychiatric and neurological disorders (Sarup et al., 2003).

E. THE GABA RECEPTORS

Receptors for GABA are divided into two main classes: $GABA_A$ receptors, which are members of the ligand-gated ion channel superfamily and $GABA_B$ metabotropic receptors which belong to G protein-linked receptors. Both types of GABA receptors are abundant in the brain, and found in almost all types of neurons and many populations of glial cells. On neurons, $GABA_A$ and $GABA_B$ receptors can be situated at both synaptic and extrasynaptic sites.

1. The Structure and Pharmacology of $GABA_A$ Receptor

In response to GABA binding, the $GABA_A$ receptor-channel opens up, and anions flow inward or outward through the channel-pore, depending on the electrochemical gradients of the permeant ions. $GABA_A$ receptors carry primarily chloride ions but bicarbonate can also permeate, although less efficiently (Bormann et al., 1987).

$GABA_A$ receptors are heteropentameric complexes of subunit proteins coded by several related genes (Macdonald and Olsen, 1994). At present, $\alpha(1–6)$, $\beta(1–3)$, $\gamma(1–3)$, $\varepsilon(1)$, $\pi(1)$, $\theta(1)$, $\rho(1–3)$, and $\delta(1)$ subunits have been identified, and some further variations are produced by alternative splicing. This multiplicity provides a large variety of potential subunit-combinations. However, there are certain preferred subunit combinations, as not all subunits can efficiently co-assemble (McKernan and Whiting, 1996). The variety is further limited by the availability of the subunits, which is restricted by the temporal and spatial pattern of subunit expression (Owens and Kriegstein, 2002).

Depending on the subunit composition, receptors with distinct electrophysiological properties will emerge, which may display also different ligand binding and gating properties (Bohme et al., 2004). The cellular distribution and turn over of the receptors are also influenced by the subunit composition (Luscher and Keller, 2004). The characteristic spatial-temporal patterns of the subunit expression well correlates with the qualitative differences observed in the properties of GABA$_A$ receptors (Owens *et al.*, 1999) (see Box 3).

Box 3

GABA(A) receptor functions are further modulated by a variety of compounds, among them a number of clinically important drugs, such as benzodiazepines, barbiturates, steroids, anesthetics, and convulsants (Macdonald and Olsen, 1994). Many clinically relevant effects occur through interactions with distinct allosteric binding sites on GABA$_A$ receptors. The effects of these chemicals depend on the subunit composition of the receptors. Synaptic and extrasynaptic receptors are composed by different subunits and, accordingly, display different sensitivity to several drugs. Differences in the subunit composition are reflected by the non-equal effects of benzodiazepine site binding ligands on the phasic and tonic actions of GABA (Farrant and Nusser, 2005).

Generally, two α, two β, and a γ subunits produce the native GABA(A) receptors. The α and β subunits are responsible for ligand binding, while γ δ, ε, or π subunits play important roles in membrane-positioning of the receptors. The γ subunits were shown to be responsible for the accumulation of the receptors at the active postsynaptic sites (Essrich *et al.*, 1998). Receptors containing a $\gamma2$ subunit in association with $\alpha1$, $\alpha2$, or $\alpha3$ subunits ($\alpha1\beta2/3$ $\gamma2$, $\alpha2\beta2/3$ $\gamma2$, or $\alpha3\beta2/3$ $\gamma2$) are the predominant receptor subtypes in the mature postsynaptic densities (Farrant and Nusser, 2005). It seems that in some receptor complexes the δ, ε, or π subunits can replace the γ, whereas the θ subunit can stand instead of a β subunit (Sieghart *et al.*, 1999). ρ subunits can form either homo- or heteromeric channels, which display special properties. These channels are insensitive for bicuculline and were previously thought to compose a distinct type of receptors, called GABA$_C$ (Barnard *et al.*, 1998).

2. The Expression of GABA$_A$ Receptors During Development

GABA$_A$ receptor mediated responses can be recorded from the very early stages of neuronal development. With the advancement of neural differentiation, the subunit-composition and the cellular distribution of GABA$_A$ receptors, as well as the ionic homeostasis of the cells, are changing. As a consequence, GABA$_A$ mediated responses will characteristically change in the course of development.

Expression of different GABA$_A$ receptor subunits shows a unique regional and temporal profile during development (Box 4). In the rat forebrain, $\alpha4$, $\beta1$, $\gamma1$ subunit proteins are present in proliferating neuroepithelial cells in the ventricular zone (Ma and Barker, 1995). During migration from the germinative zone to the cortical plate, the subunit pattern changes and an $\alpha3$, $\beta2/3$, $\gamma3$ combination becomes dominant (Maric et al., 2001). It seems that there is a shift in the subunit expression when the cells change from the proliferative to a postmitotic stage. In the fetal nervous system (E17–20 in rat), the $\alpha5$, $\beta2/3$, and $\gamma2$ subunits were distributed almost uniformly, while the expression of $\alpha2$, $\alpha3$ varied markedly among various regions (Serafini et al., 1998). It seems, that in young differentiating neurons, most GABA$_A$ receptors are composed of combinations of $\alpha2/3/5$-, $\beta2/3$-, and $\gamma2/3$ (Ma and Barker, 1995). Some of the subunits (as $\alpha5$, $\gamma1$ and $\gamma3$) are transiently expressed at a higher level in the course of embryonic and early postnatal development.

Box 4

$\alpha4$, $\beta1$, $\gamma1$ transcripts are expressed in the ventricular zone, but they have been detected in zones of postmitotic differentiating neurons, as well. δ subunits, which are expressed mainly postnatally and at extrasynaptic sites, have been described at low levels also in the cortical plate (Laurie et al., 1992). $\alpha5$ transcripts were detected in the embryonic brain, with a prominent peak in early postnatal development and a marked decreased afterwards (Poulter and Brown, 1999). The $\alpha6$ subunits are expressed predominantly in cerebellum, and mainly extrasynaptically (Mody, 2001). Among the γ subunits the $\gamma1$ and $\gamma3$ gene expression level decreased during development, while $\gamma2$ remained mostly constant (Laurie et al., 1992).

Neurotransmitter spillover and non-synaptic release are thought to underlie the activation of extrasynaptic receptors in adults (Semyanov et al., 2004). Continuous activation of extrasynaptic GABA$_A$ causes tonic signaling (Mody, 2001). The endogenous GABA level (0.5–3 μM) is sufficiently high to generate tonic conductance in certain regions (such as cerebellum and hippocampus) of the adult brain (Kaneda et al., 1995). The receptors that mediate tonic inhibition have higher affinity for GABA than the receptors mediating phasic inhibition (Stell and Mody, 2002). δ subunit-containing receptors are present exclusively at extrasynaptic membranes (Nusser et al., 1998). The contribution of δ subunit results in non-desensitization upon prolonged activation and 50-fold higher affinities for GABA (Saxena and Macdonald, 1994). The $\alpha4$ subunit was found mostly extrasynaptically in hippocampal granule cells (Sun et al., 2004). The $\alpha5$ was suggested as a non-aggregating, extrasynaptic subunit, because it was homogenously distributed over the cell surfaces (Sassoe-Pognetto et al., 2000).

During postnatal development, the receptors containing $\alpha2$ subunits are gradually replaced by receptors containing $\alpha1$ subunits (Fritschy et al., 1994). The increased $\alpha1$ participation in the receptors is responsible for the fast inhibitory synaptic currents that are normally observed from the postnatal periods of development (Vicini et al., 2001).

In many regions of the developing CNS, the spatial distribution of $GABA_A$ receptors does not follow strictly the potential GABA-sources (e.g., the distribution of GABA-containing cells). As an example, $GABA_A$ receptors were detected in the germinative layer of fetal rats, while GABA-containing neurons or neurites were demonstrated in more superficial layers in the developing forebrain (Cobas et al., 1991). Such "mismatch" indicates the paracrine action of GABA, what is expected in periods preceding synaptogenesis.

In contrast to trans-synaptic, neurocrine signaling, paracrine GABA receptors are not accumulated into dense patches, but are roughly evenly distributed along extended areas of the cell surfaces. Detection of GABA at low paracrine-range (10^{-5}–10^{-7}M) requires high affinity and non-desensitizing receptors. During early embryonic development, $GABA_A$ receptors have higher affinity to GABA (Hevers and Luddens, 1998) and slower decay kinetics than the postsynaptic receptors (Hollrigel and Soltesz, 1997). In many respects, these early paracrine receptors are similar to receptors working in the extrasynaptic membranes of mature neurons.

3. Excitation Instead of Inhibition

In the early phases of neuronal development, $GABA_A$ receptor activation causes depolarization in the cells instead of hyperpolarization. In immature neurons, due to a relatively high intracellular Cl^- concentration, the resting membrane potential is more negative than the reversal potential of chloride (Cherubini et al., 1991). Upon opening of GABA-gated Cl^- channels, Cl^- will diffuse out of the cells. The high chloride concentration is due to the function of sodium- and potassium-coupled co-transporter (NKCC1) and the lack of potassium-chloride co-transporter (KCC2) (Clayton et al., 1998). Thus, the opening of GABA-gated chloride channels results in a net chloride outflow and depolarization. Through depolarization, and via voltage sensitive L-type calcium channels, the activation of $GABA_A$ receptors increases the cytosolic $[Ca^{2+}]$ (Yuste and Katz, 1991). As development proceeds, neuronal $[Cl^-]_I$ decreases in conjunction with the onset of the function of KCC2, the Cl^--extruding neuronal transporter (Rivera et al., 1999). Therefore, the $GABA_A$ reversal potential becomes more negative, and Cl^- will flow into the cells through open anion channels. This shift will result in the establishment of the hyperpolarizing GABA effect (Owens et al., 1996).

F. THE $GABA_B$ RECEPTORS

$GABA_B$ receptors were described in the early 1980s (Bowery et al., 1980), when chloride-independent and bicuculline insensitive GABA responses were found. The GABA-mediated responses through these receptors were antagonized by baclofen. Through G protein coupled reactions, the activation of $GABA_B$ receptors can result in a decrease of cAMP level and in the modification of various ion channels (Fig. 1) (Bettler *et al.*, 2004). Signaling through $GABA_B$ receptors can reduce the presynaptic neurotransmitter release by inhibiting voltage dependent Ca^{2+} channels and decreasing $[Ca^{2+}]_I$. Activation of post-synaptic $GABA_B$ receptors activates inwardly rectifying potassium channels (GIRK) and can hyperpolarize the postsynaptic membrane (Luscher *et al.*, 1997) depending on the type of GIRKs (Cruz *et al.*, 2004). Novel data indicate that $GABA_B$ receptors also can act as calcium-sensors (Galvez *et al.*, 2000). On Purkinje cells, they can associate with mGluR1 metabotropic glutamate receptors and enhance neuronal metabotropic glutamate signaling on an outside Ca-level dependent, but G protein independent way and in the absence of GABA (Tabata *et al.*, 2004).

The $GABA_B$ receptors occur both pre- and postsynaptically, but are located mostly at extrasynaptic sites (Bettler *et al.*, 2004). The dominant extrasynaptic localization suggests that the $GABA_B$ receptor signaling mediates mainly non-synaptic GABA effects. In the developing nervous tissue, $GABA_B$ activation seems to counteract to $GABA_A$ signaling and reduces or prevents the depolarization and Ca-increase elicited by GABA through $GABA_A$ receptors. The presence of both types of receptors on the same cell provides a sensitive control of GABA action depending on the composition of the receptors and on the extracellular GABA-level.

1. *The Structure of $GABA_B$ Receptors*

The main $GABA_B$ receptor subunits (termed $GABA_BR1$ and R2 and $GABA_{BL}$) are encoded by distinct genes (Nakayasu *et al.*, 1993). From the $GABA_B1$ gene several isoforms can be translated (referred as 1a-1g), and splice variants of $GABA_{B2}$ have been also described (Bettler *et al.*, 2004). In the brain, $GABA_{B1a}$ and $GABA_{B1b}$ are the most abundant $GABA_{B1}$ isoforms in many species including human (Kaupmann *et al.*, 1998b). The various subunits provide functional receptor heterogeneity, with different pharmacological sensitivities and preferential subcellular locations.

The functional receptors contain two different subunits (Fig. 1). The $GABA_{B1}$ subunit is essential for ligand binding, whereas the $GABA_{B2}$ subunit is required for G-protein coupling and for efficient trafficking to the cell membrane (Kaupmann *et al.*, 1998a) (Box 5). The $GABA_{BL}$ can not form functional receptors with either $GABA_{B1}$ or $GABA_{B2}$ (Calver *et al.*, 2003).

Results obtained from studies on knockout mice demonstrated that $GABA_{B1}$ subunits are needed for all $GABA_B$ receptor-mediated responses in the CNS (Schuler *et al.*, 2001). Mice lacking $GABA_{B2}$ subunits, on the other hand, exhibit epileptiform seizures and behavioral symptoms reminiscent of $GABA_{B1}$ knockout mice (Thuault *et al.*, 2004) indicating the necessity of both subunits for normal

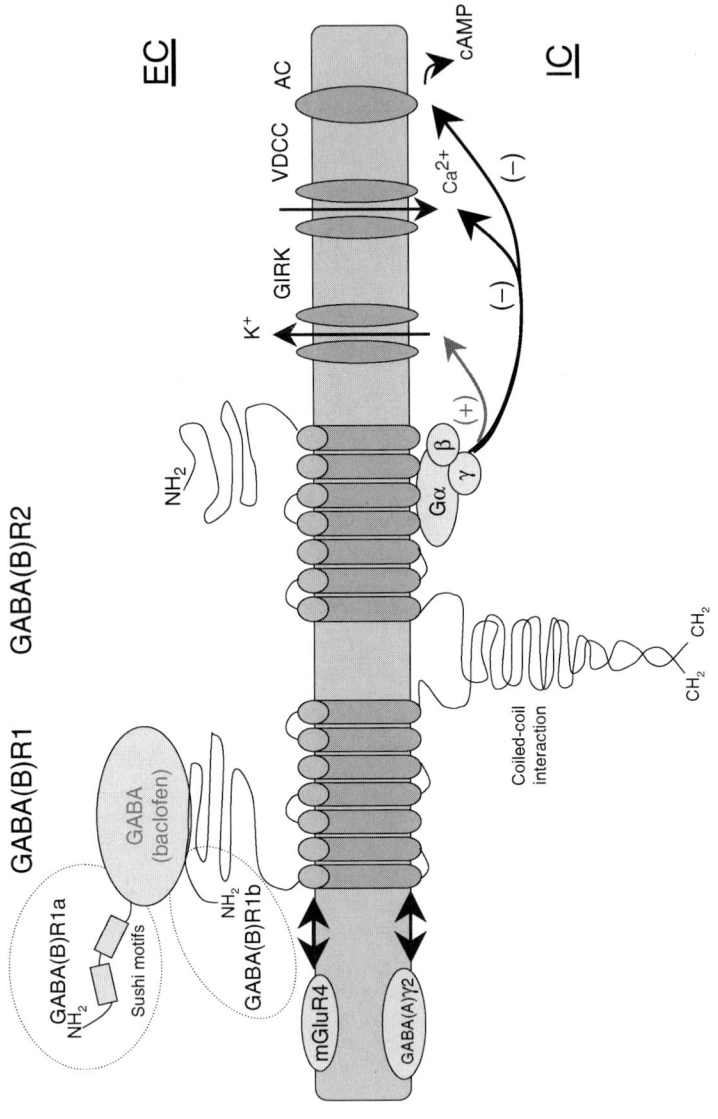

FIG. 1. The GABA$_B$ receptor dimer is composed of GABA$_B$R1 and GABA$_B$R2 subunits. GABA$_B$R1 contains the ligand binding site, while the G-protein complex associates to GABA$_B$R2. Through G-protein-mediated coupling, the activated receptor can regulate inward rectifying K-channels (GIRK), voltage dependent Ca-channels (VDDC), and adenylate cyclase (AC) enzymes. The N-terminal (extracellular) domain of GABA$_B$R1a contain, Sushi-sequences, while GABA$_B$R1b does not. Several GABA$_B$R1 subunits can associate with different receptors, among them are GABA$_A$γ2 and mGluR4. (See Color Insert.)

GABA$_B$ signaling. The knockout phenotypes provided further evidence for the importance of GABA$_B$ mediated functions in psychiatric disorders.

Box 5

There is some evidence that association of GABA$_{B2}$ with the GABA$_{B1}$ subunit is not absolutely necessary for GABA$_B$ signaling (Gassmann *et al.*, 2004). Also, the broader expression of the GABA$_{B1}$ subunit in comparison to GABA$_{B2}$ may suggest that GABA$_{B1}$ could be functional in the absence of GABA$_{B2}$ (Clark *et al.*, 2000). Besides composing homomer dimers, GABA$_{B1}$ subunits can co-assemble with multiple receptor proteins, among them with GABA$_A\gamma2$ and mGluR4. For the date, however, no conclusive evidence on the function of chimera receptors has been obtained (Sullivan *et al.*, 2000).

---

2. *The Expression of GABA$_B$ Receptor During Development*

Despite the functional requirement for both GABA$_{B1}$ and GABA$_{B2}$, the expression of the two subunits seems to be controlled by independent mechanisms, during embryonic development (Martin *et al.*, 2004). Some more important roles for GABA$_{B1}$ over GABA$_{B2}$ are indicated by the earlier and higher expression of GABA$_{B1}$ in the course of development (Kim *et al.*, 2003) (Box 6).

Box 6

GABA$_{B1}$ receptor mRNA but not GABA$_{B2}$ was detected at E11 in the rat hippocampal formation and cerebral cortex (Kim *et al.*, 2003). At E14 both transcripts were expressed, although GABA$_{B1}$ was detected more abundantly. In the prenatal period, both subunits were found in the subplate and the cortical plate, while Cajal-Retzius cells and tangentially migrating cells in the lower intermediate zone contained only GABA$_{B1}$ (Lopez-Bendito *et al.*, 2002). At E17, the two subunits displayed characteristic, region-specific distribution (Behar *et al.*, 2001).

In the rat brain GABA$_{B1a}$ is the prevalent isoform at birth, whereas GABAB$_{1b}$ protein is more abundant in adults (Fritschy *et al.*, 1999). GABA$_{B1a}$ and also GABA$_{B1b}$ proteins were demonstrated in the ventricular zone and cortical plate of the developing rat cortex at E17 (Behar *et al.*, 2001).

The expression of both proteins has a peak in early postnatal development and declines to the adult level thereafter. Receptor binding studies in rat brain also revealed a transient peak during early postnatal development (Turgeon and Albin, 1994).

In comparison to $GABA_A$, *functional* $GABA_B$ receptors appear in later phases of embryonic development. In the neocortex of developing rat brain, $GABA_B$ receptor activity was detected at embryonic day E17 both in ventricular zone and cortical plate (Behar *et al.*, 2001). On differentiating hypothalamic neurons a tonic inhibitory effect of $GABA_B$ receptor signaling was shown as early as E15. The activation of $GABA_B$ receptors depressed both the release of GABA and the $GABA_A$ receptor-mediated Ca2+ rises (Obrietan and van den Pol, 1999).

III. GABA as a Developmental Signal

To date, a growing body of evidence indicates that GABA plays an important regulatory role in the development of the nervous system. The transient GABA production in many regions of the developing CNS underlines the functional importance of GABA in the embryonic and early postnatal neural differentiation (Verney, 2003). Developing neural cells release GABA into their microenvironment by mechanisms still not properly understood (Demarque *et al.*, 2002). While in the mature CNS GABA acts mainly as a synaptic neurotransmitter and elicits phasic responses, in the developing nervous tissue GABA acts as an autocrine/paracrine signal molecule and its main roles are executed through tonic signaling. In young differentiating neurons, most important effects of GABA are mediated through $GABA_A$ receptors and result in membrane depolarization, and in an increased $[Ca^{2+}]_I$, which in turn can stimulate multiple cellular processes (Ohbayashi *et al.*, 1998). $GABA_B$ receptor-mediated effects, which can cause hyperpolarization and reduce $[Ca^{2+}]_I$, appear in later phases of tissue genesis, and seem to balance GABA(A) signaling. The GABA-evoked phasic hyperpolarizations through $GABA_A$ receptors are delayed until the late phases of the maturation of neuronal circuits.

The effects of GABA had been demonstrated in the entire period of neural tissue genesis, from proliferation, through migration and differentiation of neuronal precursors, up to the synapse formation and circuit refinement by maturing neurons.

A. EFFECTS OF GABA ON THE PROLIFERATION OF NEURAL PROGENITORS

In vertebrate central nervous system, the majority of neurons and all macroglia cells (astrocytes and oligodendrocytes) derive from specialized "germinative" zones (Box 7), which consist of proliferating neuroectodermal progenitor cells (Bayer and Altman, 1987).

Box 7

The early-generated neurons, including some transient "organizer" cells as the Cajal-Retzius cells in the cerebral cortex (Marin-Padilla, 1998), and the projection-type (Golgi-I) large neurons as the spinal motoneurons, or the V-VI layer cortical pyramidal cells, derive from the primary germinative zones (ventricular zone, VZ). With aging, the rate of cell production in the primary germinative zones decreases and the zone transforms to a transport epithelium, the ependymal layer. In most regions, a new layer of proliferating cells forms at the internal face of the transport epithelium, the *subventricular zone* (SVZ) or *secondary germinative zone*. In comparison to the primary germinative layer, SVZ generates less and smaller-size projection-type neurons. The vast majority of SVZ-derived neuronal precursors develop into interneurons. By the perinatal age, SVZ progenitors produce less neuronal and more macroglial precursors.

The *primary germinative zone* is a layer of proliferating neuroepithelial cells, and compose the entire neural tube in early phases of neurogenesis. In the primitive neural tube, neuroepithelial cells span through the total thickness of the tube wall. These radial neuroepithelial cells or embryonic neural stem cells (Gotz *et al.*, 2002) can divide symmetrically and generate identical progenies, and as a result, expand the germinative zone. Alternatively, they can go through asymmetric mitoses, which result in equal distribution of genomic DNA, but unequal distribution of cytoplasmic regulatory components. As a consequence they give rise to two different daughter cells: one, which maintains a mother-like proliferating cell phenotype, while the other is primed for a different cell fate. The later ceases contact with the ventricular wall, situates out from the germinative zone, and stops dividing. These "postmitotic" cells are the earliest neuronal precursors (Jacobson, 1978). With the advancement of embryonic development, the germinative zone gets restricted to a layer lining the brain ventricles and the central canal of the spinal cord.

At the ventricle wall of the future striatum, secondary proliferative zones appear early (E10.5 in rat) and form the primordia of the medial and lateral ganglionic eminences (MGE and LGE, respectively) (Garcia-Verdugo *et al.*, 1998). Precursors of the cortical GABAergic interneurons are generated in the ganglionic eminences (Anderson *et al.*, 1999) and migrate from here to their cortical destination (Kriegstein and Noctor, 2004). The cell production capacity at these sites persists through adulthood. A layer of proliferating cells also persists beneath the granular layer of the dentate gyrus, (hippocampal subgranular layer) and produces granule cell-precursors for the hippocampus throughout the whole life-span (Kaplan and Hinds, 1977). At large areas along the ventricle walls, however, the spontaneous cell-proliferation capacity is exhausted soon after birth. In these areas, severe loss of neural tissue cells can induce some renewal of cell productive capability.

Along the ventricle wall, the mitotic activity and cell-release occur in well-recognizable, ordered, spatial-temporal patterns (Sidman and Rakic, 1973). There are intermitting waves in cell production, and the cell-generation waves are differently timed in adjacent areas (Nornes and Das, 1974). The patterns of the cell production argue for potent regulatory mechanisms, but their nature is not understood.

Cell proliferation can be initiated and is regulated by various growth factors. Epidermal growth factor (EGF) and basic fibroblast growth factor (FGF2) are strong mitogens for neural progenitor cells in different stages of tissue development (Martens et al., 2000). The progression through the cell cycle needs regulated oscillation of $[Ca^{2+}]_I$, which can be evoked by membrane depolarization. The proliferation stimulatory effect of depolarization had been demonstrated in the developing CNS (Cone, 1980).

GABA is a factor, which can elicit membrane depolarization, can evoke increase in the intracellular free Ca-level, can stimulate the release of growth factors and can modulate their effects (Marty et al., 1996). Before the formation of the blood-brain barrier and full functioning of the GABA uptake system, a relatively high GABA-level can be postulated in the environment of proliferating cells. In addition, GABA-containing cells were detected directly above the germinative zone (Cobas et al., 1991), which might release enough GABA to stimulate adjacent cells.

Functional GABA$_A$ receptors had been demonstrated on cells in the germinative zones at various stages of neural development (Ma et al., 1998). LoTurco and co-workers provided the first evidence that the presence of GABA can directly influence the DNA synthesis in proliferating ventricular zone cells. (LoTurco et al., 1995). Activation of GABA$_A$ receptors in the ventricular zone of rat embryos older than E15 significantly reduced the DNA synthesis and resulted in a reduced number of proliferating cells (LoTurco et al., 1995). Comparative studies on cells in different germinative layers of E13-E14 rat forebrain, however, indicated that GABA decreased the cell production in the subventricular zone, but could stimulate the proliferation in the ventricular zone (Haydar et al., 2000). In respect of proliferation stimulatory or inhibitory actions of GABA, some non-congruent results were obtained on postnatal progenitors (Nguyen et al., 2003). The available data (LoTurco et al., 1995) suggest that the proliferation modulating effects of GABA highly depend on the differentiation stage of the targeted cells.

Despite the experimental evidence on the proliferation modulating effects of GABA, GAD 65/GAD 67 double knocked out mouse embryos with a GABA-level less than 0.1% of the normal, did not show severe loss or gain of neurons. While such animals die soon after birth, they failed to display anatomical brain malformations at birth (Ji et al., 1999). The observation shows that even if GABA can modulate the proliferation of neural progenitors, this capability is shared with a number of other factors and can be replaced in the course of *in vivo* neurogenesis.

B. EFFECTS OF GABA ON MIGRATING NEURONAL PRECURSORS

Postmitotic neuronal precursors reach their destination by migrating through shorter or longer distances. The migrating cells move along apparently strictly defined routes (Box 8).

Box 8

Postmitotic precursors born in the primary germinative layer leave the zone either by translocation toward the future pial surface through their apical process or by migrating along neighboring radial cells (Kriegstein and Noctor, 2004). Upon leaving the germinative zone, the early-born cells loose mitotic capability and develop into neurons. Cortical interneuron-precursors derived from the subventricular zones may take more complicated routes. Most of them migrate along some tangential guidance points (more or less parallel with the ventricular wall) before they change direction and move radially to their cortical destination.

Path finding is primarily executed by interactions between cell adhesion molecules on the surface of the migrating cell and attachment-points provided by surfaces of adjacent cells and by components of the extracellular matrix. For stable attachment, spreading, or migration, an initial chemical contact has to be established, which in turn can initiate complex cell responses involving recruitment of further adhesion molecules, cytoskeletal activation, incorporation and retraction of membrane-units, and redistribution of several protoplasmic components.

The cell will move toward fields, which provide stronger attachment and produce signals which enhance metabolism and $[Ca^{2+}]_I$. As subpopulations of cells carry different adhesion molecules and deposit different extracellular matrix components, delicate spatial-temporal maps of potential migratory routes are outlined.

The level of free Ca^{2+} is a sensitive intracellular regulator of the formation of the leading lamellipodia and generation of contractile forces needed for active cell displacement. Depolarizing agents, among them GABA, can cause Ca-influx through voltage-dependent Ca-channels, and as a consequence, can modulate the speed and direction of migration. Both too low and too high local $[Ca^{2+}]_I$, however, can prevent locomotion.

In the developing CNS, several GABA sources have been identified, which can provide local GABA and also GABA-gradients for modulation of migration. GABA-containing cells above the germinative zone might produce enough

GABA to influence the migration of postmitotic cells from the proliferative zone. In the marginal zone of the developing cerebral cortex, Cajal-Retzius cells produce GABA, beside the major migration-regulating compound, reelin (D'Arcangelo *et al.*, 1995). In the developing cortical plate, immigrating interneuron-precursors release GABA, which is thought to influence the locomotion of later arriving precursors. Ensheathing glial cells and also the migrating neuronal precursors in the rostral migratory route release GABA (Bolteus and Bordey, 2004). The available data indicate that GABA is an important constituent of the environment of migrating cells.

Migrating neuronal precursors express $GABA_A$ receptors and can respond to GABA by depolarization (Barker *et al.*, 1998). *In vitro*, femtomolar concentrations of GABA were shown to evoke directed displacement, while micromolar quantities caused enhanced but random motility in different developing neural cells (Behar *et al.*, 1996).

In contrast, studies on neurosecretory neurons migrating from the olfactory placode to the basal forebrain indicated that GABA could restrict the cell motility (Fueshko *et al.*, 1998). Activation of $GABA_A$ receptors inhibited the migration of GnRH cells (Heger *et al.*, 2003). Blocking $GABA_A$ receptor signaling, however, disrupted the migratory route and resulted in an ectopic localization of GnRH neurons in the brain (Bless *et al.*, 2000).

The migration inhibiting effect of GABA was also demonstrated on postnatal forebrain neural progenitors (Bolteus and Bordey, 2004). Application of 10 μM GABA reduced, while bicuculline increased the speed of cell migration in the anterior SVZ and in the rostral migratory stream. In higher (100 μM), desensitizing concentration, GABA increased the motility by auto-blocking the GABA-elicited inward anion-flow. The results clearly demonstrated that $GABA_A$ signaling could decrease or increase the speed of migration depending on the local GABA-concentration (Bolteus and Bordey, 2004). Moreover, GABA-induced depolarization can elicit further GABA-release, and the outcome of this GABA-loop may determine the inhibition or relief from inhibition of motility. In this system, IP_3-sensitive Ca-stores were suggested as a source of free Ca^{2+}, rather then depolarization-evoked influx of Ca^{2+} (Bolteus and Bordey, 2004).

GABA can modulate the migration of neuronal precursors via $GABA_B$ receptor-mediated effects, as well. Behar and co-workers (Behar *et al.*, 2000) demonstrated that while migration from the ventricular zone to the intermediate zone of E18 rat forebrain was modulated via $GABA_A$ receptors signaling, the migration toward the cortical plate was regulated by $GABA_B$-mediated processes. Involvement of $GABA_B$ signaling in the migration toward the cortical plate was also demonstrated by the accumulation of tangentially migrating cells in the ventricular zone of E15 rat embryos in the presence of $GABA_B$ antagonists

(Lopez-Bendito *et al.*, 2003). As $GABA_B$ signaling decreases depolarization and $[Ca^{2+}]_I$, one can assume that it can prevent the over-activation and consequent disruption of the cytoskeletal locomotory machine.

In spite of much *in vitro* and *in vivo* evidence on the migration-modulating effects of GABA, the cytoarchitecture of the GABA-deficient neonatal brain failed to show severe malformations (Ji *et al.*, 1999). The observation shows that the vital processes of correct cell migration are governed by redundant regulatory mechanisms, which can compensate for the shortage of a single factor.

C. EFFECTS OF GABA ON GROWTH CONE MOTILITY AND NEURONAL PROCESS ELONGATION

During migration, neuronal precursors have transient leading lamellipodia and pulled tails, as do any other migrating tissue cells. The *neuron-specific polarization*, namely the formation of a permanent process (future axon) on a defined part of the cell (neuritogenesis), starts usually after the settlement of the cell (Goffinet, 1984). The mechanisms resulting in the selection of one of the minor processes as the rudiment of the future axon (Dotti *et al.*, 1988) and in *the initial outgrowth* of the axon are not fully understood. Growth factors, mainly the members of the neurotrophin family, NGF (Greene *et al.*, 1984) and BDNF (Avwenagha *et al.*, 2003), were shown to promote the initial outgrowth of neurites. GABA administration was demonstrated to evoke growth factor release and also to provoke process outgrowth from cells of superior cervical ganglia (Wolff *et al.*, 1978).

The *elongation of the neurite* is persecuted by adhesion-dependent advancing of its leading edge, the growth cone, and by the orchestrated incorporation of axonal constituents right behind the growth cone. The process elongation will sensibly respond to a variety of intra- and extracellular signals and will be the subject of multiple regulations (Xiang *et al.*, 2002) (Box 9). Changes in the local concentrations of cAMP, cGMP, and Ca^{2+} were all shown to play determining roles in the growth cone responses to environmental signals indicating the involvement of multiple surface receptors in both the advancement and collapse of a growth cone (Cooper, 2002).

The advancement, halt, turn, or collapse of a growth cone well correlate with the spatial-temporal pattern of its internal free calcium concentration ($[Ca^{2+}]_I$) (Henley and Poo, 2004). The repulsive attachment-signals will either evoke an over-range increase in $[Ca^{2+}]_I$ or prevent the elevation of Ca-level. Besides the ECM and cell surface components, a large variety of secreted compounds, among them GABA and glutamate, can regulate growth cone advancement by interfering with the $[Ca^{2+}]_I$ (Obrietan and van den Pol, 1996).

Box 9

Growth cone advancement is regulated by evolutionary conserved receptor-ligand pairs (see Table I), which regulate the Ca-level at the site of attachment. These attachment-receptor systems interfere with each other in deciding the actual growth cone response.

Growth cone advancing requires the incorporation of membrane material, which is delivered by the fusion of exocytotic vesicles with the plasma membrane (Bray, 1973). The fusion needs defined ($\sim$0.5 μM) local $[Ca^{2+}]_I$ (Burgoyne and Morgan, 1995). The driving force generation by actin-myosin complexes also needs Ca^{2+}, and proceeds only at those sites of the cone which possess Ca^{2+} in the right concentration. These mechanisms result in turning the axon toward signal sources, which can provoke an appropriate local rise in $[Ca^{2+}]_I$ (Gomez et al., 2001).

The microtubule system expanding into the central (but not to peripheral) part of the growth cone provides the "conveyor" for the transportation of vesicles, actin monomers, and other constituents to the growth cone (Gordon-Weeks, 2004). The maintenance and transport functions of the microtubule bundles need Ca^{2+}, but the assembly is getting corrupted at high ($>$1 μM) $[Ca^{2+}]_I$. High $[Ca^{2+}]_I$ in the central part results in rapid collapse of the growth cone.

TABLE I

Growth cone receptor	Ligand	Attachment signal	Reference
DCC/Unc5	Netrins	repulsive/attractive	Wadsworth et al., 1996
Robo	Slit	repulsive	Wong et al., 2002
Neuropilins	Semaphorins	repulsive/attractive	Chen et al., 1998
Eph (Trk receptors)	Ephrins	repulsive/attractive	Himanen and Nikolov, 2003
Nogo	MAG, OMgp	repulsive	McGee and Strittmatter, 2003

The presence of functional $GABA_A$ receptors has been demonstrated on growth cones of many types of neurons and their activation results in a local increase in $[Ca^{2+}]_I$ (Fukura et al., 1996). Growth cones are known to release GABA, and can actively regulate the GABA-concentration in their vicinity (Lockerbie et al., 1985). The available data suggest that GABA acts as an autocrine factor for the majority of the growing neurites. The auto-release serves to keep the basic Ca-concentration at a slightly elevated level, which in turn makes the cone more sensitive for other signals.

Besides $GABA_A$ receptors, $GABA_B$ receptors are also present in the growth cone membrane of various types of neurons (Xiang et al., 2002). Selective stimulation of $GABA_B$ receptors by baclofen inhibits growth cone advancement (Priest and Puche, 2004) by decreasing the internal $[Ca^{2+}]_I$ (Behar et al., 2001).

GABA$_B$ receptor-mediated signaling may prevent the potential over-excitation caused by autocrine or environment-derived GABA and may help to maintain an optimal sensitivity level of the growth cone.

Through GABA$_A$ and GABA$_B$ signaling, GABA can influence the motility—advancement, turning, or collapse—of growth cones, and can regulate the elongation and path-finding of neurites. Despite the convincing experimental data, however, the *in vivo* consequences of the lack of GABA are less clear. Data obtained on animals lacking one (GAD 67) (Asada *et al.*, 1997) or both (GAD 67 and GAD 65) (Ji *et al.*, 1999) GABA-synthesizing enzyme, and born with a highly reduced brain GABA-level (7% and 0.02% of the normal, respectively), failed to demonstrate major alterations in the fiber maps of the CNS at birth.

D. The Role of GABA in Synaptogenesis and Synapse Stabilization

Elongating neuronal processes can travel for a long distance and pass or make only transient contacts with multiple potential synaptic partners. In the main period of process-elongation, developing axons produce excess numbers of collateral branches and grow into many areas from which they will diminish in later phases of functional network formation (Stanfield, 1984). While synapses are formed (and diminish) during the whole lifetime, the main periods of synaptogenesis in defined regions are confined to relatively narrow time-windows (critical periods) when most future innervating axons arrive at the potential target-area. Recordings made on various regions of the developing CNS (O'Donovan, 1999) demonstrated ongoing, periodic electrical activities and "spontaneous" Ca-spikes (Spitzer *et al.*, 2004) during the formation of synaptic circuits (Box 10). In the developing neural tissue, large depolarizing waves (also called as giant depolarizing potential, GDP) (Ben-Ari *et al.*, 1989) are propagated along the membranes of the differentiating cells and their processes, in spatially and temporally organized ways (Katz and Shatz, 1996), providing time- and space windows for synaptogenesis.

The travelling waves, by providing transient Ca-level elevations in *a* growth cone or in *a* potential future postsynaptic patch, are not enough for the initiation of synapse formation, but make the compartments sensitive for additional extracellular signals. The majority of the exogenous signals—comprising growth factors, matrix molecules, and Ca-elevating neurotransmitters (Box 11)—are produced and released by the potential synaptic partners, upon activation. The inherent Ca-waves and the pattern of exogenous signals together will decide on synapsing or passing by the actual partners.

Box 10

The waves presumably are generated by an initial fast Ca-spike, which evokes Ca-induced Ca-release, locally. In the developing spinal cord, periodic waves with amplitudes of 10–40 mV and with durations in the range of several 10 seconds can result in local Ca-oscillations reaching μM levels of $[Ca^{2+}]_I$ for short but repeated periods (Gu and Spitzer, 1997). Elevated free Ca^{2+} concentration is necessary for the formation of synaptic compartments (see below).

Coinciding activity of the pre- and postsynaptic partners is inevitable for the formation and also for the maintenance of a synapse and the establishment of the mature "process-map." Depolarization travelling along the developing neuronal processes is thought to play a pivotal role in the establishment of the topographic projections between innervating and target fields. The depolarization is thought to derive from pacemaker neurons, which are electrically coupled to a defined number of their neuron-mates. In this way, a number of neighboring cells get activated in a rhythm corresponding to the parameters of the spontaneous depolarization of the pacemaker, while others will adopt a different rhythm. Axons originating from a coupled group of neurons will "fire" and release factors at their endings with an almost identical timing. At the termini of bundles of axons, the concentration of synchronously released factors can rise to a level sufficient to transactivate an area where the potential postsynaptic partners are located. Activated "targets" will produce signal molecules and will be ready to respond to an initial contact by postsynaptic specification. In the same fasciculated axon bundle, there are neurites derived from neurons of a different coupling group. Their endings will be silent in those periods when the other coupling group activates the target-region. In the lack of synchronous activity, they will establish synapses at lower probability and get excluded from the given target field. This schedule of synchronous activity driven innervation gives an explanation of the formation of topographic projections, and is referred to as an example of the principle of "fire together, wire together."

Synaptogenesis is an interactive process between the future pre- and postsynaptic element (Jessell and Kandel, 1993) (see Fig. 2). Synapse formation starts with an initial membrane contact between the potential partners, and ends up in the formation of complex, functional molecular assemblies, including the active release-area in the presynaptic side, the signal-processing machinery (Kennedy, 2000) at the postsynaptic site, and the structured extracellular matrix in the synaptic cleft (Patthy and Nikolics, 1994). The initial membrane contact is a mutual signal for both partners, which, depending on the actual physiological status of the cell compartment, can or cannot initiate the assembling of synapse-building molecules. After the initial contact, the potential synaptic partners cease locomotory membrane movement for a period of 10 minutes (Pfenninger

Box 11

Specific cell adhesion molecules were shown to govern initial coupling and forth-coming molecular events on both sites (Biederer *et al.*, 2002). The assembly of pre- and postsynaptic specializations is beyond the scope of this chapter. We refer to the excellent reviews of Ziv and Garner (2004), respectively. Pre-assembled, multi-molecular units are delivered to both the pre- (Krueger *et al.*, 2003) and postsynaptic active zones by intracellular transport. Many synaptic proteins, however, are synthesized locally, at the nerve-endings (Schratt *et al.*, 2004). Local protein synthesis in both developing and mature postsynapses has been widely demonstrated. Indirect evidence for protein translation in the presynaptic terminals exists, but direct evidence was obtained only on invertebrate neurons (Martin, 2004).

The rate of the local protein production (Kalinovsky and Scheiffele, 2004) and the speed of delivery and incorporation of assembled units are regulated by growth factors and bioelectric activity (Sutton *et al.*, 2004). Increase in the $[Ca^{2+}]_I$ of the developing endings results in the exocytotic release of various factors, among them BDNF, which in turn can directly elevate the local protein synthesis (Schratt *et al.*, 2004) by signaling through cell surface tyrosine receptor kinases.

In the course of synapse-formation, the extracellular space between the partner-membranes widens gradually (Pfenninger and Maylie-Pfenninger, 1979) from an initially close apposition to a wide gap of about 200 nm. The synaptic cleft contains specific extracellular matrix components, and many of them play roles in the clustering of pre- and postsynaptic membrane components (Mi *et al.*, 2002).

and Maylie-Pfenninger, 1979). In this time-window, either the deposition of synapse-specific material starts or the locomotion activity of the membranes gets reactivated and the partners separate from each other.

Synapse-formation can take place only if both partners are activated and display synchronously elevated $[Ca^{2+}]_I$. The increase in the local $[Ca^{2+}]_I$ is mainly achieved by local Ca-influx through depolarization-activated Ca-channels. Besides the propagated bioelectric waves, developing neuronal membranes can be further depolarized by binding neurotransmitters, mainly GABA and glutamate through $GABA_A$ and NMDA receptors, respectively (Ben-Ari, 2002). Elements for functional synapses seem to be stored and kept ready for rapid incorporation in yet silent synapses of developing neurons (Krueger *et al.*, 2003). In developing hippocampal slices, rhythmic depolarization resulted in the formation of functioning synapses in 30 minutes (Gubellini *et al.*, 2001).

GABA seems to be the main, intrinsic depolarizing agent in the critical periods of neuronal circuit formation in many areas of the developing CNS. Using $GABA_A$ receptor antagonists, the role of GABA in generation of large

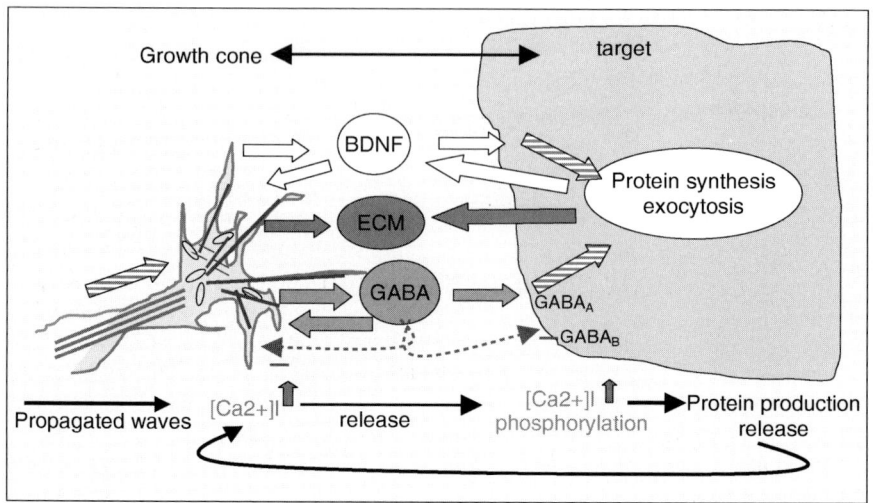

FIG. 2. Depolarization increases the intracellular Ca-level at the growth cone and elicits the release of growth factors (in most cases BDNF), extracellular matrix molecules (ECM), and GABA. BDNF through Trk receptors will stimulate local protein synthesis in both partners resulting in the production of further factors to be released. GABA also will depolarize both partners and trigger the release of various factors, among them BDNF, ECM molecules, and neurotransmitter substances, mainly GABA. The over-excitation by increasing concentrations of GABA can be prevented by the activation of GABA_B receptors, which will reduce or even stop depolarization. (See Color Insert.)

depolarizing waves was demonstrated in the areas of the developing hippocampus (Ben-Ari *et al.*, 1989), retina (Fischer *et al.*, 1998), and spinal cord (O'Donovan *et al.*, 1998). GABA-regulated Ca-spikes were recorded also in the developing cerebral cortex (Katz and Shatz, 1996) involving the early marginal zone (layer 1) containing Cajal-Retzius cells (Schwartz *et al.*, 1998).

GABA can exert these effects through both extrasynaptic and synaptic GABA_A receptors. In a developing neural circuit containing neurons and contacts at different stages of differentiation, the paracrine *versus* neurocrine modes of actions are hardly distinguishable. Depolarizing GABA signaling through either synaptic or extrasynaptic GABA_A receptors can elicit further neurotransmitter release, and creates positive feedback loops which will favor the maintenance of the activation (see also Fig. 2).

In periods and sites of massive synapse formation, GABA can act also through GABA_B receptors. Signaling through GABA_B receptors will reduce depolarization by activating GABA_B receptor-coupled K-channels. This counter-action to GABA_A signaling, together with the desensitization of GABA_A receptors by extreme GABA-concentrations, can prevent a GABA-induced over-activation and provide a mechanism for rhythm generation.

Regional analyses demonstrated that GABAergic signal-machinery develops earlier then the other neurotransmitter signaling apparatuses in a local network (Schwartz *et al.*, 1998). GABA-synapses were shown to develop prior to glutamatergic ones in the developing hippocampus and also in the cerebral cortex (Ben-Ari *et al.*, 2004). The formation of more targeted, spatially restricted neurocrine communication-apparatuses will limit the activated cells and helps to shape the local network. The targeted and coinciding activation will speed up the synapse formation between activated cells and will help their process arborization, as well. By incorporating cell-specific, pre-formed synaptic elements into synaptic specializations, GABA-induced depolarizations can elicit the formation of glutamatergic (Ben-Ari, 2002) and presumably other neurotransmitter-operated synapses. In the developing spinal cord, the frequency and the magnitude of the depolarizing waves were suggested to regulate the transmitter-phenotype-choice of the activated cells (Spitzer *et al.*, 2004).

Ongoing activity elicits fundamental changes in maturing neurons. In parallel with synapse-formation, the sets of ion-transporters change. With development of Cl^--extruding mechanisms, $GABA_A$ receptor signaling will evoke hyperpolarization instead of depolarization (Rivera *et al.*, 1999). By the period of the bulk formation of excitatory synapses, GABAergic signaling turns to inhibitory and provides a potent protection against toxic over-excitation.

GABA-signaling, with its inherent, auto-regulated biphasic potential to hypo- and hyperpolarize the target cells, and also with the mild (non-toxic) shifts it can cause in the ion-homeostasis, is a potent candidate for the role of a developmental activity-generator and neuronal circuit-organizer.

Unfortunately, animals lacking functional GABA-synthesizing enzymes, GAD65 and GAD67 (Ji *et al.*, 1999), die soon after birth, and cannot provide tools to study the relatively late events of functional network formation. Further studies should decide whether GABA-effects are dispensable in the processes of precise wiring and network formation, as well as in the determination of the neurotransmitter phenotype of differentiating neurons.

IV. Concluding Remarks

A large body of experimental evidence demonstrates that GABA-signaling possess an inherent capability for potent regulation of almost all steps of neuronal differentiation and neural tissue formation. Several data also suggest that a transient GABA-production is an inherent feature of many differentiating neuronal populations, regardless of the future neurotransmitter phenotype. The apparently normal prenatal brain development in GABA-deficient animals, however, indicates that GABA can be replaced by other regulatory factors in modulating neural cell production, neuronal precursor migration, process outgrowth, path finding, and neurite elongation. The most intriguing developmental role that has been attributed to GABA is the generation and maintenance of activity waves in the period of functional network formation. For the time being, there are no tools to approach this question at the level of the whole organism experimentally. Accumulating pathophysiological data on inherent developmental neurological, psychiatric, and mental disorders, however, may help to reach convincing conclusions.

References

Ackermann, D. (1910). Über ein neues auf bakteriellem Wege gewinnbares Aporrhegma. *Z. Physiol. Chem.* **69,** 273–281.

Al-Sarraf, H. (2002). Transport of 14C-gamma-aminobutyric acid into brain, cerebrospinal fluid and choroid plexus in neonatal and adult rats. *Brain Res. Dev. Brain Res.* **139,** 121–129.

Anderson, S., Mione, M., Yun, K., and Rubenstein, J. L. (1999). Differential origins of neocortical projection and local circuit neurons: Role of Dlx genes in neocortical interneuronogenesis. *Cereb. Cortex* **9,** 646–654.

Asada, H., Kawamura, Y., Maruyama, K., Kume, H., Ding, R. G., Kanbara, N., Kuzume, H., Sanbo, M., Yagi, T., and Obata, K. (1997). Cleft palate and decreased brain gamma-aminobutyric acid in mice lacking the 67-kDa isoform of glutamic acid decarboxylase. *Proc. Natl. Acad. Sci. USA* **94,** 6496–6499.

Attwell, D., Barbour, B., and Szatkowski, M. (1993). Nonvesicular release of neurotransmitter. *Neuron* **11,** 401–407.

Avwenagha, O., Campbell, G., and Bird, M. M. (2003). The outgrowth response of the axons of developing and regenerating rat retinal ganglion cells *in vitro* to neurotrophin treatment. *J. Neurocytol.* **32,** 1055–1075.

Awapara, J., Landua, A. J., Fuerst, R., and Seale, B. (1950). Free gamma-aminobutyric acid in brain. *J. Biol. Chem.* **187,** 35–39.

Balazs, R., Machiyama, Y., Hammond, B. J., Julian, T., and Richter, D. (1970). The operation of the gamma-aminobutyrate bypath of the tricarboxylic acid cycle in brain tissue *in vitro*. *Biochem. J.* **116,** 445–461.

Balcar, V. J., Dammasch, I., and Wolff, J. R. (1983). Is there a non-synaptic component in the K+-stimulated release of GABA in the developing rat cortex? *Brain Res.* **312,** 309–311.

Barker, J. L., Behar, T., Li, Y. X., Liu, Q. Y., Ma, W., Maric, D., Maric, I., Schaffner, A. E., Serafini, R., Smith, S. V., Somogyi, R., Vautrin, J. Y., Wen, X. L., and Xian, H. (1998). GABAergic cells and signals in CNS development. *Perspect. Dev. Neurobiol.* **5,** 305–322.

Barnard, E. A., Skolnick, P., Olsen, R. W., Mohler, H., Sieghart, W., Biggio, G., Braestrup, C., Bateson, A. N., and Langer, S. Z. (1998). International Union of Pharmacology. XV. Subtypes of gamma-aminobutyric acidA receptors: Classification on the basis of subunit structure and receptor function. *Pharmacol. Rev.* **50,** 291–313.

Bayer, S. A., and Altman, J. (1987). Directions in neurogenetic gradients and patterns of anatomical connections in the telencephalon. *Prog. Neurobiol.* **29,** 57–106.

Behar, T., Ma, W., Hudson, L., and Barker, J. L. (1994). Analysis of the anatomical distribution of GAD67 mRNA encoding truncated glutamic acid decarboxylase proteins in the embryonic rat brain. *Brain Res. Dev. Brain Res.* **77,** 77–87.

Behar, T., Schaffner, A., Laing, P., Hudson, L., Komoly, S., and Barker, J. (1993). Many spinal cord cells transiently express low molecular weight forms of glutamic acid decarboxylase during embryonic development. *Brain Res. Dev. Brain Res.* **72,** 203–218.

Behar, T. N., Li, Y. X., Tran, H. T., Ma, W., Dunlap, V., Scott, C., and Barker, J. L. (1996). GABA stimulates chemotaxis and chemokinesis of embryonic cortical neurons via calcium-dependent mechanisms. *J. Neurosci.* **16,** 1808–1818.

Behar, T. N., Schaffner, A. E., Scott, C. A., Greene, C. L., and Barker, J. L. (2000). GABA receptor antagonists modulate postmitotic cell migration in slice cultures of embryonic rat cortex. *Cereb. Cortex* **10,** 899–909.

Behar, T. N., Smith, S. V., Kennedy, R. T., McKenzie, J. M., Maric, I., and Barker, J. L. (2001). GABA$_B$ receptors mediate motility signals for migrating embryonic cortical cells. *Cereb. Cortex* **11,** 744–753.

Belhage, B., Hansen, G. H., and Schousboe, A. (1993). Depolarization by K+ and glutamate activates different neurotransmitter release mechanisms in GABAergic neurons: Vesicular versus non-vesicular release of GABA. *Neuroscience* **54,** 1019–1034.

Ben-Ari, Y. (2002). Excitatory actions of GABA during development: The nature of the nurture. *Nat. Rev. Neurosci.* **3,** 728–739.

Ben-Ari, Y., Cherubini, E., Corradetti, R., and Gaiarsa, J. L. (1989). Giant synaptic potentials in immature rat CA3 hippocampal neurones. *J. Physiol.* **416,** 303–325.

Ben-Ari, Y., Khalilov, I., Represa, A., and Gozlan, H. (2004). Interneurons set the tune of developing networks. *Trends Neurosci.* **27,** 422–427.

Benitez-Diaz, P., Miranda-Contreras, L., Mendoza-Briceno, R. V., Pena-Contreras, Z., and Palacios-Pru, E. (2003). Prenatal and postnatal contents of amino acid neurotransmitters in mouse parietal cortex. *Dev. Neurosci.* **25,** 366–374.

Bettler, B., Kaupmann, K., Mosbacher, J., and Gassmann, M. (2004). Molecular structure and physiological functions of GABA$_B$ receptors. *Physiol. Rev.* **84,** 835–867.

Biederer, T., Sara, Y., Mozhayeva, M., Atasoy, D., Liu, X., Kavalali, E. T., and Sudhof, T. C. (2002). SynCAM, a synaptic adhesion molecule that drives synapse assembly. *Science* **297,** 1525–1531.

Bless, E. P., Westaway, W. A., Schwarting, G. A., and Tobet, S. A. (2000). Effects of gamma-aminobutyric acid(A) receptor manipulation on migrating gonadotropin-releasing hormone neurons through the entire migratory route *in vivo* and *in vitro*. *Endocrinology* **141,** 1254–1262.

Bohme, I., Rabe, H., and Luddens, H. (2004). Four amino acids in the alpha subunits determine the gamma-aminobutyric acid sensitivities of GABAA receptor subtypes. *J. Biol. Chem.* **279,** 35193–35200.

Bolteus, A. J., and Bordey, A. (2004). GABA release and uptake regulate neuronal precursor migration in the postnatal subventricular zone. *J. Neurosci.* **24,** 7623–7631.

Bolteus, A. J., Garganta, C., and Bordey, A. (2005). Assays for measuring extracellular GABA levels and cell migration rate in acute slices. *Brain Res. Brain Res. Protoc.* **14,** 126–134.

Bond, R. W., Wyborski, R. J., and Gottlieb, D. I. (1990). Developmentally regulated expression of an exon containing a stop codon in the gene for glutamic acid decarboxylase. *Proc. Natl. Acad. Sci. USA* **87,** 8771–8775.

Borden, L. A. (1996). GABA transporter heterogeneity: Pharmacology and cellular localization. *Neurochem. Int.* **29,** 335–356.

Bormann, J., Hamill, O. P., and Sakmann, B. (1987). Mechanism of anion permeation through channels gated by glycine and gamma-aminobutyric acid in mouse cultured spinal neurones. *J. Physiol.* **385,** 243–286.

Bouche, N., Lacombe, B., and Fromm, H. (2003). GABA signaling: A conserved and ubiquitous mechanism. *Trends Cell. Biol.* **13,** 607–610.

Bowery, N. G., Hill, D. R., Hudson, A. L., Doble, A., Middlemiss, D. N., Shaw, J., and Turnbull, M. (1980). (-)Baclofen decreases neurotransmitter release in the mammalian CNS by an action at a novel GABA receptor. *Nature* **283**, 92–94.

Bray, D. (1973). Model for membrane movements in the neural growth cone. *Nature* **244**, 93–96.

Burgoyne, R. D., and Morgan, A. (1995). Ca2+ and secretory-vesicle dynamics. *Trends Neurosci.* **18**, 191–196.

Calver, A. R., Michalovich, D., Testa, T. T., Robbins, M. J., Jaillard, C., Hill, J., Szekeres, P. G., Charles, K. J., Jourdain, S., Holbrook, J. D., Boyfield, I., Patel, N., Medhurst, A. D., and Pangalos, M. N. (2003). Molecular cloning and characterisation of a novel GABAB-related G-protein coupled receptor. *Brain Res. Mol. Brain Res.* **110**, 305–317.

Clark, J. A., Mezey, E., Lam, A. S., and Bonner, T. I. (2000). Distribution of the GABAB receptor subunit gb2 in rat CNS. *Brain Res.* **860**, 41–52.

Clayton, G. H., Owens, G. C., Wolff, J. S., and Smith, R. L. (1998). Ontogeny of cation-Cl-cotransporter expression in rat neocortex. *Brain Res. Dev. Brain Res.* **109**, 281–292.

Cobas, A., Fairen, A., Alvarez-Bolado, G., and Sanchez, M. P. (1991). Prenatal development of the intrinsic neurons of the rat neocortex: A comparative study of the distribution of GABA-immunoreactive cells and the GABAA receptor. *Neuroscience* **40**, 375–397.

Cone, C. D., Jr. (1980). Ionically mediated induction of mitogenesis in CNS neurons. *Ann. NY Acad. Sci.* **339**, 115–131.

Conti, F., Minelli, A., and Melone, M. (2004). GABA transporters in the mammalian cerebral cortex: Localization, development and pathological implications. *Brain Res. Brain Res. Rev.* **45**, 196–212.

Conti, F., Zuccarello, L. V., Barbaresi, P., Minelli, A., Brecha, N. C., and Melone, M. (1999). Neuronal, glial, and epithelial localization of gamma-aminobutyric acid transporter 2, a high-affinity gamma-aminobutyric acid plasma membrane transporter, in the cerebral cortex and neighboring structures. *J. Comp. Neurol.* **409**, 482–494.

Cooper, H. M. (2002). Axon guidance receptors direct growth cone pathfinding: Rivalry at the leading edge. *Int. J. Dev. Biol.* **46**, 621–631.

Coyle, J. T., and Yamamura, H. I. (1976). Neurochemical aspects of the ontogenesis of cholinergic neurons in the rat brain. *Brain Res.* **118**, 429–440.

Cruz, H. G., Ivanova, T., Lunn, M. L., Stoffel, M., Slesinger, P. A., and Luscher, C. (2004). Bi-directional effects of GABAB receptor agonists on the mesolimbic dopamine system. *Nat. Neurosci.* **7**, 153–159.

D'Arcangelo, G., Miao, G. G., Chen, S. C., Soares, H. D., Morgan, J. I., and Curran, T. (1995). A protein related to extracellular matrix proteins deleted in the mouse mutant reeler. *Nature* **374**, 719–723.

Demarque, M., Represa, A., Becq, H., Khalilov, I., Ben-Ari, Y., and Aniksztejn, L. (2002). Paracrine intercellular communication by a Ca2+- and SNARE-independent release of GABA and glutamate prior to synapse formation. *Neuron* **36**, 1051–1061.

Dotti, C. G., Sullivan, C. A., and Banker, G. A. (1988). The establishment of polarity by hippocampal neurons in culture. *J. Neurosci.* **8**, 1454–1468.

Durkin, M. M., Smith, K. E., Borden, L. A., Weinshank, R. L., Branchek, T. A., and Gustafson, E. L. (1995). Localization of messenger RNAs encoding three GABA transporters in rat brain: An *in situ* hybridization study. *Brain Res. Mol. Brain Res.* **33**, 7–21.

Elliott, K. A., and Jasper, H. H. (1959). Gammaaminobutyric acid. *Physiol. Rev.* **39**, 383–406.

Engelhardt, B. (2003). Development of the blood-brain barrier. *Cell. Tissue Res.* **314**, 119–129.

Erlander, M. G., Tillakaratne, N. J., Feldblum, S., Patel, N., and Tobin, A. J. (1991). Two genes encode distinct glutamate decarboxylases. *Neuron* **7**, 91–100.

Essrich, C., Lorez, M., Benson, J. A., Fritschy, J. M., and Luscher, B. (1998). Postsynaptic clustering of major GABAA receptor subtypes requires the gamma 2 subunit and gephyrin. *Nat. Neurosci.* **1**, 563–571.

Evans, J. E., Frostholm, A., and Rotter, A. (1996). Embryonic and postnatal expression of four gamma-aminobutyric acid transporter mRNAs in the mouse brain and leptomeninges. *J. Comp. Neurol.* **376,** 431–446.

Farrant, M., and Nusser, Z. (2005). Variations on an inhibitory theme: Phasic and tonic activation of GABA$_A$ receptors. *Nat. Rev. Neurosci.* **6,** 215–229.

Fischer, K. F., Lukasiewicz, P. D., and Wong, R. O. (1998). Age-dependent and cell class-specific modulation of retinal ganglion cell bursting activity by GABA. *J. Neurosci.* **18,** 3767–3778.

Fiszman, M. L., Behar, T., Lange, G. D., Smith, S. V., Novotny, E. A., and Barker, J. L. (1993). GABAergic cells and signals appear together in the early post-mitotic period of telencephalic and striatal development. *Brain Res. Dev. Brain Res.* **73,** 243–251.

Fritschy, J. M., Meskenaite, V., Weinmann, O., Honer, M., Benke, D., and Mohler, H. (1999). GABAB-receptor splice variants GB1a and GB1b in rat brain: Developmental regulation, cellular distribution and extrasynaptic localization. *Eur. J. Neurosci.* **11,** 761–768.

Fritschy, J. M., Paysan, J., Enna, A., and Mohler, H. (1994). Switch in the expression of rat GABAA-receptor subtypes during postnatal development: An immunohistochemical study. *J. Neurosci.* **14,** 5302–5324.

Fueshko, S. M., Key, S., and Wray, S. (1998). Luteinizing hormone releasing hormone (LHRH) neurons maintained in nasal explants decrease LHRH messenger ribonucleic acid levels after activation of GABA$_A$ receptors. *Endocrinology* **139,** 2734–2740.

Fukura, H., Komiya, Y., and Igarashi, M. (1996). Signaling pathway downstream of GABAA receptor in the growth cone. *J. Neurochem.* **67,** 1426–1434.

Gadea, A., and Lopez-Colome, A. M. (2001). Glial transporters for glutamate, glycine, and GABA: II. GABA transporters. *J. Neurosci. Res.* **63,** 461–468.

Galvez, T., Urwyler, S., Prezeau, L., Mosbacher, J., Joly, C., Malitschek, B., Heid, J., Brabet, I., Froestl, W., Bettler, B., Kaupmann, K., and Pin, J. P. (2000). Ca(2+) requirement for high-affinity gamma-aminobutyric acid (GABA) binding at GABA$_B$ receptors: Involvement of serine 269 of the GABA$_B$R1 subunit. *Mol. Pharmacol.* **57,** 419–426.

Garcia-Verdugo, J. M., Doetsch, F., Wichterle, H., Lim, D. A., and Alvarez-Buylla, A. (1998). Architecture and cell types of the adult subventricular zone: In search of the stem cells. *J. Neurobiol.* **36,** 234–248.

Gassmann, M., Shaban, H., Vigot, R., Sansig, G., Haller, C., Barbieri, S., Humeau, Y., Schuler, V., Muller, M., Kinzel, B., Klebs, K., Schmutz, M., Froestl, W., Heid, J., Kelly, P. H., Gentry, C., Jaton, A. L., Van der Putten, H., Mombereau, C., Lecourtier, L., Mosbacher, J., Cryan, J. F., Fritschy, J. M., Luthi, A., Kaupmann, K., and Bettler, B. (2004). Redistribution of GABAB (1) protein and atypical GABAB responses in GABAB(2)-deficient mice. *J. Neurosci.* **24,** 6086–6097.

Goffinet, A. M. (1984). Events governing organization of postmigratory neurons: Studies on brain development in normal and reeler mice. *Brain Res.* **319,** 261–296.

Gomez, T. M., Robles, E., Poo, M., and Spitzer, N. C. (2001). Filopodial calcium transients promote substrate-dependent growth cone turning. *Science* **291,** 1983–1987.

Gordon-Weeks, P. R. (2004). Microtubules and growth cone function. *J. Neurobiol.* **58,** 70–83.

Gotz, M., Hartfuss, E., and Malatesta, P. (2002). Radial glial cells as neuronal precursors: A new perspective on the correlation of morphology and lineage restriction in the developing cerebral cortex of mice. *Brain Res. Bull.* **57,** 777–788.

Greene, L. A., Burstein, D. E., Conolly, J. L., Green, S. H., Seeley, P. J., and Shelansky, M. L. (1984). Mechanism of the promotion of neurite outgrowth by nerve growth factor. *In* "Cellular and Molecular Biology of Neuronal Development" (Black, I. B., Ed.), pp. 133–145. Plenum Press, New York.

Gu, X., and Spitzer, N. C. (1997). Breaking the code: Regulation of neuronal differentiation by spontaneous calcium transients. *Dev. Neurosci.* **19,** 33–41.

Guastella, J., Nelson, N., Nelson, H., Czyzyk, L., Keynan, S., Miedel, M. C., Davidson, N., Lester, H. A., and Kanner, B. I. (1990). Cloning and expression of a rat brain GABA transporter. *Science* **249**, 1303–1306.

Gubellini, P., Ben-Ari, Y., and Gaiarsa, J. L. (2001). Activity- and age-dependent GABAergic synaptic plasticity in the developing rat hippocampus. *Eur. J. Neurosci.* **14**, 1937–1946.

Hachiya, Y., and Takashima, S. (2001). Development of GABAergic neurons and their transporter in human temporal cortex. *Pediatr. Neurol.* **25**, 390–396.

Hansra, N., Arya, S., and Quick, M. W. (2004). Intracellular domains of a rat brain GABA transporter that govern transport. *J. Neurosci.* **24**, 4082–4087.

Haydar, T. F., Wang, F., Schwartz, M. L., and Rakic, P. (2000). Differential modulation of proliferation in the neocortical ventricular and subventricular zones. *J. Neurosci.* **20**, 5764–5774.

Hedner, T., Iversen, K., and Lundborg, P. (1982). Gamma-Aminobutyric acid concentrations in the cerebrospinal fluid of newborn infants. *Early. Hum. Dev.* **7**, 53–58.

Heger, S., Seney, M., Bless, E., Schwarting, G. A., Bilger, M., Mungenast, A., Ojeda, S. R., and Tobet, S. A. (2003). Overexpression of glutamic acid decarboxylase-67 (GAD-67) in gonadotropin-releasing hormone neurons disrupts migratory fate and female reproductive function in mice. *Endocrinology* **144**, 2566–2579.

Henley, J., and Poo, M. M. (2004). Guiding neuronal growth cones using Ca2+ signals. *Trends Cell. Biol.* **14**, 320–330.

Hevers, W., and Luddens, H. (1998). The diversity of GABAA receptors. Pharmacological and electrophysiological properties of GABAA channel subtypes. *Mol. Neurobiol.* **18**, 35–86.

Himanen, J. P., and Nikolov, D. B. (2003). Eph receptors and ephrins. *Int. J. Biochem. Cell. Biol.* **35**, 130–134.

Hollrigel, G. S., and Soltesz, I. (1997). Slow kinetics of miniature IPSCs during early postnatal development in granule cells of the dentate gyrus. *J. Neurosci.* **17**, 5119–5128.

Hsu, C. C., Davis, K. M., Jin, H., Foos, T., Floor, E., Chen, W., Tyburski, J. B., Yang, C. Y., Schloss, J. V., and Wu, J. Y. (2000). Association of L-glutamic acid decarboxylase to the 70-kDa heat shock protein as a potential anchoring mechanism to synaptic vesicles. *J. Biol. Chem.* **275**, 20822–20828.

Chen, H., He, Z., and Tessier-Lavigne, M. (1998). Axon guidance mechanisms: Semaphorins as simultaneous repellents and anti-repellents. *Nat. Neurosci.* **1**, 436–439.

Cherubini, E., Gaiarsa, J. L., and Ben-Ari, Y. (1991). GABA: An excitatory transmitter in early postnatal life. *Trends Neurosci.* **14**, 515–519.

Ikegaki, N., Saito, N., Hashima, M., and Tanaka, C. (1994). Production of specific antibodies against GABA transporter subtypes (GAT1, GAT2, GAT3) and their application to immunocytochemistry. *Brain Res. Mol. Brain Res.* **26**, 47–54.

Jacobson, M. (1978). Histogenesis and morphogenesis of the central nervous system. *In* "Developmental neurobiology" (Jacobson, M., Ed.), pp. 57–114. Plenum Press, New York.

Jelitai, M., Anderova, M., Marko, K., Kekesi, K., Koncz, P., Sykova, E., and Madarasz, E. (2004). Role of gamma-aminobutyric acid in early neuronal development: Studies with an embryonic neuroectodermal stem cell clone. *J. Neurosci. Res.* **76**, 801–811.

Jessell, T. M., and Kandel, E. R. (1993). Synaptic transmission: A bidirectional and self-modifiable form of cell-cell communication. *Cell* **72** (Suppl.), 1–30.

Ji, F., Kanbara, N., and Obata, K. (1999). GABA and histogenesis in fetal and neonatal mouse brain lacking both the isoforms of glutamic acid decarboxylase. *Neurosci. Res.* **33**, 187–194.

Jursky, F., and Nelson, N. (1996). Developmental expression of GABA transporters GAT1 and GAT4 suggests involvement in brain maturation. *J. Neurochem.* **67**, 857–867.

Jursky, F., and Nelson, N. (1999). Developmental expression of the neurotransmitter transporter GAT3. *J. Neurosci. Res.* **55**, 394–399.

Kalinovsky, A., and Scheiffele, P. (2004). Transcriptional control of synaptic differentiation by retrograde signals. *Curr. Opin. Neurobiol.* **14,** 272–279.

Kanaani, J., Lissin, D., Kash, S. F., and Baekkeskov, S. (1999). The hydrophilic isoform of glutamate decarboxylase, GAD67, is targeted to membranes and nerve terminals independent of dimerization with the hydrophobic membrane-anchored isoform, GAD65. *J. Biol. Chem.* **274,** 37200–37209.

Kaneda, M., Farrant, M., and Cull-Candy, S. G. (1995). Whole-cell and single-channel currents activated by GABA and glycine in granule cells of the rat cerebellum. *J. Physiol.* **485**(Pt 2), 419–435.

Kaplan, M. S., and Hinds, J. W. (1977). Neurogenesis in the adult rat: Electron microscopic analysis of light radioautographs. *Science* **197,** 1092–1094.

Katz, L. C., and Shatz, C. J. (1996). Synaptic activity and the construction of cortical circuits. *Science* **274,** 1133–1138.

Kaupmann, K., Malitschek, B., Schuler, V., Heid, J., Froestl, W., Beck, P., Mosbacher, J., Bischoff, S., Kulik, A., Shigemoto, R., Karschin, A., and Bettler, B. (1998a). GABA$_B$-receptor subtypes assemble into functional heteromeric complexes. *Nature* **396,** 683–687.

Kaupmann, K., Schuler, V., Mosbacher, J., Bischoff, S., Bittiger, H., Heid, J., Froestl, W., Leonhard, S., Pfaff, T., Karschin, A., and Bettler, B. (1998b). Human gamma-aminobutyric acid type B receptors are differentially expressed and regulate inwardly rectifying K+ channels. *Proc. Natl. Acad. Sci. USA* **95,** 14991–14996.

Kennedy, M. B. (2000). Signal-processing machines at the postsynaptic density. *Science* **290,** 750–754.

Kim, M. O., Li, S., Park, M. S., and Hornung, J. P. (2003). Early fetal expression of GABA(B1) and GABA(B2) receptor mRNAs on the development of the rat central nervous system. *Brain Res. Dev. Brain Res.* **143,** 47–55.

Kriegstein, A. R., and Noctor, S. C. (2004). Patterns of neuronal migration in the embryonic cortex. *Trends Neurosci.* **27,** 392–399.

Krueger, S. R., Kolar, A., and Fitzsimonds, R. M. (2003). The presynaptic release apparatus is functional in the absence of dendritic contact and highly mobile within isolated axons. *Neuron* **40,** 945–957.

Lauder, J. M., Han, V. K., Henderson, P., Verdoorn, T., and Towle, A. C. (1986). Prenatal ontogeny of the GABAergic system in the rat brain: An immunocytochemical study. *Neuroscience* **19,** 465–493.

Laurie, D. J., Wisden, W., and Seeburg, P. H. (1992). The distribution of thirteen GABAA receptor subunit mRNAs in the rat brain. III. Embryonic and postnatal development. *J. Neurosci.* **12,** 4151–4172.

Lavdas, A. A., Grigoriou, M., Pachnis, V., and Parnavelas, J. G. (1999). The medial ganglionic eminence gives rise to a population of early neurons in the developing cerebral cortex. *J. Neurosci.* **19,** 7881–7888.

Lerma, J., Herranz, A. S., Herreras, O., Abraira, V., and Martin del Rio, R. (1986). *In vivo* determination of extracellular concentration of amino acids in the rat hippocampus. A method based on brain dialysis and computerized analysis. *Brain Res.* **384,** 145–155.

Letinic, K., Zoncu, R., and Rakic, P. (2002). Origin of GABAergic neurons in the human neocortex. *Nature* **417,** 645–649.

Liu, Q. R., Lopez-Corcuera, B., Mandiyan, S., Nelson, H., and Nelson, N. (1993). Molecular characterization of four pharmacologically distinct gamma-aminobutyric acid transporters in mouse brain [corrected]. *J. Biol. Chem.* **268,** 2106–2112.

Lockerbie, R. O., Gordon-Weeks, P. R., and Pearce, B. R. (1985). Growth cones isolated from developing rat forebrain: Uptake and release of GABA and noradrenaline. *Brain Res.* **353,** 265–275.

Lopez-Bendito, G., Lujan, R., Shigemoto, R., Ganter, P., Paulsen, O., and Molnar, Z. (2003). Blockade of GABA$_B$ receptors alters the tangential migration of cortical neurons. *Cereb. Cortex* **13**, 932–942.

Lopez-Bendito, G., Shigemoto, R., Kulik, A., Paulsen, O., Fairen, A., and Lujan, R. (2002). Expression and distribution of metabotropic GABA receptor subtypes GABABR1 and GABABR2 during rat neocortical development. *Eur. J. Neurosci.* **15**, 1766–1778.

Lo Turco, J. J., Owens, D. F., Heath, M. J., Davis, M. B., and Kriegstein, A. R. (1995). GABA and glutamate depolarize cortical progenitor cells and inhibit DNA synthesis. *Neuron* **15**, 1287–1298.

Luscher, B., and Keller, C. A. (2004). Regulation of GABAA receptor trafficking, channel activity, and functional plasticity of inhibitory synapses. *Pharmacol. Ther.* **102**, 195–221.

Luscher, C., Jan, L. Y., Stoffel, M., Malenka, R. C., and Nicoll, R. A. (1997). G protein-coupled inwardly rectifying K+ channels (GIRKs) mediate postsynaptic but not presynaptic transmitter actions in hippocampal neurons. *Neuron* **19**, 687–695.

Ma, W., and Barker, J. L. (1995). Complementary expressions of transcripts encoding GAD67 and GABAA receptor alpha 4, beta 1, and gamma 1 subunits in the proliferative zone of the embryonic rat central nervous system. *J. Neurosci.* **15**, 2547–2560.

Ma, W., Behar, T., Maric, D., Maric, I., and Barker, J. L. (1992). Neuroepithelial cells in the rat spinal cord express glutamate decarboxylase immunoreactivity *in vivo* and *in vitro*. *J. Comp. Neurol.* **325**, 257–270.

Ma, W., Liu, Q. Y., Maric, D., Sathanoori, R., Chang, Y. H., and Barker, J. L. (1998). Basic FGF-responsive telencephalic precursor cells express functional GABA$_A$ receptor/Cl− channels *in vitro*. *J. Neurobiol.* **35**, 277–286.

Macdonald, R. L., and Olsen, R. W. (1994). GABA$_A$ receptor channels. *Annu. Rev. Neurosci.* **17**, 569–602.

Maddox, D. M., and Condie, B. G. (2001). Dynamic expression of a glutamate decarboxylase gene in multiple non-neural tissues during mouse development. *BMC Dev. Biol.* **1**, 1.

Maric, D., Liu, Q. Y., Maric, I., Chaudry, S., Chang, Y. H., Smith, S. V., Sieghart, W., Fritschy, J. M., and Barker, J. L. (2001). GABA expression dominates neuronal lineage progression in the embryonic rat neocortex and facilitates neurite outgrowth via GABA$_A$ autoreceptor/Cl− channels. *J. Neurosci.* **21**, 2343–2360.

Marin-Padilla, M. (1998). Cajal-Retzius cells and the development of the neocortex. *Trends Neurosci.* **21**, 64–71.

Martens, D. J., Tropepe, V., and van Der Kooy, D. (2000). Separate proliferation kinetics of fibroblast growth factor-responsive and epidermal growth factor-responsive neural stem cells within the embryonic forebrain germinal zone. *J. Neurosci.* **20**, 1085–1095.

Martin, K. C. (2004). Local protein synthesis during axon guidance and synaptic plasticity. *Curr. Opin. Neurobiol.* **14**, 305–310.

Martin, S. C., Steiger, J. L., Gravielle, M. C., Lyons, H. R., Russek, S. J., and Farb, D. H. (2004). Differential expression of gamma-aminobutyric acid type B receptor subunit mRNAs in the developing nervous system and receptor coupling to adenylyl cyclase in embryonic neurons. *J. Comp. Neurol.* **473**, 16–29.

Marty, S., Berninger, B., Carroll, P., and Thoenen, H. (1996). GABAergic stimulation regulates the phenotype of hippocampal interneurons through the regulation of brain-derived neurotrophic factor. *Neuron* **16**, 565–570.

McGee, A. W., and Strittmatter, S. M. (2003). The Nogo-66 receptor: Focusing myelin inhibition of axon regeneration. *Trends Neurosci.* **26**, 193–198.

McKernan, R. M., and Whiting, P. J. (1996). Which GABAA-receptor subtypes really occur in the brain? *Trends Neurosci.* **19**, 139–143.

Mi, R., Tang, X., Sutter, R., Xu, D., Worley, P., and O'Brien, R. J. (2002). Differing mechanisms for glutamate receptor aggregation on dendritic spines and shafts in cultured hippocampal neurons. *J. Neurosci.* **22,** 7606–7616.

Minelli, A., Alonso-Nanclares, L., Edwards, R. H., De Felipe, J., and Conti, F. (2003a). Postnatal development of the vesicular GABA transporter in rat cerebral cortex. *Neuroscience* **117,** 337–346.

Minelli, A., Barbaresi, P., and Conti, F. (2003b). Postnatal development of high-affinity plasma membrane GABA transporters GAT-2 and GAT-3 in the rat cerebral cortex. *Brain Res. Dev. Brain Res.* **142,** 7–18.

Mody, I. (2001). Distinguishing between GABA$_A$ receptors responsible for tonic and phasic conductances. *Neurochem. Res.* **26,** 907–913.

Nakayasu, H., Nishikawa, M., Mizutani, H., Kimura, H., and Kuriyama, K. (1993). Immunoaffinity purification and characterization of gamma-aminobutyric acid (GABA)B receptor from bovine cerebral cortex. *J. Biol. Chem.* **268,** 8658–8664.

Nguyen, L., Malgrange, B., Breuskin, I., Bettendorff, L., Moonen, G., Belachew, S., and Rigo, J. M. (2003). Autocrine/paracrine activation of the GABA$_A$ receptor inhibits the proliferation of neurogenic polysialylated neural cell adhesion molecule-positive (PSA-NCAM+) precursor cells from postnatal striatum. *J. Neurosci.* **23,** 3278–3294.

Nornes, H. O., and Das, G. D. (1974). Temporal pattern of neurogenesis in spinal cord of rat. I. An autoradiographic study–time and sites of origin and migration and settling patterns of neuroblasts. *Brain Res.* **73,** 121–138.

Nusser, Z., Sieghart, W., and Somogyi, P. (1998). Segregation of different GABAA receptors to synaptic and extrasynaptic membranes of cerebellar granule cells. *J. Neurosci.* **18,** 1693–1703.

Obrietan, K., and van den Pol, A. N. (1996). Growth cone calcium elevation by GABA. *J. Comp. Neurol.* **372,** 167–175.

Obrietan, K., and van den Pol, A. N. (1999). GABA$_B$ receptor-mediated regulation of glutamate-activated calcium transients in hypothalamic and cortical neuron development. *J. Neurophysiol.* **82,** 94–102.

O'Donovan, M. J. (1999). The origin of spontaneous activity in developing networks of the vertebrate nervous system. *Curr. Opin. Neurobiol.* **9,** 94–104.

O'Donovan, M. J., Chub, N., and Wenner, P. (1998). Mechanisms of spontaneous activity in developing spinal networks. *J. Neurobiol.* **37,** 131–145.

Ohbayashi, K., Fukura, H., Inoue, H. K., Komiya, Y., and Igarashi, M. (1998). Stimulation of L-type Ca2+ channel in growth cones activates two independent signaling pathways. *J. Neurosci. Res.* **51,** 682–696.

Owens, D. F., Boyce, L. H., Davis, M. B., and Kriegstein, A. R. (1996). Excitatory GABA responses in embryonic and neonatal cortical slices demonstrated by gramicidin perforated-patch recordings and calcium imaging. *J. Neurosci.* **16,** 6414–6423.

Owens, D. F., and Kriegstein, A. R. (2002). Is there more to GABA than synaptic inhibition? *Nat. Rev. Neurosci.* **3,** 715–727.

Owens, D. F., Liu, X., and Kriegstein, A. R. (1999). Changing properties of GABA$_A$ receptor-mediated signaling during early neocortical development. *J. Neurophysiol.* **82,** 570–583.

Patthy, L., and Nikolics, K. (1994). Agrin-like proteins of the neuromuscular junction. *Neurochem. Int.* **24,** 301–316.

Pfenninger, K. H., and Maylie-Pfenninger, M. F. (1979). Properties and dynamics of plasmalemmal glycoconjugates in growing neurites. *Prog. Brain. Res.* **51,** 83–94.

Poulter, M. O., and Brown, L. A. (1999). Transient expression of GABAA receptor subunit mRNAs in the cellular processes of cultured cortical neurons and glia. *Brain Res. Mol. Brain Res.* **69,** 44–52.

Priest, C. A., and Puche, A. C. (2004). GABAB receptor expression and function in olfactory receptor neuron axon growth. *J. Neurobiol.* **60,** 154–165.

Quick, M. W., Hu, J., Wang, D., and Zhang, H. Y. (2004). Regulation of a gamma-aminobutyric acid transporter by reciprocal tyrosine and serine phosphorylation. *J. Biol. Chem.* **279**, 15961–15967.

Rakic, P. (1972). Mode of cell migration to the superficial layers of fetal monkey neocortex. *J. Comp. Neurol.* **145**, 61–83.

Rivera, C., Voipio, J., Payne, J. A., Ruusuvuori, E., Lahtinen, H., Lamsa, K., Pirvola, U., Saarma, M., and Kaila, K. (1999). The K+/Cl− co-transporter KCC2 renders GABA hyperpolarizing during neuronal maturation. *Nature* **397**, 251–255.

Sabau, A., Frahm, C., Pfeiffer, M., Breustedt, J., Piechotta, A., Numberger, M., Engel, D., Heinemann, U., and Draguhn, A. (1999). Age-dependence of the anticonvulsant effects of the GABA uptake inhibitor tiagabine *in vitro. Eur. J. Pharmacol.* **383**, 259–266.

Salganicoff, L., and Derobertis, E. (1965). Subcellular distribution of the enzymes of the glutamic acid, glutamine and gamma-aminobutyric acid cycles in rat brain. *J. Neurochem.* **12**, 287–309.

Sarup, A., Larsson, O. M., and Schousboe, A. (2003). GABA transporters and GABA-transaminase as drug targets. *Curr. Drug. Targets CNS Neurol. Disord.* **2**, 269–277.

Sassoe-Pognetto, M., Panzanelli, P., Sieghart, W., and Fritschy, J. M. (2000). Colocalization of multiple GABA_A receptor subtypes with gephyrin at postsynaptic sites. *J. Comp. Neurol.* **420**, 481–498.

Saxena, N. C., and Macdonald, R. L. (1994). Assembly of GABAA receptor subunits: Role of the delta subunit. *J. Neurosci.* **14**, 7077–7086.

Semyanov, A., Walker, M. C., Kullmann, D. M., and Silver, R. A. (2004). Tonically active GABA A receptors: Modulating gain and maintaining the tone. *Trends Neurosci.* **27**, 262–269.

Serafini, R., Maric, D., Maric, I., Ma, W., Fritschy, J. M., Zhang, L., and Barker, J. L. (1998). Dominant GABA_A receptor/Cl− channel kinetics correlate with the relative expressions of alpha2, alpha3, alpha5 and beta3 subunits in embryonic rat neurones. *Eur. J. Neurosci.* **10**, 334–349.

Schratt, G. M., Nigh, E. A., Chen, W. G., Hu, L., and Greenberg, M. E. (2004). BDNF regulates the translation of a select group of mRNAs by a mammalian target of rapamycin-phosphatidylinositol 3-kinase-dependent pathway during neuronal development. *J. Neurosci.* **24**, 7366–7377.

Schuler, V., Luscher, C., Blanchet, C., Klix, N., Sansig, G., Klebs, K., Schmutz, M., Heid, J., Gentry, C., Urban, L., Fox, A., Spooren, W., Jaton, A. L., Vigouret, J., Pozza, M., Kelly, P. H., Mosbacher, J., Froestl, W., Kaslin, E., Korn, R., Bischoff, S., Kaupmann, K., van der Putten, H., and Bettler, B. (2001). Epilepsy, hyperalgesia, impaired memory, and loss of pre- and postsynaptic GABA_B responses in mice lacking GABA_{B1}. *Neuron* **31**, 47–58.

Schwartz, T. H., Rabinowitz, D., Unni, V., Kumar, V. S., Smetters, D. K., Tsiola, A., and Yuste, R. (1998). Networks of coactive neurons in developing layer 1. *Neuron* **20**, 541–552.

Sidman, R. L., and Rakic, P. (1973). Neuronal migration, with special reference to developing human brain: A review. *Brain Res.* **62**, 1–35.

Sieghart, W., Fuchs, K., Tretter, V., Ebert, V., Jechlinger, M., Hoger, H., and Adamiker, D. (1999). Structure and subunit composition of GABA_A receptors. *Neurochem. Int.* **34**, 379–385.

Soghomonian, J. J., and Martin, D. L. (1998). Two isoforms of glutamate decarboxylase: Why? *Trends Pharmacol. Sci.* **19**, 500–505.

Spitzer, N. C., Root, C. M., and Borodinsky, L. N. (2004). Orchestrating neuronal differentiation: Patterns of Ca2+ spikes specify transmitter choice. *Trends Neurosci.* **27**, 415–421.

Stanfield, B. B. (1984). Postnatal reorganization of cortical projections: The role of collateral eliminaton. *TINS* **7**, 37–40.

Stell, B. M., and Mody, I. (2002). Receptors with different affinities mediate phasic and tonic GABA_A conductances in hippocampal neurons. *J. Neurosci.* **22**, RC223.

Steward, F. C., Thomson, J. F., and Dent, C. E. (1949). γ-Aminobutyric acid. A constituent of the potato tuber. *Science* **110**, 439–440.

Sullivan, R., Chateauneuf, A., Coulombe, N., Kolakowski, L. F., Jr., Johnson, M. P., Hebert, T. E., Ethier, N., Belley, M., Metters, K., Abramovitz, M., O'Neill, G. P., and Ng, G. Y. (2000). Coexpression of full-length gamma-aminobutyric acid(B) (GABA_B) receptors with truncated receptors and metabotropic glutamate receptor 4 supports the GABA_B heterodimer as the functional receptor. *J. Pharmacol. Exp. Ther.* **293,** 460–467.

Sun, C., Sieghart, W., and Kapur, J. (2004). Distribution of alpha1, alpha4, gamma2, and delta subunits of GABAA receptors in hippocampal granule cells. *Brain Res.* **1029,** 207–216.

Sutton, M. A., Wall, N. R., Aakalu, G. N., and Schuman, E. M. (2004). Regulation of dendritic protein synthesis by miniature synaptic events. *Science* **304,** 1979–1983.

Szabo, G., Katarova, Z., and Greenspan, R. (1994). Distinct protein forms are produced from alternatively spliced bicistronic glutamic acid decarboxylase mRNAs during development. *Mol. Cell. Biol.* **14,** 7535–7545.

Tabata, T., Araishi, K., Hashimoto, K., Hashimotodani, Y., van der Putten, H., Bettler, B., and Kano, M. (2004). Ca2+ activity at GABAB receptors constitutively promotes metabotropic glutamate signaling in the absence of GABA. *Proc. Natl. Acad. Sci. USA* **101,** 16952–16957.

Taylor, J., Docherty, M., and Gordon-Weeks, P. R. (1990). GABAergic growth cones: Release of endogenous gamma-aminobutyric acid precedes the expression of synaptic vesicle antigens. *J. Neurochem.* **54,** 1689–1699.

Thuault, S. J., Brown, J. T., Sheardown, S. A., Jourdain, S., Fairfax, B., Spencer, J. P., Restituito, S., Nation, J. H., Topps, S., Medhurst, A. D., Randall, A. D., Couve, A., Moss, S. J., Collingridge, G. L., Pangalos, M. N., Davies, C. H., and Calver, A. R. (2004). The GABA(B2) subunit is critical for the trafficking and function of native GABA_B receptors. *Biochem. Pharmacol.* **68,** 1655–1666.

Turgeon, S. M., and Albin, R. L. (1994). Postnatal ontogeny of GABAB binding in rat brain. *Neuroscience* **62,** 601–613.

Valeyev, A. Y., Cruciani, R. A., Lange, G. D., Smallwood, V. S., and Barker, J. L. (1993). Cl− channels are randomly activated by continuous GABA secretion in cultured embryonic rat hippocampal neurons. *Neurosci. Lett.* **155,** 199–203.

Varju, P., Katarova, Z., Madarasz, E., and Szabo, G. (2002). Sequential induction of embryonic and adult forms of glutamic acid decarboxylase during *in vitro*-induced neurogenesis in cloned neuroectodermal cell-line, NE-7C2. *J. Neurochem.* **80,** 605–615.

Verney, C. (2003). Phenotypic expression of monoamines and GABA in the early development of human telencephalon, transient or not transient. *J. Chem. Neuroanat.* **26,** 283–292.

Vicini, S., Ferguson, C., Prybylowski, K., Kralic, J., Morrow, A. L., and Homanics, G. E. (2001). GABA_A receptor alpha1 subunit deletion prevents developmental changes of inhibitory synaptic currents in cerebellar neurons. *J. Neurosci.* **21,** 3009–3016.

Waagepetersen, H. S., Sonnewald, U., and Schousboe, A. (1999). The GABA paradox: Multiple roles as metabolite, neurotransmitter, and neurodifferentiative agent. *J. Neurochem.* **73,** 1335–1342.

Wadsworth, W. G., Bhatt, H., and Hedgecock, E. M. (1996). Neuroglia and pioneer neurons express UNC-6 to provide global and local netrin cues for guiding migrations in C. elegans. *Neuron* **16,** 35–46.

Watanabe, M., Maemura, K., Kanbara, K., Tamayama, T., and Hayasaki, H. (2002). GABA and GABA receptors in the central nervous system and other organs. *Int. Rev. Cytol.* **213,** 1–47.

Wolff, J. R., Joo, F., and Dames, W. (1978). Plasticity in dendrites shown by continuous GABA administration in superior cervical ganglion of adult rat. *Nature* **274,** 72–74.

Wong, K., Park, H. T., Wu, J. Y., and Rao, Y. (2002). Slit proteins: Molecular guidance cues for cells ranging from neurons to leukocytes. *Curr. Opin. Genet. Dev.* **12,** 583–591.

Xia, Y., Poosch, M. S., Whitty, C. J., Kapatos, G., and Bannon, M. J. (1993). GABA transporter mRNA: *In vitro* expression and quantitation in neonatal rat and postmortem human brain. *Neurochem. Int.* **22,** 263–270.

Xiang, Y., Li, Y., Zhang, Z., Cui, K., Wang, S., Yuan, X. B., Wu, C. P., Poo, M. M., and Duan, S. (2002). Nerve growth cone guidance mediated by G protein-coupled receptors. *Nat. Neurosci.* **5,** 843–848.

Xu, Q., Cobos, I., De La Cruz, E., Rubenstein, J. L., and Anderson, S. A. (2004). Origins of cortical interneuron subtypes. *J. Neurosci.* **24,** 2612–2622.

Yuste, R., and Katz, L. C. (1991). Control of postsynaptic Ca2+ influx in developing neocortex by excitatory and inhibitory neurotransmitters. *Neuron* **6,** 333–344.

Zecevic, N., and Milosevic, A. (1997). Initial development of gamma-aminobutyric acid immunoreactivity in the human cerebral cortex. *J. Comp. Neurol.* **380,** 495–506.

Ziv, N. E., and Garner, C. C. (2004). Cellular and molecular mechanisms of presynaptic assembly. *Nat. Rev. Neurosci.* **5,** 385–399.

GABAᴇʀɢɪᴄ SIGNALING IN THE DEVELOPING CEREBELLUM

Chitoshi Takayama

Department of Molecular Neuroanatomy, Hokkaido University School of Medicine
Sapporo 060–8638, Japan

In the adult central nervous system (CNS), γ-amino butyric acid (GABA) is the predominant inhibitory neurotransmitter, and it regulates glutamatergic activity. Recent studies have revealed that GABA serves as an excitatory transmitter in the immature CNS, and acts as a trophic factor for brain development. Furthermore, normal formation of GABAergic synapses is crucial for the expression of higher brain functions such as memory, learning, and motor coordination, and various psychiatric diseases such as anxiety disorders, epilepsy, schizophrenia, and autism are partly caused by the dysfunction of GABA in the developing and mature brain. These results indicate that the GABAergic roles change developmentally with special references to alterations in GABAergic transmission and signaling, and that GABA plays various roles in the expression of almost all brain functions. We morphologically investigated the developmental changes in the GABAergic transmission system and the key factors for the

INTERNATIONAL REVIEW OF
NEUROBIOLOGY, VOL. 71
DOI: 10.1016/S0074-7742(05)71003-5

63

formation of GABAergic synapses and networks. Here, we focus on the cerebellar cortex, which provides an ideal system for the investigation of brain development, and review four points: (1) The GABAergic system in the adult cerebellum, (2) GABAergic signaling before synaptogenesis—the mechanisms by which GABA exerts its effect on developing neurons, (3) Formation of GABAergic synapses—mechanisms underlying formation of functional GABAergic synapses, and (4) Functions of GABAergic synapses.

I. Introduction

In the mammalian central nervous system (CNS), GABA is the predominant neurotransmitter, and it plays various roles in the expression of brain functions by activation of ionotropic and metabotropic GABA receptors. In the adult CNS, GABA mediates inhibitory synaptic transmission, regulates glutamatergic activity, and prevents hyperexcitation (Kardos, 1999; Macdonald and Olsen, 1994; Olsen and Avoli, 1997). In the developing CNS, GABA acts as trophic factor, and induces brain morphogenesis, including changes in cell proliferation, cell migration, axonal growth, synapse formation, steroid-mediated sexual differentiation, and cell death (Barker et al., 1998; Belhage et al., 1998; Ben-Ari, 2002; Kardos, 1999; McCarthy et al., 2002; Owens and Kriegstein, 2002; Varju et al., 2001). Furthermore, during the maturation period, GABAergic transmission controls experience-dependent plasticity in the visual cortex, induces long-term potentiation, which is the electrophysiological basis of memory and learning (Ben-Ari et al., 1997; Fagiolini and Hensch, 2000; Freund and Gulyas, 1997; Hensch et al., 1998; Kardos, 1999; McBain and Maccaferri, 1997; Paulsen and Moser, 1998; Wolff et al., 1993), modulates anxiety (Nutt et al., 1990; Pratt, 1992), and generates circadian rhythms (Nutt et al., 1990; Pratt, 1992; Turek and Van Reeth, 1988; Wagner et al., 1997). Various psychiatric diseases such as epilepsy (Avoli, 2000; Baulac et al., 2001; Snead et al., 1999; Wallace et al., 2001), anxiety disorders (Freeman et al., 2002; Millan, 2003; Nutt, 2001), schizophrenia (Berry et al., 2003; Blum and Mann, 2002; Byne et al., 1999; Caruncho et al., 2004; Costa et al., 2004; Lewis et al., 2004; Wassef et al., 2003), and autism (Blatt et al., 2001; Cook et al., 1997; DeLorey et al., 1998; Dhossche et al., 2002; Dhossche, 2004; Fatemi et al., 2002; Lauritsen et al., 1999; Rolf et al., 1993) are partially caused by GABA dysfunction in the developing and mature brain.

The developmental shift in the action of GABA is based on an alteration in GABAergic transmission and signaling. During development, the GABA transmitter system changes from non-synaptic to synaptic mechanisms (Attwell et al., 1993; Fon and Edwards, 2001; Owens and Kriegstein, 2002; Taylor and Gordon-Weeks, 1991; Varju *et al.*, 2001). The subunit compositions and localization of ionotropic GABA receptors drastically change (Araki *et al.*, 1992; Fritschy *et al.*, 1994; Gambarana *et al.*, 1990; Laurie *et al.*, 1992a; Ma and Barker, 1995; Maric *et al.*, 1997). Environmental changes, such as decreasing intracellular chloride concentration, influence the response of GABA receptors (Ben-Ari, 2002; Cherubini *et al.*, 1991; Ganguly *et al.*, 2001; Owens and Kriegstein, 2002; Perkins and Wong, 1997; Rohrbough and Spitzer, 1996; Serafini *et al.*, 1998). In the first half, we review the GABAergic system in the cerebellum and the developmental changes in GABAergic signaling, and discuss how GABA exerts its effect on immature neurons during development. Establishment of GABAergic synapses is crucial for the expression of normal and higher brain functions. In the second half, we address the key factors for the formation of functional GABAergic synapses, and discuss the mechanisms underlying the formation of GABAergic synapses and networks.

II. GABAergic System in the Cerebellar Cortex

A. GABAERGIC NEURONS AND SYNAPSES

The cerebellar cortex consists of molecular, Purkinje cell, and granular layers (Ito, 1984; Llinas and Walton, 1990; Palay and Chan-Palay, 1974). Each layer contains distinct types of neurons. Stellate cells are scattered in the molecular layer, and basket cells are localized in the deep part of the molecular layer. Cell bodies of Purkinje cells are arranged in a single layer between the molecular and granular layers. Numerous granule cells occupy the granular layer and Golgi cells are localized mainly in the upper half of the granular layer. Among the five types of main neurons, four of them, Purkinje, stellate, basket, and Golgi cells, release GABA as a neurotransmitter (see Fig. 1A, B). The GABAergic neurons and the neural circuits in the cerebellar cortex are summarized in Fig. 1B (Ito, 1984; Llinas and Walton, 1990; Palay and Chan-Palay, 1974).

The Purkinje cell is a pivotal neuron in the cerebellar cortex. Each dendrite spreads out in a single vertical parasagittal plane in the molecular layer, and receives excitatory inputs from climbing and parallel fibers at the spines, and inhibitory inputs from stellate cells at the shafts. Cell bodies form GABAergic synapses with the pericellular baskets of basket cell axon collaterals. The Purkinje cell axons traverse the granular layer into the white matter, and innervate the deep cerebellar and vestibular nuclei. In addition, recurrent collateral branches arise from the third or fourth nodes of Ranvier (Cajal, 1911), ramify in the upper granular layer, and give rise to plexuses beneath the Purkinje cell bodies. In the plexus, varicosities of axon collaterals form GABAergic synapses with Purkinje

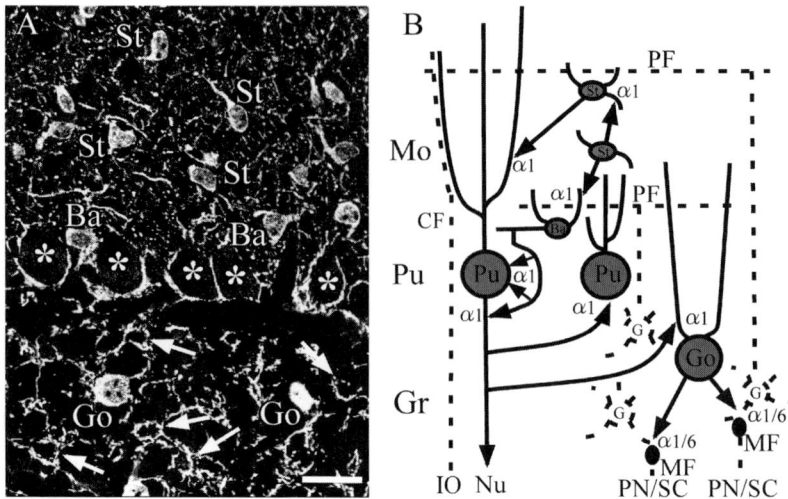

FIG. 1. GABAergic neurons (A) and the GABAergic local circuit in the cerebellar cortex (B). (A) Immunohistochemistry for GABA in the adult mouse cerebellar cortex. Cell bodies and axon terminals of stellate (St) and basket (Ba) cells are labeled in the molecular (Mo) and Purkinje cell (Pu) layers. Golgi cell bodies (Go) and their axon terminals (white arrows) are also stained in the granular layer (Gr). However, cell bodies (asterisks) and dendrites of Purkinje cells are negative. (B) Schematic illustration of the local neural circuit and GABA$_A$ receptor α subunit expression in the adult cerebellar cortex. Abbreviations and symbols. St: stellate cell, Ba: basket cell, asterisk: Purkinje cell body, Go: Golgi cell, white arrow: GABA-positive rings in the synaptic glomerulus, Mo: molecular layer, Pu: Purkinje cell layer, Gr: granular layer, PF: parallel fiber, α1: GABA$_A$ receptor α1 subunit, G: granule cell, α1/6: GABA$_A$ receptor α1 and α6 subunits, MF: mossy fiber, CF: climbing fiber, IO: inferior olivary nucleus, Nu: deep cerebellar nucleus, PN/SC: pontine nucleus and spinal cord, black arrow: GABAergic innervation and synapse, gray circle: GABAergic neuron, dotted circle: excitatory neuron. Bar: 10μm.

and Golgi cells at the cell bodies and dendrites (Palay and Chan-Palay, 1974; Takayama and Inoue, 2004a).

Each stellate and basket cell extends its dendrite only in a parasagittal plane parallel to the fan of the Purkinje cell dendrite within the molecular layer. They receive excitatory inputs from parallel and climbing fibers and inhibitory inputs from stellate cells. The stellate cell axons arborize within the molecular layer, and their varicosities make many GABAergic synaptic contacts with the dendritic shafts of Purkinje cells and other GABAergic neurons, including stellate, basket, and Golgi cells. Basket cell axons run deep in the molecular layer just above the cell bodies of Purkinje cells. Axon collaterals descend along the Purkinje cell dendrites, surround the Purkinje cell bodies, give rise to pericellular baskets, and form periaxonal plexuses, '*pinceau*,' around the initial segment of the Purkinje cell axon (Cajal, 1911). At the pericellular basket, axo-somatic synapses are formed with Purkinje cell bodies. At the *pinceau*, axo-axonic synapses are formed with the initial segments of Purkinje cells. Both synapses are GABAergic. One basket cell axon innervates about 10 Purkinje cells.

Golgi cells extend their dendrites into the molecular layer. Each Golgi cell dendrite is not confined to a single plane, but opens out into a three-dimensional ungulated field. They receive excitatory input from parallel and climbing fibers in the molecular layer and mossy fibers in the granular layer. Inhibitory inputs come from stellate cells in the molecular layer and Purkinje cell axon collaterals in the granular layer. One to three axons arise from the cell body and main dendrite, divide repeatedly, and give rise to plexuses. At the plexuses, varicosities of Golgi cell axons form GABAergic synapses with granule cell dendrites at the peripheral part of the glomeruli in the granular layer.

The pivotal neurons of the cortex, Purkinje cells, receive excitatory inputs from climbing fibers and granule cell axons, parallel fibers, and send inhibitory output to the deep cerebellar nucleus. GABAergic neurons, stellate, basket, and Golgi cells, negatively regulate above the major stream of cortical circuits at the Purkinje cell dendrites, cell bodies, and granule cell dendrites, respectively.

B. GABA ᴀɴᴅ GABA Rᴇᴄᴇᴘᴛᴏʀs

In the CNS, GABA is synthesized from glutamate by two isoforms of glutamic acid decarboxylase (GAD65 and GAD67) (Barker *et al.*, 1998; Martin and Rimvall, 1993; Varju *et al.*, 2001), and is loaded into vesicles by the vesicular GABA transporter (VGAT) (see Fig. 2) (Fon and Edwards, 2001; McIntire *et al.*, 1997; Reimer *et al.*, 1998). In response to the influx of calcium ion via a voltage-dependent calcium channel, GABA is released by the fusion of vesicles with the presynaptic membrane at the nerve terminals, and activates GABA receptors at the postsynaptic membrane. GABAergic signals are terminated by reuptake of neurotransmitter into nerve terminals or uptake into surrounding glia by plasma membrane GABA transporters (GATs) (Cherubini and Conti, 2001).

GABA receptors are classified into three groups on the basis of pharmacology and biochemistry; $GABA_A$, $GABA_B$, and $GABA_C$. Among them, fast synaptic transmission is mediated by ionotropic GABA receptors, $GABA_A$ and $GABA_C$ receptors (Bormann, 2000; Kaupmann *et al.*, 1998; Macdonald and Olsen, 1994; Mehta and Ticku, 1999). The $GABA_A$ receptor is a member of the ligand-gated ion channel receptor family, and is thought to be composed of five heteromeric subunits belonging to seven different subunit families; $\alpha 1$–6, $\beta 1$–3, $\gamma 1$–3, δ, ε, π, and θ (Macdonald and Olsen, 1994; Mehta and Ticku, 1999; Nayeem *et al.*, 1994; Olsen and Tobin, 1990; Sieghart, 1995; Sieghart *et al.*, 1999; Tretter *et al.*, 1997). Native $GABA_A$ receptors contain at least one α-, one β-, and one γ-subunit with the δ-, ε-, π-, and θ-subunits to substitute for the γ subunit

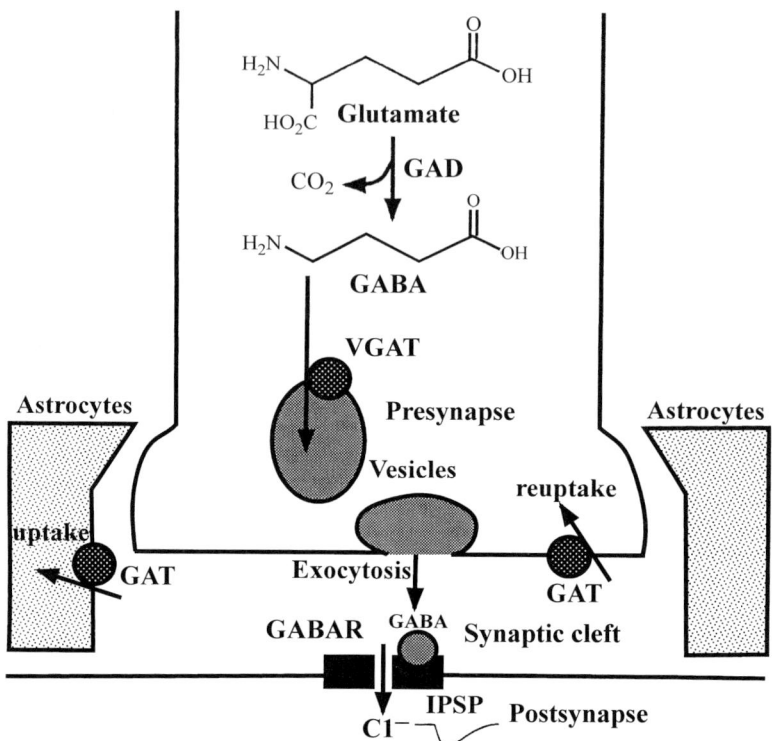

FIG. 2. Schematic illustration of GABAergic transmission in the mature GABAergic synapse. GABA is synthesized from glutamate by glutamic acid decarboxylase (GAD), and is loaded into vesicles by the vesicular GABA transporter (VGAT). GABA is released by the fusion of vesicles with the presynaptic membrane at the nerve terminals, and activates GABA receptors (GABAR) at the postsynaptic membrane. In the adult synapses, activation of $GABA_A$ receptors mediates hyperpolarization of postsynaptic membrane potential (IPSP) by the influx of chloride ion (Cl^-). GABAergic signals are terminated by uptake and reuptake of neurotransmitter into nerve terminals or uptake into surrounding glia by the plasma membrane GABA transporters (GAT). Abbreviations and symbols. GAD: glutamic acid decarboxylase, VGAT: vesicular GABA transporter, GAT: (plasma membrane) GABA transporter, GABAR: GABA receptor, IPSP: inhibitory postsynaptic potential.

(McKernan and Whiting, 1996; Pritchett *et al.*, 1989; Sieghart, 1995; Sieghart *et al.*, 1999). The subunit compositions drastically change during brain development (Araki *et al.*, 1992; Gambarana *et al.*, 1990; Laurie *et al.*, 1992a), and exhibit characteristic pharmacological and electrophysiological properties (Kardos, 1999; Luddens *et al.*, 1990; Macdonald and Olsen, 1994; Olsen and Tobin, 1990; Pritchett *et al.*, 1989; Sieghart, 1995; Vicini, 1999). GABA binding opens the pore of GABA receptors and induces influx or efflux of anions such as chloride ion (Fig. 2) (Kardos, 1999; Macdonald and Olsen, 1994; Olsen and

Tobin, 1990; Sieghart, 1995). The $GABA_C$ receptor is also an ion-channel type receptor, which is composed of only single or multiple ρ subunits. The $GABA_C$ receptor is identified as a bicuculline and baclofen insensitive GABA receptor, and is considered to be a pharmacological variant of $GABA_A$ receptors (Bormann, 2000; Bormann and Feigenspan, 1995; Mehta and Ticku, 1999). The $GABA_B$ receptor, which includes three isoforms, R1a, R1b, and R2 (Kaupmann et al., 1997; Kaupmann et al., 1998), is a metabotropic receptor, activates G proteins, negatively regulates the second messenger system, and responds to slow acting inhibition of channel and receptor functions (Bormann, 1988; Connors et al., 1988; LeVine, 1999; Nicoll, 1988).

C. $GABA_A$ Receptors in the Cerebellar Cortex

All neurons in the cerebellar cortex express the $GABA_A$ receptors. The main composition of the $GABA_A$ receptors is $\alpha1\beta2\gamma2$ (Fritschy and Mohler, 1995; Laurie et al., 1992a; Persohn et al., 1992). In addition, Purkinje cells abundantly express the $\beta3$ subunit $(\alpha1\beta2/3\gamma2)$, and granule cells abundantly express the $\alpha6$, $\beta3$, and δ subunits $(\alpha1/6\beta2/3\gamma2/\delta)$ (Fig. 1B).

D. Plasma Membrane GABA Transporters

Plasma membrane GABA transporters (GATs) are high-affinity, sodium (Na^+) and chloride (Cl^-)-dependent transporters, and GABA is co-transported with Na^+ and Cl^- (Borden, 1996; Conti et al., 2004; Gadea and Lopez-Colome, 2001; Kanner, 1994). Molecular cloning has isolated four GATs, GAT-1, GAT-2, GAT-3, and BGT-1. (Mouse GAT2, GAT3, and GAT4 are the species homolog of rat BGT-1, GAT-2, and GAT-3, respectively.) They exhibit characteristic distributions in the CNS, including the cerebellum (Durkin et al., 1995; Itouji et al., 1996; Morara et al., 1996; Ribak et al., 1996; Rosina et al., 1999). GAT-1 is mainly localized at the axon terminals containing GABAergic vesicles and is partly detected at the astrocytes (Fig. 3A, B). In contrast, GAT-3 is localized at the processes of astrocytes in the granular layer (Fig. 3C, D). GAT-2 is only localized at the leptomeningeal, ependymal cells, and choroids plexus (Conti et al., 1999).

In the adult CNS, GATs clean GABA from the synaptic cleft into the presynapse or surrounding glia (Fig. 2). In the immature brain or under abnormal condition such as ischemia and seizure, on the other hand, GATs work in reverse, releasing neurotransmitters (Attwell et al., 1993; Levi and Raiteri, 1993), in a calcium independent mechanism. It is thought that this phenomenon induces the development and protects against the effects of seizures (Phillis et al., 1994).

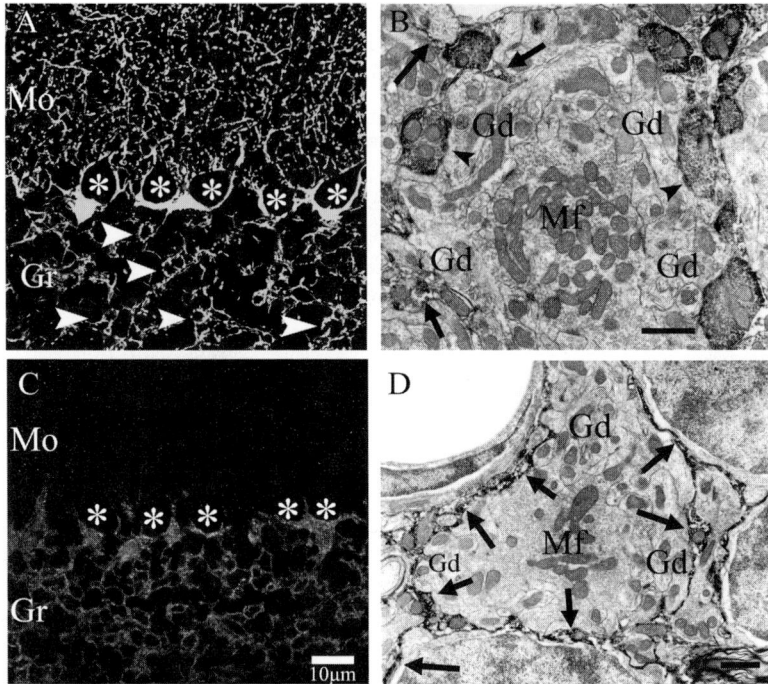

FIG. 3. Immunohistochemical localization of GAT-1 (A, B) and GAT-3 (C, D) in the adult mouse cerebellar cortex. (A and B) GAT-1-immunolabeling is localized at the axon terminals of GABAergic neurons in the molecular (Mo), Purkinje cell, and granular (Gr) layers. In the granular layer, GAT-1-immunolabeling exhibits ring-shaped profiles (white arrowheads) at the synaptic glomeruli in a light micrograph (A). An electron micrograph of the granular layer shows that the immunolabeling is detected at the axon terminals, which contain flat vesicles and form symmetric synapses (black arrows) with granule cell dendrites (Gd) (B). In addition, weak immunolabeling is also observed at the astrocytes (black arrows). (C and D) GAT-3 immunolabeling is detected in the neuropil of Purkinje cell and granular (Gr) layers in the light micrograph (C), and localize at the processes of astrocytes (black arrows) in the electron micrograph (D). Abbreviations and symbols: Mo: molecular layer, Gr: granular layer, asterisk: Purkinje cell body, white arrowhead: synaptic glomerulus, Mf: mossy fiber terminal, Gd: granule cell dendrite, black arrowhead: symmetric synapse, black arrows: immunolabeling in the process of astrocytes. Bar in the EM: 1μm.

III. GABAergic Signaling Before Synaptogenesis

A. EXTRASYNAPTIC GABA RELEASE IN THE DEVELOPING CEREBELLUM

In the developing CNS, GABA appears in GABAergic neurons long before the onset of synaptogenesis (Fairen et al., 1998; Lauder, 1993; Lauder et al., 1986; Van Eden et al., 1989) and its subcellular localization gradually changes during brain development (Behar *et al.*, 1993; McLaughlin *et al.*, 1975; Takayama and Inoue, 2004b). In the cerebellum, before GABAergic synapses are formed, GABA is distributed throughout the GABAergic neurons, including cell bodies, dendrites, axons, axon varicosities, and growth cones (Takayama and Inoue, 2004a; Takayama and Inoue, 2004c) (see Fig. 3A, 4A). VGAT, which is a membrane protein of GABAergic vesicles and transports cytosolic GABA into the vesicles (Chaudhry *et al.*, 1998; Dumoulin *et al.*, 1999; Fon and Edwards, 2001; Reimer *et al.*, 1998; Takamori *et al.*, 2000), accumulates at the axon varicosities and growth cones where GABAergic synapse are not yet formed (Fig. 3B, C, 4B, C) (Takayama and Inoue, 2004a). This indicates that GABA is distributed throughout GABAergic neurons, and vesicular GABA accumulates to the axon varicosities and growth cones. During synapse formation, GABA becomes confined to the axon terminals, and gradually disappears from axons themselves as well as dendrites. After finishing synapse formation, GABA is almost completely co-localized with VGAT at the synaptic sites where the GABA$_A$ receptor $\alpha 1$ subunit accumulates (Fig. 4D-F). This indicates that most GABA is exclusively localized in the synaptic vesicles within the axon terminals.

Physiological and biochemical studies have demonstrated that the nonvesicular form of GABA is also secreted via the plasma membrane by reverse transporter actions of GATs (see Fig. 4D) (Attwell *et al.*, 1993; Behar *et al.*, 1993; Belhage *et al.*, 1993; Gao and van den Pol, 2000; Jaffe and Vaello, 1988; Taylor *et al.*, 1990; Taylor and Gordon-Weeks, 1991; Varju *et al.*, 2001). In the developing brain, GABA could be released in two ways: exocytosis of GABAergic vesicles, and diacrine of cytosolic GABA via plasma membrane (see. Fig. 5D). It was hypothesized that cytosolic GABA might be extrasynaptically released from dendrites, axons, and cell bodies via the plasma membrane by GATs, and GABA in the vesicles might be also extrasynaptically released from axon varicosities and growth cones (Varju *et al.*, 2001).

To clarify which type of release occurs in the developing cerebellum, we examined the changes in distribution of the plasma membrane GABA transporters in the developing cerebellum (Takayama and Inoue, 2005). We could not find GAT-1 or GAT-3 in the dendrites and cell bodies in the developing GABAergic neurons (see Fig. 6). GAT-1 first appears in the granular layer and subsequently in the Purkinje and molecular layers, localizing at axons, varicosities, and terminals (Fig. 6A-F). GAT-3 appears at P10 in the deep part of the granular layer and localized in the processes of astrocytes (Fig. 6F, G). These localizations are the same as those in the adult cerebellum (Durkin *et al.*, 1995; Itouji *et al.*, 1996; Morara *et al.*, 1996; Ribak *et al.*, 1996; Rosina *et al.*, 1999). These results suggest that GABA is synthesized throughout the GABAergic neurons and transported into vesicles but is not released by diacrine. GABA in the vesicles is confined to the axon varicosities and growth cones and released by

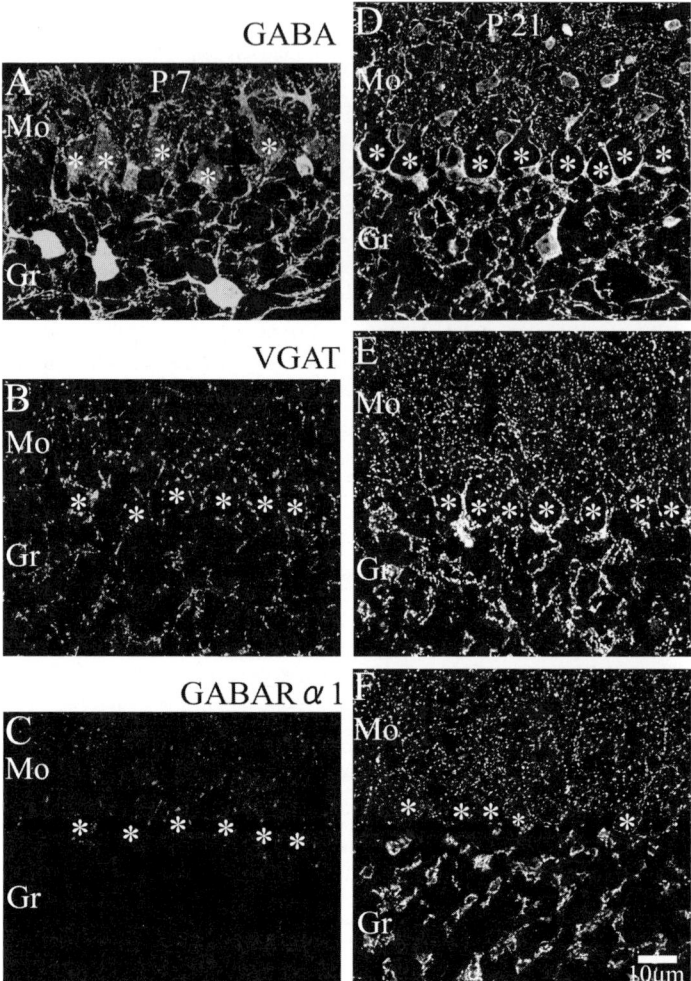

FIG. 4. Immunohistochemical localization of GABA (A, D), VGAT (B, E), and GABA$_A$ receptor α1 subunit (C, F) in the cerebellar cortex at postnatal day 7 (P7) (A-C) and P21 (D- F). At P7, GABA is localized throughout the GABAergic neurons (A). VGAT, a marker of GABAergic vesicles, is confined to the axon varicosities (B). VGAT is often localized at the axons where mature GABAergic synapses, labeled by the immunohistochemistry for GABA$_A$ receptor α1 subunit (C), are not formed in the granular layer (Gr). In contrast, at P21, the majority of GABA is confined to the terminals (D) where VGAT (E) and α1 subunit (F) are localized. Abbreviations and symbols, Mo: molecular layer, Gr: granular layer, asterisk: Purkinje cell body.

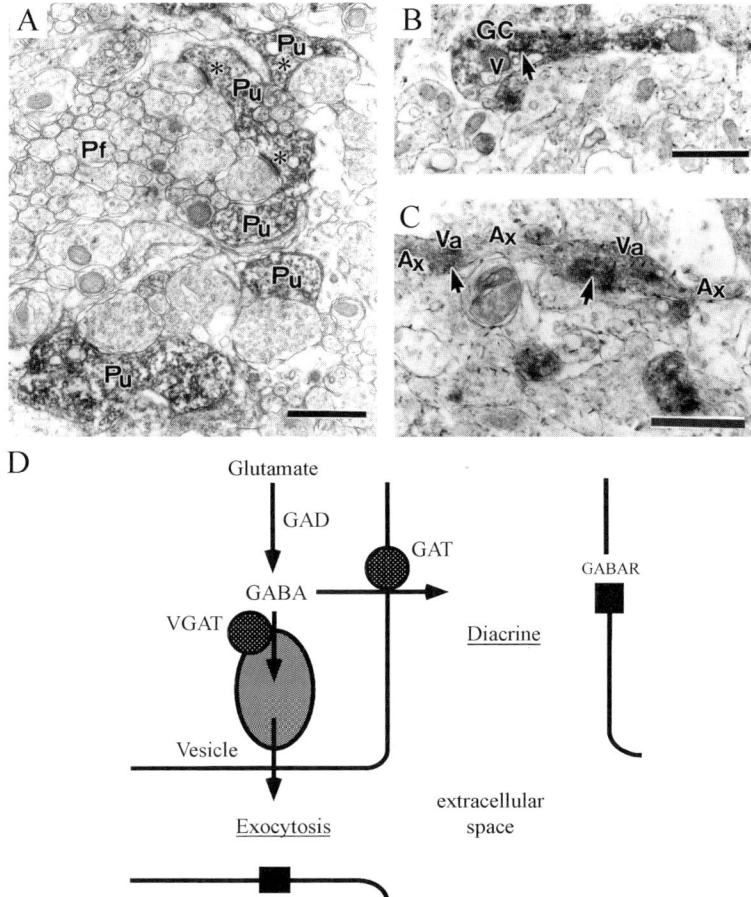

FIG. 5. Electron microscopic localization of GABA (A) and VGAT (B, C) in the immature cerebellum at P5 and the extrasynaptic GABA secretion system (D). (A-C) Electron micrographs of the immunohistochemistry for GABA (A) and VGAT (B, C) in the cerebellum at P5. GABA is distributed throughout the dendrites (A), whereas VGAT is detected at the vesicles (arrowheads) in the growth cone (GC) and axon varicosities (Va). (D) Schematic illustration of the extrasynaptic GABA-release system. Before synapse formation, GABA could be released in two ways: diacrine of cytosolic GABA by plasma membrane GABA transporters (GATs) and exocytosis of GABAergic vesicles. GABA is diffused in the extracellular space and activates GABA receptors (GABAR) on the neighboring neurons. Abbreviations and symbols. Pu: Purkinje cell dendrite, Pf: parallel fiber, asterisk: asymmetric synapse, GC: growth cone, v: vacuole, Ax: axon, Va: axon varicosity, arrowhead: synaptic vesicle, GAD: glutamic acid decarboxylase, VGAT: vesicular GABA transporter, GAT: (plasma membrane) GABA transporter, GABAR: GABA receptor. Bar in the EM=1μm.

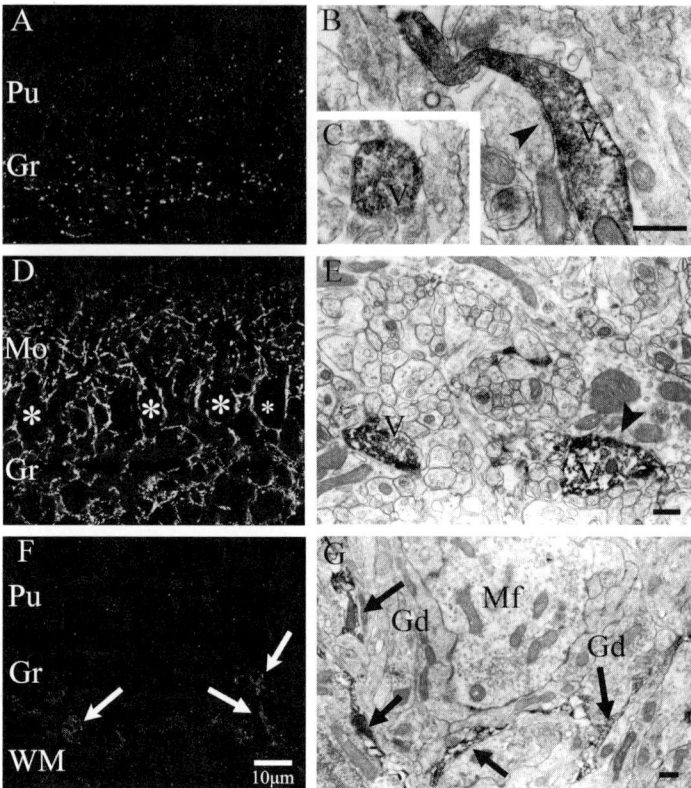

FIG. 6. Developmental expression of GAT-1 (A-E) and GAT-3 (F, G) in the cerebellar cortex. (A-E) Immunohistochemical localization of GAT-1 at P5 (A-C) and P10 (D, E). GAT-1 immunolabeling appears at P5 in the granular layer (Gr) (A), and is localized at the axons and varicosities containing vesicles (V) (B, C). At P10, the immunolabeling is also detected in the molecular (Mo) and Purkinje cell (asterisks) layers and is localized at the axon terminals of stellate cells, which often form symmetric synapses (arrowheads). (F and G) Immunohistochemical localization of GAT3 in the cerebellum at P10. GAT3-immunolabeling (white arrows) appears at P10 in the deep part of the granular layer (Gr) (F), and is localized at the processes of astrocytes (black arrows). Abbreviations and symbols, Mo: molecular layer, Gr: granular layer, asterisk: Purkinje cell body, white arrow: GAT-3 immunolabeling in the granular layer, V: GABAergic vesicle, arrowhead: synapse and synapse-like structure, black arrow: GAT3-positive process of astrocytes, Mf: mossy fiber, Gd: granule cell dendrite. Bar in the EM: 0.5μm.

exocytosis in the developing cerebellum (see Fig. 7A). The exocytosis trigger is unknown. In the mature cerebellum, on the other hand, most GABA is synthesized at the terminals, including varicosities, where synapses are formed and is released at the synapse (Fig. 7B).

Furthermore, the GABA-removing system might shift as shown in Fig. 7C-E. Before synapse formation, GABA is released from axon varicosities and growth

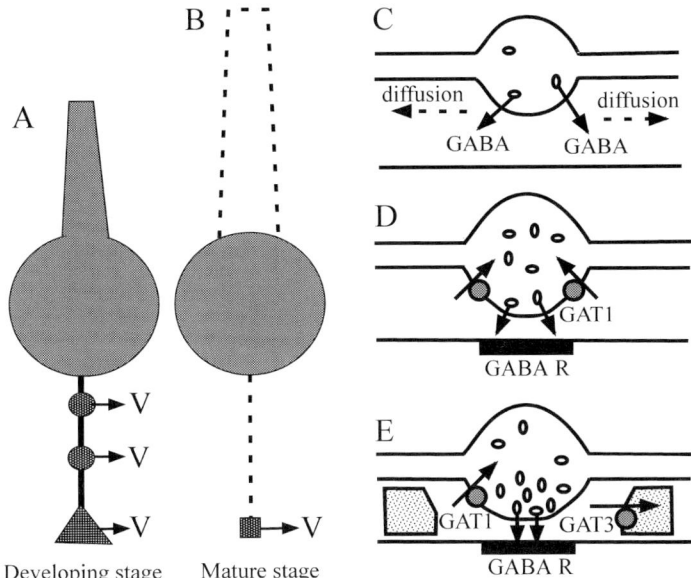

FIG. 7. Developmental changes in GABA-release (A, B) and uptake (C-E) mechanisms in the mouse cerebellum. (A and B) Schematic illustrations of developmental changes in the GABA release system. In the developing cerebellum, GABA is localized throughout the GABAergic neurons and is released by exocytosis (V) from axon varicosities (circle) and growth cones (triangles) (A). In contrast, GABA disappears from dendrites and axons in the mature brain, and is exclusively released synaptically (B). (C-E) Schematic illustrations of developmental changes in the GABA-uptake and reuptake system in the mouse cerebellar cortex. Before synapse formation, GABA is released from axon varicosities and growth cones of GABAergic neurons and disappears by diffusion (C). During synaptogenesis, GAT-1 mediates the reuptake from the extracellular space into the axon and presynapses (D). In the mature cerebellum, GABA is removed from synaptic clefts into the presynapse by GAT-1 and astrocytes by GAT-3 (E). Abbreviations. V: vesicular secretion (exocytosis), GABAR: GABA receptor.

cones of GABAergic neurons and disappears by diffusion (Fig. 6C). During synaptogenesis, GAT-1 mediates the reuptake from the extracellular space into the axons and presynapses (Fig. 7D). Finally, GABA is removed from synaptic clefts into the presynapse by GAT-1 and astrocytes by GAT-3. These results indicate that plasma membrane GATs might not be involved in the diacrine process, but only uptake of GABA from synaptic clefts in the cerebellum.

B. GABAergic Roles in the Developing Brain

During brain development, extrasynaptically released GABA diffuses in the extracellular space and activates GABA receptors on neighboring neurons.

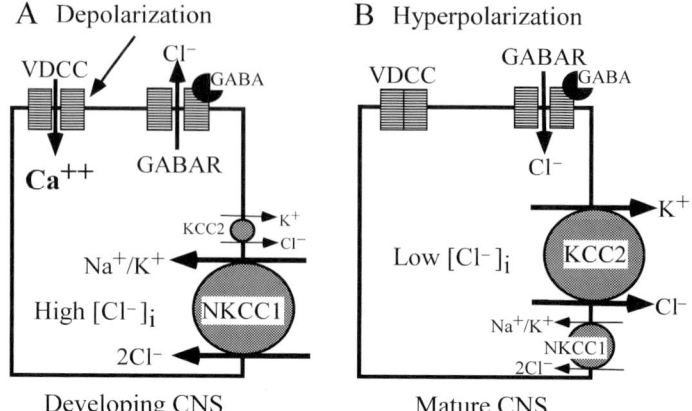

FIG. 8. Schematic illustrations of the developmental changes in GABA actions. (A) In the developing CNS, opening of GABA$_A$ receptors (GABAR) generates efflux of chloride ion (Cl$^-$) and depolarization of membrane potential, since the intracellular chloride concentration, [Cl$^-$]i, is relatively high due to the dominant action of sodium-potassium-chloride co-transporter 1 (NKCC1). GABA-inducing depolarization activates the voltage dependent calcium channel (VDCC) and mediates calcium influx (Ca^{++}). (B) In the mature CNS, GABA mediates influx of chloride ion (Cl$^-$), since potassium-chloride co-transporter 2 (KCC2) lowers the intracellular chloride concentration. Influx of chloride ion mediates hyperpolarization of membrane potential.

The activation of GABA$_A$ receptors depolarizes membrane potential, since the Cl$^-$ reversal potential of the neuronal membrane is elevated (see Fig. 8A) (Ben-Ari, 2002; Cherubini et al., 1991; Leinekugel et al., 1999; Owens and Kriegstein, 2002; Perkins and Wong, 1997; Rohrbough and Spitzer, 1996; Serafini et al., 1998). In the developing CNS, Na$^+$-K$^+$-2Cl$^-$ co-transporter 1 (NKCC1), which raises the concentration of intracellular chloride ion, [Cl$^-$]i, is predominantly expressed, and elevates the equilibrium potential of Cl$^-$. Under the high [Cl$^-$]i condition, the activation of GABA$_A$ receptors generates efflux of chloride ion and depolarization of membrane potential (Fig. 8A). In contrast, K$^+$-Cl$^-$ co-transporter 2 (KCC2), which lowers the [Cl$^-$]i, becomes a predominant chloride co-transporter in the mature CNS, and GABA induces hyperpolarization of membrane potential and inhibition of excitability (Fig. 8B).

GABA$_A$-receptor-mediated depolarization in the immature CNS activates voltage-dependent Ca^{++} channels (VDCC) and/or N-methyl-D-aspartate (NMDA) type glutamate receptors, and elevates cytosolic Ca^{++} ion (Fig. 8A) (Ben-Ari et al., 1997; Connor et al., 1987; Eilers et al., 2001; Leinekugel et al., 1995; Obrietan and van den Pol, 1996; Reichling et al., 1994; Serafini et al., 1998; Yuste and Katz, 1991). The elevation of cytosolic calcium effects various steps in CNS development such as (1) cell proliferation, (2) cell migration, and (3) neuronal maturation, including synaptogenesis (Barker *et al.*, 1998; Belhage *et al.*, 1998; Ben-Ari, 2002; Kardos, 1999; McCarthy *et al.*, 2002; Owens and Kriegstein, 2002; Varju *et al.*, 2001). GABA acts as an anti-proliferation molecule, reduces the DNA synthesis in the proliferating precursor cells, and depresses the rate of cellular proliferation by the activation of GABA$_A$ receptors, and other GABA$_A$-receptor related molecules (Haydar *et al.*, 2000; LoTurco *et al.*, 1995). GABA modulates neuronal migration by 'chemotaxis' and 'chemokinesis' at the femtomolar (10^{-15}M) and micromolar (μM) level, respectively (Behar *et al.*, 1996, 1994, 1995). Activation of GABA$_B$ and GABA$_C$ receptors promotes migration out of the proliferating layer, whereas that of the GABA$_A$ receptor slows or almost stops the movement in the cortical plates (Behar *et al.*, 2000). Furthermore, exposure of neurons to GABA or GABA$_A$ receptor agonists induces the synthesis of neuron specific molecules such as neuron-specific enolase (NSE) and neural cell adhesion molecules (NCAMs), enhances the growth rate of neuronal processes, and facilitates synapse formation by inducing the expression and targeting of GABA receptor subunits, which mediate synaptic transmission (Abraham and Schousboe, 1989; Barbin *et al.*, 1993; Belhage *et al.*, 1998; Carlson *et al.*, 1997, 1998; Elster *et al.*, 1995; Gao and van den Pol, 2000; Kim *et al.*, 1993; Meier and Jorgensen, 1986; Meier *et al.*, 1987; Mellor *et al.*, 1998; Mitchell and Redburn, 1996; Moss and Smart, 2001; Spoerri, 1988; Wolff *et al.*, 1978). In the case of interneurons and their networks, GABA might stimulate the expression of neurotrophins, such as brain derived neurotrophic factors (BDNF) and their receptors, and enhance the growth of neurons and synapses (Berninger *et al.*, 1995; Marty *et al.*, 1996; Rico *et al.*, 2002; Vicario-Abejon *et al.*, 1998).

In the cerebellum, the above change in the chloride ion concentration system could not be clarified (Eilers *et al.*, 2001; Kanaka *et al.*, 2001; Lu *et al.*, 1999; Mikawa *et al.*, 2002; Williams *et al.*, 1999). However, GABA could be also involved in morphogenesis in the cerebellum since GABA elevates the Ca^{++} ion concentration in the Purkinje and granule cells during the first two postnatal weeks (Connor *et al.*, 1987; Eilers *et al.*, 2001).

C. GABA$_A$ Receptor Expression in the Developing Cerebellum

In the CNS, the subunit compositions of GABA$_A$ receptors drastically change during brain development (Ben-Ari *et al.*, 1997; Connor *et al.*, 1987; Eilers *et al.*, 2001; Leinekugel *et al.*, 1995; Obrietan and van den Pol, 1996; Reichling *et al.*, 1994; Serafini *et al.*, 1998; Yuste and Katz, 1991). We focused on the α subunits, which may mainly reflect the functional diversity of the GABA$_A$ receptors (Kardos, 1999; Luddens *et al.*, 1990; Macdonald and Olsen, 1994; Olsen and Tobin, 1990; Pritchett *et al.*, 1989; Sieghart, 1995), and investigated the

TABLE I
CHANGES IN EXPRESSION OF THE PREDOMINANT α SUBUNITS OF THE GABA$_A$ RECEPTORS
IN THE CEREBELLAR CORTICAL NEURONS

	Proliferating stage	Migrating and differentiating stage	Matured stage
Purkinje cells	negative	$\alpha3$ subunit	$\alpha1$ subunit
Granule cells	negative	$\alpha2$ subunit	$\alpha1$ and $\alpha6$ subunits

developmental changes in expression and localization of the GABA$_A$ receptor α subunits in the cerebellum (see Table I) (Takayama and Inoue, 2004c; Takayama and Inoue, 2004d). Proliferating cells in the ventricular zone adjacent to the fourth ventricle and the upper half of the external granular layer expressed no α subunits. Since at least one α subunit is essential for functional GABA$_A$ receptors (McKernan and Whiting, 1996; Pritchett et al., 1989; Sieghart, 1995; Sieghart et al., 1999), receptor activity is absent in the proliferating zone. After finishing cell proliferation, cerebellar neurons start to express the functional GABA$_A$ receptors. Differentiating Purkinje cells express the $\alpha3$ subunit, migrating and maturating granule cells express the $\alpha2$ subunit, and both subunits disappear from the cerebellar cortex after synapse formation finishes (Takayama and Inoue, 2004d). In addition, the $\beta3$, $\gamma1$, and $\gamma3$ subunits are also abundantly expressed in the developing cerebellum (Laurie et al., 1992b). These results suggest that extrasynaptically released GABA activates GABA$_A$ receptors consisting of the restricted subunits, and may be involved in the regulation of proliferation, neuronal migration, and maturation in the cerebellum.

D. CONCLUSION FOR THIS SECTION

Before synapse formation, GABA is synthesized throughout the GABAergic neurons, transported into GABAergic vesicles at axon varicosities and growth cones, extrasynaptically released by exocytosis, and diffused within the extracellular space. Released GABA activates GABA receptors consisting of $\alpha2/3$, $\beta3$, $\gamma1/3$ subunits on the neighboring neurons, mediates the depolarization of membrane potential, and induces various types of morphogenesis.

IV. Formation of GABAergic Synapses

Synapse formation is considered to be a multi-step process (Cherubini and Conti, 2001; Moss and Smart, 2001; Vaughn, 1989). While exploring their environment, axonal growth cones lead elongating axons to their appropriate targets and make contact with dendrites and cell bodies of target neurons. Initial contact is followed by the establishment of stable synapses. In the presynapse, synaptic vesicles accumulate to the nerve terminals and dock near the active zone. In the postsynapse, GABA$_A$ receptors which mediate inhibitory synaptic transmission are targeted to and clustered at an appropriate synaptic site opposite the GABA-releasing site. At the same time, GABA$_A$ receptors, which are involved in brain morphogenesis, disappear from postsynaptic neurons (Takayama and Inoue, 2004b,d).

A. DEVELOPMENT OF GABAergic SYNAPSES IN THE CEREBELLAR CORTEX

In the mouse cerebellar cortex, the GABA$_A$ receptor $\alpha 1$ subunit protein, which is an essential subunit of mature GABA$_A$ receptors in Purkinje cells (Laurie et al., 1992a; Persohn et al., 1992), appears at P5 (Takayama and Inoue, 2004c), and symmetric synapses are detected between GAD-positive terminals and Purkinje cell dendrites at the same day (Takayama, 2005). In the mouse granular layer, on the other hand, the $\alpha 1$ and $\alpha 6$ subunit proteins, which are essential subunits for the mature GABA$_A$ receptors in the granule cells, appear in deep part at P7, and symmetric synapses are clearly discernible at P10 between GAD-positive terminals and granule cell dendrites in the synaptic glomeruli.

These results indicate that in the cerebellum excitatory synapses appear prior to the inhibitory synapses, and GABAergic synapses start to be formed on the Purkinje cell dendrites during the first postnatal week, and granule cell dendrites during the second postnatal week. The number of GABAergic synapses increase dramatically in all layers during the second and third postnatal weeks (Altman and Bayer, 1997; Jakab and Hamori, 1988; Larramendi, 1969; Takayama and Inoue, 2004c).

B. TARGET DETERMINATION

It is not fully understood how GABAergic neurons search, recognize, and determine their target neurons. To reveal how GABAergic neurons determine their targets and form synapses, we employed cerebellar mutant mice and examined the specificity of neuron-to-neuron connection in the mutant cerebellum. In the normal cerebellar cortex, five major types of neurons innervate distinct types of target neurons (Fig. 1B) (Ito, 1984; Llinas and Hillmann, 1969; Palay and Chan-Palay, 1974). The specific innervation patterns, however, are not preserved in the abnormal environment of the reeler and weaver cerebellum (see Fig. 9). Golgi cells directly innervate to Purkinje cells in the central mass of the

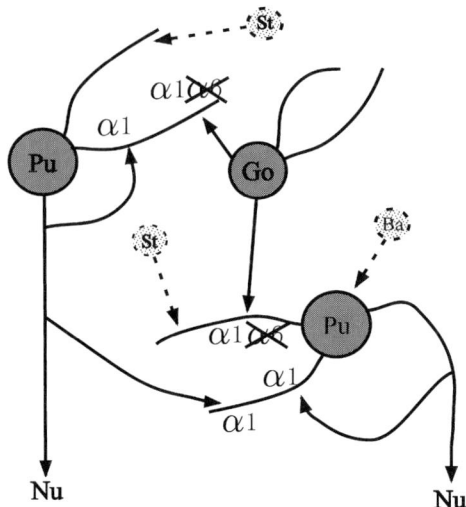

FIG. 9. Schematic illustrations of the abnormalities in the GABAergic inputs in the reeler cerebellum. In the central cerebellar mass of the reeler cerebellum, Purkinje cells (Pu) receive inhibitory inputs (arrows) from Golgi cells (Go) instead of stellate (st) and basket (ba) cells. GABA$_A$ receptors containing only the $\alpha 1$ subunit ($\alpha 1$), but not the remaining five α subunits, are localized at the GABAergic synapses on the Purkinje cells, although Golgi cells innervate them. In addition, GABAergic input from the Purkinje cell axon collaterals increased markedly. Abbreviations; Pu: Purkinje cell, Go: Golgi cell, St: stellate cell, Ba: basket cell, Nu: cerebellar nucleus.

reeler cerebellum and in the cortex of the weaver cerebellum (Caviness and Rakic, 1978; Mariani *et al.*, 1977; Rakic, 1976; Sotelo and Privat, 1978; Takayama, 1994; Wilson *et al.*, 1981). In both regions, granule cells are scarce or absent. Thus, Golgi cell axons form synapses with neighboring neurons instead of granule cells. This result indicates that targets of Golgi cells are not genetically and strictly determined, but are influenced by the environment, and that targets of GABAergic neurons plastically alter according to the environment.

C. Change in Subunit Compositions

As shown in Table I, expression of the GABA$_A$ receptor α subunits in the cerebellum developmentally changes especially during GABAergic synapse formation. While expression of the $\alpha 2$ and $\alpha 3$ subunits is decreasing, the $\alpha 1$ and $\alpha 6$ subunits appear and increase their expression (Laurie *et al.*, 1992b; Mellor *et al.*, 1998; Takayama and Inoue, 2004c,d; Tia *et al.*, 1996). Therefore, the α subunits in the GABA$_A$ receptors shift from the $\alpha 2$ and $\alpha 3$ subunits to the $\alpha 1$ and $\alpha 6$

subunits during cerebellar development. This result indicates that two pieces of evidence, the disappearance of subunits involved in morphogenesis, and the appearance of subunits which mediate inhibitory synaptic transmission, are crucial for GABAergic synapse formation.

To test the mechanism underlying the change in subunit compositions, we investigated its relationship with neuronal maturation, including migration, axonal, and dendritic extension, and formation of excitatory and inhibitory synapses using reeler mutant mice. In the reeler cerebellum, maturation of malpositioned Purkinje cells is assumed to be arrested in terms of the synaptic architecture and dendritic arborization (Caviness and Rakic, 1978; Mariani et al., 1977; Rakic, 1976; Sotelo and Privat, 1978; Takayama, 1994; Wilson et al., 1981). Parallel fibers and axons from stellate and basket cells do not innervate the Purkinje cells in the central cerebellar mass. Moreover, multiple innervations from climbing fibers remain in the adult reeler cerebellum. Instead, Purkinje cells directly form synapses with mossy fibers and Golgi cell axons. Dendrites of Purkinje cells are poorly developed and extend almost randomly. The $\alpha3$ subunit, however, is almost negative, as in the normal mature cerebellum (Fig. 10E, F), and malpositioned Purkinje cells abundantly express the $\alpha1$ subunit (Fig. 10A, B) (Frostholm et al., 1991; Takayama and Inoue, 2003). These results indicate that developmental change in subunit composition is independent of neuronal maturation such as settling in the normal neuronal position, maturation of excitatory networks. Absence of normal inhibitory synapses with stellate and basket cell axons and heterologous input from Golgi cells do not affect the developmental changes in subunit composition. Previous in vitro studies have indicated that GABAergic stimulation induces low-affinity type GABA receptor expression, which is involved in inhibitory synaptic transmission (Belhage et al., 1998; Belhage et al., 1986; Carlson et al., 1997, 1998; Elster et al., 1995; Gao and Fritschy, 1995; Kim et al., 1993; Meier et al., 1984; Mellor et al., 1998; Raetzman and Siegel, 1999; Schousboe, 1999). The change in subunit composition simultaneously occurred during GABAergic synaptogenesis (Table I). These results suggest that innervation of GABAergic fibers may be important for the change in subunit composition, even if the synapses are heterologous and ectopic, and GABAergic innervation might initiate and/or accelerate the changes in subunit composition.

D. Specific Subunit Expression

In the CNS, distinct types of subunits are expressed at distinct synapses (Fig. 1B) (Laurie et al., 1992a; Persohn et al., 1992; Wisden et al., 1992). In the normal cerebellum, GABAergic transmission between stellate cell axons and Purkinje cell dendrites is mediated by $GABA_A$ receptors containing only the $\alpha1$

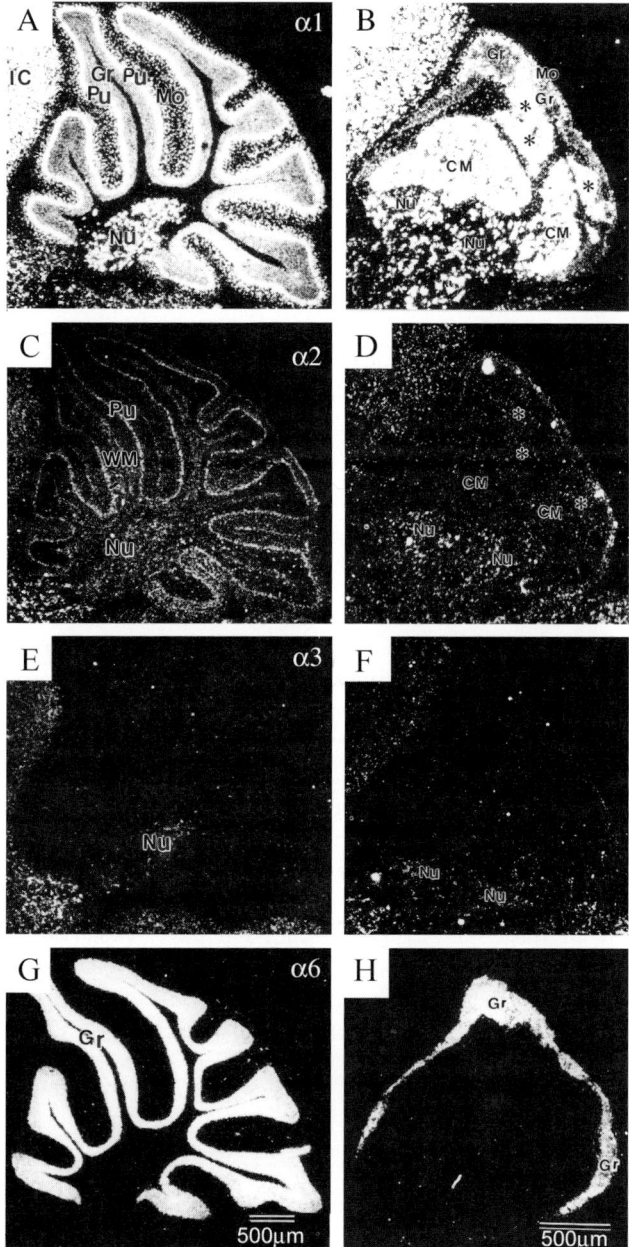

FIG. 10. Distinct expression of the GABA$_A$ receptor α1 (A, B), α2 (C, D), α3 (E, F), and α6 (G, H) subunits in the normal (A, C, E, G), and reeler (B, D, F, H) cerebella. The specific expression of α subunit mRNAs in each neuronal type was preserved in the reeler cerebellum. Furthermore,

subunit, but not the remaining five α subunits (Fig. 1B) (Laurie et al., 1992a; Persohn et al., 1992; Wisden et al., 1992, 1996). In contrast, inhibitory transmission between Golgi cell axons and granule cell dendrites is mediated by GABA$_A$ receptors containing both α1 and α6 subunits.

To test the relationship between types of presynapse and subunits in the postsynapse, we examined the expression of GABA$_A$ receptor α subunits in the reeler cerebellum. In the central cerebellar mass of the reeler cerebellum, Purkinje cells directly form synapses with Golgi cell axons (Fig. 9) (Caviness and Rakic, 1978; Mariani et al., 1977; Rakic, 1976; Sotelo and Privat, 1978; Takayama, 1994; Wilson et al., 1981). If presynaptic neurons determine the type of receptor subunits in postsynaptic neurons, GABAergic innervation from Golgi cells would induce Purkinje cells to express the α6 subunit in the central cerebellar mass. Nevertheless, Purkinje cells in the central cerebellar mass do not express the α6 or α2 subunits (Fig. 10C, D, G, H) (Takayama and Inoue, 2003). This result indicates that Golgi cell innervation does not induce expression of the α6 subunit in Purkinje cells, and suggests that postsynaptic self-autonomous mechanisms determine the types of subunits (Fig. 9).

E. Synaptic Targeting and Clustering of GABA$_A$ Receptor Proteins

Synaptic targeting and clustering of GABA$_A$ receptors are mediated by the interaction of the subunit proteins with the subsynaptic cytoskeleton, and it is thought that the diversity of subunits in the GABA$_A$ receptors is important for subcellular localization (Barnes, 2000; Moss and Smart, 2001). Most single subunits are retained within the endoplasmic reticulum (Connolly et al., 1996; Gorrie et al., 1997; Taylor et al., 2000). Specific subunits such as the γ2 subunit can lead the assembled GABA$_A$ receptors to the cell surface and synaptic site, clustering (Connolly et al., 1999) in conjunction with a range of diverse anchoring protein molecules such as gephyrin (Craig et al., 1996; Essrich et al., 1998; Kneussel et al., 1999; Sassoe-Pognetto and Fritschy, 2000), GABA$_A$-receptor associated protein (GABARAP) (Wang et al., 1999), microtubule-associated proteins, transporters, protein kinases, and so on (Moss and Smart, 2001). Furthermore, anchoring proteins such as gephyrin and GABARAP are also involved in clustering of receptor proteins (Barnes, 2000; Moss and Smart, 2001).

---

abnormal expression of α subunits was not detected, although GABAergic networks were altered and neuronal maturation is severely disturbed. Abbreviations: IC: inferior colliculus, Mo: molecular layer, Pu: Purkinje cell layer, Gr: granular layer, Nu: cerebellar nucleus, WM: white matter, asterisks: central cerebellar mass beneath the granular layer, CM: central cerebella mass under the white matter.

F. ACTIVITY-DEPENDENT SYNAPTIC REMODELING

Recent investigations revealed that GABAergic synapses are remodeled by the change in GABAergic input in auditory systems during the critical period (Kandler, 2004; Kapfer *et al.*, 2002; Kim and Kandler, 2003). Auditory experience guides subcellular localization of receptor proteins (Kapfer *et al.*, 2002), induces functional and structural elimination of inhibitory synapses during the establishment of precise topography in the GABAergic/glycinergic pathway (Kim and Kandler, 2003), and mediates aural dominance bands in the inferior colliculus (Gabriele *et al.*, 2000). In the cerebellum, the activity-dependent remodeling of GABAergic synapses has not yet been clarified, but could play roles in the formation and maturation of GABAergic synapses and networks.

G. CONCLUSION FOR THIS SECTION

GABAergic axons determine their target neurons under the influence of environmental conditions. During the formation of GABAergic synapses, axon varicosities and growth cones which contains GABAergic vesicles give rise to presynapse. GABA-release could induce the maturation of postsynapse, including expression of the mature type receptor subunits, disappearance of immature type subunits, and targeting of subunit proteins. At the postsynapse, genetically determined subunits are expressed.

V. Functions of GABAergic Synapses in the Cerebellum

The cerebellum is closely involved in learning motor skills (Ito, 1984; Llinas and Walton, 1990). GABAergic input might play important roles in cerebellar functions since GABAergic neurons regulate the neuronal activity of Purkinje cells and granule cells which organize the major stream of neural circuitry in the cerebellar cortex. Neuroanatomical analysis of the cerebellar local circuit suggests that GABAergic neurons play a role in lateral inhibition and negative feedback mechanisms on the Purkinje and granule cells. Furthermore, elimination of GABAergic input from the Golgi cells in the cerebellar granular layer caused overexcitation of granule cells resulting in severe ataxia during the acute phase (Watanabe *et al.*, 1998). Therefore, GABAergic input plays a role in the regulation of glutamatergic hyperexcitation and could be involved in motor coordination.

In addition, neuroimaging and biochemical studies indicate a dysfunction in the GABAergic system in the cerebellum of autistic patients (Dhossche, 2004;

Fatemi *et al.*, 2002). This result suggests that the GABAergic network in the cerebellum might be involved in not only motor function, but also higher brain functions.

Acknowledgments

I am grateful to Yoshihiko Ogawa and Hideki Nakamura at the Central Research Department of Hokkaido University School of Medicine for their technical assistance. I also thank Dr. Yoshiro Inoue and Dr. Takayuki Yoshida at the Department of Molecular Neuroanatomy Hokkaido University School of Medicine, Dr. Hitoshi Komuro at the Cleveland Clinic Foundation, and Miss Shoko Kogawa, a student at Hokkaido University School of Medicine, for valuable discussions.

This work was supported by the Grants-in-Aid for scientific research from the Ministry of Education, Culture, Science, Sports and Technology of Japan, the Uehara Memorial Foundation, and the Takeda Science Foundation.

References

Abraham, J. H., and Schousboe, A. (1989). Effects of taurine on cell morphology and expression of low-affinity GABA receptors in cultured cerebellar granule cells. *Neurochem. Res.* **14,** 1031–1038.

Altman, J., and Bayer, S. A. (1997). "Development of the Cerebellar System in Relation to Its Evolution, Structure, and Functions." CPC Press, Boca Raton.

Araki, T., Kiyama, H., and Tohyama, M. (1992). GABAA receptor subunit messenger RNAs show differential expression during cortical development in the rat brain. *Neuroscience* **51,** 583–591.

Attwell, D., Barbour, B., and Szatkowski, M. (1993). Nonvesicular release of neurotransmitter. *Neuron* **11,** 401–407.

Avoli, M. (2000). Epilepsy. *In* "GABA in the Nervous System: The View at Fifty Years" (D. L. Martin and R. W. Olsen, Eds.), pp. 293–316. Lippincott Williams & Wilkins, Philadelphia.

Barbin, G., Pollard, H., Gaiarsa, J. L., and Ben-Ari, Y. (1993). Involvement of GABAA receptors in the outgrowth of cultured hippocampal neurons. *Neurosci. Lett.* **152,** 150–154.

Barker, J. L., Behar, T., Li, Y. X., Liu, Q. Y., Ma, W., Maric, D., Maric, I., Schaffner, A. E., Serafini, R., Smith, S. V., Somogyi, R., Vautrin, J. Y., Wen, X. L., and Xian, H. (1998). GABAergic cells and signals in CNS development. *Perspect Dev. Neurobiol.* **5,** 305–322.

Barnes, E. M., Jr. (2000). Intracellular trafficking of GABA(A) receptors. *Life Sci.* **66,** 1063–1070.

Baulac, S., Huberfeld, G., Gourfinkel-An, I., Mitropoulou, G., Beranger, A., Prud'homme, J. F., Baulac, M., Brice, A., Bruzzone, R., and LeGuern, E. (2001). First genetic evidence of GABA(A) receptor dysfunction in epilepsy: A mutation in the gamma2-subunit gene. *Nat. Genet.* **28,** 46–48.

Behar, T., Schaffner, A., Laing, P., Hudson, L., Komoly, S., and Barker, J. (1993). Many spinal cord cells transiently express low molecular weight forms of glutamic acid decarboxylase during embryonic development. *Brain Res. Dev. Brain Res.* **72,** 203–218.

Behar, T. N., Li, Y. X., Tran, H. T., Ma, W., Dunlap, V., Scott, C., and Barker, J. L. (1996). GABA stimulates chemotaxis and chemokinesis of embryonic cortical neurons via calcium-dependent mechanisms. *J. Neurosci.* **16,** 1808–1818.

Behar, T. N., Schaffner, A. E., Colton, C. A., Somogyi, R., Olah, Z., Lehel, C., and Barker, J. L. (1994). GABA-induced chemokinesis and NGF-induced chemotaxis of embryonic spinal cord neurons. *J. Neurosci.* **14,** 29–38.

Behar, T. N., Schaffner, A. E., Scott, C. A., Greene, C. L., and Barker, J. L. (2000). GABA receptor antagonists modulate postmitotic cell migration in slice cultures of embryonic rat cortex. *Cereb. Cortex* **10,** 899–909.

Behar, T. N., Schaffner, A. E., Tran, H. T., and Barker, J. L. (1995). GABA-induced motility of spinal neuroblasts develops along a ventrodorsal gradient and can be mimicked by agonists of GABAA and GABAB receptors. *J. Neurosci. Res.* **42,** 97–108.

Belhage, B., Hansen, G. H., Elster, L., and Schousboe, A. (1998). Effects of gamma-aminobutyric acid (GABA) on synaptogenesis and synaptic function. *Perspect Dev. Neurobiol.* **5,** 235–246.

Belhage, B., Hansen, G. H., and Schousboe, A. (1993). Depolarization by K+ and glutamate activates different neurotransmitter release mechanisms in GABAergic neurons: Vesicular versus non-vesicular release of GABA. *Neuroscience* **54,** 1019–1034.

Belhage, B., Meier, E., and Schousboe, A. (1986). GABA-agonists induce the formation of low-affinity GABA-receptors on cultured cerebellar granule cells via preexisting high affinity GABA receptors. *Neurochem. Res.* **11,** 599–606.

Ben-Ari, Y. (2002). Excitatory actions of gaba during development: The nature of the nurture. *Nat. Rev. Neurosci.* **3,** 728–739.

Ben-Ari, Y., Khazipov, R., Leinekugel, X., Caillard, O., and Gaiarsa, J. L. (1997). GABAA, NMDA and AMPA receptors: A developmentally regulated 'menage a trois'. *Trends Neurosci* **20,** 523–529.

Berninger, B., Marty, S., Zafra, F., da Penha Berzaghi, M., Thoenen, H., and Lindholm, D. (1995). GABAergic stimulation switches from enhancing to repressing BDNF expression in rat hippocampal neurons during maturation *in vitro. Development* **121,** 2327–2335.

Berry, N., Jobanputra, V., and Pal, H. (2003). Molecular genetics of schizophrenia: A critical review. *J. Psychiatry Neurosci.* **28,** 415–429.

Blatt, G. J., Fitzgerald, C. M., Guptill, J. T., Booker, A. B., Kemper, T. L., and Bauman, M. L. (2001). Density and distribution of hippocampal neurotransmitter receptors in autism: An autoradiographic study. *J. Autism. Dev. Disord.* **31,** 537–543.

Blum, B. P., and Mann, J. J. (2002). The GABAergic system in schizophrenia. *Int. J. Neuropsychopharmacol.* **5,** 159–179.

Borden, L. A. (1996). GABA transporter heterogeneity: Pharmacology and cellular localization. *Neurochem. Int.* **29,** 335–356.

Bormann, J. (1988). Electrophysiology of GABAA and GABAB receptor subtypes. *Trends Neurosci.* **11,** 112–116.

Bormann, J. (2000). The 'ABC' of GABA receptors. *Trends Pharmacol. Sci.* **21,** 16–19.

Bormann, J., and Feigenspan, A. (1995). GABAC receptors. *Trends Neurosci.* **18,** 515–519.

Byne, W., Kemether, E., Jones, L., Haroutunian, V., and Davis, K. L. (1999). The neurochemistry of schizophrenia. *In* "Neurobiology of Mental Illness" (D. S. Charney, E. J. Nestler, and B. S. Bunney, Eds.), pp. 236–249. Oxford University Press, New York.

Cajal, S., and Ramon, Y. (1911). Histologie du Systeme Nerveux de l'Homme et des Vertebres.Tome 2. Paris Maloine. Reprinted by Consejo Superior de Investigaciones Cientificas, Madrid, 1955.

Carlson, B. X., Belhage, B., Hansen, G. H., Elster, L., Olsen, R. W., and Schousboe, A. (1997). Expression of the GABA(A) receptor alpha6 subunit in cultured cerebellar granule cells is developmentally regulated by activation of GABA(A) receptors. *J. Neurosci. Res.* **50,** 1053–1062.

Carlson, B. X., Elster, L., and Schousboe, A. (1998). Pharmacological and functional implications of developmentally-regulated changes in GABA(A) receptor subunit expression in the cerebellum. *Eur. J. Pharmacol.* **352,** 1–14.

Caruncho, H. J., Dopeso-Reyes, I. G., Loza, M. I., and Rodriguez, M. A. (2004). A GABA, reelin, and the neurodevelopmental hypothesis of schizophrenia. *Crit. Rev. Neurobiol.* **16,** 25–32.

Caviness, V. S., Jr., and Rakic, P. (1978). Mechanisms of cortical development: A view from mutations in mice. *Annu. Rev. Neurosci.* **1,** 297–326.

Chaudhry, F. A., Reimer, R. J., Bellocchio, E. E., Danbolt, N. C., Osen, K. K., Edwards, R. H., and Storm-Mathisen, J. (1998). The vesicular GABA transporter, VGAT, localizes to synaptic vesicles in sets of glycinergic as well as GABAergic neurons. *J. Neurosci.* **18,** 9733–9750.

Cherubini, E., and Conti, F. (2001). Generating diversity at GABAergic synapses. *Trends Neurosci.* **24,** 155–162.

Cherubini, E., Gaiarsa, J. L., and Ben-Ari, Y. (1991). GABA: An excitatory transmitter in early postnatal life. *Trends Neurosci.* **14,** 515–519.

Connolly, C. N., Krishek, B. J., McDonald, B. J., Smart, T. G., and Moss, S. J. (1996). Assembly and cell surface expression of heteromeric and homomeric gamma-aminobutyric acid type A receptors. *J. Biol. Chem.* **271,** 89–96.

Connolly, C. N., Uren, J. M., Thomas, P., Gorrie, G. H., Gibson, A., Smart, T. G., and Moss, S. J. (1999). Subcellular localization and endocytosis of homomeric gamma2 subunit splice variants of gamma-aminobutyric acid type A receptors. *Mol. Cell Neurosci.* **13,** 259–271.

Connor, J. A., Tseng, H. Y., and Hockberger, P. E. (1987). Depolarization- and transmitter-induced changes in intracellular Ca2+ of rat cerebellar granule cells in explant cultures. *J. Neurosci.* **7,** 1384–1400.

Connors, B. W., Malenka, R. C., and Silva, L. R. (1988). Two inhibitory postsynaptic potentials, and GABAA and GABAB receptor-mediated responses in neocortex of rat and cat. *J. Physiol.* **406,** 443–468.

Conti, F., Minelli, A., and Melone, M. (2004). GABA transporters in the mammalian cerebral cortex: Localization, development and pathological implications. *Brain Res. Brain Res. Rev.* **45,** 196–212.

Conti, F., Zuccarello, L. V., Barbaresi, P., Minelli, A., Brecha, N. C., and Melone, M. (1999). Neuronal, glial, and epithelial localization of gamma-aminobutyric acid transporter 2, a high-affinity gamma-aminobutyric acid plasma membrane transporter, in the cerebral cortex and neighboring structures. *J Comp. Neurol.* **409,** 482–494.

Cook, E. H., Jr., Lindgren, V., Leventhal, B. L., Courchesne, R., Lincoln, A., Shulman, C., Lord, C., and Courchesne, E. (1997). Autism or atypical autism in maternally but not paternally derived proximal 15q duplication. *Am. J. Hum. Genet.* **60,** 928–934.

Costa, E., Davis, J. M., Dong, E., Grayson, D. R., Guidotti, A., Tremolizzo, L., and Veldic, M. (2004). A GABAergic cortical deficit dominates schizophrenia pathophysiology. *Crit. Rev. Neurobiol.* **16,** 1–23.

Craig, A. M., Banker, G., Chang, W., McGrath, M. E., and Serpinskaya, A. S. (1996). Clustering of gephyrin at GABAergic but not glutamatergic synapses in cultured rat hippocampal neurons. *J. Neurosci.* **16,** 3166–3177.

De Lorey, T. M., Handforth, A., Anagnostaras, S. G., Homanics, G. E., Minassian, B. A., Asatourian, A., Fanselow, M. S., Delgado-Escueta, A., Ellison, G. D., and Olsen, R. W. (1998). Mice lacking the beta3 subunit of the GABAA receptor have the epilepsy phenotype and many of the behavioral characteristics of Angelman syndrome. *J. Neurosci.* **18,** 8505–8514.

Dhossche, D., Applegate, H., Abraham, A., Maertens, P., Bland, L., Bencsath, A., and Martinez, J. (2002). Elevated plasma gamma-aminobutyric acid (GABA) levels in autistic youngsters: Stimulus for a GABA hypothesis of autism. *Med. Sci. Monit.* **8,** PR 1–6.

Dhossche, D. M. (2004). Autism as early expression of catatonia. *Med. Sci. Monit.* **10,** RA 31–39.

Dumoulin, A., Rostaing, P., Bedet, C., Levi, S., Isambert, M. F., Henry, J. P., Triller, A., and Gasnier, B. (1999). Presence of the vesicular inhibitory amino acid transporter in GABAergic and glycinergic synaptic terminal boutons. *J. Cell. Sci.* **112**(Pt 6), 811–823.

Durkin, M. M., Smith, K. E., Borden, L. A., Weinshank, R. L., Branchek, T. A., and Gustafson, E. L. (1995). Localization of messenger RNAs encoding three GABA transporters in rat brain: An *in situ* hybridization study. *Brain Res. Mol. Brain Res.* **33,** 7–21.

Eilers, J., Plant, T. D., Marandi, N., and Konnerth, A. (2001). GABA-mediated Ca2+ signalling in developing rat cerebellar Purkinje neurones. *J. Physiol.* **536,** 429–437.

Elster, L., Hansen, G. H., Belhage, B., Fritschy, J. M., Mohler, H., and Schousboe, A. (1995). Differential distribution of GABAA receptor subunits in soma and processes of cerebellar granule cells: Effects of maturation and a GABA agonist. *Int. J. Dev. Neurosci.* **13,** 417–428.

Essrich, C., Lorez, M., Benson, J. A., Fritschy, J. M., and Luscher, B. (1998). Postsynaptic clustering of major GABAA receptor subtypes requires the gamma 2 subunit and gephyrin. *Nat. Neurosci.* **1,** 563–571.

Fagiolini, M., and Hensch, T. K. (2000). Inhibitory threshold for critical-period activation in primary visual cortex. *Nature* **404,** 183–186.

Fairen, A., Alvarez-Bolado, G., DeDiego, I., and Smith-Fernandez, A. (1998). GABA-immunoreactive cells of the cortical primordium contribute to distinctly fated neuronal populations. *Perspect Dev. Neurobiol.* **5,** 159–173.

Fatemi, S. H., Halt, A. R., Stary, J. M., Kanodia, R., Schulz, S. C., and Realmuto, G. R. (2002). Glutamic acid decarboxylase 65 and 67 kDa proteins are reduced in autistic parietal and cerebellar cortices. *Biol. Psychiatry.* **52,** 805–810.

Fon, E. A., and Edwards, R. H. (2001). Molecular mechanisms of neurotransmitter release. *Muscle Nerve* **24,** 581–601.

Freeman, M. P., Freeman, S. A., and McElroy, S. L. (2002). The comorbidity of bipolar and anxiety disorders: Prevalence, psychobiology, and treatment issues. *J. Affect Disord.* **68,** 1–23.

Freund, T. F., and Gulyas, A. I. (1997). Inhibitory control of GABAergic interneurons in the hippocampus. *Can J Physiol Pharmacol* **75,** 479–487.

Fritschy, J. M., and Mohler, H. (1995). GABAA-receptor heterogeneity in the adult rat brain: Differential regional and cellular distribution of seven major subunits. *J. Comp. Neurol.* **359,** 154–194.

Fritschy, J. M., Paysan, J., Enna, A., and Mohler, H. (1994). Switch in the expression of rat GABAA-receptor subtypes during postnatal development: An immunohistochemical study. *J. Neurosci.* **14,** 5302–5324.

Frostholm, A., Zdilar, D., Chang, A., and Rotter, A. (1991). Stability of GABAA/benzodiazepine receptor alpha 1 subunit mRNA expression in reeler mouse cerebellar Purkinje cells during postnatal development. *Brain Res. Dev. Brain Res.* **64,** 121–128.

Gabriele, M. L., Brunso-Bechtold, J. K., and Henkel, C. K. (2000). Plasticity in the development of afferent patterns in the inferior colliculus of the rat after unilateral cochlear ablation. *J. Neurosci.* **20,** 6939–6949.

Gadea, A., and Lopez-Colome, A. M. (2001). Glial transporters for glutamate, glycine, and GABA: II. GABA transporters. *J. Neurosci. Res.* **63,** 461–468.

Gambarana, C., Pittman, R., and Siegel, R. E. (1990). Developmental expression of the GABAA receptor alpha 1 subunit mRNA in the rat brain. *J. Neurobiol.* **21,** 1169–1179.

Ganguly, K., Schinder, A. F., Wong, S. T., and Poo, M. (2001). GABA itself promotes the developmental switch of neuronal GABAergic responses from excitation to inhibition. *Cell* **105,** 521–532.

Gao, B., and Fritschy, J. M. (1995). Cerebellar granule cells *in vitro* recapitulate the *in vivo* pattern of GABAA-receptor subunit expression. *Brain Res. Dev. Brain Res.* **88,** 1–16.

Gao, X. B., and van den Pol, A. N. (2000). GABA release from mouse axonal growth cones. *J. Physiol.* **523 Pt 3,** 629–637.

Gorrie, G. H., Vallis, Y., Stephenson, A., Whitfield, J., Browning, B., Smart, T. G., and Moss, S. J. (1997). Assembly of GABAA receptors composed of alpha1 and beta2 subunits in both cultured neurons and fibroblasts. *J. Neurosci.* **17,** 6587–6596.

Haydar, T. F., Wang, F., Schwartz, M. L., and Rakic, P. (2000). Differential modulation of proliferation in the neocortical ventricular and subventricular zones. *J. Neurosci.* **20,** 5764–5774.

Hensch, T. K., Fagiolini, M., Mataga, N., Stryker, M. P., Baekkeskov, S., and Kash, S. F. (1998). Local GABA circuit control of experience-dependent plasticity in developing visual cortex. *Science* **282,** 1504–1508.

Ito, M. (1984). "The Cerebellum and Neural Control." Raven Press, New York.

Itouji, A., Sakai, N., Tanaka, C., and Saito, N. (1996). Neuronal and glial localization of two GABA transporters (GAT1 and GAT3) in the rat cerebellum. *Brain Res. Mol. Brain Res.* **37,** 309–316.

Jaffe, E. H., and Vaello, M. L. (1988). Two different release mechanisms of 3H-GABA induced by glutamate in the rat olfactory bulb. *P. R. Health Sci. J.* **7,** 99–101.

Jakab, R. L., and Hamori, J. (1988). Quantitative morphology and synaptology of cerebellar glomeruli in the rat. *Anat. Embryol.* **179,** 81–88.

Kanaka, C., Ohno, K., Okabe, A., Kuriyama, K., Itoh, T., Fukuda, A., and Sato, K. (2001). The differential expression patterns of messenger RNAs encoding K-Cl cotransporters (KCC1,2) and Na-K-2Cl cotransporter (NKCC1) in the rat nervous system. *Neuroscience* **104,** 933–946.

Kandler, K. (2004). Activity-dependent organization of inhibitory circuits: Lessons from the auditory system. *Curr. Opin. Neurobiol.* **14,** 96–104.

Kanner, B. I. (1994). Sodium-coupled neurotransmitter transport: Structure, function and regulation. *J. Exp. Biol.* **196,** 237–249.

Kapfer, C., Seidl, A. H., Schweizer, H., and Grothe, B. (2002). Experience-dependent refinement of inhibitory inputs to auditory coincidence-detector neurons. *Nat. Neurosci.* **5,** 247–253.

Kardos, J. (1999). Recent advances in GABA research. *Neurochem. Int.* **34,** 353–358.

Kaupmann, K., Huggel, K., Heid, J., Flor, P. J., Bischoff, S., Mickel, S. J., McMaster, G., Angst, C., Bittiger, H., Froestl, W., and Bettler, B. (1997). Expression cloning of GABA(B) receptors uncovers similarity to metabotropic glutamate receptors. *Nature* **386,** 239–246.

Kaupmann, K., Malitschek, B., Schuler, V., Heid, J., Froestl, W., Beck, P., Mosbacher, J., Bischoff, S., Kulik, A., Shigemoto, R., Karschin, A., and Bettler, B. (1998). GABA(B)-receptor subtypes assemble into functional heteromeric complexes. *Nature* **396,** 683–687.

Kim, G., and Kandler, K. (2003). Elimination and strengthening of glycinergic/GABAergic connections during tonotopic map formation. *Nat. Neurosci.* **6,** 282–290.

Kim, H. Y., Sapp, D. W., Olsen, R. W., and Tobin, A. J. (1993). GABA alters GABAA receptor mRNAs and increases ligand binding. *J. Neurochem.* **61,** 2334–2337.

Kneussel, M., Brandstatter, J. H., Laube, B., Stahl, S., Muller, U., and Betz, H. (1999). Loss of postsynaptic GABA(A) receptor clustering in gephyrin-deficient mice. *J. Neurosci.* **19,** 9289–9297.

Larramendi, L. H. M. (1969). Analysis of synaptogenesis in the cerebellum of the mouse. *In* "Neurobiology of Cerebellar Evolution and Development" (R. Llinas, Ed.), pp. 783–843. American Medical Association, Chicago.

Lauder, J. M. (1993). Neurotransmitters as growth regulatory signals: Role of receptors and second messengers. *Trends Neurosci.* **16,** 233–240.

Lauder, J. M., Han, V. K., Henderson, P., Verdoorn, T., and Towle, A. C. (1986). Prenatal ontogeny of the GABAergic system in the rat brain: An immunocytochemical study. *Neuroscience* **19,** 465–493.

Laurie, D. J., Seeburg, P. H., and Wisden, W. (1992a). The distribution of 13 GABAA receptor subunit mRNAs in the rat brain. II. Olfactory bulb and cerebellum. *J. Neurosci.* **12,** 1063–1076.

Laurie, D. J., Wisden, W., and Seeburg, P. H. (1992b). The distribution of thirteen GABAA receptor subunit mRNAs in the rat brain. III. Embryonic and postnatal development. *J Neurosci* **12,** 4151–4172.

Lauritsen, M., Mors, O., Mortensen, P. B., and Ewald, H. (1999). Infantile autism and associated autosomal chromosome abnormalities: A register-based study and a literature survey. *J. Child Psychol. Psychiatry* **40,** 335–345.

Leinekugel, X., Khalilov, I., McLean, H., Caillard, O., Gaiarsa, J. L., Ben-Ari, Y., and Khazipov, R. (1999). GABA is the principal fast-acting excitatory transmitter in the neonatal brain. *Adv. Neurol.* **79,** 189–201.

Leinekugel, X., Tseeb, V., Ben-Ari, Y., and Bregestovski, P. (1995). Synaptic GABAA activation induces Ca2+ rise in pyramidal cells and interneurons from rat neonatal hippocampal slices. *J. Physiol.* **487**(Pt 2), 319–329.

Levi, G., and Raiteri, M. (1993). Carrier-mediated release of neurotransmitters. *Trends Neurosci* **16,** 415–419.

Le Vine, H., 3rd (1999). Structural features of heterotrimeric G-protein-coupled receptors and their modulatory proteins. *Mol. Neurobiol.* **19,** 111–149.

Lewis, D. A., Volk, D. W., and Hashimoto, T. (2004). Selective alterations in prefrontal cortical GABA neurotransmission in schizophrenia: A novel target for the treatment of working memory dysfunction. *Psychopharmacology (Berl).* **174,** 143–150.

Llinas, R., and Hillmann, D. E. (1969). Physiological and morphological organization of the cerebellar circuits in various vertebrates. *In* "Neurobiology of Cerebellar Evolution and Development" (R. Llinas, Ed.), pp. 43–73. American Medical Association, Chicago.

Llinas, R., and Walton, K. D. (1990). Cerebellum. *In* "The Synaptic Organization of the Brain" (G. M. Shepherd, Ed.), pp. 214–245. Oxford University Press, Oxford.

Lo Turco, J. J., Owens, D. F., Heath, M. J., Davis, M. B., and Kriegstein, A. R. (1995). GABA and glutamate depolarize cortical progenitor cells and inhibit DNA synthesis. *Neuron* **15,** 1287–1298.

Lu, J., Karadsheh, M., and Delpire, E. (1999). Developmental regulation of the neuronal-specific isoform of K-Cl cotransporter KCC2 in postnatal rat brains. *J. Neurobiol.* **39,** 558–568.

Luddens, H., Pritchett, D. B., Kohler, M., Killisch, I., Keinanen, K., Monyer, H., Sprengel, R., and Seeburg, P. H. (1990). Cerebellar GABAA receptor selective for a behavioural alcohol antagonist. *Nature* **346,** 648–651.

Ma, W., and Barker, J. L. (1995). Complementary expressions of transcripts encoding GAD67 and GABAA receptor alpha 4, beta 1, and gamma 1 subunits in the proliferative zone of the embryonic rat central nervous system. *J. Neurosci.* **15,** 2547–2560.

Macdonald, R. L., and Olsen, R. W. (1994). GABAA receptor channels. *Annu. Rev. Neurosci.* **17,** 569–602.

Mariani, J., Crepel, F., Mikoshiba, K., Changeux, J. P., and Sotelo, C. (1977). Anatomical, physiological and biochemical studies of the cerebellum from Reeler mutant mouse. *Philos. Trans. R. Soc. Lond. B. Biol. Sci.* **281,** 1–28.

Maric, D., Maric, I., Ma, W., Lahojuji, F., Somogyi, R., Wen, X., Sieghart, W., Fritschy, J. M., and Barker, J. L. (1997). Anatomical gradients in proliferation and differentiation of embryonic rat CNS accessed by buoyant density fractionation: Alpha 3, beta 3 and gamma 2 GABAA receptor subunit co-expression by post-mitotic neocortical neurons correlates directly with cell buoyancy. *Eur. J. Neurosci.* **9,** 507–522.

Martin, D. L., and Rimvall, K. (1993). Regulation of gamma-aminobutyric acid synthesis in the brain. *J. Neurochem.* **60,** 395–407.

Marty, S., Berninger, B., Carroll, P., and Thoenen, H. (1996). GABAergic stimulation regulates the phenotype of hippocampal interneurons through the regulation of brain-derived neurotrophic factor. *Neuron* **16,** 565–570.

McBain, C. J., and Maccaferri, G. (1997). Synaptic plasticity in hippocampal interneurons? A commentary. *Can. J. Physiol. Pharmacol.* **75,** 488–494.

McCarthy, M. M., Auger, A. P., and Perrot-Sinal, T. S. (2002). Getting excited about GABA and sex differences in the brain. *Trends Neurosci.* **25,** 307–312.

McIntire, S. L., Reimer, R. J., Schuske, K., Edwards, R. H., and Jorgensen, E. M. (1997). Identification and characterization of the vesicular GABA transporter. *Nature* **389,** 870–876.

McKernan, R. M., and Whiting, P. J. (1996). Which GABAA-receptor subtypes really occur in the brain? *Trends Neurosci.* **19,** 139–143.

McLaughlin, B. J., Wood, J. G., Saito, K., Roberts, E., and Wu, J. Y. (1975). The fine structural localization of glutamate decarboxylase in developing axonal processes and presynaptic terminals of rodent cerebellum. *Brain Res.* **85,** 355–371.

Mehta, A. K., and Ticku, M. K. (1999). An update on GABAA receptors. *Brain Res. Brain Res. Rev.* **29,** 196–217.

Meier, E., Drejer, J., and Schousboe, A. (1984). GABA induces functionally active low-affinity GABA receptors on cultured cerebellar granule cells. *J. Neurochem.* **43,** 1737–1744.

Meier, E., and Jorgensen, O. S. (1986). Gamma-aminobutyric acid affects the developmental expression of neuron-associated proteins in cerebellar granule cell cultures. *J. Neurochem.* **46,** 1256–1262.

Meier, E., Jorgensen, O. S., and Schousboe, A. (1987). Effect of repeated treatment with a gamma-aminobutyric acid receptor agonist on postnatal neural development in rats. *J. Neurochem.* **49,** 1462–1470.

Mellor, J. R., Merlo, D., Jones, A., Wisden, W., and Randall, A. D. (1998). Mouse cerebellar granule cell differentiation: Electrical activity regulates the GABAA receptor alpha 6 subunit gene. *J. Neurosci.* **18,** 2822–2833.

Mikawa, S., Wang, C., Shu, F., Wang, T., Fukuda, A., and Sato, K. (2002). Developmental changes in KCC1, KCC2 and NKCC1 mRNAs in the rat cerebellum. *Brain Res. Dev. Brain Res.* **136,** 93–100.

Millan, M. J. (2003). The neurobiology and control of anxious states. *Prog. Neurobiol.* **70,** 83–244.

Mitchell, C. K., and Redburn, D. A. (1996). GABA and GABA-A receptors are maximally expressed in association with cone synaptogenesis in neonatal rabbit retina. *Brain Res. Dev. Brain Res.* **95,** 63–71.

Morara, S., Brecha, N. C., Marcotti, W., Provini, L., and Rosina, A. (1996). Neuronal and glial localization of the GABA transporter GAT-1 in the cerebellar cortex. *Neuroreport* **7,** 2993–2996.

Moss, S. J., and Smart, T. G. (2001). Constructing inhibitory synapses. *Nat. Rev. Neurosci.* **2,** 240–250.

Nayeem, N., Green, T. P., Martin, I. L., and Barnard, E. A. (1994). Quaternary structure of the native GABAA receptor determined by electron microscopic image analysis. *J. Neurochem.* **62,** 815–818.

Nicoll, R. A. (1988). The coupling of neurotransmitter receptors to ion channels in the brain. *Science* **241,** 545–551.

Nutt, D. J. (2001). Neurobiological mechanisms in generalized anxiety disorder. *J. Clin. Psychiatry.* **62** (Suppl 11), 22–27, discussion 28.

Nutt, D. J., Glue, P., and Lawson, C. (1990). The neurochemistry of anxiety: An update. *Prog. Neuropsychopharmacol. Biol. Psychiatry.* **14,** 737–752.

Obrietan, K., and van den Pol, A. N. (1996). Growth cone calcium elevation by GABA. *J. Comp. Neurol.* **372,** 167–175.

Olsen, R. W., and Avoli, M. (1997). GABA and epileptogenesis. *Epilepsia* **38,** 399–407.

Olsen, R. W., and Tobin, A. J. (1990). Molecular biology of GABAA receptors. *FASEB J.* **4,** 1469–1480.

Owens, D. F., and Kriegstein, A. R. (2002). Is there more to GABA than synaptic inhibition? *Nat. Rev. Neurosci.* **3,** 715–727.

Palay, S. L., and Chan-Palay, V. (1974). "Cerebellar Cortex, Cytology and Organization." Springer, Berlin.

Paulsen, O., and Moser, E. I. (1998). A model of hippocampal memory encoding and retrieval: GABAergic control of synaptic plasticity. *Trends Neurosci.* **21,** 273–278.

Perkins, K. L., and Wong, R. K. (1997). The depolarizing GABA response. *Can. J. Physiol. Pharmacol.* **75,** 516–519.

Persohn, E., Malherbe, P., and Richards, J. G. (1992). Comparative molecular neuroanatomy of cloned GABAA receptor subunits in the rat CNS. *J. Comp. Neurol.* **326,** 193–216.

Phillis, J. W., Smith-Barbour, M., Perkins, L. M., and O'Regan, M. H. (1994). Characterization of glutamate, aspartate, and GABA release from ischemic rat cerebral cortex. *Brain Res. Bull.* **34,** 457–466.

Pratt, J. A. (1992). The neuroanatomical basis of anxiety. *Pharmacol. Ther.* **55,** 149–181.

Pritchett, D. B., Sontheimer, H., Shivers, B. D., Ymer, S., Kettenmann, H., Schofield, P. R., and Seeburg, P. H. (1989). Importance of a novel GABAA receptor subunit for benzodiazepine pharmacology. *Nature* **338,** 582–585.

Raetzman, L. T., and Siegel, R. E. (1999). Immature granule neurons from cerebella of different ages exhibit distinct developmental potentials. *J. Neurobiol.* **38,** 559–570.

Rakic, P. (1976). Synaptic specificity in the cerebellar cortex: Study of anomalous circuits induced by single gene mutations in mice. *Cold Spring Harb. Symp. Quant. Biol.* **40,** 333–346.

Reichling, D. B., Kyrozis, A., Wang, J., and MacDermott, A. B. (1994). Mechanisms of GABA and glycine depolarization-induced calcium transients in rat dorsal horn neurons. *J. Physiol.* **476,** 411–421.

Reimer, R. J., Fon, E. A., and Edwards, R. H. (1998). Vesicular neurotransmitter transport and the presynaptic regulation of quantal size. *Curr. Opin. Neurobiol.* **8,** 405–412.

Ribak, C. E., Tong, W. M., and Brecha, N. C. (1996). Astrocytic processes compensate for the apparent lack of GABA transporters in the axon terminals of cerebellar Purkinje cells. *Anat. Embryol. (Berl.)* **194,** 379–390.

Rico, B., Xu, B., and Reichardt, L. F. (2002). TrkB receptor signaling is required for establishment of GABAergic synapses in the cerebellum. *Nat. Neurosci.* **5,** 225–233.

Rohrbough, J., and Spitzer, N. C. (1996). Regulation of intracellular Cl- levels by Na(+)-dependent Cl- cotransport distinguishes depolarizing from hyperpolarizing GABAA receptor-mediated responses in spinal neurons. *J. Neurosci.* **16,** 82–91.

Rolf, L. H., Haarmann, F. Y., Grotemeyer, K. H., and Kehrer, H. (1993). Serotonin and amino acid content in platelets of autistic children. *Acta. Psychiatr. Scand.* **87,** 312–316.

Rosina, A., Morara, S., and Provini, L. (1999). GAT-1 developmental expression in the rat cerebellar cortex: Basket and pinceau formation. *Neuroreport* **10,** 1613–1618.

Sassoe-Pognetto, M., and Fritschy, J. M. (2000). Mini-review: Gephyrin, a major postsynaptic protein of GABAergic synapses. *Eur. J. Neurosci.* **12,** 2205–2210.

Schousboe, A. (1999). Pharmacologic and therapeutic aspects of the developmentally regulated expression of GABA(A) and GABA(B) receptors: Cerebellar granule cells as a model system. *Neurochem. Int.* **34,** 373–377.

Serafini, R., Ma, W., Maric, D., Maric, I., Lahjouji, F., Sieghart, W., and Barker, J. L. (1998). Initially expressed early rat embryonic GABA(A) receptor Cl- ion channels exhibit heterogeneous channel properties. *Eur. J. Neurosci.* **10,** 1771–1783.

Sieghart, W. (1995). Structure and pharmacology of gamma-aminobutyric acidA receptor subtypes. *Pharmacol. Rev.* **47,** 181–234.

Sieghart, W., Fuchs, K., Tretter, V., Ebert, V., Jechlinger, M., Hoger, H., and Adamiker, D. (1999). Structure and subunit composition of GABA(A) receptors. *Neurochem. Int.* **34,** 379–385.

Snead, O. C., 3rd, Depaulis, A., Vergnes, M., and Marescaux, C. (1999). Absence epilepsy: Advances in experimental animal models. *Adv. Neurol.* **79,** 253–278.

Sotelo, C., and Privat, A. (1978). Synaptic remodeling of the cerebellar circuitry in mutant mice and experimental cerebellar malformations. Study "*in vivo*" and "*in vitro*". *Acta. Neuropathol. (Berl).* **43,** 19–34.

Spoerri, P. E. (1988). Neurotrophic effects of GABA in cultures of embryonic chick brain and retina. *Synapse* **2,** 11–22.

Takamori, S., Riedel, D., and Jahn, R. (2000). Immunoisolation of GABA-specific synaptic vesicles defines a functionally distinct subset of synaptic vesicles. *J. Neurosci.* **20,** 4904–4911.

Takayama, C. (1994). Altered distribution of inhibitory synaptic terminals in reeler cerebellum with special reference to malposition of GABAergic neurons. *Neurosci. Res.* **20,** 239–250.

Takayama, C. (2005). Formation of GABAergic synapses in the cerebellum. *Cerebellum* **4,** 171–177.

Takayama, C., and Inoue, Y. (2003). Normal formation of the postsynaptic elements of GABAergic synapses in the reeler cerebellum. *Brain Res. Dev. Brain Res.* **145,** 197–211.

Takayama, C., and Inoue, Y. (2005). Developmental expression of GABA transporter-1 and 3 during formation of the GABAergic synapses in the mouse cerebellar cortex. *Brain Res. Dev. Brain Res.* **158,** 41–49.

Takayama, C., and Inoue, Y. (2004a). Extrasynaptic localization of GABA in the developing mouse cerebellum. *Neurosci. Res.* **50,** 447–458.

Takayama, C., and Inoue, Y. (2004b). Morphological development and maturation of the GABAergic synapses in the mouse cerebellar granular layer. *Brain Res. Dev. Brain Res.* **150,** 175–188.

Takayama, C., and Inoue, Y. (2004c). Morphological development and maturation of the GABAergic synapses in the mouse cerebellar granular layer. *Brain Res. Dev. Brain Res.* **150,** 177–190.

Takayama, C., and Inoue, Y. (2004d). Transient expression of GABA(A) receptor alpha2 and alpha3 subunits in differentiating cerebellar neurons. *Brain Res. Dev. Brain Res.* **148,** 169–177.

Taylor, J., Docherty, M., and Gordon-Weeks, P. R. (1990). GABAergic growth cones: Release of endogenous gamma-aminobutyric acid precedes the expression of synaptic vesicle antigens. *J. Neurochem.* **54,** 1689–1699.

Taylor, J., and Gordon-Weeks, P. R. (1991). Calcium-independent gamma-aminobutyric acid release from growth cones: Role of gamma-aminobutyric acid transport. *J. Neurochem.* **56,** 273–280.

Taylor, P. M., Connolly, C. N., Kittler, J. T., Gorrie, G. H., Hosie, A., Smart, T. G., and Moss, S. J. (2000). Identification of residues within GABA(A) receptor alpha subunits that mediate specific assembly with receptor beta subunits. *J. Neurosci.* **20,** 1297–1306.

Tia, S., Wang, J. F., Kotchabhakdi, N., and Vicini, S. (1996). Developmental changes of inhibitory synaptic currents in cerebellar granule neurons: Role of GABA(A) receptor alpha 6 subunit. *J. Neurosci.* **16,** 3630–3640.

Tretter, V., Ehya, N., Fuchs, K., and Sieghart, W. (1997). Stoichiometry and assembly of a recombinant GABAA receptor subtype. *J. Neurosci.* **17,** 2728–2737.

Turek, F. W., and Van Reeth, O. (1988). Altering the mammalian circadian clock with the short-acting benzodiazepine, triazolam. *Trends Neurosci.* **11,** 535–541.

Van Eden, C. G., Mrzljak, L., Voorn, P., and Uylings, H. B. (1989). Prenatal development of GABA-ergic neurons in the neocortex of the rat. *J. Comp. Neurol.* **289,** 213–227.

Varju, P., Katarova, Z., Madarasz, E., and Szabo, G. (2001). GABA signalling during development: New data and old questions. *Cell Tissue Res.* **305,** 239–246.

Vaughn, J. E. (1989). Fine structure of synaptogenesis in the vertebrate central nervous system. *Synapse* **3,** 255–285.

Vicario-Abejon, C., Collin, C., McKay, R. D., and Segal, M. (1998). Neurotrophins induce formation of functional excitatory and inhibitory synapses between cultured hippocampal neurons. *J. Neurosci.* **18,** 7256–7271.

Vicini, S. (1999). New perspectives in the functional role of GABA(A) channel heterogeneity. *Mol Neurobiol* **19,** 97–110.

Wagner, S., Castel, M., Gainer, H., and Yarom, Y. (1997). GABA in the mammalian suprachiasmatic nucleus and its role in diurnal rhythmicity. *Nature* **387,** 598–603.

Wallace, R. H., Marini, C., Petrou, S., Harkin, L. A., Bowser, D. N., Panchal, R. G., Williams, D. A., Sutherland, G. R., Mulley, J. C., Scheffer, I. E., and Berkovic, S. F. (2001). Mutant GABA(A) receptor gamma2-subunit in childhood absence epilepsy and febrile seizures. *Nat Genet* **28,** 49–52.

Wang, H., Bedford, F. K., Brandon, N. J., Moss, S. J., and Olsen, R. W. (1999). GABA(A)-receptor-associated protein links GABA(A) receptors and the cytoskeleton. *Nature* **397,** 69–72.

Wassef, A., Baker, J., and Kochan, L. D. (2003). GABA and schizophrenia: A review of basic science and clinical studies. *J. Clin. Psychopharmacol.* **23,** 601–640.

Watanabe, D., Inokawa, H., Hashimoto, K., Suzuki, N., Kano, M., Shigemoto, R., Hirano, T., Toyama, K., Kaneko, S., Yokoi, M., Moriyoshi, K., Suzuki, M., Kobayashi, K., Nagatsu, T., Kreitman, R. J., Pastan, I., and Nakanishi, S. (1998). Ablation of cerebellar Golgi cells disrupts synaptic integration involving GABA inhibition and NMDA receptor activation in motor coordination. *Cell* **95,** 17–27.

Williams, J. R., Sharp, J. W., Kumari, V. G., Wilson, M., and Payne, J. A. (1999). The neuron-specific K-Cl cotransporter, KCC2. Antibody development and initial characterization of the protein. *J. Biol. Chem.* **274,** 12656–12664.

Wilson, L., Sotelo, C., and Caviness, V. S., Jr. (1981). Heterologous synapses upon Purkinje cells in the cerebellum of the Reeler mutant mouse: An experimental light and electron microscopic study. *Brain Res.* **213,** 63–82.

Wisden, W., Korpi, E. R., and Bahn, S. (1996). The cerebellum: A model system for studying GABAA receptor diversity. *Neuropharmacology* **35,** 1139–1160.

Wisden, W., Laurie, D. J., Monyer, H., and Seeburg, P. H. (1992). The distribution of 13 GABAA receptor subunit mRNAs in the rat brain. I. Telencephalon, diencephalon, mesencephalon. *J. Neurosci.* **12,** 1040–1062.

Wolff, J. R., Joo, F., and Dames, W. (1978). Plasticity in dendrites shown by continuous GABA administration in superior cervical ganglion of adult rat. *Nature* **274,** 72–74.

Wolff, J. R., Joo, F., and Kasa, P. (1993). Modulation by GABA of neuroplasticity in the central and peripheral nervous system. *Neurochem. Res.* **18,** 453–461.

Yuste, R., and Katz, L. C. (1991). Control of postsynaptic Ca2+ influx in developing neocortex by excitatory and inhibitory neurotransmitters. *Neuron* **6,** 333–344.

INSIGHTS INTO GABA FUNCTIONS IN THE DEVELOPING CEREBELLUM

Mónica L. Fiszman

Instituto de Investigaciones Farmacologicas-CONICET
Ciudad de Buenos Aires 1113, Argentina

During development, GABA, the main inhibitory neurotransmitter of the nervous system is excitatory and elicits calcium-dependent trophic effects in developing neurons. Through depolarization of immature cerebellar neurons, GABA increases intracellular calcium and the phosphorylation of MAPK and CaMKII. Other depolarizing stimuli, such as exposure to 25 mM KCl, accelerate the switch of GABA from depolarizing to hyperpolarizing. Exposure to GABA and muscimol increases proliferation of immature granule cells. In addition, GABA increases the size of the dendritic arbor in early postmitotic neurons. This chapter outlines the role of GABA in the development of cerebellar granule neurons, and other central neurons, and reviews evidence that GABA and other depolarizing stimuli play a crucial role in the early development of the cerebellum.

I. Introduction

The GABAergic system is affected in many neuropathological conditions including autism (Fatemi *et al.*, 2002b). Morphological alterations, consisting of loss of granule and atrophy as well as loss of Purkinje cells (Fatemi *et al.*, 2002a;

INTERNATIONAL REVIEW OF
NEUROBIOLOGY, VOL. 71
DOI: 10.1016/S0074-7742(05)71004-7

95

Ritvo *et al.*, 1986), have been reported in the cerebellum of autistic patients. This article reviews a series of studies outlining the role of GABA in the development of the central nervous system and particularly in the development of the cerebellum. These studies provide evidence supporting the crucial role for GABA during development, the knowledge of which may be fundamental to a better understanding of the pathogenesis of autism and related diseases.

II. The Adult Cerebellum

The mammalian cerebellar cortex is an anatomically well-defined structure with a particular laminated cytoarchitecture. The cerebellum is composed of a limited number of neuronal cell types, identifiable by their position and size, and includes granule, Purkinje, Golgi, basket, and stellate neurons (Palay and Chan-Palay, 1974).

Purkinje cells are inhibitory interneurons organized in a cell layer between the granule and the molecular layers of the cerebellum. They constitute the output path of the cerebellum, and each of them receives input from just one climbing fiber axon, which originates in the inferior olive. The climbing fiber forms a very strong synapse onto the Purkinje cell, with each presynaptic spike triggering a postsynaptic spike. Climbing fibers may serve a special function other than ordinary signal transmission; an instructive signal that regulates the strength of parallel fiber–Purkinje cell synapses. Two possible actions were postulated for the synapses between climbing fibers and Purkinje cells: they may act as positive reinforcers, strengthening parallel fiber synapses, and/or as error signals, serving to weaken synapses (see review by Boyden *et al.*, 2004). Cerebellar granule cells are excitatory interneurons and their axons form the parallel fibers that project to Purkinje cells but also form synapse contacts with cells in the molecular layer. Cerebellar granule cells receive inputs from the mossy fibers coming from the vestibular nuclei. The Golgi cells are inhibitory interneurons of the cerebellum that have their cell bodies located in the granule layer. Golgi cells receive inputs from, and project their inhibitory output to granule cells, constituting a negative feedback circuit. Stellate cells are small cells lying in the outer two thirds of the molecular layer; basket cells are located in the inner third, in close proximity to the Purkinje cell layer. The basket and stellate cells receive excitatory inputs from the parallel fibers of the granule cells and the climbing fibers and direct their output by inhibiting Purkinje cells and granule cells. Anatomical and physiological studies indicate that molecular layer interneurons lie on the activated parallel fiber beam (Pouzat and Hestrin, 1997) suggesting that a Purkinje cell may be both directly excited and then inhibited via molecular layer interneurons activated by the same set of active parallel fibers (feed-forward inhibition).

The cerebellum is implicated in motor learning, a function likely accomplished by a number of plasticity mechanisms including the long-term depression (LTD) at synapses between parallel fibers and Purkinje cells. LTD is induced when parallel and climbing fiber activities occur simultaneously, whereas LTP is induced when parallel fiber activity and not climbing fiber activity occur (see review by Boyden *et al.*, 2004).

III. The Developing Cerebellum

In the rodent, all cerebellar neurons generate between the end of the second week of gestation and the first week of life. The genesis of Purkinje cells ends at the beginning of the first week of life but its dendritic tree continues to grow until day 21. The granule cells complete their genesis between 7 and 9 days after birth. These proliferating cerebellar granule cell precursors and young differentiating granule neurons migrate from the external granule layer to the internal granule layer, their final destination, guided by Bergman glial fibers (Komuro and Rakic, 1998). Basket and stellate cells reach the molecular layer postnatally, where they are innervated by granule cell axons (parallel fibers) that provide glutamatergic input (Eccles *et al.*, 1967; Palay and Chan-Palay, 1974). The synaptogenesis between parallel fibers and Purkinje and/or basket and stellate cells located in the molecular layer takes place between the first and third weeks of life (Eccles, 1967).

IV. GABA$_A$ Receptors and the Developing Cerebellum

The cerebellum is widely populated with amino acids receptors, mainly GABA and glutamate receptors, making it a very interesting preparation to study amino acids receptor pharmacology (Carlson *et al.*, 1998). The GABA$_A$ receptor system consists of 19 different subunit genes (α_{1-6}, β_{1-3}, γ_{1-3}, δ, ρ_{1-3}, ε, π, and θ) that encode the subunits of pentameric GABA$_A$ receptors (reviewed by Sieghart *et al.*, 1999; Lüscher and Keller, 2004). In granule cells from adult cerebella, high levels of α_1, α_6, β_2, β_3, γ_2, and δ mRNAs are found in comparison to low levels of the mRNAs for the α_4, β_1, and γ_3 subunits (Laurie *et al.*, 1992a). The α_6 subunit has a unique distribution in the brain, in that it is almost exclusively found in cerebellar granule cells (Laurie *et al.*, 1992a). Immunocytochemical studies have likewise shown a dense population of α_1, $\beta_{2/3}$, and δ subunits and moderate γ immunoreactivity in the granule cell layer (see review by Carlson *et al.*, 1998).

GABA$_A$ receptors subunits are expressed in the cerebellum at early stages of perinatal development and the expression pattern varies according with the stage of neuronal maturation (Laurie *et al.*, 1992b). For example, dividing precursor and pre-migratory post-mitotic cells express transcripts encoding the GABA$_A$ receptor α_2, α_3, β_3, γ_1, and γ_2 subunits (Laurie *et al.*, 1992b). Later, α_2, α_3, and γ_1 are down-regulated and replaced by the adult complement that consists predominantly of α_1, α_6, β_2, β_3, γ_2, and δ (Laurie *et al.*, 1992a,b; Thompson and Stephenson, 1994). The α_1, α_6, and δ genes are expressed only in post-migratory cerebellar granule neurons (Kuhar *et al.*, 1993; Laurie *et al.*, 1992b). Electrophysiological recording in the developing cerebellum reveals that a sequential expression of α subunits underlies changes in functional efficacy of the GABAergic network (Ortinski *et al.*, 2004; Vicini *et al.*, 2001). A 50% decrease in the decay time of miniature inhibitory postsynaptic currents (mIPSCs) between cultured cerebellar granule neurons grown for 6 days and those grown for 12 days was paralleled by the decrease of α_2 and α_3 subunits, the increase of α_1 and α_6 subunits expression, and changes in the action of selective α-subunit modulators (Ortinski *et al.*, 2004). How these electrophysiological changes correlate with developmental processes triggered by GABA in the immature cerebellum remains to be established.

V. Depolarization and Neurotrophins Are Trophic Signals

Neurotrophins are a family of peptide growth factors, including neurotrophin-3 (NT-3), NT-4/5, nerve growth factor, and brain-derived neurotrophic factor (BDNF). They play important roles in the development and maturation of the nervous system and support survival of mature neurons (Kaplan and Miller, 2000). Neurotrophins exert their trophic effects by binding to the tyrosine kinase (Trk) receptor family subtype (Chao, 1992) and triggering downstream activation of *ras*-MAP kinase (Ras-mitogen-activated protein kinase, erk) and the phosphatidylinositol-3-OH (PI3K)-protein kinase B (PKB-Akt) pathways (see review by Kaplan and Miller, 2000). Neural activity can mimic the trophic effect induced by neurotrophins promoting the differentiation and survival of central and peripheral neurons (Borodinsky *et al.*, 2002; Franklin *et al.*, 1995; Hansen *et al.*, 2001; Kaplan and Miller, 2000; Vaillant *et al.*, 2002). Intracellular cascades activated by neurotrophins are also activated by neural activity-induced calcium fluctuations (Dolmetsch *et al.*, 2001; Rosen *et al.*, 1994; Vaillant *et al.*, 1999). Furthermore, intracellular calcium fluctuations triggered by neural activity modulate gene expression, as do neurotrophins (Bradley and Finkbeiner, 2002).

The neurotrophins NT3/4 and BDNF are well known neurotrophic factors and their receptors are expressed in the developing cerebellum of humans

(Quartu et al., 2003) and rodents (Gao et al., 1995). In the rodent cerebellum neurotrophins play a critical role in the differentiation and survival of developing basket and stellate cells (Fiszman et al., 2005; Spatkowski and Schilling, 2003) and in granule cell survival (Gao et al., 1995; Shalizi et al., 2003).

Ionotropic GABA and glutamate receptors are expressed at relatively early stages of brain development (Laurie et al., 1992b; Monyer et al., 1994) and functional voltage-gated calcium channels (VGCC) are expressed even in neuro-blasts (Fiszman et al., 1990, 1993). Studies performed in different regions of the developing central nervous system (CNS) of the rodent reported that GABA, the main inhibitory neurotransmitter in the adult CNS, triggers depolarizing re-sponses through $GABA_A$ receptor activation (Ben-Ari, 2002) before the appear-ance of glutamatergic neurotransmission takes place (Ben-Ari et al., 1997). During the development of the cerebellum, cell migration requires fluctuations of intra-cellular calcium triggered by the activation of VGCC or NMDA receptors (Komuro and Rakic, 1992, 1993). In the embryonic cortex, the early presence of GABA and $GABA_A$ receptors in the marginal zone and subplate (Meinecke and Rakic, 1992) may mediate GABA-induced chemotaxis and chemokinesis. These $GABA_A$-mediated effects are calcium-dependent (Behar et al., 1996).

In hippocampal neurons in culture, exposure to glutamate increases calcium and neurite outgrowth (Wilson et al., 2000) and a glutamate-induced regulation of dendritic growth mediated by AMPA/Kainate receptors was described in motor neurons (Metzger et al., 1998). In addition, VGCC activation increases neurite outgrowth in primary cultured sympathetic (Vaillant et al., 2002), cerebellar granule (Borodinsky et al., 2002), and cortical neurons (Redmond et al., 2002).

Depolarizing responses can increase neuronal survival in the absence of added trophic factors (Chalazonitis and Fischbach, 1980; Hansen et al., 2001; Wakade et al., 1983). Activation of sodium channels promotes survival of substantia nigra neurons (Franklin et al., 1995) and sympathetic and spiral ganglion neurons respond positively to the survival promoting effect of high potassium, being in all cases due to an increase in intracellular calcium (Hansen et al., 2001; Vaillant et al., 1999). Furthermore, depolarization may promote survival by stimulating the synthesis and release of BDNF, as has been demonstrated in developing cortical neurons treated with high potassium (Ghosh et al., 1994) and a cAMP-dependent pathway mediates the survival-promoting effects of depolarization in retinal ganglion cells (Meyer-Franke et al., 1995). On the contrary, the survival-promoting effect of high potassium on cerebellar granule neurons is not due to an increased availability of BDNF (Armanino et al., 2005).

A great amount of evidence points to the fact that that survival of cerebellar cells is improved in depolarizing culture conditions mimicked by exposure to 25 mM KCl or to 50–100 micromolar concentrations of NMDA (Armanino et al., 2005; Balázs et al., 1989; Borodinsky et al., 2002; Fiszman et al., 2005; Gallo et al., 1987). The effects of NMDA and high potassium are blocked by

CaMKII blockers (Borodinsky *et al.*, 2002; Hack *et al.*, 1993) and VGCC blockers (Borodinsky *et al.*, 2002; Gallo *et al.*, 1987). Treatment of the cultures with 50 μM kainate during the first week *in vitro* leads to an increased survival of cerebellar granule cells and VGCC are involved (Balázs *et al.*, 1990a,b). However, in the presence of the same kainate concentration stellate, basket, and Golgi cells are eliminated (Damgaard *et al.*, 1996; Drejer and Schousboe, 1989), suggesting that different mechanisms are involved in the survival of granule cells and inhibitory interneurons.

VI. Establishing an *In Vitro* Model to Study the Developing Cerebellar Granule Neurons

Since granule cells constitute the vast majority of cells of the cerebellum and their mitosis takes place later than the other phenotypes, it is easy to establish an almost pure primary granule cells culture (Messer, 1977; Schousboe *et al.*, 1989). The standard procedure to culture cerebellar granule cells is to obtain and plate a cell suspension from 7- to 8-day-old mice or rats. Microarray analyses of cultured granule neurons and developing cerebellar tissue revealed that the time-course of gene expression in cultured granule neurons resembles that observed *in vivo*, confirming that these cells are a good model for the study of developmentally regulated events that take place in the intact cerebellum (Diaz *et al.*, 2002).

Taking advantage of cerebellar granule cell cultures we studied the role of neural activity on the development of cerebellar granule neurons. We focused on the neurogenesis and differentiation of cerebellar granule neurons. In addition, we were interested in the mechanism by which depolarizing responses affect the survival of early postmitotic neurons, since it is well known that these cells depend on depolarization for their survival (Gallo *et al.*, 1987). We studied the characteristics of cerebellar granule cells at different times *in vitro* and, starting from the first hours after plating, determined the time course of the proliferation of these cells by carrying out [^{3}H]-thymidine incorporation assays. We measured the complexity of the dendritic morphology by staining early postmitotic neurons with tetanus toxin fragment C labeling (a postmitotic neuronal marker, Neale *et al.*, 1993). We quantified the fractal dimension in the cells labeled with Fragment C, grown at very low density (Borodinsky and Fiszman, 2001). The long term survival of these cells was determined by measuring the number of cells after one week *in vitro* (DIV7).

During the first 24 hours *in vitro*, immature cerebellar granule cells looked rounded with very few neurites and were not stained with tetanus toxin fragment C (Borodinsky and Fiszman, 1998). Measuring [^{3}H]-thymidine incorporation in these cells revealed that they were capable of proliferating during the first

24 hours after plating (DIV1) and to a lesser extent at DIV2, reaching almost blank levels at 48 hours after plating. At the same time that a decrease in proliferation was seen (DIV2), the majority of cells became positive for tetanus toxin fragment C and acquired morphological features of differentiated pheno-types (Borodinsky and Fiszman, 1998). We therefore used the following schedule; we explored the development of cerebellar granule cells by determining $[^3H]$-thymidine incorporation at DIV1, and differentiation was determined at DIV3. Furthermore, we determined long-term survival of these cells at DIV7. $GABA_A$ agonists, high potassium, and specific blockers were added two hours after plating for the proliferation assays and at 48 hours after plating for differentiation and survival assays. GABA and muscimol (a selective $GABA_A$ agonist) effects were compared to those seen with depolarizing potassium concentrations (25 mM KCl).

VII. GABA-Mediated Depolarization and Calcium Increases

GABA, the main inhibitory neurotransmitter in the adult CNS, is excitatory during development (Ben-Ari et al., 1989; Fiszman et al., 1990, 1993; Mandler et al., 1990; Obata et al., 1978). Due to the depolarizing effect described for GABA and because $GABA_A$ receptors are already present prenatally, before synaptogenesis takes place (Laurie et al., 1992b), a paracrine trophic role of GABA during embryogenesis and early postnatal life was postulated (see reviews by Cherubini et al., 1991; Nguyen et al., 2001; Ben-Ari, 2002; Fiszman and Schousboe, 2004). Depolarization of the cell membrane by GABA in developing neurons appears to be mediated by a reversal of the chloride gradient prior to maturation (Cherubini et al., 1991; Mueller et al., 1984; Owens et al., 1996; Rohrbough et al., 1996). As a consequence of GABA-mediated depolarization, an activation of VGCC takes place with the concomitant intracellular calcium increases, as reported in the hippocampus (Gaiarsa et al., 1995), cerebral cortical neurons (Yuste et al., 1991), postnatal cerebellar granule cells explants (Connor et al., 1987), hypothalamus (Obrietan and van den Pol, 1995), and oligodendro-cytes (Kirchhoff and Kettenmann, 1992). The physiological timing switch of GABA from excitatory to inhibitory varies among brain regions, but in the rodent, it is generally complete after the second week of life (Ben-Ari, 2002; Cherubini et al., 1991). The GABA switch is accomplished in cultured cerebellar granule neurons taken from 6–8-day-old rodents and grown for seven days in appropriate culture conditions (Borodinsky et al., 2003).

In immature cerebellar neurons GABA is capable of increasing intracellular calcium at very low concentrations starting at 0.1 μM with maximal effective concentrations of 100 μM (Borodinsky et al., 2003). The effect of GABA is blocked by nifedipine, an L-subtype VGCC blocker, and by bicuculline and picrotoxin (a $GABA_A$ receptor blocker and a $GABA_A$ chloride channel blocker, respectively), confirming that the effect is mediated by $GABA_A$ receptors, chloride conductances increases, and activation of VGCC (Borodinsky et al., 2003). During the first days in vitro (DIV1-2) depolarizing potassium concentrations increase calcium and the effect is blocked by nifedipine. However, GABA was less effective in triggering calcium increases than 25 mM KCl (Borodinsky et al., 2002, 2003). This is possibly due to the fact that 25 mM KCl activates calcium channels whereas GABA effect is subject to other regulatory factors that may affect $GABA_A$-mediated responses.

The mechanism that explains GABA acting as an excitatory neurotransmitter in early development is an outward chloride gradient, resulting from a higher intracellular concentration of chloride that provides a depolarized chloride equilibrium potential (Rohrbough and Spitzer, 1996). The transporter that mediates the intracellular accumulation of chloride is the Na^{2+}-K^+-2Cl− co-transporter (NKCC1), which is highly expressed in the developing brain (Li et al., 2002; Plotkin et al., 1997). It is generally accepted that the outward chloride gradient in immature neurons is a consequence of an active chloride accumulation via NKCC1 (Ben-Ari, 2002). In more mature neurons, a mechanism that lowers the chloride concentration by extruding chloride, the exporter K^+/Cl^- co-transporter (KCC2), is also expressed. It has been demonstrated that neuronal maturation is concomitant with KCC2 expression (Rivera et al., 1999).

What are the factors that determine the switch from depolarizing to hyperpolarizing GABA? In young cultured hippocampal neurons it has been demonstrated that GABA itself, acting as a depolarizing agent, promotes the switch from depolarizing to hyperpolarizing responses, with a concomitant up-regulation of KCC2 (Ganguly et al., 2001). According to the authors, the effect is specific for GABA, since glutamate blockers failed to prevent it.

In cerebellar granule neurons cultured for 1 to 8 days in 5 mM KCl the exogenous application of 100 μM GABA induced an increase in intracellular calcium concentration with 70% of the cells being responsive (Borodinsky et al., 2003). This differs from cells plated in 25 mM KCl, where the increase in calcium induced by GABA was observed in cells from 1 to 4 DIV but not in cells cultured for 5 days or longer. At DIV5 only 20% of the cells responded to GABA with an increase in calcium. The ionic mechanism that underlies the change in the response to GABA was studied performing perforated patch recordings using amphotericin B in the recording pipette to allow access to the cell without altering its chloride and other intracellular ion concentrations. We recorded from granule neurons as early as 24 h after plating. The resting membrane potential of cells at 1–3 DIV measured in current-clamp perforated patch recordings did not differ in 5 and 25 mM KCl containing cultures. GABA applied to both neurons grown in 5 and 25 mM KCl produced a depolarization. We then compared the response induced by 100 μM GABA in voltage-clamped cerebellar granule cells at various holding potentials to determine the reversal potential (Erev). When neurons were grown in 25 mM KCl, currents evoked by GABA application to cells' voltage clamped at -45 mV were depolarizing at DIV1 and hyperpolarizing at DIV8. In cells exposed to 25 mM KCl at 1–3 DIV the Erev of GABA currents was 36 ± 1 mV. In contrast, at DIV8-9 the Erev was 47 ± 2 mV. On the other hand, when neurons were grown in 5 mM KCl for DIV6–8, currents evoked by GABA application were still depolarizing and the Erev remained unchanged from DIV1 to 8 (Borodinsky et al., 2003).

These experiments suggest that the switch of GABA responses can be induced in developing cerebellar granule cells by depolarizing potassium concentrations (Borodinsky et al., 2003), a culture condition known to increase NMDA receptor activation (Armanino et al., 2005; Fiszman et al., 2005) and VGCC activation (Borodinsky et al., 2002; Gallo et al., 1987). Furthermore, it was reported that neurotrophic factors can up-regulate KCC2 since overexpression of brain-derived neurotrophic factor (BDNF) in immature hippocampal neurons raises spontaneous activity and synaptogenesis to postnatal levels and up-regulates KCC2 (Aguado et al., 2003). Taken together, these reports suggest that the GABA switch is a multi-factorial phenomenon triggered by different conditions that promote neuronal maturation and differentiation and in the cerebellum, in particular, depolarizing conditions are one of the main regulatory factors.

VIII. Intracellular Pathways Activated Downstream GABA$_A$ Receptors

Intracellular calcium increases activate calmodulin and other related kinases, including CaMKII, which is involved mainly in neuronal plasticity (Cline, 2001). CaMK IV, which plays a crucial role in Ca^{2+}-dependent activation of CREB-dependent transcription (Bradley and Finkbeiner, 2002; Curtis and Finkbeiner, 1999) is also activated and is involved in neurite outgrowth (Redmond et al., 2002). Other kinases modulated by calcium influx such as p42/44 MAPK (erk1/2), p38-MAPK, and PI3-kinase (Dolmescht et al., 2001; Ghosh and Greenberg, 1995; Rosen and Greenberg, 1996; Rosen et al., 1994) are also involved in neurite outgrowth and plasticity (Borodinsky et al., 2002; Opazo et al., 2003; Vaillant et al., 2002).

In the developing cerebellum, GABA increases the phosphorylated forms of erk 1/2 and CaMKII as effectively as 25 mM KCl (Borodisnky et al., 2003). Therefore, GABA is capable of activating calcium-dependent and neurotrophin-related cascades. In cultured hippocampal neurons GABA can activate BDNF expression and c-fos expression through a calcium-dependent mechanism (Berninger et al., 1995) and regulate phenotypic expression of hippocampal interneurons through an increase in BDNF levels (Marty *et al.*, 1996). The effect on BDNF expression was reported to be accomplished via a MAPK-CREB-dependent mechanism in hypothalamic neurons (Obrietan *et al.*, 2002). Therefore, GABA may act directly by increasing intracellular calcium and/or indirectly by increasing availability of BDNF.

We conclude that GABA, acting as a depolarizing agent, activates intracellular cascades that mediate trophic effects similar to neurotrophins. We next explore the effects of GABA on proliferation, differentiation, and survival of cerebellar neurons.

A. GABA and Proliferation

Depolarization affects cell cycle progression by increasing proliferation in immature central and sympathetic autonomic neuroblasts (Cone, 1980; DiCicco-Bloom and Black, 1989). Amino acid neurotransmitters affect proliferation of immature neuroblasts in different directions. For example, in the embryonic neocortex, exposure to GABA, high potassium, and glutamate decrease proliferation of immature neurons (Lo Turco *et al.*, 1995). In cultured cortical progenitors, GABA$_A$ receptor activation decreases the proliferative effect induced by bFGF (Antonopoulos *et al.*, 1997). Cell cycle progression of cortical progenitors is differently affected by GABA and glutamate in ventricular and subventricular zones of the developing cortex (Haydar *et al.*, 2000). It is interesting to note that even high levels of glutamate receptor agonists are not toxic for the subventricular zone-derived neural stem/progenitor cells *in vitro* and activation of kainate receptors in a perinatal model of ischemia is trophic instead of toxic (Brazel *et al.*, 2005).

We explored the effect of GABA on immature cerebellar granule cells' proliferation. We found that GABA at micromolar concentrations induced an increase in cell proliferation that was mimicked by muscimol (Fiszman *et al.*, 1999), a selective GABA$_A$ agonist. The proliferative effect induced by GABA is mediated by GABA$_A$ receptors and chloride channels' opening since it was completely blocked by bicuculline and picrotoxin. The effect was also blocked by nifedipine and $MgCl_2$ but not by NMDA receptors blockers, suggesting that it is a calcium-dependent effect and voltage-gated calcium channel activation is required. An intracellular calcium increase triggered a downstream activation of the MAPK cascade, and the proliferative effect induced by GABA was completely blocked by PD 98059, a specific blocker of MAPKK (Fiszman *et al.*, 1999), confirming that Ras/MAPK (erk1/2) is involved in the proliferative effect induced by GABA. Moreover, VGCC activation after exposure to high potassium in the same cell preparation increases cell proliferation that is also blocked by PD 98059 (Borodinsky and Fiszman, 1998).

B. GABA and Differentiation

Cells can often generate calcium transients at very early developmental stages and excitability plays a crucial role in the differentiation of developing neurons (Spitzer et al., 2002). Early expressed glutamate receptors regulate neurite outgrowth in hippocampal neurons and embryonic rat motoneurons (Mattson et al., 1988; Metzger et al., 1998; Wilson et al., 2000) and VGCC activation increases neurite outgrowth in central and peripheral neurons (Borodinsky et al., 2002; Redmond et al., 2002; Vaillant et al., 2002). In differentiating neurospheres, intracellular calcium increases correlate positively with the appearance of the GABAergic neurotransmitter phenotype and neurite outgrowth (Ciccolini et al., 2003) and VGCC activation increase calbindin positive hippocampal pyramidal neurons (Boukhaddaou et al., 2000). Intracellular calcium oscillations modulate axon guidance and pathfinding (Gomez and Spitzer, 1999). An increased association of MAP2 with microtubules concomitant to neurite outgrowth was described in cultured sympathetic neurons treated with high potassium (Vaillant et al., 2002). Furthermore, spine outgrowth from dendrites following NMDAR activation is commonly accompanied by motile activity of the nascent spine and rapid changes in actin dynamics (Fischer et al., 1998; Maletic-Savatic et al., 1999).

There are early observations that describe GABA as a morphogen. Several years ago, many morphological changes were described after GABA exposure. For example, in vivo application of GABA in the superior cervical ganglion of the rat triggers neurite outgrowth (Wolff et al., 1978). GABA applied in vitro induces ultrastructural changes in cultured cerebellar granule cells (Hansen et al., 1984). GABA promotes neurite outgrowth in cultured hippocampal neurons (Barbin et al., 1993) and the effect is mediated by BDNF increases (Marty et al., 1996).

Since GABA activates VGCC and triggers calcium in developing cerebellar granule cells we tested the effect of GABA on the differentiation of these cells. We carried out studies in young post-mitotic cerebellar granule neurons and found that the exogenous application of GABA increases the levels of the phosphorylated forms of CaMKII and erk 1/2, at the same time increasing neurite complexity. The increase in neurite complexity is due to VGCC activation of the L-subtype similar to the effect of high potassium in the same preparation (Borodinsky et al., 2002, 2003). However, while the GABA-induced neuritogenic effect requires the activation of CaMKII and MEK1 (the kinase that activates erk 1/2), high potassium induces an increase in neuronal complexity that was mediated by CaMKII but not by MEK1 activation (Borodinsky et al., 2002, 2003). It is possible that a differential activation of downstream effectors may take place, depending on the magnitude of Ca^{2+} spikes triggered by each stimulus (Dudek and Fields, 2001). Activation of these kinases and neurite outgrowth increases were also described in rat sympathetic neurons exposed to high potassium (Vaillant et al., 2002).

C. GABA and Survival

There are several reasons to believe that GABA may increase neuronal survival. Firstly, depolarizing culture conditions promote and chronic blockade of sodium channels by TTX decreased survival of neurons *in vitro* and *in vivo* (Catsicas *et al.*, 1992; Ruitjer *et al.*, 1991). In the case of the cerebellum, activity plays a crucial role in the survival of neurons rescuing them from death induced by trophic factor deprivation (Gallo *et al.*, 1987). Secondly, GABA increases BDNF expression in developing hippocampal interneurons (Marty *et al.*, 1996) through a calcium-dependent mechanism, (Berninger *et al.*, 1995) suggesting an indirect action of GABA through neurotrophins. We expected that GABA and muscimol would increase cerebellar granule cells' survival in our experiments. However GABA$_A$ agonists failed to increase survival of cultured cerebellar granule neurons (Fiszman *et al.*, 1999). The explanation for this controversy may lie in how and when the GABA switch from depolarizing to hyperpolarizing takes place in our experimental conditions. As mentioned before, this switch is a consequence of maturation processes like neural activity and BDNF-dependent mechanisms that promote differentiation and up regulate KCC2. In the cerebellar granule neurons, activation of VGCC by GABA may induce the GABA switch as we described for cells cultured in high potassium (Borodinsky *et al.*, 2003). It is interesting to note that this high potassium culture condition is effective in promoting survival of cerebellar granule neurons by activating VGCC, (Borodinsky *et al.*, 2002) but, in contrast with previous observations in cortical neurons (Ghosh *et al.*, 1994), the effect is exerted in a BDNF-independent manner (Armanino *et al.*, 2005), suggesting that GABA has no effect on BDNF expression in cerebellar neurons. We postulate that GABA, through VGCC activation, promotes the maturation of the cerebellar granule neurons and induces the GABA switch. Therefore, GABA fails to increase survival because it becomes hyperpolarizing while cerebellar granule cells need a continuous depolarization to keep the majority of cerebellar granule cells alive.

References

Aguado, F., Carmona, M. A., Pozas, E., Aguiló, A., Martínez-Guijarro, F. J., Alcantara, S., Borrell, V., Yuste, R., Ibañez, C. F., and Soriano, E. (2003). BDNF regulates spontaneous correlated activity at early developmental stages by increasing synaptogenesis and expression of the K+/Cl⁻ co-transporter KCC2. *Development* **130,** 1267–1280.
Antonopoulos, J., Pappas, I., and Parnavelas, J. (1997). Activation of the GABA$_A$ receptor inhibits the proliferative effects of bFGF in cortical progenitor cells. *Eur. J. Neurosci.* **9,** 291–298.

Armanino, M., Gravielle, M. C., and Fiszman, M. L. (2005). NMDA receptors contribute to the survival promoting effect of high potassium in cultured cerebellar granule cells. *Int. J. Dev. Neurosci.* **23,** 545–548.

Balázs, R., Hack, N., Jørgensen, O. S., and Cotman, C. W. (1989). N-methyl-D-aspartate promotes the survival of cerebellar granule cells: Pharmacological characterization. *Neurosci. Lett.* **101,** 241–246.

Balázs, R., Hack, N., and Jørgensen, O. S. (1990a). Selective stimulation of excitatory aminoacid receptor subtypes and the survival of cerebellar granule cells in culture: Effect of kainic acid. *Neuroscience* **37,** 251–258.

Balázs, R., Hack, N., and Jørgensen, O. S. (1990b). Interactive effects involving different classes of excitatory amino-acid receptors and the survival of cerebellar granule cells in culture. *Int. J. Dev. Neurosci.* **8,** 347–359.

Barbin, G., Pollard, H., Gaiarsa, J. L., and Ben-Ari, Y. (1993). Involvement of GABA$_A$ receptors in the outgrowth of cultured hippocampal neurons. *Neurosci. Lett.* **152,** 150–154.

Behar, T., Li, Y., Tran, H. T., Ma, W., Dunlap, V., Scott, C., and Barker, J. L. (1996). GABA stimulates chemotaxis and chemokinesis of embryonic cortical neurons via calcium-dependent mechanisms. *J. Neurosci.* **16,** 1808–1818.

Ben-Ari, Y., Cherubini, E., Corradetti, R., and Gaiarsa, J. L. (1989). Giant synaptic potentials in immature rat CA3 hippocampal neurons. *J. Physiol. (Lond).* **416,** 303–325.

Ben-Ari, Y., Khazipov, R., Leinekugel, X., Callard, O., and Gaiarsa, J. (1997). GABA, NMDA and AMPA receptors: A developmentally regulated 'ménage à trois.' *TINS* **20,** 523–529.

Ben-Ari, Y. (2002). Excitatory actions of GABA during development: The nature of the nurture. *Nature Rev. Neurosci.* **3,** 728–739.

Berninger, B., Marty, S., Zafra, F., Berzaghi, M., Thoenen, H., and Lindholm, D. (1995). GABAergic stimulation switches from enhancing to repressing BDNF expression in rat hippocampal neurons during maturation *in vitro. Development* **121,** 2327–2335.

Borodinsky, L. N., and Fiszman, M. L. (1998). Extracellular potassium concentration regulates proliferation of immature cerebellar granule cells. *Dev. Brain Res.* **107,** 43–48.

Borodinsky, L. N., and Fiszman, M. L. (2001). A single-cell model to study changes in neuronal fractal dimension. *Methods* **24,** 341–345.

Borodinsky, L. N., Coso, O. A., and Fiszman, M. L. (2002). Contribution of Ca^{2+} calmodulin-dependent protein kinase II and mitogen-activated protein kinase kinase to neural activity-induced neurite outgrowth and survival of cerebellar granule cells. *J. Neurochem.* **80,** 1062–1070.

Borodinsky, L. N., O'Leary, D., Neale, J. H., Vicini, S., Coso, O. A., and Fiszman, M. L. (2003). GABA-induced neurite outgrowth of cerebellar granule cells is mediated by GABA$_A$ receptor activation, calcium influx and CaMKII and erk 1/2 pathways. *J. Neurochem.* **84,** 1141–1420.

Boyden, E. S., Katoh, A., and Raymond, J. L. (2004). Cerebellum-dependent learning: The role of multiple plasticity mechanisms. *Annu. Rev. Neurosci.* **27,** 581–609.

Bradley, J., and Finkbeiner, S. (2002). An evaluation of specificity in activity-dependent gene expression in neurons. *Prog. Neurobiol.* **67,** 469–477.

Brazel, C. Y., Nuñez, J. L., Yang, Z., and Levison, S. W. (2005). Glutamate enhances the survival and proliferation of neural progenitors derived from the subventricular zone. *Neuroscience* **131,** 55–65.

Carlson, B. X., Elster, L., and Schousboe, A. (1998). Pharmacological and functional implications of developmentally-regulated changes in GABA$_A$ receptor subunit expression in the cerebellum. *Eur. J. Pharmacol.* **352,** 1–14.

Catsicas, M., Péquinot, Y., and Clarke, P. G. H. (1992). Rapid onset of neuronal death induced by blockade of either axoplasmic transport or action potentials in afferent fibers during brain development. *J. Neurosci.* **12,** 4642–4650.

Chalazonitis, A., and Fischbach, G. D. (1980). Elevated potassium induces morphological differentiation of dorsal root ganglionic neurons in dissociated cell culture. *Dev. Biol.* **78,** 172–183.

Chao, M. V. (1992). Neurotrophin receptors: A window into neuronal differentiation. *Neuron* **9**, 583–593.

Cherubini, E., Gaiarsa, J. L., and Ben-Ari, Y. (1991). GABA an excitatory neurotransmitter in early postnatal life. *TINS* **14**, 515–519.

Ciccolini, F., Collins, T. J., Sudhoelter, J., Lipp, P., Berridge, M. J., and Bootman, M. D. (2003). Local and global spontaneous calcium events regulate neurite outgrowth and onset of GABAergic phenotype during neural precursor differentiation. *J. Neurosci.* **23**, 103–111.

Cline, H. T. (2001). Dendritic arbor development and synaptogenesis. *Curr. Opin. Neurobiol.* **11**, 118–126.

Cone, C. D., Jr. (1980). Ionically mediated induction of mitogenesis in CNS neurons. *Ann. NY Acad. Sci.* **339**, 115–131.

Connor, J. A., Tseng, H., and Hockberger, P. E. (1987). Depolarization and transmitter-induced changes in intracellular Ca^{2+} of rat cerebellar granule cells in explant cultures. *J. Neurosci.* **7**, 1384–1400.

Curtis, J., and Finkbeiner, S. (1999). Sending signals from the synapse to the nucleus: Possible roles for CaMK, Ras/erk, and SAPK pathways in the regulation of synaptic plasticity and neuronal growth. *J. Neurosci. Res.* **58**, 88–95.

Damgaard, I., Trenkner, E., Sturman, J. A., and Schousboe, A. (1996). Effect of K$^+$- and Kainate-mediated depolarization on survival and functional maturation of GABAergic and glutamatergic neurons in cultures of dissociated mouse cerebellum. *Neurochem. Res.* **21**, 267–275.

Diaz, E., Ge, Y., Yang, Y. H., Loh, K. C., Serafini, T. A., Okazaki, Y., Hayashizaki, Y., Speed, T. P., Ngai, J., and Scheiffele, P. (2002). Molecular analysis of gene expression in the developing pontocerebellar projection system. *Neuron* **36**, 417–434.

DiCicco-Bloom, E., and Black, I. B. (1989). Depolarization and insulin-like growth factor-I IGF-I differentially regulate the mitotic cycle in cultured rat sympathetic neuroblasts. *Brain Res.* **491**, 403–406.

Dolmetsch, R. E., Pajvani, U., Fife, K., Spotts, J. M., and Greenberg, M. E. (2001). Signaling to the nucleus by an L-type calcium channel-calmodulin complex through the MAP kinase pathway. *Science* **294**, 333–339.

Drejer, J., and Schousboe, A. (1989). Selection of a pure cerebellar granule cell culture by kainate treatment. *Neurochem. Res.* **14**, 751–754.

Dudek, S., and Fields, R. D. (2001). Mitogen-activated protein kinase/extracellular signal–regulated kinase activation in somatodendritic compartments: Roles of action potentials, frequency, and mode of calcium entry. *J. Neurosci.* **21**, RC122.

Eccles, J. C., Ito, M., and Szentágothai, J. (1967). "The Cerebellum as a Neuronal Machine." Springer, New York.

Fatemi, S. H., Halt, A. R., Realmuto, G., Earle, J., Kist, D. A., Thuras, P., and Merz, A. (2002a). Purkinje cell size is reduced in cerebellum of patients with autism. *Cell Mol. Neurobiol.* **22**, 171–175.

Fatemi, S. H., Halt, R. A., Stary, J. M., Kanodia, R., Schulz, S. C., and Realmuto, G. R. (2002b). Glutamic acid decarboxylase 65 and 67 kDa proteins are reduced in autistic parietal and cerebellar cortices. *Biol. Psychiatry* **52**, 805–810.

Fischer, M., Kaech, S., Knutti, D., and Matus, A. (1998). Rapid actin-based plasticity in dendritic spines. *Neuron* **20**, 847–854.

Fiszman, M. L., and Schousboe, A. (2004). Role of calcium and kinases on the neurotrophic effect induced by γ-aminobutyric acid. *J. Neurosci. Res.* **76**, 435–441.

Fiszman, M. L., Novotny, E. A., Lange, G. D., and Barker, J. L. (1990). Embryonic and early postnatal hippocampal cells respond to nanomolar concentrations of muscimol. *Dev. Brain Res.* **53**, 186–193.

Fiszman, M. L., Behar, T., Lange, G. D., Smith, S. V., Novotny, E. A., and Barker, J. L. (1993). GABAergic cells and signals appear together in the early post-mitotic period of telencephalic and striatal development. *Dev. Brain Res.* **73**, 243–251.

Fiszman, M. L., Borodinsky, L. N., and Neale, J. H. (1999). GABA induces proliferation of immature cerebellar granule cells grown *in vitro*. *Dev. Brain Res.* **115,** 1–8.

Fiszman, M. L., Barberis, A., Lu, C., Fu, Z., Erdélyi, F., Szabó, G., and Vicini, S. (2005). NMDA receptors increase the size of GABAergic terminals and enhance GABA release. *J. Neurosci.* **23,** 2024–2031.

Franklin, J. L., Sanz-Rodriguez, C., Juhasz, A., Deckwerth, T. L., and Johnson, E. M., Jr. (1995). Chronic depolarization prevents programmed death of sympathetic neurons *in vitro* but does not support growth: Requirement for Ca^{2+} influx but not Trk activation. *J. Neurosci.* **15,** 643–664.

Gaiarsa, J. L., McLean, H., Congar, P., Leinekugel, X., Khazipov, R., Tseeb, V, and Ben-Ari, Y. (1995). Postnatal maturation of gamma-aminobutyric acid -mediated inhibition in the CA3 hippocampal region of the rat. *J. Neurobiol.* **26,** 339–349.

Gallo, V., Kingsbury, A., Balázs, R., and Jørgensen, O. S. (1987). The role of depolarization in the survival and differentiation of cerebellar granule cells in culture. *J. Neurosci.* **7,** 2203–2213.

Ganguly, K., Schinder, A. F., Wong, S. T., and Poo, M. (2001). GABA promotes the developmental switch of neuronal GABAergic responses from excitation to inhibition. *Cell* **105,** 521–532.

Gao, W. Q., Zheng, J. L., and Karihaloo, M. (1995). Neurotrophin-4/5 (NT-4/5) and brain-derived neurotrophic factor (BDNF) act at later stages of cerebellar granule cell differentiation. *J. Neurosci.* **15,** 2656–2667.

Ghosh, A., Carnahan, J., and Greenberg, M. E. (1994). Requirement for BDNF in activity-dependent survival of cortical neurons. *Science* **263,** 1618–1623.

Ghosh, A., and Greenberg, M. E. (1995). Calcium signaling in neurons: Molecular mechanisms and cellular consequences. *Science* **268,** 239–247.

Gomez, T. M., and Spitzer, N. C. (1999). *In vivo* regulation of axon extension and pathfinding by growth-cone calcium transients. *Nature* **397,** 350–355.

Hack, N., Hidaka, H., Wakefield, M. J., and Balász, R. (1993). Promotion of granule cell survival by high K^+ or excitatory aminoacid treatment and Ca^{2+}/calmodulin-dependent protein kinase activity. *Neuroscience* **57,** 9–20.

Hansen, G. H., Meier, E., and Schousboe, A. (1984). GABA influences the ultrastructure composition of cerebellar granule cells during development in culture. *Int. J. Dev. Neurosci.* **2,** 247–257.

Hansen, M. R., Zha, X., Bok, J., and Green, S. H. (2001). Multiple distinct signal pathways, including an autocrine neurotrophic mechanism, contribute to the survival-promoting effect of depolarization on spiral ganglion neurons *in vitro*. *J. Neurosci.* **21,** 2256–2267.

Haydar, T. F., Wang, F., Schwartz, M. L., and Rakic, P. (2000). Differential modulation of proliferation in the neocortical ventricular and subventricular zones. *J. Neurosci.* **20,** 5764–5774.

Kaplan, D. R., and Miller, F. D. (2000). Neurotrophin signal transduction in the nervous system. *Curr. Opin. Neurobiol.* **10,** 381–391.

Kirchhoff, F., and Kettenmann, H. (1992). GABA triggers a Ca^{2+} increase in murine precursor cells of the oligodendrocyte lineage. *Eur. J. Neurosci.* **4,** 1049–1058.

Komuro, H., and Rakic, P. (1992). Selective role of N-type calcium channels in neuronal migration. *Science* **257,** 806–809.

Komuro, H., and Rakic, P. (1993). Modulation of neuronal migration by NMDA receptors. *Science* **260,** 95–97.

Komuro, H., and Rakic, P. (1998). Distinct modes of neuronal migration in different domains of developing cerebellar cortex. *J. Neurosci.* **18,** 1478–1490.

Kuhar, S. G., Feng, L., Vidan, S., Ross, M. E., Hatten, M. E., and Heintz, N. (1993). Changing patterns of gene expression define four stages of cerebellar granule neuron differentiation. *Development* **117,** 97–10.

Laurie, D. J., Seeburg, P. H., and Wisden, W. (1992a). The distribution of 13 GABA receptor subunit mRNAs in the rat brain: II. Olfactory bulb and cerebellum. *J. Neurosci.* **12,** 1063–1076.

Laurie, D. J., Wisden, W., and Seeburg, P. H. (1992b). The distribution of thirteen GABA receptor subunit mRNAs in the rat brain: III. Embryonic and postnatal development. *J. Neurosci.* **12,** 4151–4172.

Li, H., Tornberg, J., Kaila, K., Airaksinen, M. S., and Rivera, C. (2002). Patterns of cation-chloride co-transporter expression during embryonic rodent CNS development. *Eur. J. Neurosci.* **16,** 2358–2370.

LoTurco, J. J., Owens, D. F., Heath, M. J. S., Davis, M. B. E., and Kriegstein, A. R. (1995). GABA and glutamate depolarize cortical progenitor cells and inhibit DNA synthesis. *Neuron* **15,** 1287–1298.

Lüscher, B., and Keller, C. A. (2004). Regulation of GABA$_A$ receptor trafficking, channel activity, and functional plasticity of inhibitory synapses. *Pharmacol. Ther.* **102,** 195–221.

Maletic-Savatic, M., Malinow, R., and Svoboda, K. (1999). Rapid dendritic morphogenesis in CA1 hippocampal dendrites induced by synaptic activity. *Science* **283,** 1923–1927.

Mandler, R. N., Schaffner, A. E., Novotny, E. A., Lange, G. D., Smith, S. V., and Barker, J. L. (1990). Electrical and chemical excitability appear one week before birth in the embryonic rat spinal cord. *Brain Res.* **522,** 46–54.

Marty, S., Berninger, B., Carroll, P., and Thoenen, H. (1996). GABAergic stimulation regulates the phenotype of hippocampal interneurons through the regulation of brain-derived neurotrophic factor. *Neuron* **16,** 565–570.

Mattson, M. P., Don, P., and Kater, S. B. (1988). Outgrowth-regulating actions of glutamate in isolated hippocampal pyramidal neurons. *J. Neurosci.* **8,** 2087–2100.

Meinecke, D. L., and Rakic, P. (1992). Expression of GABA and GABA$_A$ receptors by neurons of the subplate zone in developing primate occipital cortex: Evidence for transient local circuits. *J. Comp. Neurol.* **317,** 91–101.

Messer, A. (1977). The maintenance and identification of mouse cerebellar granule cells in monolayer cultures. *Brain Res.* **130,** 1–12.

Metzger, F., Wiese, S., and Sendtner, M. (1998). Effect of glutamate on dendritic growth in embryonic rat motoneurons. *J Neurosci.* **18,** 1735–1742.

Meyer-Franke, A., Kaplan, M. R., Pfrieger, F. W., and Barres, B. A. (1995). Characterization of the signaling interactions that promote the survival and growth of developing retinal ganglion cells in culture. *Neuron* **15,** 805–819.

Monyer, H., Burnashev, N., Laurie, D. J., Sakmann, B., and Seeburg, P. H. (1994). Developmental and regional expression in the rat brain and functional properties of four NMDA receptors. *Neuron* **12,** 529–540.

Mueller, A. L., Taube, J. S., and Schwartzkroin, P. A. (1984). Development of hyperpolarizing inhibitory postsynaptic potentials and hyperpolarizing response to γ-aminobutyric acid in rabbit hippocampus studied *in vitro. J. Neurosci.* **4,** 860–867.

Neale, E. A., Bowers, L. M., and Smith, T. G. (1993). Early dendrite development in spinal cord cell cultures: A quantitative study. *J. Neurosci. Res.* **34,** 54–66.

Nguyen, L., Rigo, J., Rocher, V., Belachew, S., Malgrange, B., Rogister, B., Leprince, P, and Moonen, G. (2001). Neurotransmitters as early signals for central nervous system development. *Cell Tissue Res.* **305,** 187–202.

Obata, K., Oide, M., and Tanaka, H. (1978). Excitatory and inhibitory actions of GABA and glycine on embryonic chick spinal neurons in culture. *Brain Res.* **144,** 179–184.

Obrietan, K., and van den Pol, A. N. (1995). GABA neurotransmission in the hypothalamus: Developmental reversal from Ca^{2+} elevating to depressing. *J. Neurosci.* **15,** 5065–5077.

Obrietan, K., Gao, X., and van-den Pol, A. N. (2002). Excitatory actions of GABA increase BDNF expression via a MAPK-CREB-dependent mechanism. A positive feedback circuit in developing neurons. *J. Neurophysiol.* **88,** 1005–1015.

Opazo, P., Watabe, A. M., Grant, S. G. N., and O' Dell, T. J. (2003). Phosphatidylinositol 3-Kinase regulates the induction of long-term potentiation through extracellular signal-related kinase-independent mechanisms. *J. Neurosci.* **23,** 3679–3688.

Ortinski, P. I., Lu, C., Takagaki, K., Fu, Z., and Vicini, S. (2004). Expression of distinct α subunits of GABA$_A$ receptor regulates inhibitory synaptic strength. *J. Neurophysiol.* **92,** 1718–1727.

Owens, D. F., Boyce, L. H., Davis, M. B. E., and Kriegstein, A. R. (1996). Excitatory GABA responses in embryonic and neonatal cortical slices demonstrated by gramicidin perforated-patch recordings and calcium imaging. *J. Neurosci.* **16,** 6414–6423.

Palay, S. L., and Chan-Palay, V. (1974). "Cerebellar Cortex, Cytology and Organization." Springer, Berlin.

Plotkin, M. D., Snyder, E. Y., Hebert, S. C., and Delpire, E. (1997). Expression of the Na-K-2Cl cotransporter is developmentally regulated in postnatal rat brains: A possible mechanism underlying GABA's excitatory role in immature brain. *J. Neurobiol.* **33,** 781–795.

Pouzat, C., and Hestrin, S. (1997). Developmental regulation of basket/stellate cell–Purkinje cell synapses in the cerebellum. *J. Neurosci.* **17,** 9104–9112.

Quartu, M., Serra, M. P., Manca, A., Follesa, P., Ambu, R., and Del Fiaco, M. (2003). High affinity neurotrophin receptors in the human pre-term newborn, infant, and adult cerebellum. *Inter. J. Dev. Neurosci.* **21,** 309–320.

Redmond, L., Kashani, A. H., and Ghosh, A. (2002). Calcium regulation of dendritic growth via CaMKIV and CREB-mediated transcription. *Neuron* **34,** 999–1010.

Ritvo, E. R., Freeman, B. J., Scheibel, A. B., Duong, T., Robinson, H., Guthrie, D., and Ritvo, A. (1986). Lower Purkinje cell counts in the cerebella of four autistic subjects: Initial findings of the UCLA-NSAC autopsy research report. *Am. J. Psychiatry* **143,** 862–866.

Rivera, C., Voipio, J., Payne, J. A., Ruusuvuori, E., Lahtinen, H., Lamsa, K., Pirvola, U., Saarma, M., and Kaila, K. (1999). The K$^+$/Cl$^-$ co-transporter KCC2 renders GABA hyperpolarizing during neuronal maturation. *Nature* **397,** 251–255.

Rohrbough, J., and Spitzer, N. C. (1996). Regulation of intracellular Cl levels by Na^{2+}-dependent Cl$^-$ cotransport distinguishes depolarizing from hyperpolarizing GABA receptor-mediated responses in spinal neurons. *J. Neurosci.* **16,** 82–91.

Rosen, L. B., and Greenberg, M. E. (1996). Stimulation of growth factor receptor signal transduction of voltage-sensitive calcium channels. *Proc. Natl. Acad. Sci. USA* **93,** 1113–1118.

Rosen, L. B., Ginty, D.D, Weber, M. J., and Greenberg, M. E. (1994). Membrane depolarization and calcium influx stimulate MEK and MAP kinase via activation of Ras. *Neuron* **12,** 1207–1221.

Ruitjer, J. M., Baker, R. E., De Jong, B. M., and Romijn, H. J. (1991). Chronic blockade of bioelectric activity in neonatal rat cortex grown *in vitro*. Morphological effects. *Int. J. Dev. Neurosci.* **9,** 331–338.

Spatkowski, G., and Schilling, K. (2003). Postnatal dendritic morphogenesis of cerebellar basket and stellate cells *in vitro*. *J. Neurosci. Res.* **72,** 317–326.

Schousboe, A., Meier, E., Drejer, J., and Hertz, L. (1989). Preparation of primary cultures of mouse rat cerebellar granule cells. *In* "A Dissection and Tissue Culture Manual of the Nervous System." (Shahar, A., De Vellis, J., Vernadakis, A., and Haber, B., Eds.), pp. 203–206. Alan R. Liss, New York.

Shalizi, A., Lehtinen, M., Gaudilliére, B., Donovan, N., Han, J., Yoshiyuki Konishi, Y., and Bonni, A. (2003). Characterization of a neurotrophin signaling mechanism that mediates neuron survival in a temporally specific pattern. *J. Neurosci.* **23,** 7326–7336.

Sieghart, W., Fuchs, K., Tretter, V., Ebert, V., Jechlinger, M., Hoger, H., and Adamiker, D. (1999). Structure and subunit composition of GABA(A) receptors. *Neurochem. Int.* **34,** 379–385.

Spitzer, N. C., Kingston, P. A., Manning, T. J., Jr., and Conklin, M. W. (2002). Outside and in: Development of neuronal excitability. *Curr. Opin. Neurobiol.* **12,** 315–323.

Thompson, C. L., and Stephenson, F. A. (1994). GABA$_A$ receptor subtypes expressed in cerebellar granule cells: A developmental study. *J. Neurochem.* **62,** 2037–2044.

Vaillant, A. R., Mazzoni, I., Tudan, C., Boudreau, M., Kaplan, D. R., and Miller, F. D. (1999). Depolarization and neurotrophins converge on the phosphatidylinositol-3-kinase-Akt pathway to synergistically regulate neuronal survival. *J. Cell Biol.* **146,** 955–966.

Vaillant, A. R., Zanassi, P., Walsh, G. S., Aumont, A., Alonso, A., and Miller, F. D. (2002). Signaling mechanisms underlying reversible, activity-dependent dendrite formation. *Neuron* **34,** 985–998.

Vicini, S., Ferguson, C., Prybylowski, K., Kralic, J., Morrow, A. L., and Homanics, G. E. (2001). GABA$_A$ receptor alpha1 subunit deletion prevents developmental changes of inhibitory synaptic currents in cerebellar neurons. *J. Neurosci.* **21,** 3009–3016.

Wakade, A. R., Edgar, D., and Thoenen, H. (1983). Both nerve growth factor and high K+ concentrations support the survival of chick embryo sympathetic neurons. Evidence for a common mechanism of action. *Exp. Cell. Res.* **144,** 377–384.

Wilson, T. M., Kisaalita, W. S., and Keith, C. H. (2000). Glutamate-induced changes in the pattern of hippocampal dendrite outgrowth: A role for calcium-dependent pathways and the microtubule cytoskeleton. *J. Neurobiol.* **43,** 159–172.

Wolff, J. R., Joo, F., and Dames, W. (1978). Plasticity in dendrites shown by continuous GABA administration in superior cervical ganglion of adult rat. *Nature* **274,** 72–74.

Yuste, R. L., and Katz, C. (1991). Control of postsynaptic Ca^{2+} influx in developing neocortex by excitatory and inhibitory neurotransmitters. *Neuron* **6,** 333–344.

ROLE OF GABA IN THE MECHANISM OF THE ONSET OF PUBERTY IN NON-HUMAN PRIMATES

Ei Terasawa

Department of Pediatrics and Wisconsin National Primate Research Center
University of Wisconsin, Madison, Wisconsin 53715, USA

Evidence indicates that GABA is an inhibitory neurotransmitter responsible for restricting luteinizing hormone-releasing hormone (LHRH) release before the onset of puberty. LHRH neurons in the hypothalamus of female rhesus monkeys are already active during the neonatal period, but subsequently enter a dormant state in the juvenile/prepubertal period because of an elevated level of GABA in the stalk-median eminence (S-ME). The developmental reduction in tonic GABA inhibition results in an increase in LHRH release in the S-ME, triggering puberty. The reduction in GABA also appears to allow an increase in glutamate release in the S-ME and this glutamate seems to further contribute to the pubertal increase in LHRH release. These observations conducted in non-human primates, as a model for humans, provide some insights into future studies of the importance of GABAergic mechanisms in the relation between onset of puberty and neurodevelopmental disorders including autism.

I. Introduction

There are several neurological and psychiatric diseases associated with puberty and changes in GABAergic function may, at least in part, be responsible for underlying etiology. First, recent discoveries by several laboratories indicate

INTERNATIONAL REVIEW OF
NEUROBIOLOGY, VOL. 71
DOI: 10.1016/S0074-7742(05)71005-9

that an abnormality of the GABAergic neuronal system may be a cause of autism or responsible for some symptoms of autism (Cohen *et al.*, 2002; Dhossche, 2002; Hussman, 2001; Menold *et al.*, 2001; Muhle *et al.*, 2004; Nurmi *et al.*, 2003; Prosser *et al.*, 1997). Moreover, in some patients, a worsening of autistic behavior is observed in association with puberty (Gillberg, 1984; Mouridsen *et al.*, 1999). Second, it has been well documented that the onset of schizophrenia occurs between late puberty and early young adulthood (Lewis, 1997; Lewis *et al.*, 2005). A significant decrease in the GABA transporter, GAT-1, immuno-reactivity on axonal terminals of a subset of GABA neurons that innervate pyramidal cells in the frontal cortex is observed in patients with schizophrenia when compared to normal human subjects (Woo *et al.*, 1998). Third, the new onset of epileptic seizures tends to occur early in life and during the adolescent period (Appleton and Gibbs, 1998; Robertson *et al.*, 1990). Precocious puberty is also often associated with epilepsy in children (Elian, 1970; Lennox and Lennox, 1960; Mouridsen *et al.*, 1999; Shenoy and Raja, 2004). Furthermore, treatment with sodium valproic acid, a GABA agonist, delays the timing of puberty in children with seizure disorders (Cook *et al.*, 1992; Lundberg *et al.*, 1986) and in genetically epilepsy-prone mice (Snyder and Badura, 1995). It is possible that the pubertal increase in gonadal steroids may sensitize neurocircuits involved in epileptic seizures, but it is also possible that there is a common mechanism of developmental deficiency (i.e., weakened GABA inhibition in the LHRH neuronal system resulting in precocious puberty and weakened GABAergic inhibition in the brain at the pubertal age resulting in epilepsy) (Olsen and Avoli, 1997). Bourguignon and colleagues treated an 11-month-old child who exhibited severe epileptic seizures and precocious puberty with loreclezole and vigabatrin, GABA agonists. At an earlier stage traditional treatment for epilepsy with phenobarbital was not effective in this patient. However, treatment with loreclezole followed by vigabatrin not only regressed all signs of precocious puberty, but also settled seizure attacks (Bourguignon *et al.*, 1997).

This laboratory has been studying the mechanism of the onset of puberty in the rhesus monkey, as a model for humans. Specifically, results from a series of experiments suggest that the GABAergic neuronal system is, in part, responsible for the timing of puberty in primates (Terasawa, 1995, 2000). Puberty is an important developmental stage during which not only reproductive function is attained (Terasawa and Fernandez, 2001), but also the maturation of the prefrontal cortex, responsible for adolescent behaviors, occurs (Gogtay *et al.*, 2004). Recently, a concept has been proposed that the maturation of the hypothalamus, responsible for puberty, may occur independently from the maturation of the cortices, but there may be common mechanisms governing the maturation of reproductive function and behaviors (Sisk and Foster, 2004). Therefore, a series of observations from this laboratory conducted in non-human primates provide some insights into better understanding the role of GABA function in the possible relation between onset of puberty and clinical changes in autism and other neuropsychiatric disorders.

II. Developmental Changes in Luteinizing Hormone-Releasing Hormone Release

The decapeptide, luteinizing hormone-releasing hormone (LHRH, also called gonadotropin-releasing hormone or GnRH), is synthesized in the preoptic area and hypothalamus and is released into the pituitary portal circulation in a pulsatile manner at approximately 60 minute intervals in mature primates including humans (Hotchkiss and Knobil, 1994). In juveniles the pulse interval of LHRH release is much longer, 90–120 minutes (Plant, 1994). Acceleration of the pulse frequency accompanied by an increase in the pulse amplitude, hence an increase in total output of LHRH release, triggers the onset of puberty (Watanabe and Terasawa, 1989). It has also been shown that pulsatile administration of LHRH into juvenile monkeys results in precocious puberty (Wildt et al., 1980), and pulsatile administration of LHRH agonist and antagonist analogs has been used for the treatment of precocious and delayed puberty in humans (Crowley et al., 1985).

Although LHRH neurons in the preoptic area and hypothalamus are reasonably mature at birth (Terasawa and Fernandez, 2001) and release the decapeptide in a pulsatile manner shortly after birth, the adult type of secretory pattern with a higher pulse frequency (interval, ~60 minutes) is not established until puberty. This is probably due to the immaturity of transsynaptic input regulating LHRH neurons before puberty, either by (1) insufficient excitatory neuronal input to LHRH neurons, or (2) by inhibitory neuronal input to LHRH neurons suppressing activity of LHRH neurons. Electrical stimulation of the medial basal hypothalamus in prepubertal monkeys results in LHRH release of the same magnitude observed in pubertal monkeys (Claypool et al., 1990). However, the immaturity of inhibitory input in control of LHRH release is the predominant mechanism over the immaturity of excitatory input before the onset of puberty (i.e., activity of LHRH neurons during the neonatal period is elevated for the first few months after birth in rhesus monkeys and several months after birth in humans, but is then suppressed by an unknown source of inhibition until shortly before puberty) (Plant, 1994). This inhibition is central in origin, independent from suppression by the ovarian steroid hormone estrogen (Chongthammakun et al., 1993; Terasawa et al., 1983). Thus, we investigated inhibitory neurotransmitters for LHRH release before puberty.

At an early stage we examined the role of beta-endorphin, an inhibitory neuropeptide. However, we excluded it as a candidate for prepubertal inhibition of LHRH release: The release of beta-endorphin increased concomitant with the pubertal increase in LHRH release (Terasawa and Fernandez, 2001). Subsequently, we examined the role of GABA, a dominant inhibitory neurotransmitter in the hypothalamus, (Decavel and van den Pol, 1990) and have found that GABA is an important neurotransmitter responsible for the timing of puberty.

III. Developmental Changes in GABA Release

LHRH neurons release the decapeptide into portal circulation located in the median eminence (ME) and pituitary stalk (S), and appear to be controlled by presynaptic input from GABA neurons. Modulation of LHRH neurons by GABA neurons appears to occur at the cell body and dendrites as well as at the neuroterminals. As the first step to assess the role of GABA in puberty, we measured the simultaneous release of LHRH and GABA in the S-ME. The collection of samples from the S-ME located in the base of the hypothalamus in unanesthetized monkeys is a challenging task, but we have been successful in collecting hypothalamic perfusates for the detection of neurochemical substances using a push-pull perfusion method. As described previously (Terasawa, 1994), a double lumen cannula is inserted into the S-ME with aid of x-ray ventriculograms, and artificial CSF is slowly infused to the area, approximately 1 mm^3, through the push cannula while perfusates are continuously collected through the pull cannula using two peristaltic pumps calibrated at identical speeds. Using this method, we are able to measure LHRH, GABA, glutamates, beta-endorphin, neuropeptide Y, prostaglandin E_2, and catecholamines and their metabolites in various physiological conditions (Terasawa, 1994). This method is also useful in examining the effects of neurotransmitter agonists and antagonists on neurotransmitter/neuromodulator release, such as LHRH and/or GABA, by direct application through the push cannula.

Developmental changes in GABA and LHRH levels in the same samples were assessed using this method. Perfusate samples from the S-ME are collected from prepubertal monkeys at 13–21 months of age (before any sign of puberty is apparent), early pubertal monkeys at 22–28 months of age (after some signs of puberty, before menarche), and midpubertal monkeys at 34–46 months of age (after menarche, but before first ovulation). LHRH and GABA levels were measured by radioimmunoassay and HPLC with electrochemical detection, respectively. In prepubertal monkeys LHRH levels are low, whereas GABA levels measured in the same samples are high. LHRH levels significantly increase in early pubertal and midpubertal monkeys, whereas GABA levels are significantly low in both early and midpubertal monkeys (see Fig. 1; Terasawa *et al.*, 1999). These observations are similar to those reported previously (Mitsushima *et al.*, 1994). Although we were not able to obtain data from monkeys between the neonatal period and 12 months of age because they were not weaned, a scatter plot of GABA levels during the ages of 13 to 46 months (see Fig. 2) indicates that there is a clear developmental GABA decrease. Interestingly, this profile in female rhesus monkeys resembles that described for circulating GABA levels in normal children (Dhossche *et al.*, 2002).

IV. Evidence for GABA as an Inhibitory Neurotransmitter Before Puberty

A question arises as to whether high levels of GABA release in the S-ME before puberty have any physiological significance. To answer this question, two experiments were conducted. First, we examined the effect of the GABA$_A$ receptor antagonist, bicuculline, on LHRH release (Mitsushima *et al.*, 1994). Results suggest that bicuculline stimulates LHRH release in prepubertal monkeys by removing endogenous GABA inhibition, whereas exogenous GABA is not effective in suppressing LHRH release until after the onset of puberty, when endogenous GABAergic tone is reduced. The GABA$_B$ receptor blocker, saclofen, was not effective in prepubertal monkeys. Second, we examined whether lowering GABA levels in the S-ME by chronic infusion of bicuculline triggers puberty (Keen *et al.*, 1999). The average ages of menarche and first ovulation in female rhesus monkeys in our colony are approximately 30 and 45 months, respectively (Terasawa *et al.*, 1983). Pulsatile infusion of bicuculline into the third ventricle of prepubertal monkeys results in precocious menarche, which occurs 6–8 weeks after the initiation of bicuculline infusion, and in precocious first ovulation, which occurs by 30 months, the age of menarche in control females (see Fig. 3; Keen *et al.*, 1999; Richter and Terasawa, 2001). However, since the interval between menarche and first ovulation is not shortened by bicuculline infusion, additional mechanisms, such as the establishment of the stimulatory neuronal system for pulsatile LHRH release, are necessary for the pubertal transition in female primates. The mean ages of menarche and first ovulation in control monkeys receiving saline infusions are not different from the data in colony controls. The results of these two experiments indicate that tonic GABAergic inhibition in the S-ME is, at least in part, responsible for restraining the activity of LHRH neurons before the onset of puberty and reduction in GABAergic inhibition triggers the pubertal increase in LHRH release resulting in the onset of puberty.

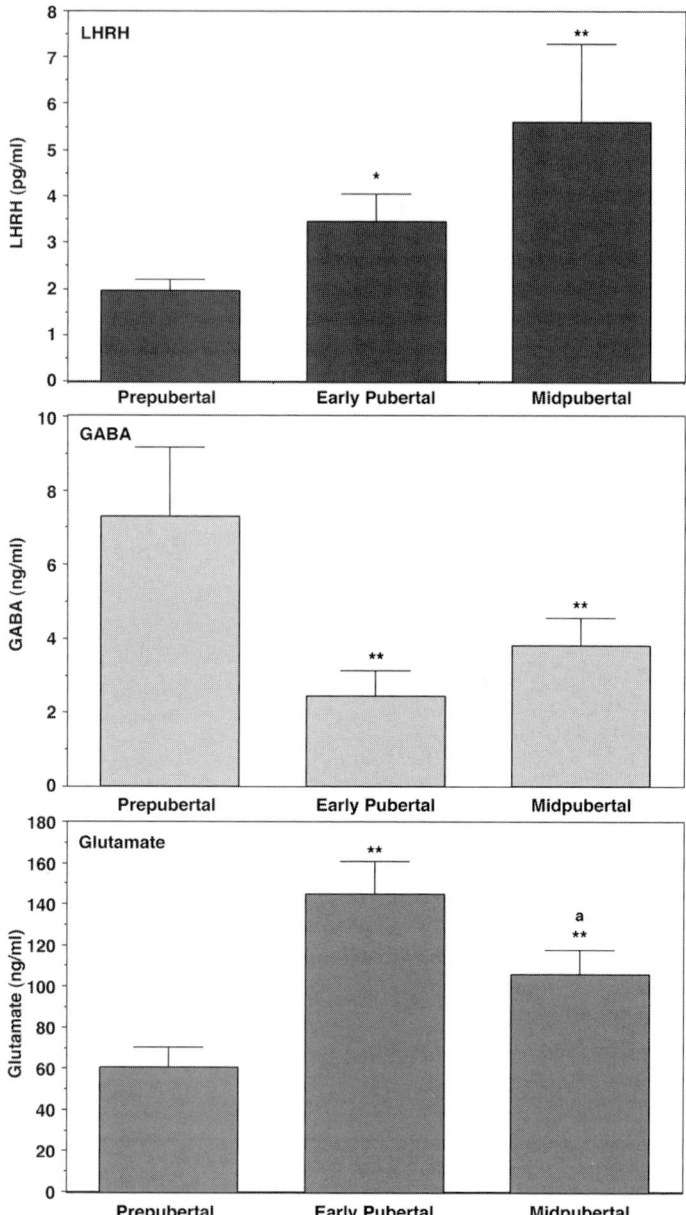

FIG. 1. Developmental changes in luteinizing hormone-releasing hormone (LHRH, top), GABA (middle), and glutamate (bottom) levels in the stalk-median eminence of the hypothalamus in female monkeys. Samples were obtained using the push-pull perfusion method. Note that GABA release in the stalk-median eminence decreases, whereas glutamate release increases, when the pubertal

V. Role of Glutamic Acid Decarboxylase in Puberty

In presynaptic neurons GABA is synthesized from glutamate by decarboxylation in the presence of glutamic acid decarboxylase (GAD), stored in vesicles, and released by exocytosis upon depolarization in the presence of extracellular Ca^{2+} (Rando et al., 1981). There are two different proteins, GAD67 and GAD65, derived from respective genes (Erlander and Tobin, 1991; Rimvall and Martin, 1993). To assess the possible involvement of GADs in puberty, we examined whether interference in GAD67 and GAD65 synthesis in prepubertal monkeys modifies the LHRH release pattern (Mitsushima et al., 1996). Infusion of antisense oligodeoxynucleotides for GAD67 and GAD65 mRNAs into the S-ME of prepubertal monkeys results in a dramatic increase in LHRH release (Kasuya et al., 1999; Mitsushima et al., 1996), presumably due to the reduction in GABA synthesis and subsequent GABA release (Terasawa et al., 1999; Mitsushima et al., 1996). Scrambled oligodeoxynucleotides for GAD67 and GAD65 mRNAs as controls did not induce any significant effect. Observations from this experiment indicate that interference in GAD67 and GAD65 synthesis is effective in reducing tonic GABAergic inhibition, resulting in an increase in LHRH release.

Developmental changes in GABA and GAD in primates are not well studied, and there are conflicting data from the postnatal period. GAD activity in the human neocortex sharply increases at birth and continues to increase until 1 year of age, after which it declines gradually until pubertal age and then slightly increases at adulthood (Diebler et al., 1979; Johnston and Coyle, 1981). Urbanski et al. (1998) reported that the distribution pattern and concentration of GAD67 and GAD65 mRNAs in hypothalamic nuclei assessed by in situ hybridization in gonadally intact juvenile ($\sim$0.6 years of age) male rhesus monkeys were not different from those in adult ($\sim$10 years of age) male monkeys, and a report by Plant and his colleagues (El Majdoubi et al., 2000) indicates that GAD mRNA levels in the basal hypothalamus of juvenile castrated male rhesus monkeys did not differ from those in adult castrated male monkeys. Although these data do not appear to support the hypothesis that GAD plays an important role in puberty, more precise developmental studies with the exact regional distribution pattern of GABA neurons in non-human primates, including in females, are needed before conclusions can be drawn.

---

increase in LHRH release occurs. The ages of prepubertal, early pubertal, and midpubertal monkeys are 13–20 months (before any sign of puberty), 21–30 months (some signs of puberty, but before menarche), and 34–46 months (after menarche, but before first ovulation), respectively. Ages of menarche and first ovulation in our colony females are $\sim$30 months and $\sim$45 months, respectively. *p $<$ 0.05 vs. prepubertal; **p $<$ 0.01 vs. prepubertal; [a]p $<$ 0.05 vs. early pubertal monkeys. Modified from Terasawa et al. (1999) with permission.

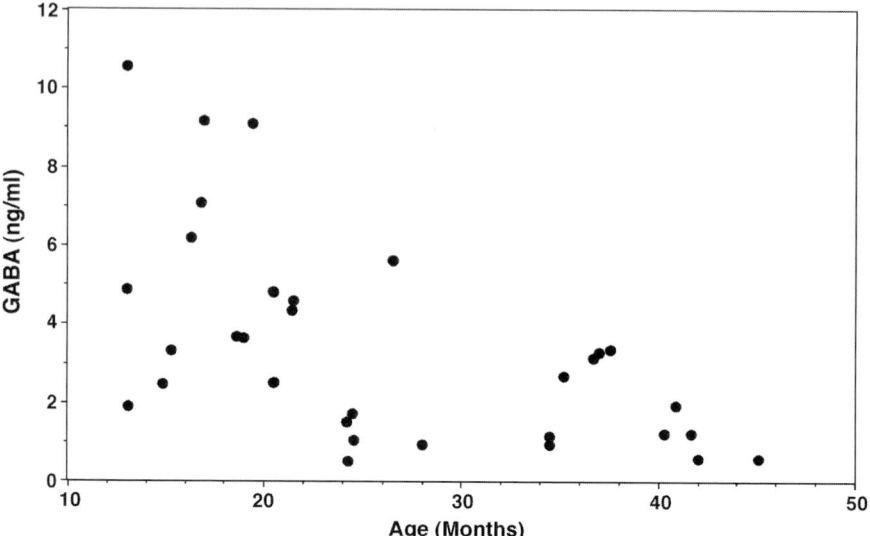

FIG. 2. A scatter plot of GABA release in the stalk-median eminence of female monkeys at ages of 13–46 months. A gradual decline of GABA levels with age is seen.

GAD67 and GAD65 exist in the enzymatically active holo form and inactive apo form, and conversion of holo-GAD67 or holo-GAD65 to and from apo-GAD67 or apo-GAD65 is determined by the presence of the cofactor, pyridoxal-5′-phosphate, which is influenced by physiological states as well as by experimental conditions (Erlander and Tobin, 1991; Kaufman et al., 1991). We do not have any data showing how the ratio of the active and inactive form changes during development.

VI. Other Possible Factors Associated with Developmental Changes in GABAergic Function

A. GABA Transporters

Elevated local concentrations of GABA in the synaptic cleft are actively removed by GABA transporters (GAT), located on presynaptic terminals and surrounding glial cells, where the neurotransmitters are recycled. Four GABA transporters (GAT-1, GAT-2, GAT-3, and BGT-1), classified as the Na^+ and Cl^--coupled transporter family, have been described (Guastella et al., 1990;

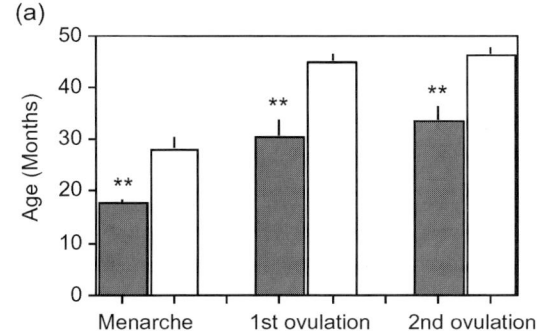

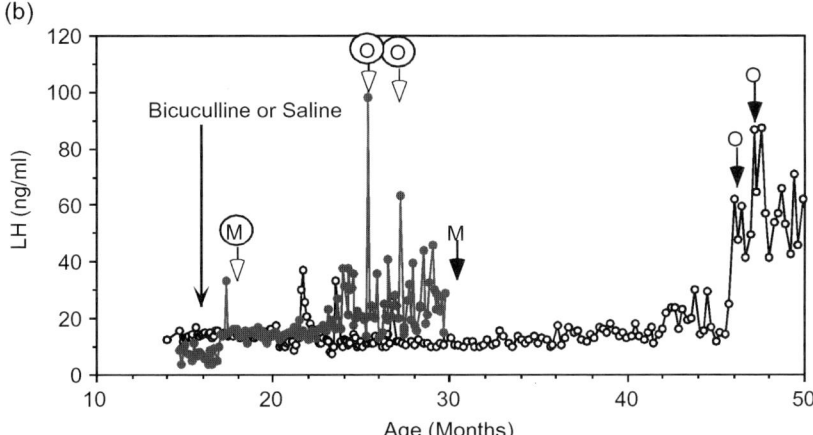

FIG. 3. Reducing GABA neurotransmission with the GABA$_A$ receptor blocker, bicuculline, advances puberty in female rhesus monkeys. (a) The age of menarche and first ovulation in female rhesus monkeys treated by chronic infusion of bicuculline occur at a significantly earlier age than that in controls. Bicuculline treated group, filled bars; saline-treated control group, open bars. **P < 0.01. (b) Representative examples showing LH concentration in a bicuculline treated (filled circles) monkey and a saline-treated control (open circles) monkey. The dramatically earlier onset of puberty is shown by menarche (M) and first and second ovulations (O). Modified from Keen *et al.* (1999) and Richter and Terasawa (2001) with permission.

Nelson *et al.*, 1990). GAT-1 is the predominant GABA transporter in the mammalian brain (Borden, 1996) and the GAT-1 transcript contains an estrogen responsive element (Herbison *et al.*, 1995). Thus, it is possible that the reduction of GABA concentration in the S-ME at the onset of puberty might be due to a developmentally regulated increase in GABA transporter activity. At this time there is no information on the role of GAT activity in puberty.

B. GABA$_A$ Receptors and Their Subunit Composition

The developmental pattern of each GABA$_A$ receptor subunit is complex and differs from region to region or neuron to neuron, which ensures the functional heterogeneity of GABA input. Nonetheless, it has been consistently reported that, in general, α_2 subunit expression is very high before birth to shortly after birth, and decreases gradually toward adult levels, whereas α_1 expression is minimal in prenates and then gradually increases after birth until adulthood (Brooks-Kayal and Pritchett, 1993; Fritschy and Mohler, 1995; Fritschy et al., 1994; Hendrickson et al., 1994; Laurie et al., 1992). In fact, it appears that GABA$_A$ receptors containing α_1 subunits gradually replace GABA$_A$ receptors containing the α_2 subunit during postnatal maturation in the rat, monkey, and human brain, and that the increase in the α_1 subunit is an indication of brain maturation (i.e., the onset of synaptic GABA inhibition, whereas β subunits do not generally undergo developmental changes) (Brooks-Kayal and Pritchett, 1993; Hendrickson et al., 1994; Laurie et al., 1992). An example of developmental changes in subunit composition in association with function has been shown: Changes in GABA$_A$ receptor subunit composition in hippocampal neurons preceded the onset of epilepsy by weeks in epileptic rats (Brooks-Kayal et al., 1998).

It is, therefore, possible that developmental changes in the GABA$_A$ subunit composition in neuronal cells may occur prior to the onset of puberty. Analysis of the literature provides support for the hypothesis that GABA disinhibition of LHRH neurons through GABA$_A$ receptors at the onset of puberty in female monkeys (Mitsushima et al., 1994) is due to changes in GABA$_A$ receptor subunit composition. For example, a report using single cell RT-PCR (Sim et al., 2000) suggests that the pattern of LHRH neurons expressing GABA subunits in the POA and medial septum of sexually immature mice at neonatal and juvenile ages is more heterogeneous than that in adults, and it becomes homogeneous when the mice mature. Interestingly, the same authors have reported that sensitivity to GABA in LHRH neurons of juvenile mice is lower than that in adult mice and that the pattern of the dose response curve to GABA in prepubertal LHRH neurons is more heterogeneous than that in adult LHRH neurons (Sim et al., 2000). At this time, we have little knowledge of developmental changes in the GABA$_A$ subunit composition of LHRH neurons in non-human primates.

VII. The Pubertal Reduction in GABAergic Inhibition is Followed by an Increase in Glutamatergic Tone

Glutamate is profoundly involved in pulsatile LHRH release *in vivo* and *in vitro* through NMDA and kainate receptors. NMDA stimulates release of LH and

LHRH in adult rats and monkeys *in vivo* (Bourguignon *et al.*, 1995; Brann and Mahesh, 1997; Olney *et al.*, 1976; Price *et al.*, 1978; van den Pol *et al.*, 1994; Wilson *et al.*, 1982), and glutamate, NMDA, and kainate all stimulate LHRH/LH release in sexually immature monkeys (Gay and Plant, 1987; Medhamurthy *et al.*, 1990), rats (Bourguignon *et al.*, 1989; Brann and Mahesh, 1992; Cicero *et al.*, 1988), sheep (I'Anson *et al.*, 1993), and fetal sheep (Bettendorf *et al.*, 1999) *in vitro* and *in vivo*. Moreover, stimulation of NMDA receptors results in precocious puberty in rats and monkeys (Plant *et al.*, 1989; Urbanski and Ojeda, 1990), whereas administration of the NMDA receptor blockers, MK-801 or 2-amino-5-phosphonovaleric acid (AP-5), delays the timing of puberty in rats (MacDonald and Wilkinson, 1990; Meiji-Roelofs *et al.*, 1991; Urbanski and Ojeda, 1990; Wu *et al.*, 1990). In contrast, the non-NMDA receptor antagonist, 6,7-dinitro-quinoxaline-2,3-dione (DNQX), fails to change the timing of puberty (Brann and Mahesh, 1994). The excitatory action of glutamate on LHRH release may occur not only through NMDA receptors, but also through metabolic receptors. Therefore, the developmental changes in NMDA and kainate receptors are integrated parts of the mechanism of the onset of puberty.

We have measured glutamate release in the S-ME using the push-pull perfusion method (Terasawa *et al.*, 1999). Glutamate levels during the prepubertal period are very low, but increase strikingly during the early pubertal period, and remain high during the midpubertal period, although midpubertal levels decline slightly from early pubertal levels (Fig. 1). However, this observation from monkeys at different ages (cross-sectional study) does not provide the exact timing of glutamate increase during puberty. Fortunately, there is an indication that the pubertal elevation in glutamate release may occur promptly following GABA reduction. The results of the antisense GAD67 infusion experiment in prepubertal monkeys suggest that the reduction in GABA release induced by the antisense GAD67 treatment is followed by an increase in glutamate release for several hours (Terasawa *et al.*, 1999).

Sensitivity to glutamatergic stimulation increases after the onset of puberty. For example, (1) infusion of NMDA into the S-ME at 10 μM-100 μM stimulates LHRH release in pubertal monkeys, whereas only 100 μM NMDA results in LHRH release in prepubertal monkeys (Claypool *et al.*, 2000), and (2) *i.v.* injection of NMDA at 10 mg/kg results in LHRH responses with a longer duration in pubertal monkeys than in prepubertal monkeys (Claypool *et al.*, 2000). Although this increase in the responsiveness of LHRH neurons to NMDA after puberty may, in part, be due to an increase in circulating estrogen, glutamatergic tone is more elevated after the onset of puberty.

VIII. Conclusions

The mechanism of the onset of puberty is complex. As previously discussed, LHRH neurons are reasonably mature at birth and are already active during the neonatal period. However, in primates "central inhibition" suppresses pulsatile LHRH release during the juvenile period. Studies from this laboratory suggest that the GABAergic neuronal system appears to be a substrate for "central inhibition" in primates. When approaching puberty, this GABA inhibition is removed or diminished, and an increase in LHRH release occurs. Subsequently, increases in stimulatory input from glutamatergic neurons as well as new stimulatory input from norepinephrine and NPY neurons (which we did not discuss here, see Terasawa and Fernandez, 2001) and inhibitory input from β-endorphin neurons to the LHRH neuronal system become active to establish the adult type of regulatory mechanism for pulsatile LHRH release. This pubertal increase in LHRH release results in a cascade of events during puberty, such as increases in synthesis and release of gonadotropins, and increases in steroidogenesis and gametogenesis, followed by the appearance of secondary sexual characteristics.

The most important question still remains: What determines the timing to remove "GABA inhibition"? Because many genes in the brain are turned on or turned off to establish a complex series of events occurring during puberty, the timing of spontaneous puberty must be regulated by a master gene or genes, as part of a series of developmental events. We expect that future studies will include a search for genes determining events to remove "GABA inhibition" and genes which ultimately trigger the onset of puberty in primates.

Understanding the mechanism of the onset of puberty in detail is very important, as puberty is associated with onset of or changes in many neurodevelopmental disorders, including autism, schizophrenia, and affective disorder. At this time it is unclear whether the series of events in the hypothalamus with puberty also occur in the higher brain regions where cognitive function is controlled. Future studies to assess whether similar events in the hypothalamus also occur in the higher brain regions that are involved in autism and other disorders will be useful for treatment strategies.

Acknowledgments

The author thanks Kim Keen for her proofreading of the manuscript. This work is supported by NIH grants 2R01HD11355, 2R01HD1543, and RR00167.

References

Appleton, R., and Gibbs, J. (1998). "Epilepsy in Childhood and Adolescence." Martin Dunitz Publishers, London.

Bettendorf, M., de Zegher, F., Albers, N., Hart, C. S., Kaplan, S. L., and Grumbach, M. M. (1999). Acute N-methyl-D,L-aspartate administration stimulates the luteinizing hormone releasing hormone pulse generator in the ovine fetus. *Horm. Res.* **51,** 25–30.

Borden, L. A. (1996). GABA transporter heterogeneity: Pharmacology and cellular localization. *Neurochem. Int.* **29,** 335–356.

Bourguignon, J. P., Gérard, A., and Franchimont, P. (1989). Direct activation of gonadotropin-releasing hormone secretion through different receptors to neuroexcitatory amino acids. *Neuroendocrinology* **49,** 402–408.

Bourguignon, J. P., Gérard, A., Gonzalez, M. L. A., Purnelle, G., and Franchimont, P. (1995). Endogenous glutamate involvement in pulsatile secretion of gonadotropin-releasing hormone: Evidence from the effect of glutamine and developmental changes. *Endocrinology* **136,** 911–916.

Bourguignon, J.-P., Jaeken, J., Gérard, A., and deZegher, F. (1997). Amino acid neurotransmission and initiation of puberty: Evidence from nonketotic hyperglycinemia in a female infant and gonadotropin-releasing hormone secretion by rat hypothalamic explants. *J. Clin. Endocrinol. Metab.* **82,** 1899–1903.

Brann, D. W., and Mahesh, V. B. (1992). Excitatory amino acid neurotransmission. Evidence for a role in neuroendocrine regulation. *Trends Endocrinol. Metab.* **3,** 122–126.

Brann, D. W., and Mahesh, V. B. (1994). Excitatory amino acids: Function and significance in reproduction and neuroendocrine regulation. *Front Neuroendocrinol.* **15,** 3–49.

Brann, D. W., and Mahesh, V. B. (1997). Excitatory amino acids: Evidence for a role in the control of reproduction and anterior pituitary hormone secretion. *Endocr. Rev.* **18,** 678–700.

Brooks-Kayal, A. R., and Pritchett, D. B. (1993). Developmental changes in human γ-aminobutyric acid A receptor subunit composition. *Ann. Neurol.* **34,** 687–693.

Brooks-Kayal, A. R., Shumate, M. D., Jin, H., Rikhter, T. Y., and Coulter, D. A. (1998). Selective changes in single cell GABA$_A$ receptor subunit expression and function in temporal lobe epilepsy. *Nat. Med.* **4,** 1166–1172.

Chongthammakun, S., Claypool, L. E., and Terasawa, E. (1993). Ovariectomy increases *in vivo* LHRH release in pubertal, but not prepubertal, female rhesus monkeys. *J. Neuroendocrinol.* **5,** 41–50.

Cicero, T. J., Meyer, E. R., and Bell, R. D. (1988). Characterization and possible opioid modulation of N-methyl-D-aspartic acid induced increases in serum luteinizing hormone levels in the developing male rat. *Life Sci.* **42,** 1725–1732.

Claypool, L. E., Kasuya, E., Saitoh, Y., Marzban, F., and Terasawa, E. (2000). N-methyl-D, L-aspartate induces the release of luteinizing hormone-releasing hormone in prepubertal and pubertal female rhesus monkeys as measured by *in vivo* push-pull perfusion in the stalk-median eminence. *Endocrinology* **141,** 219–228.

Claypool, L. E., Watanabe, G., and Terasawa, E. (1990). Effects of electrical stimulation of medial basal hypothalamus on the *in vivo* release of luteinizing hormone-releasing hormone in the prepubertal and peripubertal monkey. *Endocrinology* **127,** 3014–3022.

Cohen, B. I. (2002). Use of GABA-transaminase agonist for treatment of infantile autism. *Med. Hypotheses* **59,** 115–116.

Cook, J. S., Bale, J. F., and Hoffman, R. P. (1992). Pubertal arrest associated with valproic acid therapy. *Pediatr. Neurol.* **8,** 229–231.

Crowley, W. F., Filicori, M., Spratt, D. T., and Santoro, N. F. (1985). The physiology of gonadotropin-releasing hormone (GnRH) secretion in men and women. *Recent Prog. Horm. Res.* **41,** 473–531.

Decavel, C., and van den Pol, A. N. (1990). GABA: A dominant neurotransmitter in the hypothalamus. *J. Comp. Neurol.* **302,** 1019–1037.

Dhossche, D., Applegate, H., Abraham, A., Maertens, P., Bland, L., Bencsath, A., and Martinez, J. (2002). Elevated plasma gamma-aminobutyric acid (GABA) levels in autistic youngsters: Stimulus for a GABA hypothesis of autism. *Med. Sci. Monit.* **8,** PR1–PR6.

Diebler, M. F., Farkas-Bargeton, E., and Wehrle, R. (1979). Developmental changes of enzymes associated with energy metabolism and the synthesis of some neurotransmitters in discrete areas of human neocortex. *J. Neurochem.* **32,** 429–435.

El Majdoubi, M., Sahu, A., Ramaswamy, S., and Plant, T. M. (2000). Neuropeptide Y: A hypothalamic brake restraining the onset of puberty in primates. *Proc. Natl. Acad. Sci. USA* **97,** 6179–6184.

Elian, M. (1970). EEG, epilepsy, and precocious puberty. *Electroencephalogr. Clin. Neurophysiol.* **28,** 642.

Erlander, M. G., and Tobin, A. J. (1991). The structural and functional heterogeneity of glutamic acid decarboxylase: A review. *Neurochem. Res.* **16,** 215–226.

Fritschy, J. M., and Mohler, H. (1995). GABA$_A$-receptor heterogeneity in the adult rat brain: Differential regional and cellular distribution of seven major subunits. *J. Comp. Neurol.* **359,** 154–194.

Fritschy, J. M., Paysan, J., Enna, A., and Mohler, H. (1994). Switch in the expression of rat GABA$_A$-receptor subtypes during postnatal development: An immunohistochemical study. *J. Neurosci.* **14,** 5302–5324.

Gay, V. L., and Plant, T. M. (1987). N-Methyl-D,L-aspartate (NMDA) elicits hypothalamic GnRH release in prepubertal male rhesus monkeys (*Macaca mulatta*). *Endocrinology* **120,** 2289–2296.

Gillberg, C. (1984). Autistic children growing up: Problems during puberty and adolescence. *Dev. Med. Child. Neurol.* **26,** 125–129.

Gogtay, N., Giedd, J. N., Lusk, L., Hayashi, K. M., Greenstein, D., Vaituzis, A. C., Nugent, T. F., III, Herman, D. H., Clasen, L. S., Toga, A. W., Rapoport, J. L., and Thompson, P. M. (2004). Dynamic mapping of human cortical development during childhood through early adulthood. *Proc. Natl. Acad. Sci. USA* **101,** 8174–8179.

Guastella, J., Nelson, N., Nelson, H., Czyzyk, L., Keynan, S., Miedel, M. C., Davidson, N., Lester, H. A., and Kanner, B. I. (1990). Cloning and expression of a rat brain GABA transporter. *Science* **249,** 1303–1306.

Hendrickson, A., March, D., Richards, G., Erickson, A., and Shaw, C. (1994). Coincidental appearance of the α1 subunit of the GABA$_A$-receptor and the type I benzodiazepine receptor near birth in macaque monkey visual cortex. *Int. J. Dev. Neurosci.* **12,** 299–314.

Herbison, A. E., Augwood, S. J., Simonian, S. X., and Chapman, C. (1995). Regulation of GABA transporter activity and mRNA expression by estrogen in rat preoptic area. *J. Neurosci.* **15,** 8302–8309.

Hotchkiss, J., and Knobil, E. (1994). The menstrual cycle and its neuroendocrine control. *In* "The Physiology of Reproduction" (E. Knobil and J. D. Neill, Eds.), 2nd ed, pp. 711–749. Raven Press, New York.

Hussman, J. P. (2001). Suppressed gabaergic inhibition as a common factor in suspected etiologies of autism. *J. Autism Dev. Disord.* **31,** 247–248.

I' Anson, H., Herbosa, C. G., Ebling, F. J., Wood, R. I., Bucholtz, D. C., Mieher, C. D., Foster, D. L., and Padmanabhan, V. (1993). Hypothalamic versus pituitary stimulation of luteinizing hormone secretion in the prepubertal female lamb. *Neuroendocrinology* **57,** 467–475.

Johnston, M. V., and Coyle, J. T. (1981). Development of central neurotransmitter systems. *Ciba. Found. Symp.* **86,** 251–271.

Kasuya, E., Nyberg, C. L., Mogi, K., and Terasawa, E. (1999). A role of γ-aminobutyric acid (GABA) and glutamate in control of puberty in female rhesus monkeys: Effect of an antisense oligodeoxynucleotide for GAD67 messenger ribonucleic acid and MK801 on luteinizing hormone-releasing hormone release. *Endocrinology* **140,** 705–712.

Kaufman, D. L., Houser, C. R., and Tobin, A. J. (1991). Two forms of the γ-aminobutyric acid synthetic enzyme glutamate decarboxylase have distinct intraneuronal distributions and cofactor interactions. *J. Neurochem.* **56,** 720–723.

Keen, K. L., Burich, A. J., Mitsushima, D., Kasuya, E., and Terasawa, E. (1999). Effects of pulsatile infusion of the GABA_A receptor blocker bicuculline on the onset of puberty in female rhesus monkeys. *Endocrinology* **140,** 5257–5266.

Laurie, D. J., Wisden, W., and Seeburg, P. H. (1992). The distribution of thirteen GABA_A receptor subunit mRNAs in the rat brain. III. Embryonic and postnatal development. *J. Neurosci.* **12,** 4151–4172.

Lennox, W. G., and Lennox, M. A. (1960). "Epilepsy and Related Disorders." Little, Brown and Company, Boston.

Lewis, D. A. (1997). Development of the prefrontal cortex during adolescence—insights into vulnerable neural circuits in schizophrenia. *Neuropsycopharmacol.* **16,** 385–398.

Lewis, D. A., Hashimoto, T., and Volk, D. W. (2005). Cortical inhibitory neurons and schizophrenia. *Nat. Rev. Neurosci.* **6,** 312–324.

Lundberg, B., Nergardh, A., Ritzen, E., and Samuelson, K. (1986). Influence of valproic acid on the gonadotropin releasing hormone test in puberty. *Acta. Pediatr. Scand.* **75,** 787–792.

MacDonald, M. C., and Wilkinson, M. (1990). Peripubertal treatment with N-methyl-D-aspartic acid (NMDA) or neonatally with monosodium glutamate (MSG) accelerates sexual maturation in female rats, an effect reversed by MK801. *Neuroendocrinology* **52,** 143–149.

Martin, D. L., and Rimvall, K. (1993). Regulation of γ-aminobutyric acid synthesis in the brain. *J. Neurochem.* **60,** 395–407.

Medhamurthy, R., Dichek, H. L., Plant, T. M., Bernardini, I., and Cutler, G. B., Jr. (1990). Stimulation of gonadotropin secretion in prepubertal monkeys after hypothalamic excitation with aspartate and glutamate. *J. Clin. Endocrinol. Metab.* **71,** 1390–1392.

Meiji-Roelofs, H. M. A., Kramer, P., and van Leeuwen, E. C. M. (1991). The N-methyl-D-aspartate receptor antagonist MK-801 delays the onset of puberty and may acutely block the first spontaneous LH surge and ovulation in the rat. *J. Endocrinol.* **131,** 435–441.

Menold, M. M., Shao, Y., Wolpert, C. M., Donnelly, S. L., Raiford, K. L., Martin, E. R., Ravan, S. A., Abramson, R. K., Wright, H. H., Delong, G. R., Cuccaro, M. L., Pericak-Vance, M. A., and Gilbert, J. R. (2001). Association analysis of chromosome 15 GABA receptor subunit genes in autistic disorder. *J. Neurogenet.* **15,** 245–259.

Mitsushima, D., Hei, D. L., and Terasawa, E. (1994). GABA is an inhibitory neurotransmitter restricting the release of luteinizing hormone-releasing hormone before the onset of puberty. *Proc. Natl. Acad. Sci. USA* **91,** 395–399.

Mitsushima, D., Marzban, F., Luchansky, L. L., Burich, A. J., Keen, K. L., Durning, M., Golos, T. G., and Terasawa, E. (1996). Role of glutamic acid decarboxylase in the prepubertal inhibition of the luteinizing hormone-releasing hormone release in prepubertal female monkeys. *J. Neurosci.* **16,** 2563–2573.

Mouridsen, S. E., Rich, B., and Isager, T. (1999). Epilepsy in disintegrative psychosis and infantile autism: A long-term validation study. *Dev. Med. Child. Neurol.* **41,** 110–114.

Muhle, R., Trentacoste, S. V., and Rapin, I. (2004). The genetics of autism. *Pediatrics* **113,** e472–e486.

Nelson, H., Mandiyan, S., and Nelson, N. (1990). Cloning of the human brain GABA transporter. *FEBS Lett.* **269,** 181–184.

Nurmi, E. L., Dowd, M., Tadevosyan-Leyfer, O., Haines, J. L., Folstein, S. E., and Sutcliffe, J. S. (2003). Exploratory subsetting of autism families based on savant skills improves evidence of genetic linkage to 15q11-q13. *J. Am. Acad. Child Adolesc. Psychiatry* **42,** 856–863.

Olney, J. W., Cicero, T. J., Meyer, E. R., and De Gubareff, T. (1976). Acute glutamate-induced elevations in serum testosterone and luteinizing hormone. *Brain Res.* **12**, 420–424.

Olsen, R. W., and Avoli, M. (1997). GABA and epileptogenesis. *Epilepsia* **38**, 399–407.

Plant, T. M. (1994). Puberty in primates. *In* "The Physiology of Reproduction" (E. Knobil and J. D. Neill, Eds.), pp. 453–485. Raven Press, New York.

Plant, T. M., Gay, V. L., Marshall, G. R., and Arslan, M. (1989). Puberty in monkeys is triggered by chemical stimulation of the hypothalamus. *Proc. Natl. Acad. Sci. USA* **86**, 2506–2510.

Price, M. T., Olney, J. W., and Cicero, T. J. (1978). Acute elevations of serum luteinizing hormone induced by kainic acid, N-methyl-aspartic acid, or homosysteic acid. *Neuroendocrinology* **26**, 352–358.

Prosser, J., Hughes, C. W., Sheikha, S., Kowatch, R. A., Kramer, G. L., Rosenbarger, N., Trent, J., and Petty, F. (1997). Plasma GABA in children and adolescents with mood, behavior, and comorbid mood and behavior disorders: A preliminary study. *J. Child Adolesc. Psychopharmacol.* **7**, 181–199.

Rando, R. R., Bangerter, F. W., and Farb, D. H. (1981). The inactivation of γ-aminobutyric acid transaminase in dissociated neuronal cultures from spinal cord. *J. Neurochem.* **36**, 985–990.

Richter, T. A., and Terasawa, E. (2001). Inhibitory neural mechanisms underlying changes in LHRH release at the onset of puberty in the rhesus monkey. *Trends Endocrinol. Metab.* **12**, 353–359.

Rimvall, K., and Martin, D. L. (1994). The level of GAD67 protein is highly sensitive to small increases in intraneuronal gamma-aminobutyric acid levels. *J. Neurochem.* **62**, 1375–1381.

Robertson, C. M., Morrish, D. W., Wheler, G. H., and Grace, M. G. (1990). Neonatal encephalopathy: An indicator of early sexual maturation in girls. *Pediatr. Neurol.* **6**, 102–108.

Shenoy, S. N., and Raja, A. (2004). Hypothalamic hamartoma with precocious puberty. *Pediatr. Neurosurg.* **40**, 249–252.

Sim, J. A., Skynner, M. J., Pape, J. R., and Herbison, A. E. (2000). Late postnatal reorganization of GABA$_A$ receptor signaling in native GnRH neurons. *Eur. J. Neurosci.* **12**, 3497–3504.

Sisk, C. L., and Foster, D. L. (2004). The neural basis of puberty and adolescence. *Nat. Neurosci.* **7**, 1040–1047.

Snyder, P. J., and Badura, L. L. (1995). Chronic administration of sodium valproic acid slows pubertal maturation in inbred DBA/2J mice: Skeletal, histological, and endocrinological evidence. *Epilepsy Res.* **20**, 203–211.

Terasawa, E. (1994). *In vivo* measurement of pulsatile release of neuropeptides and neurotransmitters in rhesus monkeys using push-pull perfusion. *In* "Methods in Neurosciences: Pulsatility in Neuroendocrine System" (J. E. Levine, Ed.), pp. 184–202. Academic Press, New York.

Terasawa, E. (1995). Mechanisms controlling the onset of puberty in primates: The role of GABAergic neurons. *In* "The Neurobiology of Puberty" (T. M. Plant and P. A. Lee, Eds.), pp. 139–151. Journal of Endocrinology Limited, Bristol.

Terasawa, E. (2000). The control of the onset of puberty by neurotransmitters in female rhesus monkeys. *In* "The 5th International Conference on the Control of the Onset of Puberty" (J. P. Bourguignon and T. M. Plant, Eds.), pp. 131–143. Elsevier Science, Amsterdam.

Terasawa, E., and Fernandez, D. L. (2001). Neurobiological mechanisms of the onset of puberty in primates. *Endocr. Rev.* **22**, 111–151.

Terasawa, E., Luchansky, L. L., Kasuya, E., and Nyberg, C. L. (1999). An increase in glutamate release follows a decrease in gamma aminobutyric acid and the pubertal increase in luteinizing hormone releasing hormone release in female rhesus monkeys. *J. Neuroendocrinol.* **11**, 275–282.

Terasawa, E., Nass, T. E., Yeoman, R. R., Loose, M. D., and Schultz, N. J. (1983). Hypothalamic control of puberty. *In* "Neuroendocrine Aspects of Reproduction" (R. L. Norman, Ed.), pp. 149–182. Academic Press, New York.

Urbanski, H. F., and Ojeda, S. R. (1990). A role for N-methyl-D-aspartate (NMDA) receptors in the control of LH secretion and initiation of female puberty. *Endocrinology* **126**, 1774–1776.

Urbanski, H. F., Rodrigues, S. M., Garyfallou, V. T., and Kohama, S. G. (1998). Regional distribution of glutamic acid decarboxylase (GAD65 and GAD67) mRNA in the hypothalamus of male rhesus before and after puberty. *Mol. Brain. Res.* **57,** 86–91.

van den Pol, A. N., Kogelman, L., Ghost, P., Liljelund, P., and Blackstone, C. (1994). Developmental regulation of the hypothalamic metatropic glutamate receptor mGluR1. *J. Neurosci.* **14,** 3816–3834.

Watanabe, G., and Terasawa, E. (1989). *In vivo* release of luteinizing hormone-releasing hormone (LHRH) increases with puberty in the female rhesus monkey. *Endocrinology* **125,** 92–99.

Wildt, L., Marshall, G., and Knobil, E. (1980). Experimental induction of puberty in the infantile rhesus monkey. *Science* **207,** 1373–1375.

Wilson, R. C., and Knobil, E. (1982). Acute effects of N-methyl-DL-aspartate on the release of pituitary gonadotropins and prolactin in the adult female rhesus monkey. *Brain Res.* **248,** 177–179.

Woo, T. U., Whitehead, R. E., Melchitzky, D. S., and Lewis, D. A. (1998). A subclass of prefrontal gamma-aminobutyric acid axon terminals are selectively altered in schizophrenia. *Proc. Natl. Acad. Sci. USA* **95,** 5341–5346.

Wu, F. C. W., Howe, D. C., and Naylor, A. M. (1990). N-methyl-D-aspartate (NMDA) receptor antagonism by D-2-amino-5-phosphonovaleric acid delays onset of puberty in female rat. *J. Neuroendocrinol.* **2,** 627–631.

RETT SYNDROME: A ROSETTA STONE FOR UNDERSTANDING THE MOLECULAR PATHOGENESIS OF AUTISM

Janine M. LaSalle, Amber Hogart, and Karen N. Thatcher

Medical Microbiology and Immunology and Rowe Program in Human Genetics
School of Medicine, University of California, Davis, California 95616, USA

Autism is a common but complex disorder, showing high heritability but elusive genetic etiology. One approach to "deciphering" the etiology of autism is to investigate overlapping pathways with other autism spectrum-disorders with known genetic causes. Rett syndrome, an X-linked disorder caused by mutations in *MECP2*, and Angelman syndrome, an imprinted disorder caused by maternal 15q11-13 deficiency, share many features with autism. Furthermore, maternal 15q11-13 duplications in autism and rare *MECP2* mutations in autism and Angelman syndrome point to genetic overlap between these three disorders. Recent studies on human postmortem brain samples and *Mecp2*-deficient mouse models have demonstrated overlapping pathways in the molecular pathogenesis of these disorders. Although originally described as a transcriptional repressor of methylated genes, new roles for MeCP2 in chromatin organization have recently emerged. In addition, the recent demonstration of MeCP2 as a regulator of both *UBE3A* (the Angelman gene) and *GABRB3* (encoding a $GABA_A$ receptor subunit) expression within 15q11-13 has revealed some interesting insights into the genetic and epigenetic pathways common to all three disorders.

The discovery of mutations in *MECP2*, the gene encoding methyl-CpG binding protein 2, as the cause of Rett syndrome (RTT) in 1999 (Amir *et al.*, 1999) has opened up an entirely new line of investigations about the role of DNA methylation and chromatin in the development of the mammalian brain. Furthermore, of the five pervasive developmental disorders (PDD) subtypes, including autistic disorder, RTT, Asperger disorder, disintegrative disorder, and PDD Not Otherwise Specified (PDD-NOS), RTT is the only one with a known genetic cause (Zoghbi, 2003). Could the emerging information from the rapidly progressing field of MeCP2 research therefore be applied to autism in a "Rosetta Stone" approach to understanding the complex genetics of autism? The Rosetta stone was a small but significant stone tablet containing the same message written in Greek and two different Egyptian scripts that was essential to deciphering Egyptian hieroglyphics. Could the discovery of *MECP2* mutations in RTT be as significant a finding for the field of autism?

I. Overlapping Syndromes: Rett Syndrome, Angelman Syndrome, and Autism

Autism is a neurodevelopmental disorder characterized by severe impairments in social interaction and communication. Symptoms generally appear around 1 to 3 years of age and are characterized by stereotyped mannerisms, abnormal preoccupations, lack of pragmatic language and imaginative play, impaired eye gaze, and impaired joint attention (Volkmar and Pauls, 2003). Males with autism outnumber females by approximately 4 to 1, and the frequency in the population has increased dramatically in recent decades to around 1 in 500 children (Services, 2003). Whether the increased reported prevalence is due to increased diagnosis is under debate, but may suggest that environmental factors affect the incidence of autism (Blaxill, 2004). Neonatal or early childhood exposures to mercury, pesticides, or infections have all been suggested as potential environmental triggers of autism (Lawler *et al.*, 2004).

Twin studies, however, have provided compelling evidence for a genetic origin of autism, as the concordance rate is 70–90% for monozygotic twins and 0–10% for dizygotic twins (Zoghbi, 2003). Yet pinpointing the genetic cause of autism remains challenging since many linkage scans show conflicting results for multiple loci across the human genome (Polleux and Lauder, 2004). Recent estimates have suggested as many as 10 genes may contribute to the risk of developing autism (Muhle et al., 2004). The complexity of the disorder is further complicated by the large range of phenotypes grouped under the single title of autism-spectrum disorders that may be linked more by aberrant biochemical pathways than by specific genetic loci. Research involving autism spectrum disorders with known genetic causes is therefore being used as an approach to understand the molecular pathogenesis of autism and the normal development of the human brain.

Angelman syndrome (AS) is a rare autism-spectrum disorder (approximately 1/20,000) in which children display severe motor problems, mental retardation, ataxia, hypotonia, seizures, absence of language, and inappropriate laughter (Lalande, 1996). In a recent study, 42% of AS patients also were found to meet the diagnostic criteria for autism (Peters et al., 2004). AS is caused by maternal deletions of 15q11-13, deletions or point mutations in UBE3A, paternal disomy of chromosome 15, or maternal methylation defects. Prader-Willi syndrome (PWS) is a distinct neurodevelopmental disorder caused by paternal deficiency of 15q11-13 (Lalande, 1996).

Rett syndrome (RTT) is more common than AS, with an incidence of approximately 1 in 10,000 children (Zoghbi, 2003). RTT is characterized by mental retardation, deceleration of head growth, loss of purposeful hand movements, ataxia, loss of language, autistic features, seizures, and respiratory dysfunction. RTT infants appear normal at birth and develop normally until 6 to 18 months when they exhibit a progressive loss of any previous milestones, including any previously acquired language skills. RTT is an X-linked dominant disorder in which approximately 80% of RTT patients have detectable mutations in MECP2 within Xq28 (Amir et al., 2000; Bienvenu et al., 2000). Although RTT occurs almost exclusively in females, a hemizygous mutation in MECP2 is compatible with life, as rare cases of males with MECP2 mutations have been reported (Schanen, 2001). The phenotype of males with MECP2 mutations is much more severe, with death in infancy or early childhood. The almost exclusive occurrence of RTT in females has been explained by a bias of de novo MECP2 mutations in the paternal X chromosome (Girard et al., 2001).

Autism, RTT, and AS are all characterized by loss or impairment of language, stereotyped behaviors, and a high frequency of seizures and sleep abnormalities. RTT and AS both cause severe mental retardation, while approximately 70% of autistic patients are also mentally retarded (Polleux and Lauder, 2004). AS is usually apparent at birth, while the onset of both RTT and autism follows a period of apparently normal infant development that is often followed by a period of apparent regression in skills.

Genetic overlap has previously been suggested between AS and autism because the most common cytogenetic abnormality (1–3% of autism cases) is a maternal duplication of 15q11-13 (Schroer *et al.*, 1998). In addition, the 15q11-13 region has emerged as a candidate autism locus from multiple linkage and association studies (Buxbaum *et al.*, 2002; Kim *et al.*, 2002; McCauley *et al.*, 2004; Nurmi *et al.*, 2001, 2003). In one study, approximately 2% of AS patients had mutations in *MECP2* (Watson *et al.*, 2001) while in another three RTT patients had apparent small deletions within 15q11-13 (Renieri *et al.*, 2003). Several cases of infantile autism with a *MECP2* mutation have been described (Beyer *et al.*, 2002; Lam *et al.*, 2000), however, two large studies have determined that mutations or polymorphisms in the coding region of *MECP2* occur at a low frequency in autistic patients (Beyer *et al.*, 2002; Vourc'h *et al.*, 2001).

The combined phenotypic and genotypic evidence has set the stage for the emerging hypothesis that these three autism spectrum disorders share an overlapping molecular pathogenesis. This review will summarize and discuss the recent evidence supporting this hypothesis by linking MeCP2, expression of genes within 15q11-13, and neuronal nuclear organization in an important molecular pathway that is defective in all three disorders.

II. The 15q11-13 Imprinted Gene Cluster

A. Overview of Imprinting

Genomic imprinting, or preferential gene expression based on parent of origin, was discovered over two decades ago after mice with exclusive maternal or paternal genomes failed to develop (McGrath and Solter, 1984). At least 70 imprinted genes have been identified in mammals and as many as 100–200 imprinted genes are estimated to exist in our genomes (Murphy and Jirtle, 2003). There are several hallmark features of imprinted genes, including CpG islands, repetitive elements near or within the CpG islands, and clustering of maternally and paternally expressed genes (Reik and Walter, 2001). Clustering of imprinted genes likely allows coordinated expression of the differentially expressed genes within the domain and suggests that common control elements regulate imprinting for all genes in the same cluster (Verona *et al.*, 2003). The ~4 Mb imprinted gene cluster on human 15q11-13 is one of the most complex and least understood imprinted loci in the human genome. Imprinted genes within 15q11-13 are implicated in both PWS and AS, clinically distinct disorders that are caused by paternal or maternal deficiencies in 15q11-13, respectively. Due to the significance of this region to human disease much effort has been spent investigating the mechanisms governing gene expression within this complex locus.

B. PHENOTYPES OF PRADER-WILLI (PWS) AND ANGELMAN SYNDROMES

PWS and AS are both characterized by developmental, neurological, and behavioral abnormalities, although the phenotypes are distinct. PWS patients have hypotonia and growth defects early in life but later in childhood exhibit neurological and behavioral phentotypes such as mild mental retardation, obsessive-compulsive behavior, and obesity due to an insatiable appetite (Lalande, 1996; Nicholls and Knepper, 2001). In contrast, AS patients have severe mental retardation, lack of speech, ataxia, aggressive behavior, and most notably excessive inappropriate laughter. Most commonly these syndromes are caused by a *de novo* deletion of the 15q11-13 chromosomal locus or uniparental disomy of chromosome 15. In AS patients, maternal mutations in the imprinted gene *UBE3A* are also known to cause the disease (Kishino *et al.*, 1997; Rougeulle *et al.*, 1997; Vu and Hoffman, 1997). The protein product of *UBE3A* is a ubiquitin E3A ligase that functions to ubiquitinate proteins thus marking them for degradation via the proteosome (Jiang and Beaudet, 2004). Although *Ube3a* has been shown to affect p53 levels in the brain (Jiang *et al.*, 1998), a paucity of downstream targets have complicated understanding how deficits in *UBE3A* lead to the AS phenotype. No single gene has been identified to independently cause PWS, however the paternally expressed *NDN*, encoding necdin, is considered a strong candidate based on a similar phenotype of the *Ndn*-deficient mouse model (Gerard *et al.*, 1999; Lee *et al.*, 2005; Ren *et al.*, 2003). In a small percentage of patients, PWS or AS is caused by aberrant imprinting and gene silencing due to a reversal of the parental methylation status at the imprinting control region (ICR) (Buiting *et al.*, 2003).

C. 15q11-13 IMPRINTING MECHANISM

The 15q11-13 imprinted cluster contains both maternally and paternally imprinted genes including the paternally expressed *MKRN3*, *MAGEL2*, *NDN*, and *SNURF-SNRPN* transcripts and maternally expressed *UBE3A* and *ATP10A* transcripts (see Fig. 1 for more detail). Preferential gene expression within this ~4 Mb region strictly depends on the 4.3 kb imprinting control region (ICR) that overlaps with the *SNRPN* promoter (Sutcliffe *et al.*, 1994). On the maternal chromosome this locus exhibits heavy CpG methylation, while the paternal chromosome is almost completely unmethylated (Glenn *et al.*, 1993). Deletion of the paternal ICR has been observed in PWS patients, demonstrating that this PWS-ICR is required for paternal expression in the 15q11-13 region (Buiting *et al.*, 1995; Sutcliffe *et al.*, 1994). A 900 bp region 35 kb centromeric to the PWS-ICR is termed the AS-ICR because of its maternal deletion in AS patients (Buiting *et al.*, 1995). The AS-ICR establishes the maternal imprint switch during

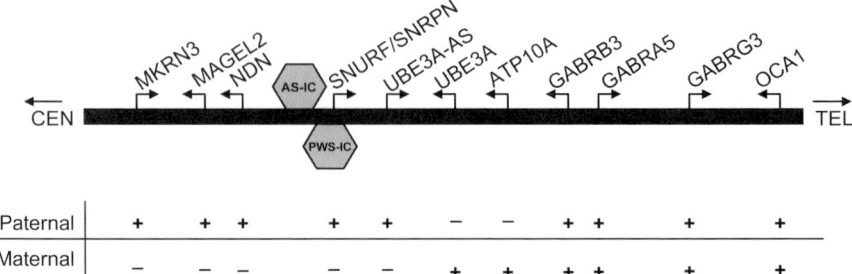

FIG. 1. A schematic map of parental imprinting of transcripts within 15q11-13. The roughly 6 MB region of 15q11-13 that is duplicated in autism patients is shown. Arrows are depicting the direction of each transcript and the expression patterns on the maternal and paternal chromosomes are shown in the lower table. The imprinting control regions for the paternal (PWS-ICR) and maternal (AS-ICR) chromosomes are shown.

early development, however the mechanism involved remains unclear (Buiting *et al.*, 2001; Dittrich *et al.*, 1996). The paternally expressed genes upstream of *SNRPN* exhibit differential methylation, with heavy methylation on the silent maternal chromosome. Unlike the paternally expressed transcripts, *UBE3A* does not exhibit differential methylation but instead is thought to be regulated by a paternally expressed antisense transcript, *UBE3A-AS*, which originates from the *SNRPN* promoter as part of an extensive transcription unit (Runte *et al.*, 2001). The exact termination point of the *UBE3A-AS* transcript is currently unknown and the possibility remains that this transcript also extends through *ATP10A* presumably to regulate imprinted expression of this gene as well. *UBE3A* appears to be imprinted exclusively in postnatal neurons in certain regions of the brain and the antisense transcript is only expressed in regions where imprinting of the paternal *UBE3A* occurs (Yamasaki *et al.*, 2003). Exactly how the ICR regulates imprinted gene expression throughout this entire 4 Mb domain remains unclear.

III. MeCP2: A "Rosetta Stone" for Autism?

A. BIOCHEMICAL PROPERTIES OF MeCP2

MECP2 encodes the founding member of a family of methyl CpG binding domain (MBD) proteins (Jorgensen and Bird, 2002). Methylated cytosine in the context of a CpG dinucleotide is associated with heterochromatin and transcriptionally silent genomic DNA, but until the discovery of MBD proteins, how this epigenetic mark resulted in a silent chromatin state was unclear. MeCP2 was originally isolated from rat brain nuclear extracts based on its ability to bind selectively to methylated CpGs (Lewis et al., 1992). Consistent with a role in transcriptional silencing, MeCP2 was found to colocalize with nuclear heterochromatin (Nan et al., 1996). The MeCP2 MBD domain has since been used to identify four other MBD proteins (MBD1-4) (Hendrich et al., 1999) and has identified MECP2 among other vertebrates such as frog, chicken, and zebrafish (Coverdale et al., 2004; Jones et al., 2001; Weitzel et al., 1997). In addition to having a conserved MBD, MeCP2 also contains a transcriptional repression domain (TRD) that has been shown to interact with the histone deacetylase complex to direct transcriptional repression of target genes (Jones et al., 1998; Nan et al., 1998). A third domain of a putative binding site for WW-domain proteins (such as splicing factors) has also been described (Buschdorf and Stratling, 2004). MECP2 mutations in RTT patients are found throughout the gene, but tend to fall into three categories: MBD missense mutations, TRD truncation mutations, and complex rearrangements and deletions in or near the C-terminal WW-binding domain (Zoghbi, 2003).

Although MeCP2 binds specifically to methylated DNA, its action on chromatin usually involves association with additional chromatin modifying enzymes involved in inactivating genes or compacting nucleosomes. MeCP2 interacts with the transcriptional repressor Sin3A and histone deacetylase (HDAC) through the TRD (Jones et al., 1998). In addition, MeCP2 associates with histone methyltransferase activity that directs the silencing histone H3 K9 methylation mark (Fuks et al., 2003). MeCP2 can also interact with the maintenance DNA methyltransferase, Dnmt1, when bound to hemimethylated DNA, providing a mechanism for maintaining inactive chromatin states (Kimura and Shiota, 2003). More recently, MeCP2 was also found to associate with Brahma, a factor normally associated with a chromatin remodeling complex that uses ATP to change nucleosome positions (Harikrishnan et al., 2005). These recently described partners for MeCP2 activity are all consistent with its role in inactive heterochromatin, but are diverging from the originally described role for MeCP2 as a simple transcriptional repressor.

B. MOUSE MODELS

Three different groups have established mouse models of either Mecp2 deficiency or mutation that recapitulate many of the features of RTT (Chen et al., 2001; Guy et al., 2001; Shahbazian et al., 2002a). Two groups used Cre-lox recombination to create Mecp2-null mice that show no apparent abnormalities at birth, but developed neurological symptoms around 4–6 weeks of age and died around 10 weeks (Chen et al., 2001; Guy et al., 2001). Conditional deletion of

Mecp2 only in post-mitotic neurons resulted in a similar phenotype (Chen *et al.*, 2001; Guy *et al.*, 2001). Another mouse model, the $Mecp2^{308/y}$ truncation mutant mouse, also showed a delayed phenotype with several important features of RTT, including motor impairments, seizures, hyopoactivity, and repetitive stereotyped forepaw clasping (Shahbazian *et al.*, 2002a). The $Mecp2^{308/y}$ mouse was also shown to have specific defects in social behavior, making this a potential genetic animal model for autism (Moretti *et al.*, 2005). These mouse models have established that MeCP2 is required for the normal functioning of the mammalian brain and that *Mecp2* mutation is sufficient to cause both RTT and autistic-like phenotypes in mice.

Mecp2 deficiency specifically targeted to post-mitotic neurons is sufficient to cause the neurological phenotypes exhibited in *Mecp2*-null mice (Chen *et al.*, 2001; Guy *et al.*, 2001). Likewise, the phenotypic abnormalities presented in *Mecp2*-null mice have been rescued by expression of transgenic *Mecp2* as a fusion protein with Tau in post-mitotic neurons (Luikenhuis *et al.*, 2004). Interestingly, expression of the same transgenic construct in wild-type mice, resulting in a two-fold increase in expression of *Mecp2*, caused severe motor dysfunction (Luikenhuis *et al.*, 2004). A different transgenic mouse model with 2-fold overexpression of *MECP2* under its own promoter also showed severe neurological abnormalities and death in adult mice (Collins *et al.*, 2004). The intolerance for too much or too little MeCP2 in mice has exposed the delicate balance of MeCP2 that is required for proper neuronal maturation during postnatal brain development.

C. Developmental and Tissue-Specific Regulation of MeCP2 Expression

Although the earliest investigation of *MECP2* expression by Northern blot analyses suggested ubiquitous expression in all tissues and stages of development (Coy *et al.*, 1999; D'Esposito *et al.*, 1996; Reichwald *et al.*, 2000), multiple recent studies using immunohistochemistry, immunofluorescence, or immunoblot analyses of MeCP2 protein expression have revealed a complex and highly controlled developmental expression pattern in the postnatal brain. In mouse, rat, macaque, and human, MeCP2 expression is heterogeneous in all regions of the brain, with highest expression in mature neurons and lowest expression in glia (Akbarian *et al.*, 2001; Balmer *et al.*, 2003; Braunschweig *et al.*, 2004; Cassel *et al.*, 2004; Cohen *et al.*, 2003; LaSalle *et al.*, 2001; Shahbazian *et al.*, 2002b). Increases in MeCP2 expression follow the developmental maturation of the CNS, with the earliest developing structures in the spinal cord and hindbrain showing highest MeCP2 expression before the cerebral cortex and hippocampus (Braunschweig *et al.*, 2004; Mullaney *et al.*, 2004; Shahbazian *et al.*, 2002b). In rodents, the olfactory bulb also shows high MeCP2 expression (Cassel *et al.*, 2004; Cohen *et al.*, 2003). Olfactory receptor neurons have been a useful experimental model for determining that increased MeCP2 expression occurs prior to synaptogenesis (Cohen *et al.*, 2003). Neuronal cell lines can also be induced to increase MeCP2 expression with maturational differentiation, consistent with the involvement in pre-synaptogenic events (Mullaney *et al.*, 2004). All of these studies have led to the emerging view that increased MeCP2 expression is a marker of mature neurons that coincides with synaptogenesis in the postnatal brain.

While some of the developmental regulation of MeCP2 is controlled at a transcriptional level (Samaco *et al.*, 2004), the discrepancies between RNA and protein studies for tissue-specific developmental regulation of MeCP2 (Shahbazian *et al.*, 2002b) suggests that multiple post-transcriptional mechanisms play a major role in regulating this dynamically expressed protein. At least four different isoforms exist for *MECP2* transcripts due to alternative splicing and alternative polyadenylation. Alternative splicing of *MECP2/Mecp2* exon 2 results in two different transcripts that encode proteins with different N-terminal amino acid sequences due to different translation start sites (Kriaucionis and Bird, 2004; Mnatzakanian *et al.*, 2004). Both MeCP2_e1 (*MECP2B*, *Mecp2*□B&A□(B) and MeCP2e2 (*MECP2A*, *Mecp2*□B&B□(B) contain the three known functional domains and appear to localize similarly to nuclear heterochromatin (Kriaucionis and Bird, 2004). MeCP2_el is more efficiently translated and more highly expressed in brain is more efficiently translated and more highly expressed in brain, but MeCP2_e2 appears sufficient to rescue the RTT phenotype in the *Mecp2*-null mouse when over-expressed (Luikenhuis *et al.*, 2004). Although a few mutations in *MECP2* exon 1 have been described in RTT patients (Mnatzakanian *et al.*, 2004; Ravn *et al.*, 2005), these are a rare cause of RTT (Amir *et al.*, 2005; Evans *et al.*, 2005). The functional significance of the two different N-termini of MeCP2 for the pathogenesis of RTT is still unclear.

The relevance of the two alternatively polyadenylated forms of *MECP2* is also not well understood, although there appears to be both tissue specific and developmental regulation of the process. *MECP2* has an extraordinarily large 3′ untranslated region (3′ UTR) that can add 8.5 kb to the 1.5 kb coding region (Coy *et al.*, 1999; Reichwald *et al.*, 2000). Both 10 kb and 1.9 kb bands are observed by Northern blot, but the long *MECP2* transcript is more prevalent in brain and fetal tissues (Coy *et al.*, 1999). No difference in stability was observed between the two transcripts (Reichwald *et al.*, 2000). In single cell *in situ* analyses of the two transcripts, the ratio of long 3′ UTR transcript to total *MECP2* was reduced with increasing postnatal age, but both transcripts showed higher levels in mature neurons (Samaco *et al.*, 2004).

In addition to variant transcripts of *MECP2*, some evidence for post-translational modifications that may regulate activity of MeCP2 is also emerging. Phosphorylation of MeCP2 was associated with its dissociation from the *BDNF* promoter following activity dependent stimulation of primary neurons (Chen *et al.*, 2003). The combined results of expression, molecular, and biochemical studies of MeCP2 all suggest that MeCP2 is an intricately regulated protein with multiple levels of control to ensure the correct timing and function in the developing brain. The mouse studies have also demonstrated that the correct dosage of MeCP2 is critical to normal brain development and suggest that multiple *trans* acting factors and pathways could potentially regulate MeCP2 expression.

D. MeCP2 Expression Abnormalities in Autism-Spectrum Neurodevelopmental Disorders

Since *MECP2* is an X-linked gene, RTT females are mosaic for mutant and wild-type expressing cells due to X chromosome inactivation (XCI). In some cases of X-linked dominant disorders skewing towards inactivation of the mutant X chromosome occurs (Hatakeyama *et al.*, 2004). Thus far this does not appear to be the case with *MECP2* mutations in the majority of human females (Zoghbi *et al.*, 1990). Some cases have been reported where a phenotypically normal female is a carrier of mutant *MECP2*, although these cases are relatively rare, as ~90% of RTT mutations are sporadic (Schanen, 2001). In the *Mecp2*-deficient mouse models, however, nonrandom XCI ratios favoring inactivation of the mutant allele were observed in two independent studies (Braunschweig *et al.*, 2004; Young and Zoghbi, 2004). This skewing toward wild-type expression may help to explain why the heterozygous female mice do not develop neurological impairments until adulthood, unlike the onset of RTT in infancy.

In RTT individuals and $Mecp2^{-/+}$ female mice, not only is the overall expression of MeCP2 lower, but the wild-type expressing neurons are unable to reach a normal high level of MeCP2 expression (Braunschweig *et al.*, 2004). This indicates that *MECP2* mutation negatively impacts even wild-type expressing neurons in the mosaic RTT brain, potentially by not providing the appropriate local environment for proper neuronal synaptogenesis. Despite having reduced brain size, RTT post-mortem brain tissue is relatively normal. This reduction is due to a decrease in the size of neuronal cell bodies as well as a higher packing density of neurons within several brain regions including the prefrontal, post-frontal, and anterior temporal regions (Subramaniam *et al.*, 1997). In addition, RTT brains have more tightly arranged minicolumn structures, similar to autism brain (Casanova *et al.*, 2002, 2003).

To test the hypothesis that abnormal MeCP2 expression could result from multiple genetic and developmental abnormalities, MeCP2 expression levels were investigated in RTT and other autism-spectrum post-mortem cerebral cortex samples by quantitative immunofluorescence and laser scanning cytometry (Samaco *et al.*, 2004). Although a defect in MeCP2 expression was expected in cases of RTT with known *MECP2* mutations, MeCP2 expression abnormalities were also found in samples from autism, AS, and PWS compared to age-matched controls. Expression was examined at the RNA level and protein levels and a complex pattern of dysregulation was revealed, suggesting that multiple pathways regulating MeCP2 expression could be aberrant in different autism cases. Although some reports of *MECP2* mutations have been found in a few cases of idiopathic autism (Beyer *et al.*, 2002; Lam *et al.*, 2000), the autism individuals reported lacked coding mutations in *MECP2* (Samaco *et al.*, 2004).

E. MeCP2 Target Genes

Although much effort has been placed in determining the downstream targets for MeCP2 regulation, to date few genes have been found. Expression profiling analysis of RTT and control human brain reported less than 5% of genes significantly altered, with more genes up-regulated than down-regulated (Colantuoni *et al.*, 2001). Microarray analysis of transcriptional differences between *Mecp2*-null and wild-type mouse brain samples also found very subtle effects on gene expression (Tudor *et al.*, 2002). Several careful studies have been successful in identifying genes that are subtly affected by aberrant levels of MeCP2. Two laboratories independently showed that the brain derived neurotropic factor (BDNF) gene is directly regulated by MeCP2 (Chen *et al.*, 2003; Martinowich *et al.*, 2003). A recent study by Horike *et al.* (2005) cloned and sequenced MeCP2 binding sites to screen for MeCP2 target genes. This strategy identified a cluster of imprinted DLX genes that are regulated by MeCP2. Although several other imprinted genes such as *SNRPN* and *H19* have been examined and do not show aberrant expression in the absence of MeCP2 (Balmer *et al.*, 2002; Samaco *et al.*, 2005), *DLX5* showed an apparent loss of imprinting in RTT lymphoblast cultures (Horike *et al.*, 2005).

Other reports (Jiang *et al.*, 2004; Makedonski *et al.*, 2005; Samaco *et al.*, 2005) have speculated that MeCP2 controls gene expression at another imprinted locus in humans and mice, the PWS/AS imprinted cluster at 15q11-13. This locus, illustrated in Fig. 1, contains genes that exhibit parent-of-origin specific gene expression, as discussed in the previous section. Deficiency in MeCP2 results in a decrease in gene expression of two genes within this locus, *Ube3a* and *Gabrb3*, although it does not affect expression of a paternally expressed transcript, *Snrpn* (Samaco *et al.*, 2005). MeCP2 has been shown to physically bind to the imprinting control region (ICR) on the methylated maternal allele of the *Snrpn* promoter (Makedonski *et al.*, 2005; Samaco *et al.*, 2005; Thatcher *et al.*, 2005), however MeCP2 deficiency does not result in maternal *Snrpn* transcription (Balmer *et al.*, 2002; Samaco *et al.*, 2005). Curiously, MeCP2 was not found to be associated with the promoter of either *Ube3a* or *Gabrb3* although expression of these genes is significantly reduced in the absence of MeCP2 (Samaco *et al.*, 2005). Although Makedonski (2005) reported aberrant maternal expression of the *Ube3a-as* transcript in *Mecp2*-deficient mice, the loss of imprinting was not observed in a different mouse strain cross and no difference in the expression level of *Ube3a-as* was observed (Samaco *et al.*, 2005). Presumably other factors in addition to the *Ube3a-as* can control the level of expression of *Ube3a*.

Surprisingly, reduced expression in both RTT and *Mecp2*-deficient mouse brain was also observed for *GABRB3/Gabrb3*, located ~1 MB distal to *UBE3A/ Ube3a* (Samaco *et al.*, 2005). Although no direct binding of MeCP2 to the *GABRB3 CpG* island has yet been reported, regulation of this gene by MeCP2 may resemble that of *Dlx5* (Horike *et al.*, 2005) or *Igf2/H19* loci (Murrell *et al.*, 2004), with long-range effects due to chromatin loop structure.

IV. GABRB3 Expression in Normal Development and Autism

A. The GABA$_A$ Receptor Gene Family

GABRB3 and two other GABA$_A$ receptor subunit genes (*GABRA5* and *GABRG3*) present on 15q11-13 are attractive candidate genes for autism due to evidence for positive linkage, common occurrence of 15q11-13 cytological abnormalities, functional relevance to autistic features, and recent molecular evidence of high frequency expression defects in autism brain. GABRB3 is one of 19 different mammalian subunits that compose the pentameric GABA$_A$ receptor (Barnard *et al.*, 1998). This transmembrane receptor, upon binding of GABA$_A$ forms a chlorine channel that allows rapid membrane depolarization and resulting inhibitory synaptic neurotransmission in the brain. Although there are seven different classes of receptor subunits, the α, β, and γ subunits comprise the majority of GABA$_A$ receptors (Barnard *et al.*, 1998). There are six different α subunits, four β subunits, and three γ subunits which are used in different stoichiometries to form the majority of GABA$_A$ receptors (Pirker *et al.*, 2000). The GABA$_A$ receptor genes exhibit an interesting distribution in the human genome in that all are present in clusters consisting of at least one of each of the three main subunits (McLean *et al.*, 1995). GABA$_A$ receptor clusters are present on chromosomes 4p11-13, 5q34-q35, 6q14-q21, 15q11-13, and Xq28. These clusters are hypothesized to have originated from gene duplication events from an ancestral GABA$_A$ receptor cluster (McLean *et al.*, 1995).

Despite the large variety of subunits not all GABA$_A$ receptor subunits are expressed equally. Extensive studies of the mRNA and protein distributions in rat brain have shown that $\alpha 1$ and $\alpha 2$ as well as all three β subunits are widely distributed throughout the brain (Laurie *et al.*, 1992; Pirker *et al.*, 2000). These data indicate that these constituents are likely present in almost all varieties of GABA$_A$ receptors, while other subunits including the $\gamma 3$ subunit are restricted to very few brain areas (Pirker *et al.*, 2000). The expression patterns of the GABA$_A$ receptor subunits are consistent with the phenotypes of mouse knockouts of the syntenic 15q11-13 GABA$_A$ receptor cluster on murine 7qB4. A deletion that included the mouse *p* locus (pink eye dilution, the mouse homologue to the *OCA2* gene in human 15q11-13) through the GABA$_A$ receptor cluster was mostly lethal at birth, with rare severely neurologically affected survivors (Nicholls *et al.*, 1993). Mice with deletions encompassing only the *Gabrg3* or both *Gabrg3* and *Gabra5* chromosomal regions were phenotypically normal, indicating that the lethality seen in the larger deletions was due to absence of *Gabrb3* (Culiat *et al.*, 1994). The importance of *Gabrb3* was further illustrated in mice with a disruption of the promoter and exons 1–3 of *Gabrb3* that exhibited a high frequency of neonatal lethality and reduced lifespan (Homanics *et al.*, 1997). In addition, the phenotypic manifestations of *Gabrb3* disruption included cleft palate, presumably contributing to the early fatalities, and neurological abnormalities including hyperactivity, hypersensitive behavior, and seizures (DeLorey and Olsen, 1999; DeLorey *et al.*, 1998). The vast difference in the severe phenotypic manifestation of removal of the *Gabrb3* subunit and the normal phenotype in mice with deletions of the *Gabra5* and *Gabrg3* subunits indicates that *Gabrb3* is an important constituent of GABA$_A$ receptors in the brain.

B. IMPRINTING STATUS OF THE 15q11-13 GABA$_A$ RECEPTOR CLUSTER

Early mouse studies involving deletions of the GABA$_A$ receptor cluster on mouse chromosome 7 (syntenic to human 15q11-13) demonstrated that expression of these genes is biallelic (Nicholls *et al.*, 1993). In addition, analysis of a single nucleotide polymorphism (SNP) in interspecies hybrid and wild-type mice showed biallelic expression of *Gabrb3* in mouse brain (Nicholls *et al.*, 1993; Yamasaki *et al.*, 2003). However, several studies have suggested that parental origin affects *GABRB3* expression. Since polymorphisms within coding genes are rare in human GABA$_A$ receptor genes, determining parent-specific expression can be difficult. A study by Meguro *et al.* (1997) utilized microcell-mediated chromosome transfer to produce hybrid mice with a single human chromosome 15. Cell lines containing maternal or paternal human chromosome 15 were analyzed for expression of known imprinted genes and found to maintain proper paternal-only expression of *SNRPN* and *IPW*. Expression of the three GABA$_A$ receptor genes in this system only occurred when the human chromosome was of paternal origin (Meguro *et al.*, 1997). Although this system is artificial, two recent reports from actual human samples corroborate preferential paternal expression of the 15q11-13 GABA$_A$ receptor genes (Bittel *et al.*, 2003, 2005). These two studies conducted 15q11-13 gene expression profiles of lymphoblasts from individuals with AS and PWS, arising from 15q11-13 deletions or uniparental disomy. Comparing cell lines with AS and PWS, *GABRB3* and *GABRA5* showed paternal expression bias (Bittel *et al.*, 2003, 2005). These results pose interesting considerations for the 15q11-13 GABA$_A$ receptor cluster containing candidate genes for autism. Most linkage studies presume equal parental inheritence, however if the gene is only expressed from one allele then significant findings may be masked. Recently, an association study was conducted for GABA$_A$ receptor genes and alcohol dependence (Song *et al.*, 2003). In this study, upon considering a model of paternal expression of *GABRB3* and *GABRA5*, significant association was found, while no significant findings were reported for maternal transmission. Such considerations may be important in analyzing *GABRB3* contribution to autism as well as in searches for candidate genes in autism pedigrees.

C. Genetic Linkage and Association Studies of *GABRB3* and 15q11-13 in Autism

Since autism is a complex trait with potential oligogenic or polygenic inheritance, genome-wide scans for susceptibility alleles have yielded inconsistent results. To date multiple genome-wide scans have been completed and the results regarding 15q11-13 are mixed (Wassink and Piven, 2000). Early studies of linkage disequilibrium supported a role for chromosome 15 in autism (Cook *et al.*, 1998) although this locus has not been consistently identified in all scans (Martin *et al.*, 2000). Although genome-wide approaches with widely distributed markers have failed to find highly significant linkage evidence for 15q11-13, more recent association studies with dense mapping over smaller regions, as well as use of phenotypic subtypes have increased the evidence for autism susceptibility genes in 15q11-13 (McCauley *et al.*, 2004; Nurmi *et al.*, 2003; Shao *et al.*, 2003). One study used dense SNP coverage across the maternally expressed regions of 15q11-13 and found significant evidence for transfer of susceptibility in two SNPs around *ATP10A* upstream of *GABRB3* (Fig. 1) (Nurmi *et al.*, 2003). A later investigation specifically analyzed the GABA$_A$ receptor cluster on 15q11-13 with a dense coverage of markers and found marginally significant evidence for association with *GABRB3* and *GABRA5* (McCauley *et al.*, 2004). Perhaps the most significant evidence for linkage within 15q11-13 was gathered by using a subtype of autism families with the characteristic of insistence on sameness (IS) (Shao *et al.*, 2003). The LOD score for the *GABRB3* locus was increased from 1.45 (insignificant) to a highly significant score of 4.71 by narrowing the analysis down to only families sharing high scores for the IS character. Since autism is a complex disorder with multiple etiologies, the use of phenotypic subtypes, or endophenotyping, may be necessary to remove much of the noise generated from genome scans for linkage.

D. Cytogenetic Abnormalities of 15q11-13 in Autism

Despite the variable support for autism candidate genes by linkage and association studies, the cytogenetic support for 15q11-13 genes' involvement in autism is compelling. Duplications of maternal 15q11-13 are the most common cytogenetic abnormalities reported in autism, occurring in 1–3% of autistic individuals (Cook et al., 1997). Duplications of 15q11-13 are the reciprocal meiotic products of a maternal deletion that results in the related neurodevelopmental disorder Angelman syndrome (AS). AS individuals often exhibit comorbid autism (Peters et al., 2004), which further indicates that abnormalities in genes in 15q11-13 are associated with autism. Deletions of 15q11-13 are the most common cause of AS and result in a more severe phenotypic manifestation of the disease than mutations in UBE3A. The phenotypic disparity in AS patients with large deletions and UBE3A mutations indicates that the GABA$_A$ receptor genes contribute to the phenotype of this disorder (Moncla et al., 1999). Since autism 15q11-13 abnormalities usually arise from maternal duplications it is unclear how gene expression is modified. One report has indicated that expression of UBE3A in a lymphoblast cell line with maternal 15q11-13 duplication is increased (Herzing et al., 2002), however the expression of other genes within the duplicated region, including the GABA$_A$ receptor genes, has not yet been examined. Thus, although 15q11-13 abnormalities are clearly associated with autism, the consequences of the duplication and the effect on gene expression still remain unclear.

E. Abnormal Expression of GABRB3 in Autism Brain

Based on evidence for the involvement of 15q11-13 genes in autism disorder, molecular studies have been conducted to examine expression of genes within this region (Jiang et al., 2004; Samaco et al., 2005). Jiang et al. (2004) analyzed UBE3A protein levels in a panel of autism post-mortem cerebellum and cerebral cortex samples, and found reduced levels of UBE3A in some but not all autism samples compared to controls. This result was also confirmed by Samaco et al. (2005) using an independent method of quantitative immunofluorescence and laser scanning cytometry. Additionally, expression of GABRB3 was analyzed and significant reduction was found in more than 50% of autism brain samples (Samaco et al., 2005). Other related neurodevelopmental disorder samples including AS and RTT were also analyzed and found to have lower levels of GABRB3 expression (Samaco et al., 2005). Although this result is expected for AS patients with a deletion of maternal 15q11-13, reduced GABRB3 levels in RTT, a genetically distinct condition, is surprising. In addition to sharing a defect in GABRB3 expression, several of the same post-mortem autism brain samples have been shown to exhibit defects in MeCP2 expression (Samaco *et al.*, 2004). Direct evidence for a causal effect of MeCP2 on 15q11-13 genes was obtained by examining gene expression of *Ube3a* and *Gabrb3* in *Mecp2* null mice. Both *Ube3a* and *Gabrb3* were found to be reduced in MeCP2 deficient mice, although the levels of reduction were not as significant as AS and autism human samples (Samaco *et al.*, 2005). Furthermore, an independent study confirmed that UBE3A expression is reduced in RTT brain samples and *Mecp2* deficient mice (Makedonski *et al.*, 2005).

Despite affecting the level of expression of *UBE3A* and *GABRB3*, MeCP2 does not affect the expression level of the paternally expressed *SNRPN* (Balmer *et al.*, 2002). Even more surprising is the result that MeCP2 binds to the methylated *SNRPN* promoter on the maternal allele of the PWS-ICR (Makedonski *et al.*, 2005; Samaco *et al.*, 2005; Thatcher *et al.*, 2005), but not to the promoters of *Ube3a* or *Gabrb3* (Samaco *et al.*, 2005). Although a model involving MeCP2 regulation of *UBE3A* through loss of imprinting of the maternal *UBE3A-AS* transcript has been proposed (Makedonski *et al.*, 2005), this model fails to explain how MeCP2 regulates other genes, such as *GABRB3*, within the region. Thatcher *et al.* (2005) have reported that homologous pairing occurs at chromosome 15 at the *GABRB3* locus during normal development in human cerebral cortex. Interestingly, this pairing is absent in autism, RTT, and AS samples. Perhaps a new model that incorporates emerging hypotheses about higher order chromatin and nuclear organization is justified in light of these apparently paradoxical new findings. Thus, elucidation of the mechanism used by MeCP2 to regulate gene expression of 15q11-13 may help explain a pathway important in the complex etiology of autism.

V. The Dynamics of Chromatin and Nuclear Organization within the Maturing Neuron

A. THE DYNAMIC INTERPHASE NEURONAL NUCLEUS AND GLOBAL TRANSCRIPTIONAL REGULATION

Chromosomal organization is most commonly considered in a metaphase state, and chromosomal movements are best studied in the context of cells undergoing mitosis or meiosis. The organization of chromosomes in interphase is often disregarded. Recent studies using tools to determine nuclear organization, however, have revealed the highly organized and dynamic chromosome architecture within the interphase nucleus. Non-cycling differentiated cells can vary in shape, nuclear volume, size and number of nucleoli, and nuclear protein distribution even more than their replicating counterparts. In addition, the organization and distribution of condensed heterochromatin is similarly maintained within a given cell lineage but can differ significantly between cell types (Leitch, 2000).

Nuclei can undergo highly dynamic changes in their DNA and chromatin organization even after initial differentiation. This is particularly evident in the mammalian neuronal nucleus which undergoes a highly ordered pattern of chromatin re-organization during neuronal maturation (Manuelidis, 1990; Martou and De Boni, 2000). In immature neurons the nucleus is small and heterochromatic and contains multiple nucleoli. As the neuron matures its nucleus increases greatly in size and a large single nucleolus begins to form (Manuelidis, 1990). The nucleus becomes highly euchromatic with the hetero-chromatin centralized around the nucleolus and in a complex web-like network throughout the nucleus. In murine neurons the heterochromatin becomes concentrated into two large heterochromatic foci located on either side of the large single nucleolus. These changes in neuronal nuclear organization with maturational differentiation coincide with an increase in levels of transcription and the need for a large nucleolus to transcribe sufficient ribosomal RNA to efficiently translate the new transcripts.

Dynamic chromosome movements have been previously implicated in neuronal maturation and neurodevelopmental disorders. Global nuclear changes in Purkinje neurons located in the cerebellum have been experimentally linked to increased dendritic differentiation and synaptic maturation (Martou and De Boni, 2000). In addition, changes in X chromosome positioning within the human neuronal nucleus have been shown to coincide with epilepsy and changes in cellular activities (Borden and Manuelidis, 1988). Interestingly, addition of gamma-aminobutyric acid (GABA) to neurons has been shown to significantly alter the arrangement of heterochromatin at chromosome centromeres, suggesting a new transcriptional state (Holowacz and De Boni, 1991).

B. MULTIPLE LAYERS OF TRANSCRIPTIONAL REGULATION

Gene regulation is a highly ordered process involving multiple layers of regulation, from DNA methylation, histone modifications, and nucleosome assembly on linear DNA to spatial positioning of chromosomes in nuclear three dimensional space. Linear gene positioning and clustering with respect to repetitive DNA and other genes can often dictate a gene's transcriptional activity. CpG methylation often regulates vertebrate gene transcription (Bird, 1984) and serves an important role in regulation of imprinted genes (Li et al., 1993). Nucleosome compaction of linear DNA can be modified by methylation, acetylation, phosphorylation, and ubiquitination of histone tails (Jenuwein and Allis, 2001). Modifications of these histones create a "histone code" that alters nucleosome structure leading to changes in DNA accessibility to transcriptional machinery, transcription factors, and other nuclear proteins (Jenuwein and Allis, 2001). Nucleosomes serve to compact the expansive amounts of DNA found in each cell, acting at the primary level of chromatin structure. A secondary conformation is made of chromatin loops anchored to a nuclear matrix scaffold. Chromatin loops have now been detected by new molecular techniques in mouse and Drosophila, and range in size from 15–400kb (Byrd and Corces, 2003; Chambeyron and Bickmore, 2004).

All of these local modifications to single genes or coordinately regulated gene clusters are part of a larger global organization of the nucleus that affects genome-wide expression. Chromosome organization appears to be an ordered arrangement in the interphase nucleus as seen by fluorescent *in situ* hybridization (FISH) of whole chromosomes (Cremer *et al.*, 1993). Each chromosome appears to maintain fairly distinct chromosome territories with only partial overlap between chromosomes (Kosak and Groudine, 2004). Reviews have speculated that transcriptional regulation is one determinant of a highly organized interphase nucleus (Kosak and Groudine, 2004; Leitch, 2000). Precise nuclear positioning of a chromosomal domain may facilitate the expression of the genes within that region by making it more or less accessible to transcription factors and enable *cis* or *trans* interactions with enhancers. A gene's activity can often be correlated to its nuclear position and distance from constitutive heterochromatin (Brown *et al.*, 1997, 1999, 2001; Francastel *et al.*, 1999; Wang *et al.*, 2004). Initial studies showed that active genes were more susceptible to DNAse I degradation than inactive genes, providing evidence that active genes are in a more open chromatin structure (Weintraub and Groudine, 1976). As a recent example, the murine *Hoxb* gene has been shown by FISH to move out of its chromosome territory upon transcriptional activation and back inside its territory when inactive (Chambeyron and Bickmore, 2004). Active genes from different chromosomes have also been shown to colocalize in specific areas of the nucleus, indicating nuclear regions of increased transcriptional activity and movement outside of chromosome territories (Osborne *et al.*, 2004).

Originally, transcription machinery was thought to be assembled at each active gene throughout the genome regardless of the gene's location. However, recent data suggest that transcription may be localized to discrete areas of the nucleus known as transcription factories where transcribed genes are localized (Jackson *et al.*, 1993, 1998). Evidence for transcription factories includes staining patterns for RNA polymerase II, the primary polymerase of gene transcription, that show punctuate foci staining in multiple cell types pinpointing discrete areas of transcriptional activity (Iborra *et al.*, 1996). The number of RNA polymerase II foci within a single nucleus are too few to account for only one gene being transcribed at each focus (Grande *et al.*, 1997; Zaidi *et al.*, 2002). Active genes from multiple chromosomes which have been shown to localize together by FISH have also been shown to colocalize with a single focus of RNA polymerase II (Osborne *et al.*, 2004).

Transcription factories are thought to interact with clusters of active genes looping out of their chromosome territories, resulting in "active chromatin hubs." One of the best characterized active chromatin hubs, the human β-globin locus, has been found to have a dynamic loop structure which brings active genes in the region into direct spatial orientation with the locus control region (LCR) (de Laat and Grosveld, 2003). Hypersensitive sites are located close to the LCR while the inactive genes are looped out away from the LCR and hypersensitive sites. Deletions of the LCR and specific hypersensitive sites showed that these regions are critical for loop formation (Patrinos *et al.*, 2004). The region is stabilized by *cis*-regulatory elements and high levels of histone acetylation protecting against heterochromatic silencing. Interestingly, in brain, where none of the genes in the β-globin locus are active, the region forms a linear structure in the interphase nucleus (Tolhuis *et al.*, 2002).

A similar active chromatin hub and looping structure may facilitate imprinted gene expression since most imprinted genes occur in clusters and often appear to be regulated by an ICR or differentially methylated region (DMR) that controls gene expression over long distances. These features of long-range control and association with hypersensitive sites make ICRs similar to the LCR of the β-globin locus. PWS and BWS ICRs, for example, regulate gene imprints bi-directionally over several megabases (Soejima and Wagstaff, 2005). Chromatin loop structure is determined by the methylation status, histone modifications, or proteins interacting with the ICR or DMR and can dictate the expression pattern of each gene within the region. Evidence to support this model has been shown at the *Igf2/H19* imprinted locus in mouse. The DMRs in this region interact to produce distinct maternal and paternal chromatin loops that regulate the differential gene expression of each homolog (Murrell *et al.*, 2004).

The combined concepts of transcription factories and active chromatin hubs create a dynamic model by which active gene regions loop out of heterochromatin to transcription factories to be brought into proximity with not only RNA polymerase II, but possibly many other nuclear proteins and enzymes which can facilitate interactions that regulate transcription. This exciting concept of dynamic chromosomal changes in the context of a global nuclear architecture opens a new realm of possibilities to investigate gene regulation during cellular differentiation and maturation.

C. A ROLE FOR MeCP2 IN TRANSCRIPTIONAL REGULATION AND CHROMATIN ORGANIZATION

MeCP2, classically defined as a methylated DNA binding protein that re-presses gene expression, now appears to have multiple roles at different levels of gene regulation. In the simplest transfected constructs, MeCP2 has been shown *in vitro* to repress genes with methylated CpG promoters through its transcription repression domain (TRD) (Nan *et al.*, 1997). MeCP2 has also been shown to interact with histone deacetylase 1 (HDAC) (Jones *et al.*, 1998; Nan *et al.*, 1998) and histone methyltransferase (Fuks *et al.*, 2003) and is therefore thought to be involved in increased histone methylation which generally leads to further DNA compaction and silencing. In addition, MeCP2 has the ability to bind non-methylated DNA when present in high concentrations and to compact naked DNA independently of other factors (Georgel *et al.*, 2003).

Recently Bowen *et al.* (2004) proposed both enzymatic and structural models for MeCP2's involvement in DNA silencing and compaction. The enzymatic model predicts that MeCP2 binds to methylated CpGs and recruits histone modifying enzymes to repress transcription. The structural model suggests a role for MeCP2 in chromatin compaction independent of histone modifiers. Bowen *et al.* (2004) suggest a mixture of the two models, particularly in brain, where higher levels of MeCP2 may even cause MeCP2 to interact with itself or with nucleosomes to create conformational chromatin changes. The staining pattern of MeCP2 in neurons specifically to areas of heterochromatin in the nucleus supports MeCP2 as a factor involved in both gene silencing and heterochromatin condensation (see Fig. 2).

MeCP2 expression is particularly high in mature neurons which have increased transcription and euchromatin (Balmer *et al.*, 2003). This relationship may be a function of the high level of chromatin organization in the mature neuron that requires careful packaging and silencing of specific regions of chromatin while allowing high levels of transcription from others. MeCP2 may also serve to silence genes that inhibit terminal differentiation of neurons. Additionally, MeCP2 may contribute to global chromatin organization by mediating the formation of chromatin loops, and may be of particular importance in the transcriptional regulation of complex clusters of imprinted genes, many of which are expressed in brain. Evidence for this has recently been shown in mouse brain where MeCP2 creates a 11 kb silent-chromatin loop at the *DLX* imprinted region (Horike *et al.*, 2005). The silent loop is absent in *Mecp2* deficient male mice where expression of *DLX5* is two times greater than in wild-type mice. The MeCP2 mediated silent chromatin loop is associated with methylated histone H3 K9, while the active loop, found in the *Mecp2* deficient mice is associated with acetylation of histone H3 K9 and K14.

While no loop structures have yet been defined for the imprinted 15q11-13 locus, three recent papers have demonstrated that MeCP2 binds to the maternal methylated allele of the *SNRPN/Snrpn* promoter in human and mouse brain and human neuroblastoma cells (Makedonski *et al.*, 2005; Samaco *et al.*, 2005; Thatcher *et al.*, 2005). This region is also associated with dimethylation of histone H3 K9 on the maternal allele, and is hyper-acetylated on the paternal allele (Soejima and Wagstaff, 2005). While MeCP2 was not found to be associated with

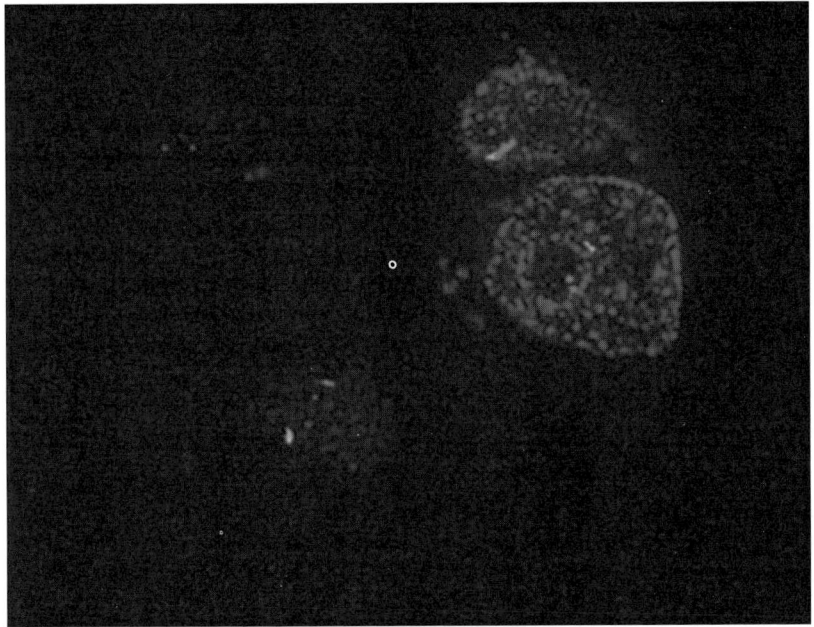

FIG. 2. Homologous pairing of 15q11-13 loci within MeCP2 high expressing neurons in human brain. A normal human cerebral cortical section stained for MeCP2 (blue), chromosome 15 centromere (green), and 15q11-13 (*SNRPN*, red). Two nuclei with high levels of MeCP2 expression (top right) show homologous association of 15q11-13 domains around the nucleolus. In contrast, a neuronal nucleus with low MeCP2 expression (bottom left) shows no homologous pairing. (See Color Insert.)

either the promoter for *Ube3a* or *Gabrb3* (Samaco *et al.*, 2005), MeCP2 may bind at other promoters, perhaps at *NDN* or *MKRN3* which are differentially methylated (Soejima and Wagstaff, 2005). Alternatively, MeCP2 may bind to intergenic or intronic regions throughout 15q11-13 to produce maternal or paternal specific chromatin loop structures, thus regulating transcription of the entire region. Interactions of MeCP2 binding sites along 15q11-13 may even facilitate the interaction of separate gene clusters on different chromosomes.

E. MeCP2 and Homologous Pairing of 15q11-13

Chromatin movement in the neuronal interphase nucleus may facilitate the ongoing maturation and growth of mammalian neurons by orienting genes for transcription that promote increased dendritic branching and synaptogenesis. This dynamic organization may also regulate imprinted genes, many of which are important for proper brain development.

Specific homologous pairing of chromosomes has been previously described in neuronal nuclei and Sertoli cells including association of 9q12 and 1q12 with the nucleolus (Manuelidis, 1990; Manuelidis and Borden, 1988) and somatic pairing of homologous chromosomes 1 and 17 (Arnoldus *et al.*, 1989, 1991). The phenomenon of homologous pairing in interphase nuclei is generally limited to non-cycling differentiated cells (Leitch, 2000), however, transient somatic pairing of 15q11-13 and the centromere of chromosome 15 has been described in late S-phase of cycling human lymphocytes. Lymphocytes derived from PWS and AS patients with 15q11-13 deletions showed a deficiency in this pairing at the centromere (LaSalle and Lalande, 1996), demonstrating that homologous pairing of chromosome 15 is dependent on having both maternal and paternal copies of 15q11-13. Chromosome 15 is one of five human acrocentric chromosomes whose p-arms carry the genes for rRNA. These acrocentric chromosomes come together to form a single large nucleolus in mature neurons.

Evidence for homologous pairing of the imprinted 15q11-13 locus has been observed in the frontal cortex of human brain and in a differentiating human neuroblastoma cell line (Thatcher *et al.*, 2005). Using DNA FISH probes for *GABRB3* and the centromere of chromosome 15, a significant increase was observed in the homologous association of this region from infancy to adulthood in human frontal cortex that was not seen for another acrocentric chromosome. A similar increase in chromosome 15 homologous pairing was observed in the SH-SY5Y neuroblastoma cell line following induced differentiation in culture. Approximately 90% of all nuclei in the human frontal cortex exhibiting chromosome 15 pairing were large mature neuronal nuclei expressing high levels of MeCP2 (Fig. 2, unpublished data). Homologous 15q11-13 pairing was deficient in RTT, autism, PWS, and AS brain samples suggesting a similar pathway that is aberrant in all four disorders. Experimental evidence for the involvement of MeCP2 in 15q11-13 homologous pairing was obtained by transfecting the SH-SY5Y cells with an excess of a MeCP2 decoy, a methylated 22 bp DNA fragment which binds MeCP2, decreasing its binding to endogenous DNA within the nucleus. Homologous pairing was significantly blocked by the MeCP2 decoy, implicating MeCP2 in the mechanism (Thatcher *et al.*, 2005). Jugloff *et al.* (2005) showed that primary mouse cortical neurons over-expressing MeCP2 show increased axonal length and dendritic branching, perhaps due to altered chromatin conformation induced by higher levels of MeCP2 creating a nucleus transcriptionally poised for neuronal maturational changes.

The combination of aberrant MeCP2 expression in autism, PWS, AS, and RTT brain, decreased *UBE3A* and *GABRB3* expression in RTT and autism brain, and binding of MeCP2 to the PWS-IC provides a compelling link between MeCP2 and the imprinted region 15q11-13 and suggests a mechanism linking all of these disorders together.

VI. A Higher Order Model for MeCP2 in Regulating Gene Expression within 15q11-13

The findings supporting a role for MeCP2 in regulating the PWS/AS region and the *DLX* imprinted region have suggested that MeCP2 does not simply act in *cis* to repress methylated genes. Evidence for chromatin loop changes in *Mecp2*-deficient mouse brain in the *DLX* locus (Horike *et al.*, 2005) suggests that MeCP2 may form a major structural basis for chromatin loop formation. Higher order nuclear organization models have suggested areas of high transcriptional activity, so recruitment to these regions by MeCP2 and associated factors could have a major impact on the expression of genes such as *UBE3A*, *ATP10A*, *GABRB3*, and potentially *GABRA5* and *GABRG3*.

Although much work lies ahead to define the other binding sites for each parental allele across this 2–3 MB region, a speculative model is shown in Fig. 3A that diagrams how MeCP2 binding sites may serve to colocalize the promoters of *UBE3A*, *ATP10A*, and *GABRB3* with the imprinting control regions (ICR) on either the paternal or maternal chromosome (pink for maternal and blue for paternal ICR). Just as for the more simple loop structures of *H19* and *Igf2* (Murrell *et al.*, 2004), allele-specific methylation patterns would result in different chromatin loop structures on each parental allele causing *SNRPN* to be silent on the maternal and *UBE3A* and *ATP10A* to be reduced on the paternal chromosome. While the loss of MeCP2 is not sufficient to change the imprinted expression patterns, it could result in the inaccessibility of the promoters of *Ube3a* and *Gabrb3* to the ICRs, thus reducing expression on both parental alleles.

While the model in Fig. 3A would be consistent with the experimental results for both mouse and human, human 15q11-13 has the added complexity that it is on an acrocentric chromosome that is tightly linked to nucleolar organization and undergoes homologous pairing during neuronal maturation (Thatcher *et al.*, 2005; Fig. 2). The model in Fig. 3B therefore incorporates the nuclear and nucleolar changes that were discussed in the previous section. During neuronal maturational differentiation, MeCP2 expression increases and MeCP2-associated heterochromatic regions fuse and form around the outer perimeter of the larger nucleolus. As the p arms of chromosome 15 contain repeated rDNA genes, the 15q11-13 regions are recruited to this region as well as other acrocentric chromosomes. However, the 15q11-13 region has an additional "glue," perhaps MeCP2 itself or the combination of MeCP2 and other proteins or regulatory RNAs, that allows a tighter interaction than other acrocentric chromosomes (LaSalle and Lalande, 1996; Thatcher *et al.*, 2005). It is predicted that both alleles of *UBE3A* and *GABRB3* would benefit from this interaction simply by being brought into closer proximity to the ICR and active chromatin hub on the opposing chromosome. In this way, homologous pairing would be required for the optimal level of expression of *GABRB3* and *UBE3A* without invoking a

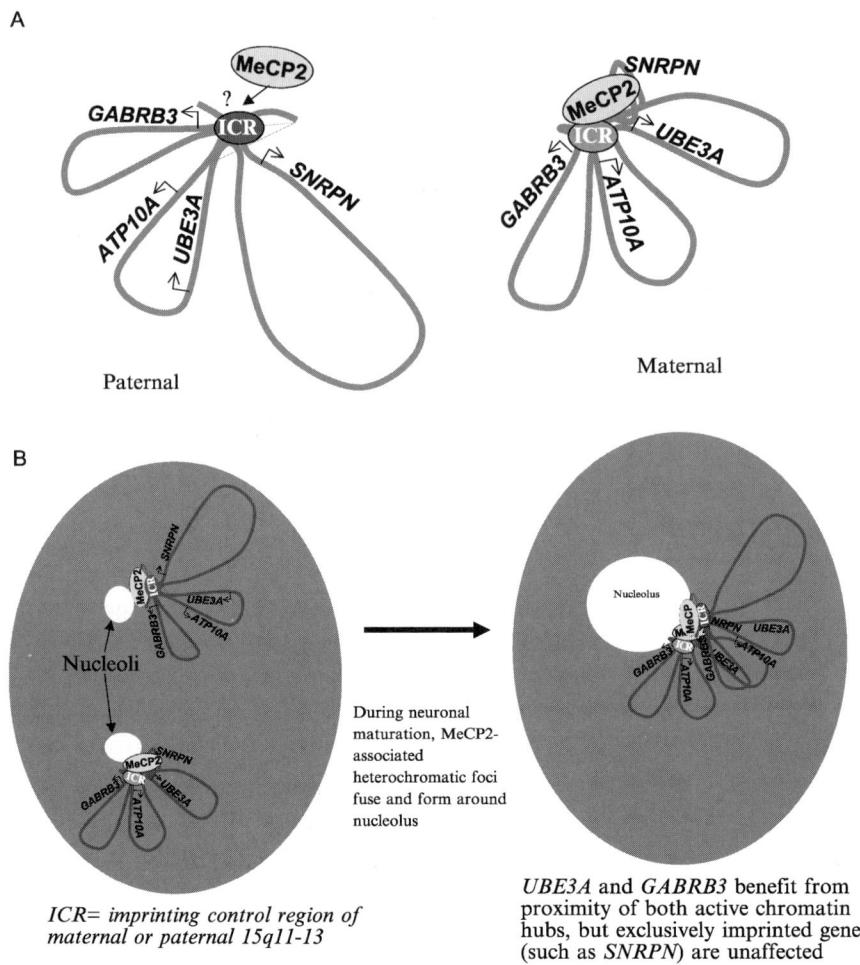

FIG. 3. A higher order model for MeCP2 in regulating gene expression within 15q11-13. (A) Chromatin loop structures are proposed that would change the gene expression patterns on the paternal and maternal 15q11-13 regions. Differential MeCP2 binding sites would be determined by differentially methylated regions (DMRs). One known binding site for MeCP2 is present in the maternal methylated allele of the *SNRPN* promoter, but presumably MeCP2 could also bind to sites of paternal-specific methylation. The chromatin loop structures could position the imprinting control regions (ICR) for either the paternal (PWS-ICR, blue) or maternal (AS-ICR, pink) ICR near the promoters of *Gabrb3* and *Ube3a*, thus explaining how MeCP2 deficiency can alter expression of these genes without necessarily binding at the promoters. (B) In the context of a neuronal nucleus undergoing maturational differentiation, the 15q11-13 chromatin loop structures are brought into close proximity by the formation of a large nucleolus (as in Fig. 2). MeCP2 colocalizes with the peri-nucleolar heterochromatin, bringing the associated chromatin loops together. *UBE3A* and *GABRB3* expression would benefit by being brought into the active chromatin hub of the opposite parental chromosome. (See Color Insert.)

requisite change in the imprinted expression. The loss of MeCP2 function through mutation in RTT could result in reduced 15q11-13 pairing by a combination of aberrant chromatin looping of 15q11-13 and the loss of MeCP2-mediated inter-actions of homologous chromosomes. This model could also explain the more significant defects in UBE3A and GABRB3 expression in human RTT and autism samples compared to *Mecp2*-deficient mice (Samaco *et al.*, 2005), since syntenic mouse chromosome 7 does not show homologous pairing (Thatcher *et al.*, 2005).

The recent data showing autism brain samples having significantly reduced homologous pairing of 15q11-13 and reduced expression of *GABRB3* and *UBE3A* compared to age-matched controls raise interesting implications for understand-ing the etiology of autism. The samples examined also had significantly reduced levels of MeCP2, suggesting that the defects in pairing and 15q11-13 gene expres-sion may be simply downstream effects of impaired neurodevelopment. Alterna-tively, the autism samples may have genetic or environmental factors reducing MeCP2 expression that could mimic the *MECP2* mutant state in RTT. Curiously, the autism samples showed more significant defects in homologous pairing than the RTT, AS, and PWS samples that have known genetic defects. In addition, the GABRB3 expression defects in autism were also more significant than those observed in RTT or *Mecp2*-deficient mouse brain. Could defects in other genetic and epigenetic pathways regulating GABRB3 expression be responsible for this finding? The linkage and association data using markers around *GABRB3* and within 15q11-13 are supportive of this possibility (Buxbaum *et al.*, 2002; Kim *et al.*, 2002; McCauley *et al.*, 2004; Nurmi *et al.*, 2001, 2003).

VII. Future Directions and Remaining Questions

Clearly, much more work lies ahead in understanding the epigenetics and genetics of gene regulation within 15q11-13 and its relevance to autism. Additional autism brain samples should be screened for expression defects and compared with normal controls and other neurodevelopmental disorders with known genetic etiologies with and without comorbid autism. First and foremost, the coding regions and upstream regulatory regions of *GABRB3*, *UBE3A*, and *MECP2* should be examined for genetic variants that may explain the expression defects in these genes in autism brain samples. Such genetic causes may be infrequent or heterogeneous in different autism families, as has been seen for the *NLGN* genes (Jamain *et al.*, 2003; Gauthier *et al.*, 2005; Vincent *et al.*, 2004). Alternatively, inheritance of polymorphic variants in or around these genes may be one of multiple susceptibility alleles that are predicted in the etiology of autism.

Further understanding of genetic and epigenetic pathways that regulate MeCP2 and GABRB3 expression is also expected to continue to shed light on the molecular pathogenesis of autism. The MeCP2 binding sites and allele-specific methylation patterns across the entire 15q11-13 region need to be defined, as well as the proposed chromatin loop structures. Similar to the approaches used for determining chromatin loops in the relatively simpler regions of *H19/Igf2* and *Dlx5* loci (Horike *et al.*, 2005; Murrell *et al.*, 2004), the chromatin loops on both parental chromosomes across the entire 15q11-13 region could be defined. Although the imprinting status of transcription has been defined for much of the 15q11-13 region, the question of whether the $GABA_A$ receptor genes may show preferential allelic expression in human brain remains open. Perhaps proper expression of *GABRB3* depends on *trans* interactions between oppositely imprinted maternal and paternal 15q11-13 regions during homologous pairing that occur during neuronal maturation.

Mouse models are expected to continue to play an important role in understanding the syntenic chromosome 7qB4 region and the regulation of *Gabrb3* and *Ube3a*. Gene targeting knockouts of both genes have been made and could be useful in future experiments (Homanics *et al.*, 1997; Jiang *et al.*, 1998; Miura *et al.*, 2002). To truly mimic the expression level defects observed in autism, however, perhaps knock-down models should be attempted by transgenic or RNAi approaches. Testing the behavioral phenotype of these mouse models for relevance to autism would be important. In addition, testing the effect of double knock-down mice for *Mecp2* and *Gabrb3* or *Ube3a* may prove to be informative in understanding the combined effect of expression defects of these genes on behavioral phenotype.

Another potential avenue to pursue would be genetic and environmental interactions on MeCP2 and 15q11-13 gene regulation. As MeCP2 expression is tightly linked to neuronal maturation, pharmacological agents and other compounds that either enhance or reduce neuronal stimulation could affect MeCP2 expression levels. In addition, environmental agents such as mercury or pesticides that inhibit MeCP2 or GABRB3 function could be involved in triggering autism in genetically susceptible individuals.

In conclusion, we have discussed the results of the beginning stages of a "Rosetta stone" approach to understanding how distinct genetic defects may contribute to a common pathway for regulating nuclear organization, chromatin structure, and gene expression in the developing brain.

Acknowledgments

The authors thank R. Nagarajan, S. Peddada, and D. Yasui for helpful discussions and critical review of the manuscript; the Rett Syndrome Research Foundation, the U.C. Davis M.I.N.D. Institute, and the N.I.H. (1R01HD/NS41462) for support of research programs.

References

Akbarian, S., Chen, R. Z., Gribnau, J., Rasmussen, T. P., Fong, H., Jaenisch, R., and Jones, E. G. (2001). Expression pattern of the Rett syndrome gene MeCP2 in primate prefrontal cortex. *Neurobiol. Dis.* **8,** 784–791.

Amir, R. E., Fang, P., Yu, Z., Glaze, D. G., Percy, A. K., Zoghbi, H. Y., Roa, B. B., and Van den Veyver, I. B. (2005). Mutations in exon 1 of MECP2 are a rare cause of Rett syndrome. *J. Med. Genet.* **42,** e15.

Amir, R. E., Van den Veyver, I. B., Schultz, R., Malicki, D. M., Tran, C. Q., Dahle, E. J., Philippi, A., Timar, L., Percy, A. K., Motil, K. J., Lichtarge, O., Smith, E. O., Glaze, D. G., and Zoghbi, H. Y. (2000). Influence of mutation type and X chromosome inactivation on Rett syndrome phenotypes. *Ann. Neurol.* **47,** 670–679.

Amir, R. E., Van den Veyver, I. B., Wan, M., Tran, C. Q., Francke, U., and Zoghbi, H. Y. (1999). Rett syndrome is caused by mutations in X-linked *MECP2*, encoding methyl- CpG-binding protein 2. *Nat. Genet.* **23,** 185–188.

Arnoldus, E. P., Noordermeer, I. A., Peters, A. C., Raap, A. K., and Van der Ploeg, M. (1991). Interphase cytogenetics reveals somatic pairing of centromeres in normal human brain tissue, but no chromosome loss. *Cytogenet. Cell. Genet.* **56,** 214–216.

Arnoldus, E. P. J., Peters, A. C. B., Bots, G. T. A. M., Raap, A. K., and van der Ploeg, M. (1989). Somatic pairing of chromosome 1 centromeres in interphase nuclei of human cerebellum. *Hum. Genet.* **83,** 231–234.

Balmer, D., Arredondo, J., Samaco, R. C., and LaSalle, J. M. (2002). MECP2 mutations in Rett syndrome adversely affect lymphocyte growth, but do not affect imprinted gene expression in blood or brain. *Hum. Genet.* **110,** 545–552.

Balmer, D., Goldstine, J., Rao, Y. M., and LaSalle, J. M. (2003). Elevated methyl-CpG-binding protein 2 expression is acquired during postnatal human brain development and is correlated with alternative polyadenylation. *J. Mol. Med.* **81,** 61–68.

Barnard, E. A., Skolnick, P., Olsen, R. W., Mohler, H., Sieghart, W., Biggio, G., Braestrup, C., Bateson, A. N., and Langer, S. Z. (1998). International Union of Pharmacology. XV. Subtypes of gamma-aminobutyric acidA receptors: Classification on the basis of subunit structure and receptor function. *Pharmacol. Rev.* **50,** 291–313.

Beyer, K. S., Blasi, F., Bacchelli, E., Klauck, S. M., Maestrini, E., and Poustka, A., and International Molecular Genetic Study of Autism, C. (2002). Mutation analysis of the coding sequence of the MECP2 gene in infantile autism. *Hum. Genet.* **111,** 305–309.

Bienvenu, T., Carrie, A., de Roux, N., Vinet, M. C., Jonveaux, P., Couvert, P., Villard, L., Arzimanoglou, A., Beldjord, C., Fontes, M., Tardieu, M., and Chelly, J. (2000). *MECP2* mutations account for most cases of typical forms of rett syndrome. *Hum. Mol. Genet.* **9,** 1377–1384.

Bittel, D. C., Kibiryeva, N., Talebizadeh, Z., and Butler, M. G. (2003). Microarray analysis of gene/transcript expression in Prader-Willi syndrome: Deletion versus UPD. *J. Med. Genet.* **40,** 568–574.

Bittel, D. C., Kibiryeva, N., Talebizadeh, Z., Driscoll, D. J., and Butler, M. G. (2005). Microarray analysis of gene/transcript expression in Angelman syndrome: Deletion versus UPD. *Genomics* **85,** 85–91.

Blaxill, M. F. (2004). What's going on? The question of time trends in autism. *Public Health Rep.* **119,** 536–551.

Borden, J., and Manuelidis, L. (1988). Movement of the X chromosome in epilepsy. *Science* **242,** 1687–1691.

Bowen, N. J., Palmer, M. B., and Wade, P. A. (2004). DNA damage repair and transcription chromosomal regulation by MeCP2: Structural and enzymatic considerations. *Cell. Mol. Life Sci.* **61,** 2163–2167.

Braunschweig, D., Simcox, T., Samaco, R. C., and LaSalle, J. M. (2004). X-Chromosome inactivation ratios affect wild-type MeCP2 expression within mosaic Rett syndrome and Mecp2−/+ mouse brain. *Hum. Mol. Genet.* **13,** 1275–1286.

Brown, K. E., Amoils, S., Horn, J. M., Buckle, V. J., Higgs, D. R., Merkenschlager, M., and Fisher, A. G. (2001). Expression of alpha- and beta-globin genes occurs within different nuclear domains in haemopoietic cells. *Nat. Cell. Biol.* **3,** 602–606.

Brown, K. E., Baxter, J., Graf, D., Merkenschlager, M., and Fisher, A. G. (1999). Dynamic repositioning of genes in the nucleus of lymphocytes preparing for cell division. *Mol. Cell* **3,** 207–217.

Brown, K. E., Guest, S. S., Smale, S. T., Hahm, K., Merkenschlager, M., and Fisher, A. G. (1997). Association of transcriptionally silent genes with Ikaros complexes at centromeric heterochromatin. *Cell* **91,** 845–854.

Buiting, K., Barnicoat, A., Lich, C., Pembrey, M., Malcolm, S., and Horsthemke, B. (2001). Disruption of the bipartite imprinting center in a family with Angelman syndrome. *Am. J. Hum. Genet.* **68,** 1290–1294.

Buiting, K., Gross, S., Lich, C., Gillessen-Kaesbach, G., el-Maarri, O., and Horsthemke, B. (2003). Epimutations in Prader-Willi and Angelman syndromes: A molecular study of 136 patients with an imprinting defect. *Am. J. Hum. Genet.* **72,** 571–577.

Buiting, K., Saitoh, S., Gross, S., Dittrich, B., Schwartz, S., Nicholls, R. D., and Horsthemke, B. (1995). Inherited microdeletions in the Angelman and Prader-Willi syndromes define an imprinting centre on human chromosome 15. *Nat. Genet.* **9,** 395–400.

Buschdorf, J. P., and Stratling, W. H. (2004). A WW domain binding region in methyl-CpG-binding protein MeCP2: Impact on Rett syndrome. *J. Mol. Med.* **82,** 135–143.

Buxbaum, J. D., Silverman, J. M., Smith, C. J., Greenberg, D. A., Kilifarski, M., Reichert, J., Cook, E. H., Fang, Y., Song, C. Y., and Vitale, R. (2002). Association between a GABRB3 polymorphism and autism. *Mol. Psychiatry* **7,** 311–316.

Byrd, K., and Corces, V. G. (2003). Visualization of chromatin domains created by the gypsy insulator of *Drosophila. J. Cell. Biol.* **162,** 565–574.

Casanova, M. F., Buxhoeveden, D., Switala, A., and Roy, E. (2003). Rett syndrome as a minicolumnopathy. *Clin. Neuropathol.* **22,** 163–168.

Casanova, M. F., Buxhoeveden, D. P., Switala, A. E., and Roy, E. (2002). Minicolumnar pathology in autism. *Neurology* **58,** 428–432.

Cassel, S., Revel, M. O., Kelche, C., and Zwiller, J. (2004). Expression of the methyl-CpG-binding protein MeCP2 in rat brain. An ontogenetic study. *Neurobiol. Dis.* **15,** 206–211.

Chambeyron, S., and Bickmore, W. A. (2004). Chromatin decondensation and nuclear reorganization of the HoxB locus upon induction of transcription. *Genes. Dev.* **18,** 1119–1130.

Chen, R. Z., Akbarian, S., Tudor, M., and Jaenisch, R. (2001). Deficiency of methyl-CpG binding protein-2 in CNS neurons results in a Rett-like phenotype in mice. *Nat. Genet.* **27,** 327–331.

Chen, W. G., Chang, Q., Lin, Y., Meissner, A., West, A. E., Griffith, E. C., Jaenisch, R., and Greenberg, M. E. (2003). Derepression of BDNF transcription involves calcium-dependent phosphorylation of MeCP2. *Science* **302,** 885–889.

Cohen, D. R., Matarazzo, V., Palmer, A. M., Tu, Y., Jeon, O. H., Pevsner, J., and Ronnett, G. V. (2003). Expression of MeCP2 in olfactory receptor neurons is developmentally regulated and occurs before synaptogenesis. *Mol. Cell. Neurosci.* **22,** 417–429.

Colantuoni, C., Jeon, O. H., Hyder, K., Chenchik, A., Khimani, A. H., Narayanan, V., Hoffman, E. P., Kaufmann, W. E., Naidu, S., and Pevsner, J. (2001). Gene expression profiling in postmortem rett syndrome brain: Differential gene expression and patient classification. *Neurobiol. Dis.* **8,** 847–865.

Collins, A. L., Levenson, J. M., Vilaythong, A. P., Richman, R., Armstrong, D. L., Noebels, J. L., David Sweatt, J., and Zoghbi, H. Y. (2004). Mild overexpression of MeCP2 causes a progressive neurological disorder in mice. *Hum. Mol. Genet.* **13,** 2679–2689.

Cook, E. H., Jr., Courchesne, R. Y., Cox, N. J., Lord, C., Gonen, D., Guter, S. J., Lincoln, A., Nix, K., Haas, R., Leventhal, B. L., and Courchesne, E. (1998). Linkage-disequilibrium mapping of autistic disorder, with 15q11-13 markers. *Am. J. Hum. Genet.* **62,** 1077–1083.

Cook, E. H., Jr., Lindgren, V., Leventhal, B. L., Courchesne, R., Lincoln, A., Shulman, C., Lord, C., and Courchesne, E. (1997). Autism or atypical autism in maternally but not paternally derived proximal 15q duplication. *Am. J. Hum. Genet.* **60,** 928–934.

Coverdale, L. E., Martyniuk, C. J., Trudeau, V. L., and Martin, C. C. (2004). Differential expression of the methyl-cytosine binding protein 2 gene in embryonic and adult brain of zebrafish. *Brain Res. Dev. Brain Res.* **153,** 281–287.

Coy, J. F., Sedlacek, Z., Bachner, D., Delius, H., and Poustka, A. (1999). A complex pattern of evolutionary conservation and alternative polyadenylation within the long 3′-untranslated region of the methyl- CpG-binding protein 2 gene (MeCP2) suggests a regulatory role in gene expression. *Hum. Mol. Genet.* **8,** 1253–1262.

Cremer, T., Kurz, A., Zirbel, R., Dietzel, S., Rinke, B., Schrock, E., Speicher, M. R., Mathieu, U., Jauch, A., Emmerich, P., *et al.* (1993). Role of chromosome territories in the functional of the cell nucleus. *Cold Spring Harb. Symp. Quant. Biol.* **58,** 777–792.

Culiat, C. T., Stubbs, L. J., Montgomery, C. S., Russell, L. B., and Rinchik, E. M. (1994). Phenotypic consequences of deletion of the gamma 3, alpha 5, or beta 3 subunit of the type A gamma-aminobutyric acid receptor in mice. *Proc. Natl. Acad. Sci. USA* **91,** 2815–2818.

D'Esposito, M., Quaderi, N. A., Ciccodicola, A., Bruni, P., Esposito, T., D'Urso, M., and Brown, S. D. (1996). Isolation, physical mapping, and northern analysis of the X-linked human gene encoding methyl CpG-binding protein, MECP2. *Mamm. Genome* **7,** 533–535.

de Laat, W., and Grosveld, F. (2003). Spatial organization of gene expression: The active chromatin hub. *Chromosome Res.* **11,** 447–459.

DeLorey, T. M., Handforth, A., Anagnostaras, S. G., Homanics, G. E., Minassian, B. A., Asatourian, A., Fanselow, M. S., Delgado-Escueta, A., Ellison, G. D., and Olsen, R. W. (1998). Mice lacking the beta3 subunit of the GABAA receptor have the epilepsy phenotype and many of the behavioral characteristics of Angelman syndrome. *J. Neurosci.* **18,** 8505–8514.

DeLorey, T. M., and Olsen, R. W. (1999). GABA and epileptogenesis: Comparing gabrb3 gene-deficient mice with Angelman syndrome in man. *Epilepsy. Res.* **36,** 123–132.

Dittrich, B., Buiting, K., Korn, B., Rickard, S., Buxton, J., Saitoh, S., Nicholls, R. D., Poustka, A., Winterpacht, A., Zabel, B., and Horsthemke, B. (1996). Imprint switching on human chromosome 15 may involve alternative transcripts of the SNRPN gene. *Nature Genetics* **14,** 163–170.

Evans, J. C., Archer, H. L., Whatley, S. D., Kerr, A., Clarke, A., and Butler, R. (2005). Variation in exon 1 coding region and promoter of MECP2 in Rett syndrome and controls. *Eur. J. Hum. Genet.* **13,** 124–126.

Francastel, C., Walters, M. C., Groudine, M., and Martin, D. I. (1999). A functional enhancer suppresses silencing of a transgene and prevents its localization close to centrometric heterochromatin. *Cell* **99,** 259–269.

Fuks, F., Hurd, P. J., Wolf, D., Nan, X., Bird, A. P., and Kouzarides, T. (2003). The methyl-CpG-binding protein MeCP2 links DNA methylation to histone methylation. *J. Biol. Chem.* **278,** 4035–4040.

Gauthier, J., Bonnel, A., St-Onge, J., Karemera, L., Laurent, S., Mottron, L., Fombonne, E., Joober, R., and Rouleau, G. A. (2005). NLGN3/NLGN4 gene mutations are not responsible for autism in the Quebec population. *Am. J. Med. Genet. B. Neuropsychiatr. Genet.* **132,** 74–75.

Georgel, P. T., Horowitz-Scherer, R. A., Adkins, N., Woodcock, C. L., Wade, P. A., and Hansen, J. C. (2003). Chromatin compaction by human MeCP2. Assembly of novel secondary chromatin structures in the absence of DNA methylation. *J. Biol. Chem.* **278,** 32181–32188.

Gerard, M., Hernandez, L., Wevrick, R., and Stewart, C. L. (1999). Disruption of the mouse necdin gene results in early post-natal lethality. *Nat. Genet.* **23,** 199–202.

Girard, M., Couvert, P., Carrie, A., Tardieu, M., Chelly, J., Beldjord, C., and Bienvenu, T. (2001). Parental origin of *de novo* MECP2 mutations in Rett syndrome. *Eur. J. Hum. Genet.* **9,** 231–236.

Glenn, C. C., Porter, K. A., Jong, M. T., Nicholls, R. D., and Driscoll, D. J. (1993). Functional imprinting and epigenetic modification of the human SNRPN gene. *Hum. Mol. Genet.* **2,** 2001–2005.

Grande, M. A., van der Kraan, I., de Jong, L., and van Driel, R. (1997). Nuclear distribution of transcription factors in relation to sites of transcription and RNA polymerase II. *J. Cell Sci.* **110** (Pt. 15), 1781–1791.

Guy, J., Hendrich, B., Holmes, M., Martin, J. E., and Bird, A. (2001). A mouse Mecp2-null mutation causes neurological symptoms that mimic Rett syndrome. *Nat. Genet.* **27,** 322–326.

Harikrishnan, K. N., Chow, M. Z., Baker, E. K., Pal, S., Bassal, S., Brasacchio, D., Wang, L., Craig, J. M., Jones, P. L., Sif, S., and El-Osta, A. (2005). Brahma links the SWI/SNF chromatin-remodeling complex with MeCP2-dependent transcriptional silencing. *Nat. Genet.* **37,** 254–264.

Hatakeyama, C., Anderson, C. L., Beever, C. L., Penaherrera, M. S., Brown, C. J., and Robinson, W. P. (2004). The dynamics of X-inactivation skewing as women age. *Clin. Genet.* **66,** 327–332.

Hendrich, B., Abbott, C., McQueen, H., Chambers, D., Cross, S., and Bird, A. (1999). Genomic structure and chromosomal mapping of the murine and human Mbd1, Mbd2, Mbd3, and Mbd4 genes. *Mamm. Genome.* **10,** 906–912.

Herzing, L. B. K., Cook, E. H., and Ledbetter, D. H. (2002). Allele-specific expression analysis by RNA-FISH demonstrates preferential maternal expression of UBE3A and imprint maintenance within 15q11-q13 duplications. *Hum. Mol. Genet.* **11,** 1707–1718.

Holowacz, T., and De Boni, U. (1991). Arrangement of kinetochore proteins and satellite DNA in neuronal interphase nuclei: Changes induced by gamma-aminobutyric acid (GABA). *Exp. Cell Res.* **197,** 36–42.

Homanics, G. E., DeLorey, T. M., Firestone, L. L., Quinlan, J. J., Handforth, A., Harrison, N. L., Krasowski, M. D., Rick, C. E., Korpi, E. R., Makela, R., Brilliant, M. H., Hagiwara, N., Ferguson, C., Snyder, K., and Olsen, R. W. (1997). Mice devoid of gamma-aminobutyrate type A receptor beta3 subunit have epilepsy, cleft palate, and hypersensitive behavior. *Proc. Natl. Acad. Sci. USA* **94,** 4143–4148.

Horike, S., Cai, S., Miyano, M., Cheng, J. F., and Kohwi-Shigematsu, T. (2005). Loss of silent-chromatin looping and impaired imprinting of DLX5 in Rett syndrome. *Nat. Genet.* **37,** 31–40.

Iborra, F. J., Pombo, A., Jackson, D. A., and Cook, P. R. (1996). Active RNA polymerases are localized within discrete transcription "factories" in human nuclei. *J. Cell Sci.* **109**(Pt. 6), 1427–1436.

Jackson, D., Hassan, A., Errington, R., and Cook, P. (1993). Visualization of focal sites of transcription within human nuclei. *EMBO J.* **12,** 1059–1065.

Jackson, D. A., Iborra, F. J., Manders, E. M., and Cook, P. R. (1998). Numbers and organization of RNA polymerases, nascent transcripts, and transcription units in HeLa nuclei. *Mol. Biol. Cell* **9,** 1523–1536.

Jamain, S., Quach, H., Betancur, C., Rastam, M., Colineaux, C., Gillberg, I. C., Soderstrom, H., Giros, B., Leboyer, M., Gillberg, C., and Bourgeron, T. (2003). Mutations of the X-linked genes encoding neuroligins NLGN3 and NLGN4 are associated with autism. *Nat. Genet.* **34,** 27–29.

Jenuwein, T., and Allis, C. D. (2001). Translating the histone code. *Science* **293,** 1074–1080.

Jiang, Y. H., Armstrong, D., Albrecht, U., Atkins, C. M., Noebels, J. L., Eichele, G., Sweatt, J. D., and Beaudet, A. L. (1998). Mutation of the Angelman ubiquitin ligase in mice causes increased cytoplasmic p53 and deficits of contextual learning and long-term potentiation. *Neuron.* **21,** 799–811.

Jiang, Y. H., and Beaudet, A. L. (2004). Human disorders of ubiquitination and proteasomal degradation. *Curr. Opin. Pediatr.* **16,** 419–426.

Jiang, Y. H., Sahoo, T., Michaelis, R. C., Bercovich, D., Bressler, J., Kashork, C. D., Liu, Q., Shaffer, L. G., Schroer, R. J., Stockton, D. W., Spielman, R. S., Stevenson, R. E., and Beaudet, A. L. (2004). A mixed epigenetic/genetic model for oligogenic inheritance of autism with a limited role for UBE3A. *Am. J. Med. Genet.* **131,** 1–10.

Jones, P. L., Veenstra, G. J., Wade, P. A., Vermaak, D., Kass, S. U., Landsberger, N., Strouboulis, J., and Wolffe, A. P. (1998). Methylated DNA and MeCP2 recruit histone deacetylase to repress transcription. *Nat. Genet.* **19,** 187–191.

Jones, P. L., Wade, P. A., and Wolffe, A. P. (2001). Purification of the MeCP2/histone deacetylase complex from *Xenopus laevis. Methods Mol. Biol.* **181,** 297–307.

Jorgensen, H. F., and Bird, A. (2002). MeCP2 and other methyl-CpG binding proteins. *Ment. Retard. Dev. Disabil. Res. Rev.* **8,** 87–93.

Jugloff, D. G., Jung, B. P., Purushotham, D., Logan, R., and Eubanks, J. H. (2005). Increased dendritic complexity and axonal length in cultured mouse cortical neurons overexpressing methyl-CpG-binding protein MeCP2. *Neurobiol. Dis.* **19,** 18–27.

Kim, S. J., Herzing, L. B., Veenstra-Vander Weele, J., Lord, C., Courchesne, R., Leventhal, B. L., Ledbetter, D. H., Courchesne, E., and Cook, E. H., Jr. (2002). Mutation screening and transmission disequilibrium study of ATP10C in autism. *Am. J. Med. Genet.* **114,** 137–143.

Kimura, H., and Shiota, K. (2003). Methyl-CpG-binding protein, MeCP2, is a target molecule for maintenance DNA methyltransferase, Dnmt1. *J. Biol. Chem.* **278,** 4806–4812.

Kishino, T., Lalande, M., and Wagstaff, J. (1997). UBE3A/E6-AP mutations cause Angelman syndrome. *Nat. Genet.* **15,** 70–73.

Kosak, S. T., and Groudine, M. (2004). Gene order and dynamic domains. *Science* **306,** 644–647.

Kriaucionis, S., and Bird, A. (2004). The major form of MeCP2 has a novel N-terminus generated by alternative splicing. *Nucleic. Acids Res.* **32,** 1818–1823.

Lalande, M. (1996). Parental imprinting and human disease. *Ann. Rev. Genet.* **30,** 173–195.

Lam, C., Yeung, W., Ko, C., Poon, P., Tong, S., Chan, K., Lo, I., Chan, L., Hui, J., Wong, V., Pang, C., Lo, Y., and Fok, T. (2000). Spectrum of mutations in the MECP2 gene in patients with infantile autism and Rett syndrome. *J. Med. Genet.* **37,** E41.

LaSalle, J., Goldstine, J., Balmer, D., and Greco, C. (2001). Quantitative localization of heterologous methyl-CpG-binding protein 2 (MeCP2) expression phenotypes in normal and Rett syndrome brain by laser scanning cytometry. *Hum. Mol. Genet.* **10,** 1729–1740.

LaSalle, J., and Lalande, M. (1996). Homologous association of oppositely imprinted chromosomal domains. *Science* **272,** 725–728.

Laurie, D. J., Seeburg, P. H., and Wisden, W. (1992). The distribution of 13 GABAA receptor subunit mRNAs in the rat brain. II. Olfactory bulb and cerebellum. *J. Neurosci.* **12,** 1063–1076.

Lawler, C. P., Croen, L. A., Grether, J. K., and Van de Water, J. (2004). Identifying environmental contributions to autism: Provocative clues and false leads. *Ment. Retard. Dev. Disabil. Res. Rev.* **10,** 292–302.

Lee, S., Walker, C. L., Karten, B., Kuny, S. L., Tennese, A. A., O'Neill, M. A., and Wevrick, R. (2005). Essential role for the Prader-Willi syndrome protein necdin in axonal outgrowth. *Hum. Mol. Genet.* **14,** 627–637.

Leitch, A. R. (2000). Higher levels of organization in the interphase nucleus of cycling and differentiated cells. *Microbiol. Mol. Biol. Rev.* **64,** 138–152.

Lewis, J. D., Meehan, R. R., Henzel, W. J., Maurer-Fogy, I., Jeppesen, P., Klein, F., and Bird, A. (1992). Purification, sequence, and cellular localization of a novel chromosomal protein that binds to methylated DNA. *Cell* **69,** 905–914.

Li, E., Beard, C., and Jaenisch, R. (1993). Role for DNA methylation in genomic imprinting. *Nature* **366,** 362–365.

Luikenhuis, S., Giacometti, E., Beard, C. F., and Jaenisch, R. (2004). Expression of MeCP2 in postmitotic neurons rescues Rett syndrome in mice. *Proc. Natl. Acad. Sci. USA* **101,** 6033–6038.

Makedonski, K., Abuhatzira, L., Kaufman, Y., Razin, A., and Shemer, R. (2005). MeCP2 deficiency in Rett syndrome causes epigenetic aberrations at the PWS/AS imprinting center that affects UBE3A expression. *Hum. Mol. Genet.* **14,** 1049–1058.

Manuelidis, L. (1990). A view of interphase chromosomes. *Science* **250,** 1533–1540.

Manuelidis, L., and Borden, J. (1988). Reproducible compartmentalization of individual chromosome domains in human CNS cells revealed by in situ hybridization and three-dimensional reconstruction. *Chromosoma* **96,** 397–410.

Martin, E. R., Menold, M. M., Wolpert, C. M., Bass, M. P., Donnelly, S. L., Ravan, S. A., Zimmerman, A., Gilbert, J. R., Vance, J. M., Maddox, L. O., Wright, H. H., Abramson, R. K., DeLong, G. R., Cuccaro, M. L., and Pericak-Vance, M. A. (2000). Analysis of linkage disequilibrium in gamma-aminobutyric acid receptor subunit genes in autistic disorder. *Am. J. Med. Genet.* **96,** 43–48.

Martinowich, K., Hattori, D., Wu, H., Fouse, S., He, F., Hu, Y., Fan, G., and Sun, Y. E. (2003). DNA methylation-related chromatin remodeling in activity-dependent BDNF gene regulation. *Science* **302,** 890–893.

Martou, G., and De Boni, U. (2000). Nuclear topology of murine, cerebellar Purkinje neurons: Changes as a function of development. *Exp. Cell Res.* **256,** 131–139.

McCauley, J. L., Olson, L. M., Delahanty, R., Amin, T., Nurmi, E. L., Organ, E. L., Jacobs, M. M., Folstein, S. E., Haines, J. L., and Sutcliffe, J. S. (2004). A linkage disequilibrium map of the 1-Mb 15q12 GABA(A) receptor subunit cluster and association to autism. *Am. J. Med. Genet.* **131,** 51–59.

McGrath, J., and Solter, D. (1984). Completion of mouse embryogenesis requires both the maternal and paternal genomes. *Cell* **37,** 179–183.

McLean, P. J., Farb, D. H., and Russek, S. J. (1995). Mapping of the alpha 4 subunit gene (GABRA4) to human chromosome 4 defines an alpha 2-alpha 4-beta 1-gamma 1 gene cluster: Further evidence that modern GABAA receptor gene clusters are derived from an ancestral cluster. *Genomics* **26,** 580–586.

Meguro, M., Mitsuya, K., Sui, H., Shigenami, K., Kugoh, H., Nakao, M., and Oshimura, M. (1997). Evidence for uniparental, paternal expression of the human GABAA receptor subunit genes, using microcell-mediated chromosome transfer. *Hum. Mol. Genet.* **6,** 2127–2133.

Miura, K., Kishino, T., Li, E., Webber, H., Dikkes, P., Holmes, G. L., and Wagstaff, J. (2002). Neurobehavioral and electroencephalographic abnormalities in Ube3a maternal-deficient mice. *Neurobiol. Dis.* **9,** 149–159.

Mnatzakanian, G. N., Lohi, H., Munteanu, I., Alfred, S. E., Yamada, T., MacLeod, P. J., Jones, J. R., Scherer, S. W., Schanen, N. C., Friez, M. J., Vincent, J. B., and Minassian, B. A. (2004). A previously unidentified MECP2 open reading frame defines a new protein isoform relevant to Rett syndrome. *Nat. Genet.* **36,** 339–341.

Moncla, A., Malzac, P., Voelckel, M. A., Auquier, P., Girardot, L., Mattei, M. G., Philip, N., Mattei, J. F., Lalande, M., and Livet, M. O. (1999). Phenotype-genotype correlation in 20 deletion and 20 non-deletion Angelman syndrome patients. *Eur. J. Hum. Genet.* **7,** 131–139.

Moretti, P., Bouwknecht, J. A., Teague, R., Paylor, R., and Zoghbi, H. Y. (2005). Abnormalities of social interactions and home-cage behavior in a mouse model of Rett syndrome. *Hum. Mol. Genet.* **14,** 205–220.

Muhle, R., Trentacoste, S. V., and Rapin, I. (2004). The genetics of autism. *Pediatrics* **113,** e472–e486.

Mullaney, B. C., Johnston, M. V., and Blue, M. E. (2004). Developmental expression of methyl-CpG binding protein 2 is dynamically regulated in the rodent brain. *Neuroscience* **123,** 939–949.

Murphy, S. K., and Jirtle, R. L. (2003). Imprinting evolution and the price of silence. *Bioessays* **25,** 577–588.

Murrell, A., Heeson, S., and Reik, W. (2004). Interaction between differentially methylated regions partitions the imprinted genes Igf2 and H19 into parent-specific chromatin loops. *Nat. Genet.* **36,** 889–893.

Nan, X., Campoy, F. J., and Bird, A. (1997). MeCP2 is a transcriptional repressor with abundant binding sites in genomic chromatin. *Cell* **88,** 471–481.

Nan, X., Ng, H. H., Johnson, C. A., Laherty, C. D., Turner, B. M., Eisenman, R. N., and Bird, A. (1998). Transcriptional repression by the methyl-CpG-binding protein MeCP2 involves a histone deacetylase complex. *Nature* **393,** 386–389.

Nan, X., Tate, P., Li, E., and Bird, A. (1996). DNA methylation specifies chromosomal localization of MeCP2. *Mol. Cell. Biol.* **16,** 414–421.

Nicholls, R. D., Gottlieb, W., Russell, L. B., Davda, M., Horsthemke, B., and Rinchik, E. M. (1993). Evaluation of potential models for imprinted and nonimprinted components of human chromosome 15q11-q13 syndromes by fine-structure homology mapping in the mouse. *Proc. Natl. Acad. Sci. USA* **90,** 2050–2054.

Nicholls, R. D., and Knepper, J. L. (2001). Genome organization, function, and imprinting in Prader-Willi and Angelman syndromes. *Annu. Rev. Genomics Hum. Genet.* **2,** 153–175.

Nurmi, E. L., Amin, T., Olson, L. M., Jacobs, M. M., McCauley, J. L., Lam, A. Y., Organ, E. L., Folstein, S. E., Haines, J. L., and Sutcliffe, J. S. (2003). Dense linkage disequilibrium mapping in the 15q11-q13 maternal expression domain yields evidence for association in autism. *Mol. Psychiatry* **8,** 570, 624–634.

Nurmi, E. L., Bradford, Y., Chen, Y.-H., Hall, J., Arnone, B., Gardiner, M. B., Hutcheson, H. B., Gilbert, J. R., Pericak-Vance, M. A., Copeland-Yates, S. A., Michaelis, R. C., Wassink, T. H., Santangelo, S. L., Sheffield, V. C., Piven, J., Folstein, S. E., Haines, J. L., and Sutcliffe, J. S. (2001). Linkage disequilibrium at the Angelman syndrome gene UBE3A in autism families. *Genomics* **77,** 105–113.

Osborne, A. R., Zhang, H., Fejer, G., Palubin, K. M., Niesen, M. I., and Blanck, G. (2004). Oct-1 maintains an intermediate, stable state of HLA-DRA promoter repression in Rb-defective cells: An Oct-1-containing repressosome that prevents NF-Y binding to the HLA-DRA promoter. *J. Biol. Chem.* **279,** 28911–28919.

Patrinos, G. P., de Krom, M., de Boer, E., Langeveld, A., Imam, A. M., Strouboulis, J., de Laat, W., and Grosveld, F. G. (2004). Multiple interactions between regulatory regions are required to stabilize an active chromatin hub. *Genes. Dev.* **18,** 1495–1509.

Peters, S. U., Beaudet, A. L., Madduri, N., and Bacino, C. A. (2004). Autism in Angelman syndrome: Implications for autism research. *Clin. Genet.* **66,** 530–536.

Pirker, S., Schwarzer, C., Wieselthaler, A., Sieghart, W., and Sperk, G. (2000). GABA(A) receptors: Immunocytochemical distribution of 13 subunits in the adult rat brain. *Neuroscience* **101,** 815–850.

Polleux, F., and Lauder, J. M. (2004). Toward a developmental neurobiology of autism. *Ment. Retard. Dev. Disabil. Res. Rev.* **10,** 303–317.

Ravn, K., Nielson, J., and Schwartz, M. (2005). Mutations found within exon 1 of MECP2 in Danish patients with Rett syndrome. *Clin. Genet.* **67,** 532–533.

Reichwald, K., Thiesen, J., Wiehe, T., Weitzel, J., Poustka, W. A., Rosenthal, A., Platzer, M., Stratling, W. H., and Kioschis, P. (2000). Comparative sequence analysis of the MECP2-locus in human and mouse reveals new transcribed regions. *Mamm. Genome.* **11,** 182–190.

Reik, W., and Walter, J. (2001). Genomic imprinting: Parental influence on the genome. *Nat. Rev. Genet.* **2,** 21–32.

Ren, J., Lee, S., Pagliardini, S., Gerard, M., Stewart, C. L., Greer, J. J., and Wevrick, R. (2003). Absence of Ndn, encoding the Prader-Willi syndrome-deleted gene necdin, results in congenital deficiency of central respiratory drive in neonatal mice. *J. Neurosci.* **23,** 1569–1573.

Renieri, A., Pescucci, C., Longo, I., Ariani, F., Meloni, I., Zappella, M., Russo, L., Giordano, T., Neri, G., and Gurrieri, F. (2003). Rett patients with both MECP2 mutations and 15q11-13 rearrangements. *Am. J. Hum. Genet.* **73,** 6.

Rougeulle, C., Glatt, H., and Lalande, M. (1997). The Angelman syndrome candidate gene, UBE3A/E6-AP, is imprinted in brain [letter] [In Process Citation]. *Nat. Genet.* **17,** 14–15.

Runte, M., Huttenhofer, A., Gross, S., Kiefmann, M., Horsthemke, B., and Buiting, K. (2001). The IC-SNURF-SNRPN transcript serves as a host for multiple small nucleolar RNA species and as an antisense RNA for UBE3A. *Hum. Mol. Genet.* **10,** 2687–2700.

Samaco, R. C., Hogart, A., and LaSalle, J. M. (2005). Epigenetic overlap in autism-spectrum neurodevelopmental disorders: MECP2 deficiency causes reduced expression of UBE3A and GABRB3. *Hum. Mol. Genet.* **14,** 483–492.

Samaco, R. C., Nagarajan, R. P., Braunschweig, D., and LaSalle, J. M. (2004). Multiple pathways regulate MeCP2 expression in normal brain development and exhibit defects in autism-spectrum disorders. *Hum. Mol. Genet.* **13,** 629–639.

Schanen, C. (2001). Rethinking the fate of males with mutations in the gene that causes Rett syndrome. *Brain Dev.* **23**(Suppl. 1), S144–S146.

Schroer, R. J., Phelan, M. C., Michaelis, R. C., Crawford, E. C., Skinner, S. A., Cuccaro, M., Simensen, R. J., Bishop, J., Skinner, C., Fender, D., and Stevenson, R. E. (1998). Autism and maternally derived aberrations of chromosome 15q. *Am. J. Med. Genet.* **76,** 327–336.

Services, C. D. o. D. (2003). "Autism Spectrum Disorders, Changes in the California Caseload, An Update: 1999 to 2002." California Health and Human Services, Sacramento California Health and Human Services.

Shahbazian, M., Young, J., Yuva-Paylor, L., Spencer, C., Antalffy, B., Noebels, J., Armstrong, D., Paylor, R., and Zoghbi, H. (2002a). Mice with truncated MeCP2 recapitulate many Rett syndrome features and display hyperacetylation of histone H3. *Neuron* **35,** 243–254.

Shahbazian, M. D., Antalffy, B., Armstrong, D. L., and Zoghbi, H. Y. (2002b). Insight into Rett syndrome: MeCP2 levels display tissue- and cell-specific differences and correlate with neuronal maturation. *Hum. Mol. Genet.* **11,** 115–124.

Shao, Y., Cuccaro, M. L., Hauser, E. R., Raiford, K. L., Menold, M. M., Wolpert, C. M., Ravan, S. A., Elston, L., Decena, K., Donnelly, S. L., Abramson, R. K., Wright, H. H., DeLong, G. R., Gilbert, J. R., and Pericak-Vance, M. A. (2003). Fine mapping of autistic disorder to chromosome 15q11-q13 by use of phenotypic subtypes. *Am. J. Hum. Genet.* **72,** 539–548.

Soejima, H., and Wagstaff, J. (2005). Imprinting centers, chromatin structure, and disease. *J. Cell. Biochem.* **95,** 226–233.

Song, J., Koller, D. L., Foroud, T., Carr, K., Zhao, J., Rice, J., Nurnberger, J. I., Jr., Begleiter, H., Porjesz, B., Smith, T. L., Schuckit, M. A., and Edenberg, H. J. (2003). Association of GABA(A) receptors and alcohol dependence and the effects of genetic imprinting. *Am. J. Med. Genet. B. Neuropsychiatr. Genet.* **117,** 39–45.

Subramaniam, B., Naidu, S., and Reiss, A. L. (1997). Neuroanatomy in Rett syndrome: Cerebral cortex and posterior fossa. *Neurology* **48,** 399–407.

Sutcliffe, J. S., Nakao, M., Christian, S., Orstavik, K. H., Tommerup, N., Ledbetter, D. H., and Beaudet, A. L. (1994). Deletions of a differentially methylated CpG island at the SNRPN gene define a putative imprinting control region. *Nat. Genet.* **8,** 52–58.

Thatcher, K., Peddada, S., Yasui, D., and LaSalle, J. M. (2005). Homologous pairing of 15q11-13 imprinted domains in brain is developmentally regulated but deficient in Rett and autism samples. *Hum. Mol. Genet.* in revision.

Tolhuis, B., Palstra, R. J., Splinter, E., Grosveld, F., and de Laat, W. (2002). Looping and interaction between hypersensitive sites in the active beta-globin locus. *Mol. Cell* **10,** 1453–1465.

Tudor, M., Akbarian, S., Chen, R. Z., and Jaenisch, R. (2002). Transcriptional profiling of a mouse model for Rett syndrome reveals subtle transcriptional changes in the brain. *Proc. Natl. Acad. Sci. USA* **99,** 15536–15541.

Verona, R. I., Mann, M. R., and Bartolomei, M. S. (2003). Genomic imprinting: Intricacies of epigenetic regulation in clusters. *Annu. Rev. Cell Dev. Biol.* **19,** 237–259.

Vincent, J. B., Kolozsvari, D., Roberts, W. S., Bolton, P. F., Gurling, H. M., and Scherer, S. W. (2004). Mutation screening of X-chromosomal neuroligin genes: No mutations in 196 autism probands. *Am. J. Med. Genet. B. Neuropsychiatr. Genet.* **129,** 82–84.

Volkmar, F. R., and Pauls, D. (2003). Autism. *Lancet* **362,** 1133–1141.

Vourc'h, P., Bienvenu, T., Beldjord, C., Chelly, J., Barthelemy, C., Muh, J. P., and Andres, C. (2001). No mutations in the coding region of the Rett syndrome gene MECP2 in 59 autistic patients. *European J. Hum. Genet.* **9,** 556–558.

Vu, T., and Hoffman, A. (1997). Imprinting of the Angelman syndrome gene, UBE3A, is restricted to brain. *Nature Genet.* **17,** 12–13.

Wang, Y., Fischle, W., Cheung, W., Jacobs, S., Khorasanizadeh, S., and Allis, C. D. (2004). Beyond the double helix: Writing and reading the histone code. *Novartis Found Symp.* **259,** 3–17; discussion 17–21, 163–169.

Wassink, T. H., and Piven, J. (2000). The molecular genetics of autism. *Curr. Psychiatry Rep.* **2,** 170–175.

Watson, P., Black, G., Ramsden, S., Barrow, M., Super, M., Kerr, B., and Clayton-Smith, J. (2001). Angelman syndrome phenotype associated with mutations in MECP2, a gene encoding a methyl CpG binding protein. *J. Med. Genet.* **38,** 224–228.

Weintraub, H., and Groudine, M. (1976). Chromosomal subunits in active genes have an altered conformation. *Science* **193,** 848–856.

Weitzel, J. M., Buhrmester, H., and Stratling, W. H. (1997). Chicken MAR-binding protein ARBP is homologous to rat methyl-CpG- binding protein MeCP2. *Mol. Cell. Biol.* **17,** 5656–5666.

Yamasaki, K., Joh, K., Ohta, T., Masuzaki, H., Ishimaru, T., Mukai, T., Niikawa, N., Ogawa, M., Wagstaff, J., and Kishino, T. (2003). Neurons but not glial cells show reciprocal imprinting of sense and antisense transcripts of Ube3a. *Hum. Mol. Genet.* **12,** 837–847.

Young, J. I., and Zoghbi, H. Y. (2004). X-chromosome inactivation patterns are unbalanced and affect the phenotypic outcome in a mouse model of rett syndrome. *Am. J. Hum. Genet.* **74,** 511–520.

Zaidi, S. K., Sullivan, A. J., van Wijnen, A. J., Stein, J. L., Stein, G. S., and Lian, J. B. (2002). Integration of Runx and Smad regulatory signals at transcriptionally active subnuclear sites. *Proc. Natl. Acad. Sci. USA* **99,** 8048–8053.

Zoghbi, H. Y. (2003). Postnatal neurodevelopmental disorders: Meeting at the synapse? *Science* **302,** 826–830.

Zoghbi, H. Y., Percy, A. K., Schultz, R. J., and Fill, C. (1990). Patterns of X chromosome inactivation in the Rett syndrome. *Brain Dev.* **12,** 131–135.

GABAᴇʀɢɪᴄ CEREBELLAR SYSTEM IN AUTISM: A NEUROPATHOLOGICAL AND DEVELOPMENTAL PERSPECTIVE

Gene J. Blatt

Department of Anatomy and Neurobiology, Boston University School of Medicine
Boston, Massachusetts 02118, USA

I. Introduction
II. The GABAergic System in the Cerebellum in Autism
   A. Effects on Purkinje Cells
   B. GAD Expression in Cerebellar Cortex is Altered in Autism
   C. GABAergic Interneurons in Cerebellar Cortex in Autism
   D. Implications of the Timing of Purkinje Cell Decrease in Autism
   E. Neuropathology of the Deep Cerebellar Nuclei in Autism
III. Functional Consequences of Altered Cerebellar Function in Autism
IV. Future and Ongoing Neuropathological Experiments to Further Understand How Cerebellar Circuitry is Altered in Autism
   References

Alterations in the GABAergic system in autism only recently became evident affecting specific receptors in the hippocampus, GAD protein levels in parietal cortex and cerebellum and a decreased number of GABAergic Purkinje cells evident in a subset of autistic cases. A candidate gene for the GABA-Aβ3 subunit has remained attractive on the chromosome 15q11-q13 region in a subset of individuals. Current studies are revealing valuable data toward understanding how cerebellar circuitry may be altered in autism as better markers and techniques become available. In particular, immunostaining of Purkinje cells demonstrated that previous Nissl studies may have overlooked agonal effects on the efficacy of tissue sections bringing into question results of previous studies, some of which also did not control for age. It appears that the cerebellum in autism shows variability in neuropathology similar to that described for other affected brain areas in autism. Decreased Purkinje cells are present in about half of cases in the posterior lateral cerebellar cortex yet normal cytoarchitecture persists throughout the cerebellum indicating completion of migration of both Purkinje cell and granule cell populations. Despite the Purkinje cell decrease, normal density of GABAergic interneurons, basket and stellate cells, are present in the molecular layer and neurons are also preserved in the principal olive that projects to the posterior lobe in the lateral hemisphere. This suggests that the Purkinje cell loss probably occurs at about 32 weeks of gestation or later but most likely no later that early postnatal development. Current studies have demonstrated that Purkinje cells are deficient in GAD67 possibly affecting GABA synthesis.

INTERNATIONAL REVIEW OF
NEUROBIOLOGY, VOL. 71
DOI: 10.1016/S0074-7742(05)71007-2

Altered input to the deep cerebellar nuclei is discussed and how recent studies strongly link the cerebellum as a modulator of both motor and cognitive function thus potentially implicating it in a variety of autistic behaviors.

I. Introduction

The GABAergic system in autism has received increased attention in the recent literature largely due to: (1) reports of decreased numbers of cerebellar GABAergic Purkinje cells especially in the posterior lobe (Bailey *et al.*, 1998; Bauman and Kemper, 1985, 1994; Kemper and Bauman, 1998; Ritvo *et al.*, 1986; Whitney *et al.*, 2004); (2) decreased levels of key synthesizing enzymes (GAD65 and GAD67) in the cerebellum and parietal cortex (Fatemi *et al.*, 2002) and recently, decreased GAD67 localized to Purkinje cells (Yip *et al.*, 2005a); (3) neuropathology in the deep cerebellar nuclei, many cells of which are GABAergic (Bauman and Kemper, 1985, 1994; Kemper and Bauman, 1998); (4) decreased density of GABA-A receptors and benzodiazepine binding sites in specific subfields of the hippocampal formation (Blatt *et al.*, 2001) and increased packing density of parvalbumin-labeled GABAergic interneurons in the CA3 and CA1 subfields (Lawrence *et al.*, 2005); (5) the most common chromosomal abnormality in autism is an alteration(s) in chromosome 15q11-q13, a region that contains three GABA-A receptor subunit candidate genes for autism (Schroer *et al.*, 1998; Shao *et al.*, 2003) including the GABA-Aβ3 subunit receptor gene (Buxbaum *et al.*, 2002; Martin *et al.*, 2000; Shao *et al.*, 2003); and (6) elevated plasma GABA in autistic children aged 5–15 years (Dhossche *et al.*, 2002). This chapter will primarily focus on the cerebellar GABAergic system in autism since it has repeatedly been described by researchers as the most consistent neuropathological finding in autism. Caution should be taken however to make such a claim as more advanced techniques are redefining what we know about cerebellar pathology in autism. Similar to other neuropathological abnormalities in the autistic brain, results among cases are variable and not complete. This chapter will also address the functional implications of upsetting the delicate balance of excitatory-inhibitory inputs to key cerebellar GABAergic neurons as well as address the current and future path of investigation of this important inhibitory system in autism research.

II. The GABAergic System in the Cerebellum in Autism

A. EFFECTS ON PURKINJE CELLS

The initial study reporting reduced numbers of cerebellar Purkinje cells (PCs) in autism reporting on 4 cases age 3 to 33 years of age was by Williams *et al.* (1980). General PC loss was reported and most patients had histories of seizures.

Ritvo *et al.* (1986) reported the density of PCs in the cerebellar hemisphere and vermis of 4 autistic males, all without history of seizure activity, ranging in age from 10 to 22 years with a mean age of 17.5 years compared to a control group ranging in age from 3 to 13 years with a mean age of 7.75 years. The disparity in age of the two groups makes interpretation of these results difficult. These authors found significantly lower PC counts in the autistic cases but failed to account for the age-related change in cerebellar size (Hall *et al.*, 1975). Hall *et al.* (1975) showed that a normal reduction in the number of PCs per measured cortical distance normally occurs with age as the brain grows. For example, Hall *et al.* (1975) found that in males 10 years and under the mean PC count was 5.15 PCs/mm and ranged from 4.40 to 5.95 PCs/mm whereas males between the ages of 20 and 50 years had a mean PC count of 4.47 PCs/mm with a range of 3.40 to 6.92 PCs/mm. This difference could in part account for the lower PC count in the autistic group reported by Ritvo *et al.* (1986), as recalculation of Ritvo's data expressed in PCs/mm demonstrated that their findings are in agreement with Hall's reported normal variability in PC number with age (i.e., a mean of 5.6 PC/mm (range 5.0 to 6.3) in the younger group and a mean of 4.2 PC/mm (range 3.8 to 4.6) in the older group.

In other studies, nine autistic brains (aged 6–54 years) were serially sectioned by Kemper and Bauman with results reported in the 1980s and 1990s in a series of three papers (Bauman and Kemper, 1985, 1994; Kemper and Bauman, 1998). These authors reported qualitative reductions (some severe) of PCs in the posterior lateral cerebellar cortex with lesser effects in the adjacent archicerebellar cortex. Normal numbers of PCs were observed in the anterior lobe (Bauman and Kemper, 1985) and vermis (Bauman and Kemper, 1985; Kemper and Bauman, 1998). There was a pallor of staining in Nissl sections in the granule layer in four cases suggesting possible reduction or alteration in granule cells.

More recently, Whitney *et al.* (2004) quantified PCs in six autistic and four age-matched control brains in the posterior lateral neocerebellar cortex aged 13–54 years. With the investigator "blind" to the identification of the study groups, three cases with marked reductions in Nissl stained PCs (about 90% decrease) were identified. When the group code was broken, two of these three cases were controls! Further examination revealed that when the immunostain calbindin, a calcium-binding protein, was used on adjacent sections, the "missing" PCs in the 2 controls appeared but not in the autistic case! After extensive testing of Nissl stain and standard H & E preparation it was concluded that some brains must undergo agonal changes, even with short post-mortem intervals (PMI: the PMI in one case was only 5 hours) and these changes can directly affect the ability to count PCs using stereological techniques. When calbindin was used, $99^{+}\%$ of PCs immunostained. Quantitative results revealed that three autistic cases had PC density in the normal range while the other three were reduced. Specifically, the calbindin PC counts in the control brains ranged from 4.0 to 5.4 PCs/mm with mean SD of 4.7 +/− 0.8 PCs/mm while cell counts in the autistic brains spanned a wider range, from 0.5 to 5.8 PCs/mm with mean SD of 3.5 ± 1.8 PCs/mm (not significant by t-test, $p = 0.25$; Whitney et al., 2004). Thus, reduction in PC density only appears to be in a subset of autistic brains and unless a specific immuno marker such as calbindin is used in conjunction with other stains the results may not be entirely accurate.

B. GAD EXPRESSION IN CEREBELLAR CORTEX IS ALTERED IN AUTISM

Glutamic acid decarboxylase (GAD) 65 and 67 kDa proteins, the key synthesizing enzymes for GABA, are reduced in both the cerebellar cortex and parietal cortex in autistic subjects (Fatemi et al., 2002). In their study, cerebellar samples were taken from different cerebellar regions for Western Blot biochemical analysis, so results are difficult to interpret with regard to any one cerebellar area or which cell types might be affected. From seven young adult autistics aged 19–30 compared to nine age-matched controls, the mean GAD65 values in cerebellum were reduced by 51% and GAD67 reduced by 61%, both highly significant ($p < 0.02$ and $p < 0.03$, t-test respectively). Although the brain samples varied by location it was nevertheless an important finding because this suggests that the ability to synthesize GABA may be altered in the autistic cerebellum.

Yip et al. (2005a) addressed the issue as to whether it is the Purkinje cell that has a major reduction in GAD. In posterior lateral cerebellar cortex from 16 cases, eight autistics aged 16–30 years and eight controls aged 19–43 years, these investigators used in situ hybridization to measure GAD67 mRNA in a quantitative study. Yip et al. (2005a) found that there is a 40% decrease in GAD67 mRNA in PCs in the autistic cases relative to controls and this difference was found whether there was a decrease in PCs or not. Since every PC contains GAD67, PC counts were made in the same 16 cases and most autistic cases contained normal numbers of PCs per length of posterior lateral cerebellar cortex (i.e., most counts were in the low normal range). This provides further evidence that the PC decrease in autism is limited to a subset of cases, not a consistent neuropathological finding in autism and, that the majority of the GAD67 decrease reported in the cerebellum by Fatemi et al. (2002) is indeed found in Purkinje cells.

C. GABAERGIC INTERNEURONS IN CEREBELLAR CORTEX IN AUTISM

With the quantitative assessment of decreased cerebellar PCs in a subset of autistic cases, the question arises as to the state of GABAergic basket and stellate cells in the molecular layer of the cerebellar cortex. These interneurons intimately innervate PCs; the basket cells send horizontal axons in the lower third of the molecular layer (parallel to the pial surface) forming axonal plexuses around the soma of each PC as a "nest." In contrast, stellate cells are present mainly in the outer two-thirds of the molecular layer and innervate PC dendrites. Both cell types immunostain with the calcium binding protein parvalbumin enabling quantitative density counts possible. Whitney *et al.* (2005) performed a stereo-logical count in the posterior lateral cerebellar cortex in the same lobular location in the same 10 cases where PC counts were reported (Whitney *et al.*, 2004). In this way, the density of basket and stellate cells were calculated per PC from each control and autistic case. Results from the study demonstrated that there was no statistical difference in the density of either basket or stellate cells between control and autistic cases (i.e., even though there was moderate or severe PC loss in three out of six autistic cases, there was no change in GABAergic interneuron populations in the molecular layer) (Whitney *et al.*, 2005).

D. IMPLICATIONS OF THE TIMING OF PURKINJE CELL DECREASE IN AUTISM

Another corollary observation is that in the autistic cerebellum, the adult laminar pattern of the cerebellar cortex is similar to controls (i.e., migration of the external granular layer is complete with no obvious ectopia). An exception may be PCs where Bailey *et al.* (1998) and colleagues have reported ectopic cells in autistic cases but these can also occasionally be found in controls. In any case, it is very difficult microscopically to tell the difference between an autistic and control cerebellar cortex, in any region or location, unless there is a decrease of PCs. If some Purkinje cells in autistic cases were never generated and thus did not migrate to their target location in the cerebellar cortex then it would be expected that there would be a major disruption in the cortical laminar pattern such as in the formation of the internal granular layer and, a likely loss of both granule cells and interneurons. Because the cerebellar cortical laminar pattern is not disturbed in autism, it is likely that PCs were generated, completed their migration to the PC layer, and then subsequently died. If the PCs died just after they established their position in the PC layer, before the elaboration of much of their dendritic tree, then it is likely that there would be a loss of interneurons, especially stellate cells, unable to make stabilizing synapses with their major targets. But because there is no difference in the density of GABAergic interneurons in the cerebellar molecular layer in autism, at least in the most affected area of reported PC loss, this infers that basket and stellate cells were able to make stabilizing synaptic contacts prior to the PC death. Therefore, from the results of Whitney *et al.*'s (2004, 2005) studies, the timing of the PC loss in autism is most likely not before 32 weeks of gestational age but may extend into early postnatal development since the completion of migration of the external granular layer in humans is not complete until about 18–20 months.

Bauman and Kemper (1985) reported a pallor of staining in the granular layer in some but not all of their original eight cases. It could be that in those cases, the Purkinje cell loss may have occurred earlier rather than later in this range (i.e., late prenatal period affecting the survival of some granule cells whereas cases with normal Nissl stained granule cells but with PC loss may represent a loss in the early postnatal period). Although this is speculation, there is additional evidence that suggests PC loss during this approximate time period and that is the preservation of inferior olivary neurons that project excitatory olivocerebellar climbing fibers directly to Purkinje cells.

In a stereological study of adult autistic and control brainstems, Thevarkunnel *et al.* (2005) reported normal numbers of neurons in the principal olive in autistic cases, the subnucleus that directly projects to the lateral cerebellar hemisphere including the posterior lobe. Again, the preservation of inferior olivary neurons, similar to the GABAergic interneruons in the molecular layer, have in common that their primary target neuron is the Purkinje cell. Olivocerebellar climbing fibers (CFs) arrive early in cerebellar development when PCs are still in clusters awaiting final migration to the PC layer (Sotelo *et al.*, 1984). Permanent stabilizing synapses are made when PCs are in their final position and CF continues to grow as the PC elaborates its extensive dendritic arbor in a mostly two-dimensional plane. The formation of the olivocerebellar projection is thought to be regulated by positional information shared by pre- and post-synaptic neurons. Sotelo and Chedotal (1997) suggested the cell adhesion molecule BEN/SCI/DM-GRASP, isolated from chick embryos, could be one of the target recognition molecules controlling the development of the olivocerebellar projection to innervate its target PC thus forming a topographic map. Postnatal loss of PCs in spontaneous genetic mutations in mice results in inferior olivary cell loss as a secondary consequence with some cells surviving probably due to their sustaining collaterals to the deep cerebellar nuclei (Blatt and Eisenman, 1985a,b). In mutant mice, the PC loss is usually regional and continuous along folia and each CF innervates only two to four PCs increasing the likelihood that a particular CF would lose all of its intended targets in the PC layer. In marked contrast, each CF in the human innervates up to ten PCs and the PC decrease observed in autism results in clusters of remaining PCs (Bailey *et al.*, 1998; Whitney *et al.* 2004, 2005) with the clusters variable from section to section (Whitney *et al.*, 2004, 2005). Therefore, the likelihood of all intended target PCs missing for a given CF is low presumably contributing to the survival of the inferior olivary cells. The timing of events allowing this survival is likely at a similar temporal period as the elaboration of the PC dendritic tree (i.e., between 32 weeks prenatally into the early postnatal period). Interestingly, one case that Whitney *et al.* (2004, 2005) counted had a severe PC loss in the posterior lateral cerebellar cortex (95% decrease) but had a normal number of principal olive cells (Thevarkunnel *et al.*, 2005) although the counts were in the low normal range. This suggests that if the PC loss is large enough some inferior olivary cells may also die but not in sufficient numbers to place it out of the low normal range. It is unknown whether a contributing factor to inferior olive cell survival in autism is their sustaining collaterals to the deep cerebellar nuclei (DCN) because there is a report in the literature that these nuclei also undergo neuropathological changes in autism (Bauman and Kemper, 1985, 1994; Kemper and Bauman, 1998).

E. Neuropathology of the Deep Cerebellar Nuclei in Autism

Age-dependent changes were observed in the deep cerebellar nuclei (Bauman and Kemper, 1985, 1994; Kemper and Bauman, 1998) such that brains from young autistic individuals exhibited unusually large neurons in all four DCN with older individuals (>22 years) showing unusually small and pale neurons with qualitative observation of diminished neuronal numbers in the fastigial, globose, and emboliform nuclei. It must be stressed that as of the date of this publication, there have not been any quantitative studies of the DCN in autism so it is unknown how the PC loss in some autistic cases affects cell number in the DCN. From qualitative observations, the dentate nucleus appears the least affected (Bauman and Kemper, 1985), which is surprising since one of its main inputs is from PC axons emanating from the posterior lateral hemisphere, the area considered most affected with decreases of PCs. The presumed preservation of dentate neurons may indicate a reorganization of PC inputs and/or CF collaterals. Any disruption in the delicate inhibitory-excitatory balance of inputs to either PCs or DCN cells potentially can have profound consequences to the output of the DCN and presumably affect functionality of the nucleo-olivocerebellar system and/or the dentatorubrothalamic tract (Brodal, 1978) projecting back to cerebral cortical areas. Some DCN cells are GABAergic and it has been demonstrated by de Zeeuw et al. (1994, 1997, 1998) that these inhibitory neurons project back to the inferior olive to modulate its activity. Since firing of inferior olivary neurons is synchronous via coupled electrical synapses, disruption in the GABAergic feedback of the DCN could potentially interfere with the ability of inferior olivary neurons to generate coherent rhythmic outputs thus slowing overall cognitive processing speed (Welsh, 2002; Welsh et al., 2005). Alterations within this circuit due to possible miswiring (heterotopic innervation), hyperinnervation of CFs to its PC and/or DCN targets, or innervation of fewer targets due to cell loss would also have the potential of disrupting the modulatory effects of cerebellar function.

III. Functional Consequences of Altered Cerebellar Function in Autism

The classical literature has described the main function of the cerebellum as a motor structure, modulating the output of motor cortex to produce fine motor coordination and skills. Although there are motor signs in autism such as repetitive and stereotypic movements of the body, limbs and fingers, hype-ractivity, abnormal muscle tone and reflexes, and usual gait patterns (see Bauman, 1999, for comprehensive review), autistic individuals do not demonstrate classic cerebellar signs such as marked ataxia and it has thus been difficult for investigators to correlate PC deficits or DCN alterations with the severity of autism or to specific autistic behaviors based only on the consideration of motor criteria.

In normal children, cerebellar hemisphere tumor resection yields a quite different pattern of clinical deficits. Children demonstrated impairments in exec-utive functions, auditory and visual learning, time-based attention tasks, expres-sive language, and visuospatial abilities (Levisohn et al., 2000; Riva and Giorgi, 2000). Clinical research indicates that discrete cerebellar lesions, in otherwise healthy children, cause behavioral and/or cognitive impairments. In autism, however, cerebellar pathology is likely acquired during critical developmental period(s), during a time when the brain is capable of constructing alternate innervation patterns (Sugihara et al., 2003; Zagrebelsky et al., 1997). The apparent lack of correlation of clinical and neuropathological data may be secondary to severe behavioral problems in some individuals that impede testing of cognitive functions attributable to the cerebellar damage. Future research elucidating the correlation between cerebellar pathology and autism phenotype will likely re-quire consistent and detailed neuropsychological testing of individuals with autism.

What are the neuroanatomical substrates within the cerebellar circuitry that may in part underlie cognitive changes observed in the autistic phenotype? Much progress has recently been made in tract tracing and lesion-behavior studies in the rhesus monkey. In a large number of manuscripts, Schmahmann and Pandya (1997; also see Schmahmann, 2001a,b, for review) mapped out the cerebral cortical projections to the cerebellar cortex in the rhesus monkey and found that the pons receives widespread afferents from prefrontal, posterior parietal, para-striate, and limbic cortices in a highly organized pattern. In turn, segregated loops of cerebellar-thalamo-cortical projections have the potential of participat-ing in a wide variety of motor, cognitive, and emotional behaviors, in a diverse stream of information flow. This new revolutionary view of cerebellar function actually has its roots in Ito's theories (see Ito, 1993, for review) but has been largely overlooked. In autism, it is more likely that cerebellar abnormalities may better correlate with functional and high order behavioral alterations rather than classical motor disturbances. We are in the early stages in deciphering how the cerebellum works contrary to Eccles et al.'s (1967) original view that the simplicity of its cytoarchitecture and simple physiological characteristics would make it one of the first understood structures in the human brain. Within the simple cytoarchitecture, however, lies very organized streams of informa-tion processing affecting a diverse number of sensorimotor, associative, and limbic functions.

Schmahmann *et al.* (2004) utilized MRI targeted electrolytic lesions of the dentate nucleus bilaterally in rhesus monkeys and assessed them on conceptual set shifting tasks, a modified version of the Wisconsin card sorting task. This test determines the monkey's ability to learn a rule based on abstract principles and then to "shift set" as the rule shifts from one concept condition to another. Statistically significant greater preservation was found when the monkeys' shifted from one abstraction to another. This test has been shown to be sensitive to prefrontal cortical damage, and suggests that the cerebellum is a critical modulator of prefrontal systems mediating executive function (Schmahmann *et al.*, 2004). In autism, altered GABAergic PC input to the DCN seen in at least half of examined cases with decreased numbers of PCs, would have direct effects on the normal functioning of the DCN and thus may be an underlying substrate, perhaps among many, that may contribute to the behavioral phenotype.

IV. Future and Ongoing Neuropathological Experiments to Further Understand How Cerebellar Circuitry is Altered in Autism

First, a detailed examination of the pontine nuclei in autistic individuals is needed to determine whether the "motor" or "cortical association" input region(s) are affected either in cell number, cell size, and/or distribution. This study should be correlated with a detailed analysis of the cerebellar granular layer. Next, olivocerebellar CFs that use glutamate and aspartate as neurotransmitters need to be marked to determine whether its distribution their altered in the cerebellar cortex especially in those cases with decreased numbers of Purkinje cells. In our laboratory, Yip *et al.* (2005b) is using the neurofilament marker peripherin specific to CFs and to two mossy fiber pathways, spinocerebellar and vestibulocerebellar tracts. She found that CFs do indeed innervate PCs and extend into the molecular layer in autistic cases and noticed some differences in the thickness and pattern of CF distribution to PCs and in its collaterals to the dentate nucleus (Yip *et al.*, 2005b). Another study underway in our laboratory is determining the density and distribution of GABA receptor types in the cerebellar cortex and dentate nucleus. This should provide important insights into the functionality of the GABAergic system especially since there is reduced GAD67 mRNA in Purkinje cells (Yip *et al.*, 2005a). Finally, a detailed quantitative analysis of all four deep cerebellar nuclei is needed to determine the extent of any cell loss and investigate the possible cause(s) of the age-related change in cell size first recognized by Bauman and Kemper (1985). Indications thus far suggest that cerebellar circuitry is disrupted in autism possibly at many levels and pathways. Attention to correlates with altered cognitive functions should lead to a better understanding of the role(s) of the cerebellum in autism. Identification of candidate genes in autism may also yield much needed information regarding the developmental timing of events influencing connectivity. Autism may be the result of action of many genes acting at different but precise times during pre- and early postnatal development in the cerebellum and at other sites in the brain.

Acknowledgments

The author would like to acknowledge Drs. Eisenman, Rosene, Pandya, Bauman, and Kemper who have made and continue to make an invaluable impact on me as a scientist. The author also acknowledges the following research grant support: NIH NICHD HD 39459-04 "Olivocerebellar circuitry in autism" and a grant from the John and Lisa Hussman Foundation. The author also acknowledges past support from the National Alliance for Autism Research (NAAR). Finally, the author wishes to thank Dr. Jean-Jacques Soghomonian and Lin Nguyen for their assistance with the *in situ* hybridization of GAD67 and, Dr. Jane Yip, Dr. Elizabeth Whitney, Sandy Thevarkunnel, Rita Marcon, Yuri Lawrence, Marissa Simms, Claudia Persico, and Gillian VanSluytman for their work in the laboratory. We are also grateful to the Harvard Brain Tissue Resource Center and The Autism Tissue Program for their tissue support for our studies.

References

Bailey, A., Luthert, P., Dean, A., Harding, B., Janota, I., Montgomery, M., Rutter, M., and Lantos, P. (1998). A clinicopathological study of autism. *Brain* **121,** 889–905.

Bauman, M. L. (1999). Motor dysfunction in autism. *In* "Movement Disorders in Neurology and Neuropsychiatry" (A. B. Joseph and R. R. Young, Eds.), 2nd ed., pp. 601–605. Blackwell Science, Malden, MA.

Bauman, M. L., and Kemper, T. L. (1985). Histoanatomic observations of the brain in early infantile autism. *Neurology* **35,** 866–874.

Bauman, M. L., and Kemper, T. L. (1994). Neuroanatomic observations of the brain in autism. Ch. 7. *In* "The Neurobiology of Autism" (M. L. Bauman and T. L. Kemper, Eds.), pp. 119–145. Johns Hopkins University Press, Baltimore.

Blatt, G. J., and Eisenman, L. M. (1985a). A qualitative and quantitative light microscopic study of the inferior olivary complex of normal, reeler, and weaver mutant mice. *J. Comp. Neurol.* **232,** 117–128.

Blatt, G. J., and Eisenman, L. M. (1985b). A qualitative and quantitative light microscopic study of the inferior olivary complex of the adult staggerer mutant mouse. *J. Neurogenet.* **2,** 51–66.

Blatt, G. J., Fitzgerald, C. M., Guptill, J. T., Booker, A. B., Kemper, T. L., and Bauman, M. L. (2001). Density and distribution of hippocampal neurotransmitter receptors in autism: An autoradiographic study. *J. Autism Devel. Disord.* **31,** 537–543.

Brodal, A. (1978). "Neurological Anatomy: In Relation to Clinical Medicine." Oxford University Press, London.

Buxbaum, J. D., Silverman, J. M., Smith, C. J., Greenberg, D. A., Kilifarski, M., Reichert, J., Cook, E., Fang, Y., Song, C. Y., and Vitale, R. (2002). Association between a GABARB3 polymorphism and autism. *Mol. Psychiatry* **7,** 311–316.

de Zeeuw, C. I., Gerrits, N. M., Voogd, J., Leonard, C. S., and Simpson, J. I. (1994). The rostral dorsal cap and ventrolateral outgrowth of the rabbit inferior olive receive a GABAergic input from dorsal group Y and the ventral dentate nucleus. *J. Comp. Neurol.* **341,** 420–432.

de Zeeuw, C. I., Van Alphen, A. M., Hawkins, R. K., and Ruigrok, T. J. (1997). Climbing fibre collaterals contact neurons in the cerebellar nuclei that provide a GABAergic feedback to the inferior olive. *Neurosci.* **80,** 981–986.

de Zeeuw, C. I., Simpson, J. I., Hoogenraad, C. C., Galjart, N., Koekkoek, S. K., and Ruigrok, T. J. (1998). Microcircuitry and function of the inferior olive. *Trends in Neurosci.* **21,** 391–400.

Dhossche, D., Applegate, H., Abraham, A., Maertens, P., Bland, L., Bencsath, A., and Martinez, J. (2002). Elevated plasma gamma-aminobutyric acid (GABA) levels in autistic youngsters: Stimulus for a GABA hypothesis of autism. *Med. Sci. Monit.* **8,** 1–6.

Eccles, J. C., Ito, M., and Szentagothai, J. (1967). "The Cerebellum as a Neuronal Machine." Springer-Verlag, New York.

Fatemi, S. H., Halt, A. R., Stary, J. M., Kanodia, R., Schulz, S. C., and Realmuto, G. R. (2002). Glutamic acid decarboxylase 65 and 67 kDa proteins are reduced in autistic parietal and cerebellar cortices. *Biol. Psychiatry* **52,** 805–810.

Hall, T. C., Miller, A. K. H., and Corsellis, J. A. N. (1975). Variations in the human Purkinje cell population according to age and sex. *Neuropathol. Appl. Neurobiol.* **1,** 267–292.

Ito, M. (1993). Movement and thought: Identical control mechanisms by the cerebellum. *Trends Neurosci.* **16,** 448–450.

Kemper, T. L., and Bauman, M. L. (1998). Neuropathology of infantile autism. *J. Neuropath. Exper. Neurol.* **57,** 645–652.

Lawrence, Y., Kemper, T. L., Bauman, M. L., and Blatt, G. J. (2005). Increased density of parvalbumin labeled hippocampal interneurons in autism. *Internat. Meeting for Autism Res. Abstr.* (IMFAR) **4,** 45–46.

Levisohn, L., Cronin-Golomb, A., and Schmahmann, J. D. (2000). Neuropsychological consequences of cerebellar tumour resection in children: Cerebellar cognitive affective syndrome in a paediatric population. *Brain* **123,** 1041–1050.

Martin, E. R., Menold, M. M., Wolpert, C. M., Bass, M. P., Donnelly, S. L., Ravan, S. A., Zimmerman, A., Gilbert, J. R., Vance, J. M., Maddox, L. O., Wright, H. H., Abramson, R. K., DeLong, G. R., Cuccaro, M. I., and Pericak-Vance, M. A. (2000). Analysis of linkage disequilibrium in GABA receptor subunit genes in autistic disorders. *Am. J. Med. Genet.* **96,** 43–48.

Ritvo, E. R., Freeman, B. J., Schiebel, A. B., Robinson, D. T., Guthrie, D., and Ritvo, A. (1986). Lower Purkinje cell counts in the cerebella of four autistic subjects: Initial findings of the UCLA-NSAC autopsy research report. *Amer. J. Psychiatry* **146,** 862–866.

Riva, D., and Giorgi, C. (2000). The cerebellum contributes to higher functions during development; evidence from a series of children surgically treated for posterior fossa tumours. *Brain* **123,** 1051–1061.

Schmahmann, J. D. (2001a). The cerebrocerebellar system: Anatomic substrates of the cerebellar contribution to cognition and emotion. *Int. Rev. Psychiatry* **13,** 247–260.

Schmahmann, J. D. (2001b). The cerebellar cognitive affective syndrome: Clinical correlations of the dysmetria of thought hypothesis. *Int. Rev. Psychiatry* **13,** 313–322.

Schmahmann, J. D., and Pandya, D. N. (1997). The cerebrocerebellar system. *Int. Rev. Neurobiol.* **41,** 31–60.

Schmahmann, J. D., Killiany, R. J., Moore, T. L., DeMong, C., MacMore, J. P., and Moss, M. B. (2004). Cerebellar dentate nucleus lesions impair flexibility but not motor function in monkeys. *Soc. Neurosci. Abstr.* **34,** 254.12.

Schroer, R. J., Phelan, M. C., Michaelis, R. C., Crawford, E. C., Skinner, S. A., Cuccaro, M., Simensen, R. J., Bishop, J., Skinner, C., Fender, D., and Stevenson, R. E. (1998). Autism and maternally derived aberrations of chromosome 15q. *Amer. J. Med. Genet.* **76,** 327–336.

Shao, Y., Cuccaro, M. L., Hauser, E. R., Raiford, K. L., Menold, M. M., Wolpert, C. M., Ravan, S. A., Elston, L., Decena, K., Donnelly, S. L., Abramson, R. K., Wright, H. H., DeLong, G. R., Gilbert, J. R., and Pericak-Vance, M. A. (2003). Fine mapping of autistic disorder to chromosome 15q11-q13 by use of phenotypic subtypes. *Amer. J. Human Genet.* **72,** 539–548.

Sotelo, C., and Chedotal, A. (1997). Development of the olivocerebellar projection. *Perspect. Devel. Neurobiol.* **5,** 57–67.

Sotelo, C., Bourrat, F., and Triller, A. (1984). Postnatal development of the inferior olivary complex in the rat. II. Topographic organization of the immature olivocerebellar projection. *J. Comp. Neurol.* **222,** 177–199.

Sugihara, I., Lohof, A. M., Letellier, M., Mariani, J., and Sherrard, R. M. (2003). Post-lesion transcommissural growth of olivary climbing fibres creates functional synaptic microzones. *Eur. J. Neurosci.* **18,** 3027–3036.

Thevarkunnel, S., Bauman, M. L., Kemper, T. L., and Blatt, G. J. (2005). Stereological study of the number and size of neurons in the principal olive in autism. *Internat. Meeting for Autism Res. Abstr.* (IMFAR) **4,** 48.

Welsh, J. P. (2002). Functional significance of climbing fiber synchrony: A population coding and behavioral analysis. *Ann. N.Y. Acad. Sci.* **978,** 188–204.

Welsh, J. P., Ahn, E. S., and Placantonakis, D. G. (2005). Is autism due to brain desynchronization? *Int. J. Devl. Neuroscience* **23,** 253–263.

Whitney, E. R., Kemper, T. L., Bauman, M. L., and Blatt, G. J. (2004). Calcium binding proteins in cerebellar Purkinje cells in the autistic cerebellum. *Soc. Neurosci. Abstr.* **34,** 116.11.

Whitney, E. R., Kemper, T. L., Bauman, M. L., and Blatt, G. J. (2005). Quantitative analysis of cerebellar basket and stellate cells in autism. *Internat. Meeting for Autism Res. Abstr.* (IMFAR) **4,** 49.

Williams, R. S., Hauser, S. L., Purpura, D. P., DeLong, G. R., and Swisher, C. N. (1980). Autism and mental retardation: Neuropathologic studies performed in four retarded persons with autistic behavior. *Arch. Neurol.* **37,** 749–753.

Yip, J., Soghomonian, J.-J., Nguyen, L., and Blatt, G. J. (2005a). GAD67 mRNA decrease in cerebellar Purkinje cells in autism: An *in situ* hybridization study. *Soc. Neurosci. Abstr.* **35**(In Press).

Yip, J., Marcon, R., Kemper, T. L., Bauman, M. L., and Blatt, G. J. (2005b). The olivocerebellar projection in autism: Using the intermediate filament protein peripherin as a marker for climbing fibers. *Internat. Meeting for Autism Res. Abstr.* (IMFAR) **4,** 49.

Zagrebelsky, M., Strata, P., Hawkes, R., and Rossi, F. (1997). Reestablishment of the olivocerebellar projection map by compensatory transcommissural reinnervation following unilateral transection of the inferior cerebellar peduncle in the inferior cerebellar peduncle in the newborn rat. *J. Comp. Neurol.* **379,** 283–299

REELIN GLYCOPROTEIN IN AUTISM AND SCHIZOPHRENIA

S. Hossein Fatemi

Department of Psychiatry, University of Minnesota Medical School
Minneapolis, Minnesota 55455, USA

Reelin glycoprotein is a serine protease with important roles in embryogenesis and in adult life. Reelin mutations or deficiency of the protein product could cause abnormal cortical development and/or Reelin-induced signal transduction impairment in brain. Reelin abnormalities in several neuropsychiatric disorders, such as autism and schizophrenia, may provide mechanistic explanations for etiologies of these disorders.

I. Reelin Gene and Protein

Reelin glycoprotein is a serine protease (Quattrocchi *et al.*, 2002) with dual roles in mammalian brain: embryologically, it guides neurons and radial glial cells to their final positions in brain (Forster *et al.*, 2002; Goffinet, 1992); during adult life, it participates in a signaling cascade which may serve synaptic plasticity, memory processing, and cognition (Fatemi, 2005; Weeber *et al.*, 2002). Reelin gene (Reln) is localized to chromosome 7 in man (DeSilva *et al.*, 1997). The Reln gene codes for protein products which on SDS-PAGE range from 410 to 330, 180kDa and several smaller fragments in man (Fatemi *et al.*, 2002a, 2004, 2005a; Ignatova *et al.*, 2004; Smallheiser *et al.*, 2000) and in rodents (D'Arcangelo *et al.*, 1995; Ogawa *et al.*, 1995).

INTERNATIONAL REVIEW OF
NEUROBIOLOGY, VOL. 71
DOI: 10.1016/S0074-7742(05)71008-4

179

II. Reelin Signaling Cascade

Reelin protein binds several receptors including apolipoprotein E receptor 2 (ApoER2), very-low-density lipoprotein receptor (VLDLR), and $\propto 3\beta 1$ integrin to initiate a signaling cascade which underlies downstream biochemical events leading to synaptic plasticity (see Fig. 1). Binding of Reelin to its receptors, specifically ApoER2 and VLDLR, induces clustering of these receptors and oligomerization of the adaptor protein, disabled-1 (Dab-1) (D'Arcangelo *et al.*, 1999; Dulabon *et al.*, 2000; Hiesberger *et al.*, 1999; Strasser *et al.*, 2004). This is then followed by tyrosine phosphorylation of Dab-1 which causes actin polymerization and final ubiquitination of Dab-1 protein, steps involved in mechanisms behind cell migration and synaptic plasticity (Beffert *et al.*, 2005; Suetsugu *et al.*, 2004). Phosphorylation of Dab-1 can also act as a substrate for inhibition of the level of glycogen synthase-kinase 3β (GSK-3β) and modulation of pathways for cell survival and growth (Beffert *et al.*, 2002).

III. The Reeler Mouse

The significance of Reelin's role in embryogenesis of brain became evident following the discovery of a Reln gene mutation nearly half a century ago (Falconer, 1951). The Reelin mutant mouse, which carries an autosomal recessive mutation in Reln, exhibited ataxia and a reeling gait. Examination of the brain in these animals showed inverted cortical lamination, abnormal positioning of neurons, and aberrant orientation of neuronal cell bodies and nerve fibers (Falconer, 1951; Goffinet, 1979). The mutant mice also exhibited cerebellar hypoplasia with associated lack of foliation (Caviness and Sidman, 1973). Ectopic expression of Reelin in the Reeler mouse rescues cerebellar development and corrects ataxia in the mutant mouse (Magdaleno *et al.*, 2002). The homozygous mutant mouse does not produce Reelin. The heterozygous Reeler mouse has a 50% reduction in Reelin protein and mRNA with decreases in dendritic spine density, neuropil hypoplasticity, and decreased GABA turnover (Carboni *et al.*, 2004). Behaviorally, the heterozygous Reeler mouse exhibits decreased prepulse inhibition, a phenomen observed in subjects with autism and schizophrenia (McAlonan *et al.*, 2002; Meincke *et al.*, 2004; Tueting *et al.*, 1999). Several recent reports using various prenatal insults, e.g., viral infection in midterm pregnant mice (Fatemi *et al.*, 1999, 2002b), and 5 methoxytryptamine exposure in E17 pregnant rats (Janusonis *et al.*, 2004) cause reductions in levels of brain/blood Reelin levels and result in abnormal corticogenesis in the offspring.

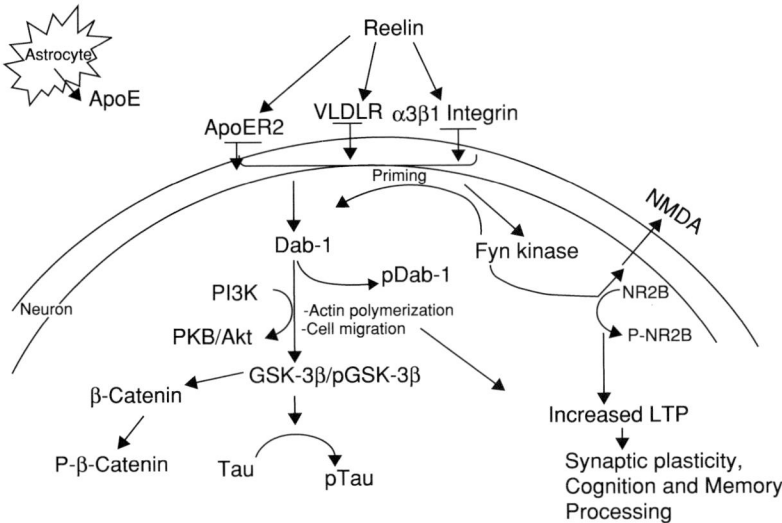

FIG. 1. The Reelin signaling system and cognition. Extracellular Reelin glycoprotein is secreted by Cajal-Retzius cells and certain cortical and hippocampal GABAergic cells and cerebellar granule cells. Reelin can bind its receptors ApoER2, VLDLR, and $\alpha3\beta1$ integrin directly, initiating the signaling system in the effector cells (i.e., cortical pyramidal cells). Reelin induction of the cascade leads to clustering of the receptors causing dimerization/oligomerization of Dab-1 protein and activation of Src-tyrosine kinase family/Fyn-kinase leading to tyrosine phosphorylation of Dab-1 protein in a positive-feedback loop. Interaction between Dab-1, N-WASP, and ARP 2/3 complex causes formation of microspikes or filopodia which are important in processes of cell migration and synaptic plasticity. Finally, phosphorylation of a subpopulation of Dab-1 molecules causes degradation of Dab-1 via ubiquitination, resulting in termination of Reelin signaling cascade. Downstream effector proteins involved in Reelin signaling path include phosphatidylinositol-3-kinase (PI3K) and protein kinase B (PKB/Akt), which further impact on three other important molecules, glycogen synthase kinase (GSK-3β), β-catenin, and tau. The latter proteins can modulate pathways, affecting cell proliferation, apoptosis, and neurodegeneration respectively. Finally, Reelin has a direct effect on enhancement of long term potentiation (LTP), via direct involvement of its receptors VLDLR and ApoER2. Alternately, tyrosine phosphorylation of NR2B subunit of NMDA receptor by Fyn kinase is essential for induction of LTP and modulation of synaptic plasticity, potentially converging on Reelin's role in cognition and memory processing (Fatemi *et al.*, 2001, 2005b).

IV. Reelin in Autism

Analogous defects involving the Reelin signaling system appear to be present in several neuropsychiatric disorders (e.g., schizophrenia (Fatemi *et al.*, 2000, 2005a; Guidotti *et al.*, 2000; Impagnatiello *et al.*, 1998) and autism (Fatemi *et al.*, 2001, 2005b)). The findings of Reelin defects are more robust in autism and are supported by two positive genetic linkage studies (Persico *et al.*, 2001;

TABLE I
mRNA Levels for Reelin, VLDLR, Dab-1, and GSK3*

Area	Gene of interest (GOI)	GOI relative to age matched control	P Value
Area 9	Reelin	−4.7	p < 0.035
	VLDLR	+14.2	p < 0.01
	DAB1	−5.4	p < 0.01
	GSK3	+1.9	NS
Cerebellum	Reelin	−3.9	p < 0.01
	VLDLR	+2.8	p < 0.04
	DAB1	−3.4	p < 0.001
	GSK3	+1	NS

*Fatemi, S. H., et al. (2005b).

Zhang et al., 2002). Autism, which is a severe childhood disorder of the brain, is characterized by impairments in communication, social skills, and repetitive behavior (Kanner, 1943). Despite a large body of evidence supporting a genetic cause for autism (Folstein and Rosen-Sheidley, 2002), environmental causes are potentially responsible for some cases of autism (Rodier, 2000). A controversial issue is the involvement of Reln in causation of autism. Of the six genetic linkage studies, four deny a link (Bonora et al., 2003; Devlin et al., 2004; Krebs et al., 2002; Li et al., 2004) while two are supportive (Persico et al., 2001; Zhang et al., 2002). Additionally, four recent biochemical reports, however, show reductions in brain (Fatemi et al., 2001, 2005b) and blood (Fatemi et al., 2002a; Lugli et al., 2003). Reelin levels in subjects with autism (Fatemi et al., 2005b) showed that Reelin protein and mRNA species were reduced significantly in area 9 and cerebellum of autistic subjects vs. age and postmortem intervaled-matched controls (see Table I). The reductions in Reelin levels accompanied significant increases in mRNA levels of Reelin receptor VLDL-R in frontal and cerebellar cortices of autistic subjects (Table I). Surprisingly, levels of Dab-1 mRNA were also reduced significantly in the same brain sites in autistic subjects (Table I) implying involvement of the Reelin signaling cascade in the autistic pathology (see Fig. 2).

V. Reelin in Schizophrenia

Schizophrenia, a neurodevelopmental disorder, which in contrast to autism affects youth in puberty and is manifested by presence of hallucinations, disorganized behavior, and fragmentation of thought (Kraeplin, 1923), may also share

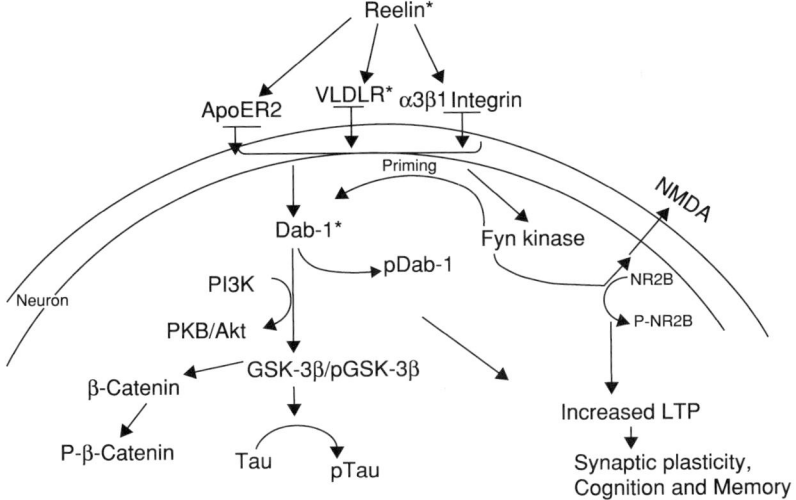

FIG. 2. The role of Reelin signaling system in autism. Normally, extracellular Reelin is secreted by Cajal-Retzius cells and certain GABAergic cells to bind its receptors VLDLR, ApoER2, and $\alpha 3\beta 1$ integrin on effector cells. Following binding of Reelin to its receptors, Dab-1 protein is oligomerized and phosphorylated. In autistic brain, Reelin signaling system appears to be impaired in 3 steps (marked by *); (1) Reelin ligand is not produced adequately as evident by reductions in mRNA and protein levels in superior frontal cortex and cerebellum; (2) Reelin receptor VLDLR mRNA is upregulated potentially in response to reduced levels of its ligand, Reelin; (3) Dab-1 mRNA is also reduced potentially due to reduction in levels of Reelin which normally activates Dab-1 phosphorylation via a positive-feedback loop. Alternatively, Dab-1 levels may be reduced in response to increases in levels of VLDLR acting via a negative-feedback loop. Alterations in levels of Reelin, its receptor VLDLR, and adaptor protein Dab-1 interfere with the Reelin signaling system affecting LTP, synaptic plasticity, cognition, and memory, modalities involved in autism (Fatemi *et al.*, 2005b).

Reelin abnormalities with autism (Fatemi, 2005). Investigation of the postmortem brains using a multitude of techniques showed downregulation of Reelin protein and mRNA in the prefrontal cortex (Guidotti *et al.*, 2000), and decreased Reelin protein in hippocampus (Fatemi *et al.*, 2000) and cerebellum (Fatemi *et al.*, 2005a) of subjects with schizophrenia. Mechanistically, it appears that hypermethylation of the promoter for Reln may be partially responsible for the observed decreases in Reelin in schizophrenia (Abdolmaleky *et al.*, 2005; Costa *et al.*, 2003). Two genetic linkage studies have not been able to show a significant linkage between Reelin polymorphisms and schizophrenia (Akahane *et al.*, 2002; Chen *et al.*, 2002). The biochemical data also support involvement of Reelin abnormalities in mood disorders (Fatemi *et al.*, 2000, 2005a; Guidotti *et al.*, 2000) regardless of presence of psychosis.

VI. Conclusions

In summary, increasing biochemical evidence points to involvement of Reelin glycoprotein in a number of psychiatric disorders including schizophrenia, autism, and mood disorders. The disparity seen in levels of Reelin production in man appears to be similar to the same scenario seen in various animal models which affect Reelin production, leading to production of cognitive deficits in rodents (Fatemi, 2005). Future correlative studies of human postmortem brains in schizophrenia and autism and pertinent animal models of mental disorders may expand our knowledge of the role of Reelin in cognition.

Acknowledgments

The work of the author has been supported by Stanley Medical Research Institute, March of Dimes, National Institute of Child Health and Human Development, The Jonty Foundation, and the Kunin Fund of St. Paul Foundation. I am grateful for technical assistance by Mr. T. Folsom and Ms. T. Reutiman and secretarial assistance by Ms. Laurie Iversen.

References

Abdolmaleky, H. M., Cheng, K. H., Russo, A., Smith, C. L., Faraone, S. V., Wilcox, M., Shafa, R., Glatt, S. J., Nguyen, G., Ponte, J. F., Thiagalingam, S., and Tsuang, M. T. (2005). Hypermethylation of the reelin (RELN) promoter in the brain of schizophrenic patients: A preliminary report. *Am. J. Med. Genet. B. Neuropsychiatr. Genet.* **134,** 60–66.

Akahane, A., Kunugi, H., Tanaka, H., and Nanko, S. (2002). Association analysis of polymorphic CGG repeat in 5' UTR of the reelin and VLDLR genes with schizophrenia. *Schizophr. Res.* **58,** 37–41.

Beffert, U., Morfini, G., Bock, H. H., Reyna, H., Brady, S. T., and Herz, J. (2002). Reelin-mediated signaling locally regulates protein kinase B/Akt and glycogen synthase kinase 3β. *J. Biol. Chem.* **51,** 49958–49964.

Bonora, E., Beyer, K. S., Lamb, J. A., Parr, J. R., Klauck, S. M., Benner, A., Paolucci, M., Abbott, A., Ragoussis, I., Poustka, A., Bailey, A. J., and Monaco, A. P. (2003). International Molecular Genetic Study of Autism (IMGSAC). Analysis of reelin as a candidate gene for autism. *Mol. Psychiatry* **10,** 885–892.

Carboni, G., Tueting, P., Tremolizzo, L., Sugaya, I., Davis, J., Costa, E., and Guidotti, A. (2004). Enhanced dizocilpine efficacy in heterozygous reeler mice relates to GABA turnover downregulation. *Neuropharmacology* **46,** 1070–1081.

Caviness, V. S., Jr., and Sidman, R. L. (1973). Time of origin of corresponding cell classes in the cerebral cortex of normal and mutant Reeler mice: An autoradiographic analysis. *J. Comp. Neurol.* **148,** 141–152.

Chen, M. L., Chen, S. Y., Huang, C. H., and Chen, C. H. (2002). Identification of a single nucleotide polymorhism at the 5' promoter region of human reelin gene and association study with schizophrenia. *Mol. Psychiatry* **7,** 447–448.

Costa, E., Grayson, D. R., and Guidotti, A. (2003). Epigenetic downregulation of GABAergic function in schizophrenia; potential for pharmacological intervention? *Mol. Interv.* **3,** 220–229.

D'Arcangelo, G., Miao, G. G., Chon, S. C., Soares, H. D., Morgan, J. I., and Curran, T. (1995). A protein related to extracellular matrix proteins detected in the mouse mutant reeler. *Nature* **374,** 719–723.

D'Arcangelo, G., Homayouni, R., Keshvara, L., Rice, D. S., Sheldon, M., and Curran, T. (1999). Reelin is a ligand for lipoprotein receptors. *Neuron* **24,** 471–479.

DeSilva, U., D' Arcangelo, G., Braden, V. V., Chen, J., Miao, G. G., Currant, T., and Green, E. D. (1997). The human reelin gene: Isolation, sequencing, and mapping on chromosome 7. *Genome Res.* **7,** 157–164.

Devlin, B., Bennett, P., Dawson, G., Figlewicz, D. A., Grigorenko, E. L., McMahon, W., Minshew, N., Pauls, D., Smith, M., Spence, M. A., Rodier, P. M., Stodgell, C., and Schellenberg, G. D. (2004). Alleles of a reelin CGG repeat do not convey liability to autism in a sample from the CPEA network. *Am. J. Med. Genet. B. Neuropsychiatr. Genet.* **126,** 46–50.

Dulabon, L., Olson, E. C., Taglienti, M. G., Eisenhuth, S., McGrath, B., Walsh, C. A., Kreidberg, J. A., and Anton, E. S. (2000). Reelin binds alpha 3 beta 1 integrin and inhibits neuronal migration. *Neuron* **27,** 33–44.

Falconer, D. S. (1951). Two new mutants, Trembler and 'Reeler', with neurological actions in the house mouse. *J. Genetics* **50,** 182–201.

Fatemi, S. H., Emamian, E. S., Kist, D., Sidwell, R. W., Nakajima, K., Akhter, P., Shier, A., Sheikh, S., and Bailey, K. (1999). Defective corticogenesis and reduction in Reelin immunoreactivity in cortex and hippocampus of prenatally infected neonatal mice. *Mol. Psychiatry* **4,** 145–154.

Fatemi, S. H., Earle, J. A., and McMenomy, T. (2000). Reduction in Reelin immunoreactivity in hippocampus of subjects with schizophrenia, bipolar disorder and major depression. *Mol. Psychiatry* **5,** 654–663.

Fatemi, S. H., Kroll, J. L., and Stary, J. M. (2001). Altered levels of Reelin and its isoforms in schizophrenia and mood disorders. *Neuroreport* **12,** 3209–3215.

Fatemi, S. H., Stary, J. M., and Egan, E. A. (2002a). Reduced blood levels of reelin as a vulnerability factor in pathophysiology of autistic disorder. *Cell. Mol. Neurobiol.* **22,** 139–52.

Fatemi, S. H., Earle, J. A., Kanodia, R., Kist, D. A., Emamian, E. S., Patterson, P. H., Shi, L., and Sidwell, R. W. (2002b). Prenatal viral infection leads to pyramidal cell atrophy and macrocephaly in adulthood: Implications for genesis of autism and schizophrenia. *Cell. Mol. Neurobiol.* **22,** 25–33.

Fatemi, S. H. (2005). Reelin glycoprotein: Structure, biology and roles in health and disease. *Mol. Psychiatry* **10,** 251–257.

Fatemi, S. H., Stary, J. M., Araghi-Niknam, M., and Egan, E. (2005a). GABAergic dysfunction in schizophrenia and mood disorders as reflected by decreased levels of Reelin and GAD 65 & 67 kDa proteins in cerebellum. *Schizophr. Res.* **72,** 109–122.

Fatemi, S. H., Snow, A. V., Stary, J. M., Araghi-Niknam, M., Reutiman, T. J., Lee, S., Brooks, A. I., and Pearce, D. A. (2005b). Reelin signaling is impaired in autism. *Biol. Psychiatry* **57,** 777–787.

Folstein, S. E., and Rosen-Sheidley, B. (2002). Genetics of autism: Complex etiology for a heterogeneous disorder. *Nature Rev. Genetics* **2,** 943–955.

Forster, E., Tielsch, A., Saum, B., Weiss, K. H., Johanssen, C., Graus-Porta, D., Muller, U., and Frotscher, M. (2002). Reelin, disabled 1, and beta 1 integrins are required for the of the radial glial scaffold in the hippocampus. *Proc. Natl. Acad. Sci. USA* **99,** 13178–13183.

Goffinet, A. M. (1979). An early developmental defect in the cerebral cortex of the Reelor mouse. *Anat. Embryol.* **157,** 205–218.

Goffinet, A. M. (1992). The reeler gene: A clue to brain development and evolution. *Int. J. Dev. Biol.* **36,** 101–107.

Guidotti, A., Auta, J., Davis, J. M., Di-Giorgi-Gerevini, V., Dwivedi, Y., Grayson, D. R., Impagnatiello, F., Pandey, G., Pesold, C., Sharma, R., Uzunov, D., and Costa, E. (2000). Decrease in Reelin and glutamic acid decarboxylase 67 (GAD 67) expression in schizophrenia and bipolar disorder: A postmortem brain study. *Arch. Gen. Psychiatry* **57,** 1061–1069.

Hiesberger, T., Trommsdorff, M., Howell, B. W., Goffinet, A., Mumby, M. C., Cooper, J. A., and Herz, J. (1999). Direct biding of Reelin to VLDL receptor and ApoE receptor 2 induces tyroene phosphorylation of disabled-1 and modulates tau phosphorylation. *Neuron* **24,** 481–489.

Ignatova, N., Sindic, C. J. M., and Goffinet, A. M. (2004). Characterization of the various farms of the Reelin protein in the cerebrospinal fluid of normal subjects and in neurological diseases. *Neurobiol. Disease* **15,** 326–330.

Impagnatiello, F., Guidotti, A. R., Pesold, C., Dwivedi, Y., Caruncho, H., Pisu, M. G., Uzunov, D. P., Smalheiser, N. R., Davis, J. M., Pandey, G. N., Pappas, G. D., Tueting, P., Sharma, R. P., and Costa, E. (1998). A decrease of Reelin expression as a putative vulnerability factor in schizophrenia. *Proc. Natl. Acad. Sci. USA* **95,** 15718–15723.

Janusonis, S., Gluncic, V., and Rakic, P. (2004). Early serotonergic projections to Cajal-Retzius cells: Relevance for cortical development. *J. Neuroscience* **24,** 1652–1659.

Kanner, L. (1943). Autistic disturbances of affective contact. *J. Nerv. Child* **2,** 217–250.

Kraeplin, E. (1923). "Psychiatrie," (8th edition). Lepizig, Barth.

Krebs, M. O., Betancur, C., Leroy, S., Bourdel, M. C., Gillberg, C., and Leboyer, M. (2002). Paris Autism Research International Sibpair (PARIS) study. Absence of association between a polymorphic GGC repeat in the 5′ untranslated region of the reelin gene and autism. *Mol. Psychiatry* **7,** 801–804.

Li, J., Nguyen, L., Gleason, C., Lotspeich, L., Spiker, D., Risch, N., and Myers, R. M. (2004). Lack of evidence for an association between WNT2 and RELN polymorphisms and autism. *Am. J. Med. Genet. B. Neuropsychiatr. Genet.* **126,** 51–57.

Lugli, G., Krueger, J. M., Davis, J. M., Persico, A. M., Keller, F., and Smalheiser, N. R. (2003). Methodological factors influencing measurement and processing of plasma reelin in humans. *BMC Biochemistry* **4,** 9.

Magdaleno, S., Keshvara, L., and Curran, T. (2002). Rescue of ataxia and prepalte splitting by ectopic expression of Reelin in reeler mice. *Neuron* **33,** 573–586.

McAlonan, G. M., Daly, E., Kumari, V., Critchley, H. D., van Amelsvoort, T., Suckling, J., Simmons, A., Sigmundsson, T., Greenwood, K., Russell, A., Schmitz, N., Happe, F., Howlin, P., and Murphy, D. G. (2002). Brain anatomy and sensorimotor gating in Asperger's syndrome. *Brain* **125,** 1594–606.

Meincke, U., Light, G. A., Geyer, M. A., Braff, D. L., and Gouzoulis-Mayfrank, E. (2004). Sensitization and habituation of the acoustic startle reflex in patients with schizophrenia. *Psychiatry Res.* **126,** 51–61.

Ogawa, M., Miyata, T., Nakajima, K., Yagyu, K., Seike, M., Ikenaka, K., Yamamoto, H., and Mikoshiba, K. (1995). The reeler gene-associated antigen on Cajal–Retzius neurons is a crucial molecule for laminar organization of cortical neurons. *Neuron* **14,** 899–912.

Persico, A. M., D'Agruma, L., Maiorano, N., Totaro, A., Militerni, R., Bravaccio, C., Wassink, T. H., Schneider, C., Melmed, R., Trillo, S., Montecchi, F., Palermo, M., Pascucci, T., Puglisi-Allegra, S., Reichelt, K. L., Conciatori, M., Marino, R., Quattrocchi, C. C., Baldi, A., Zelante, L., Gasparini, P., and Keller, F. (2001). Reelin gene alleles and haplotypes as a factor predisposing to autistic disorder. *Mol. Psychiatry* **6,** 150–159. Collaborative Linkage, Study of Autism.

Quattrocchi, C. C., Wannenes, F., Persico, A. M., Ciafre, S. A., D'Arcangelo, G., Farace, M. G., and Keller, F. (2002). Reelin is a serine protease of the extracellular matrix. *J. Biol. Chem.* **277,** 303–309.

Rodier, P. (2000). The early origins of autism. *Sci. Am.* **282,** 56–63.

Smalheiser, N. R., Costa, E., Guidotti, A., Impagnatiello, F., Auta, J., Lacor, P., Kriho, V., and Pappas, G. D. (2000). Expression of reelin in adult mammalian blood, liver, pituitary pars intermedia, and adrenal chromaffin cells. *Proc. Natl. Acad. Sci. USA* **97,** 1281–1286.

Strasser, V., Fasching, D., Hauser, C., Mayer, H., Bock, H. H., Hiesberger, T., Herz, J., Weeber, E. J., Sweatt, J. D., Pramatarova, A., Howell, B., Schneider, W. J., and Nimpf, J. (2004). Receptor clustering is involved in reelin signaling. *Mol. Cell. Biol.* **24,** 1378–1386.

Suetsugu, S., Tezuka, T., Morimura, T., Hattori, M., Mikoshiba, K., Yamamoto, T., and Takenawa, T. (2004). Regulation of actin cytoskeleton by mDab1 through N-WASP and ubiquitination of mDab1. *Biochem. J.* **384,** 1–8.

Tueting, P., Costa, E., Dwivedi, Y., Guidotti, A., Impagnatiello, F., Manev, R., and Pesold, C. (1999). The phenotypic characteristics of heterozygous reeler mouse. *NeuroReport* **10,** 1327–1334.

Weeber, E. J., Beffert, U., Jones, C., Christian, J. M., Forster, E., Sweatt, J. D., and Herz, J. (2002). Reelin and ApoE receptors cooperate to enhance hippocampal synaptic plasticity and learning. *J. Biol. Chem.* **277,** 39944–39952.

Zhang, H., Liu, X., Zhang, C., Mundo, E., Macciardi, F., Grayson, D. R., Guidotti, A. R., and Holden, J. J. (2002). Reelin gene alleles and susceptibility for autism spectrum disorders. *Mol. Psychiatry* **7,** 1012–1017.

IS THERE A CONNECTION BETWEEN AUTISM, PRADER-WILLI SYNDROME, CATATONIA, AND GABA?

Dirk M. Dhossche,* Yaru Song,[†] and Yiming Liu[†]

*Department of Psychiatry & Human Behavior
University of Mississippi Medical Center
Jackson, Mississippi 39216, USA
[†]Department of Chemistry, Jackson State University
Jackson, Mississippi 39217, USA

Research advances in autism, Prader-Willi syndrome, and catatonia allow new inferences about their mutual relations. Catatonia has been described in both people with autism and Prader-Willi syndrome, but remains unaccounted. Although autism and Prader-Willi syndrome have been considered entities in their own right, clinical observations suggest common ground and point to GABA dysfunction as a putative shared risk factor. The knowledge base on GABA abnormalities in autism, Prader-Willi syndrome, and catatonia is limited but expanding, and is summarized in this chapter. Preliminary findings that plasma GABA is a familial marker of autism are presented. Future research avenues are outlined.

I. Introduction

The interest in neurosis sparked by psychoanalytical ideas and the segregation of the
"feeble-minded" into special institutions may be the potent factors in the decline of
catatonia. Likewise, the impact of Kanner's delineation of autism, and research into the
stereotypies of mental retardation has tended to mute the earlier enthusiasm concerning the
interrelationship between the functional psychoses and mental retardation. In fact, the
clinical and research "hole" resembles a form of diagnostic sterilization reminiscent of the
operative practices of the eugenics period. Reviewing the whole process in the light of
critical science and historical understanding in services shows up areas of underdevelopment
in services, research, and clinical acumen. Yet the potential for genuinely new discoveries is
there provided we understand what really has gone before.

T. Turner, M.D. (1989)

Historically, autism, childhood schizophrenia, mental retardation, and other
early-onset developmental disorders were thought to be connected (Earl, 1934;
Turner, 1989). In 1943, Kanner isolated autism from other childhood disorders.
This may have caused premature closure of the search for commonalities between
autism, psychosis, mental retardation, and other early-onset neurodevelopmental
disorders. However, research advances in these disorders offer an opportunity for
re-evaluation. New findings reveal phenomenological, biochemical, and genetic
areas of overlap. GABA dysfunction may provide a common risk factor in autism,
Prader-Willi syndrome, and catatonia (Dhossche, 2004). In this chapter, evidence
that GABA abnormalities are present in autism, Prader-Willi syndrome, and
catatonia is reviewed. Implications for future research are described.

II. GABA in Autism

The charm of GABA lies in nature's choice of this simple molecule, made from the common
metabolic soil of glutamic acid, for the all-important role as major controller of the infinitely
complex machinery of the brain, allowing it to operate in the manner best described as freedom
without license. Try as one might, one cannot come up with a better choice for the job.

Eugene Roberts (2000)

There are many theories about putative core deficit(s) in autism, but no
definitive findings. Many psychological, physiological, biochemical, immunological, and genetic differences between autistic people and control groups have been
reported (Rapin & Katzman, 1998). Some differences have been replicable;
others have been attributed to the cognitive deficits that are prevalent in autistic
people (Rutter, 1983). No comprehensive theory on autism has been clearly
articulated in which psychomotor deficits play a central role, with the possible

exception of Robin Allott's contribution (2001) linking autism and the Motor Theory of Language.

GABA hypotheses of autism have recently been formulated (Dhossche *et al.*, 2002; Hussman, 2001). Hussman (2001) speculates that autism reflects dysfunction in a single factor (i.e., decreased GABA inhibition, shared in common by many systems). In his model, suppression of GABAergic inhibition results in excessive stimulation of glutamate specialized neurons and loss of sensory gating. Inhibition of GABA may become defective through multiple etiological factors. Loss of inhibitory control may cause deterioration in the quality of sensory information due to the failure to suppress competing "noise," resulting in compensatory restrictions in sensory input to a narrow, repetitive, or controllable scope.

Criteria for a comprehensive theory of autism have been formulated (Dhossche *et al.*, 2002). Any viable theory must consider the protean nature of symptoms, course, and outcome in autistic people, and should account for: (a) the early onset of clinical abnormalities; (b) worsening of symptoms around puberty in a considerable number of patients; (c) association between autism and epilepsy; and (d) genetic transmission of the disorder.

The evidence that central GABA dysfunction can account for these key features is briefly reviewed:

a. GABA is the main inhibitory neurotransmitter in the mature brain as perhaps 25–40% of all terminals contain GABA. In early development, GABA has an excitatory trophic role affecting neuronal wiring, plasticity of neuronal network, and neural organization (Roberts, 2000). Interference with the trophic role of GABA may affect development of neuronal wiring, plasticity of neuronal network, and neural organization (Christie *et al.* 2002; Herlenius and Lagercrantz, 2001; Varju *et al.*, 2001). For example, in mice, developmental changes in inhibitory synaptic currents in cerebellar neurons are determined primarily by developmental changes in $GABA_A$ receptor subunit expression. Overall, the effects of abnormal trophic GABA function could account for the brain abnormalities reported so far in autistic people (Courchesne *et al.*, 1988, 2001).

b. Decreased GABA inhibition in the hypothalamus is considered an important trigger for onset of puberty (Genazzani *et al.*, 2000; Mitushima *et al.*, 1994). Adaptive changes in GABA function at the onset of or during puberty may worsen or induce disorders associated with underlying abnormalities of GABA function. Increased rates of seizure disorders (Deykin and MacMahon, 1979; Gillberg, 1991b), catatonia (Wing and Shah, 2000), and worsening of autistic symptoms or overall behavioral deterioration (Gillberg, 1991a; Gillberg and Schaumann, 1981) have been reported in autistic people around and after puberty.

c. GABA has been strongly implicated in epilepsy (Petroff *et al.*, 1996). About 30% of autistic people develop some type of epilepsy (Gillberg, 1991b).

d. Genetic studies have implicated the proximal long arm of chromosome 15 in autism (Bakker *et al.*, 2003; Borgatti *et al.*, 2001a; Cook *et al.*, 1997; Lauritsen *et al.*, 1999) and catatonia (Stöber *et al.*, 2000, 2002). Three $GABA_A$ receptor subunits genes (GABRB3, GABRA5, and GABRG3) are located at the chromosome location that includes the PWS/AS region. Animal and human studies have suggested a role for these genes in the phenotypic expression of Prader-Willi syndrome (Ebert *et al.*, 1997) and Angelman syndrome (DeLorey *et al.*, 1998; Odano *et al.*, 1996). GABRB3 has been associated with autism in several studies (Cook *et al.*, 1998; Menold *et al.*, 2001; Nurmi *et al.*, 2003), especially in patients with increased levels of "insistence on sameness" (a composite score of difficulties with minor changes in personal routine or environment, resistance to trivial changes in environment, and compulsions/rituals) (Shao *et al.*, 2003) and savant skills (Nurmi *et al.*, 2003).

Empirical support for GABA dysfunction in autism is limited. Elevated plasma GABA levels in autistic children were found in a case-report (Cohen, 1999) and case-series (Dhossche *et al.*, 2002). Reduced ^{3}[H]-flunitrazepam labeled benzodiazepine bindings sites and ^{3}[H]-muscimol labeled $GABA_A$ receptors have been reported in the hippocampus of autistic people (Blatt *et al.*, 2001), providing direct evidence of abnormal benzodiazepine-GABA receptor complexes in this brain region. Glutamic acid decarboxylase (GAD) is the enzyme responsible for normal conversion of glutamate to GABA in the brain. GAD exists in two isoenzymes, GAD_{65} and GAD_{67}, that are products of two independently regulated genes (Soghomonian and Martin, 1998). In a postmortem study (Fatemi *et al.*, 2002), brain levels of both isoenzymes were reduced approximately by 50% in the parietal and cerebellar cortices of autistic patients. GAD deficiency in autism may be due to or associated with brain abnormalities in levels of glutamate/gamma amino butyric acid, or transporter/receptor density.

III. Plasma GABA as Marker for Autism

GABA is present in plasma at very low concentrations. Its role is unknown. Its correlation with GABA in brain or peripheral organs is also not clear. There is evidence that plasma GABA levels are to some extent under genetic control (Petty *et al.*, 1999). Others have observed that blood plasma levels seem fairly constant among various mammalian species (i.e., between 500–1200 pmoles/ml) (Ferkany *et al.*, 1978). This suggests non-dietary sources of plasma GABA.

Plasma GABA levels may reflect global or local aspects of brain GABA activity. However, there are no studies that have correlated plasma GABA levels and localized brain GABA levels. A peripheral source of plasma GABA has been postulated but has not been demonstrated, at least not in normal physiological conditions (Petty et al., 1987). GABA does not enter the brain easily through the brain blood barrier (van Gelder and Elliot, 1958). GABA is eliminated from the brain interstitial fluid to the circulating blood across the brain blood barrier (Kakee et al., 2001). This efflux mechanism is inhibited by probenecid and valproic acid (Löscher and Frey, 1982). There is active transport of GABA into the CSF (Löscher and Frey, 1982).

Relations between plasma, CSF, and brain GABA levels are unclear. Correlations between plasma and CSF GABA levels and correlations between changes in plasma and CSF GABA levels under specific pharmacological conditions (Ferkany et al., 1978; Löscher and Schmidt, 1981) or (patho)-physiological states (Adinoff et al., 1995) have been reported in some studies, but not all (Berrettini et al., 1982; Schmidt and Löscher, 1982).

The simultaneous assessment of levels of brain, CSF, and plasma GABA levels in future studies may clarify their correlations in different disorders and under different conditions. However, it cannot be automatically assumed that central GABA measures (CSF and brain) perform better than plasma GABA levels as a marker of GABAergic function in specific disorders. This has been demonstrated in a study in schizophrenia (van Kammen et al., 1998) where plasma GABA levels, but not CSF GABA levels, correlated with specific brain morphology.

There are a few studies that have measured plasma GABA in human diseases. Low levels of GABA were found in CSF of patients with Alzheimer's disease (Jimenez-Jimenez et al., 1998) and in plasma of patients with bipolar disorder (Petty et al., 1993) and mood disorders (Petty et al., 1992). Low post-trauma GABA plasma levels have been suggested as a predictive factor in the development of acute posttraumatic stress disorder (Vaiva et al., 2004). Elevated GABA levels were observed in a small sample of autistic children (Dhossche et al., 2002).

Measuring plasma GABA measurements is difficult because of the low concentrations and the need to process samples quickly. It has been estimated that about 30% of whole blood GABA is present in plasma (Ferkany et al., 1978). The remaining fraction is bound to formed elements, mostly homocarnosine i.e., a dipeptide. CSF GABA increases during storage at room temperature, probably because of GABA release from homocarnosine (Grossman et al., 1980).

A sensitive assay is critical for adequate measurement of plasma GABA. In the past, various methods have been developed for the determination of GABA. These methods were based on techniques including HPLC, GC-MS, CE-MS, CE/laser induced fluorescence detection, electrochemical sensor, and spectrophotometry. Most of these methods, particularly the GC-MS based methods, required time-consuming sample pretreatments such as liquid-liquid or solid phase extraction. HPLC coupled to mass spectrometry (MS) with an electrospray ionization (ESI) source has become a popular analytical technique. However, the sensitivity of an HPLC-MS combination is often lackluster compared with fluorescence detection, particularly when standard-size HPLC columns (4mm ID) are used. This problem can be simply caused by sample dilution within the relatively large volume comprised by the column and tubing. Therefore, capillary HPLC-MS hyphenation is gaining research interest.

Our group has developed a sensitive HPLC-ESI-MS/MS method for the determination of GABA in physiological fluid samples, as described in a recent paper (Song *et al.*, 2005). This new method is about 500 times more sensitive than previously reported HPLC–MS methods and, therefore, well suited to quantify trace levels of GABA present in physiological fluids. A linear calibration curve from 10 to 250 ng/Ml GABA with an *r*2 value of 0.9994 was obtained. Detection limit was estimated to be 5.00 ng/mL GABA (S/N = 3). Human plasma and CSF samples were analyzed. The concentrations of GABA were found to be 98.6 ± 33.9 ng/mL (mean ± S.D., $n = 12$), and 44.3 ± 10.0 ng/mL ($n = 6$) in plasma and CSF, respectively. The method is based on capillary liquid chromatography (LC)/tandem mass spectrometry (MS/MS) using deuterium-labeled GABA (gamma-aminobutyric acid-2,2-D_2, GABA-d_2) as internal standard. Further details of the method can be obtained from the paper.

Elevated plasma GABA may be a marker of autism (Fig. 1). In replication of an earlier study (Dhossche *et al.*, 2002), our research group has found increased plasma GABA levels in a larger sample of prepubertal autistic children (N = 24, ages 3–12). Children with autism had higher plasma GABA levels than control children (without autism or any other neuropsychiatric conditions) (T-test,

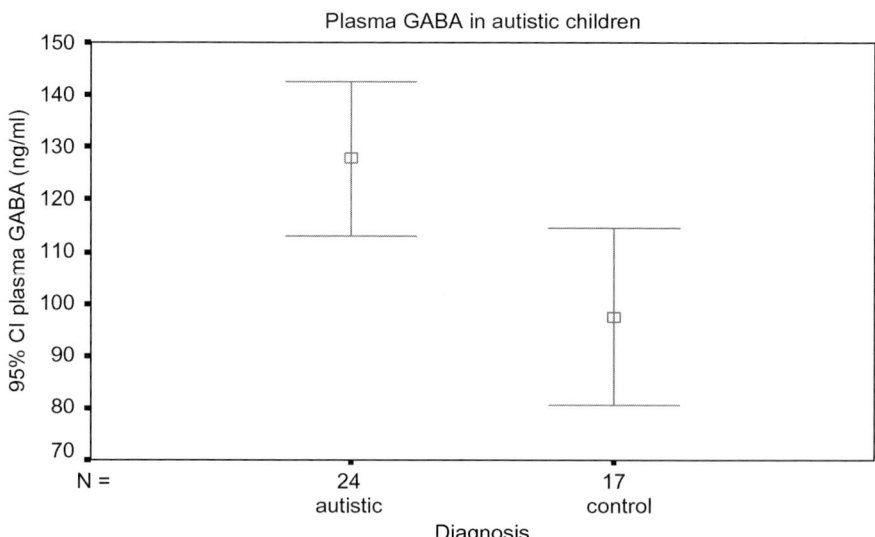

FIG. 1. Plasma GABA levels in autistic children versus controls (p = 0.008).

t = 2.8, df = 39, p = 0.008). Age and sex were not different between diagnostic groups. Although psychotropic medications and certain antiepileptic medication may alter plasma GABA levels (Kemph *et al.*, 1993; Lichtshtein *et al.*, 1978; Löscher and Schmidt, 1981; Shiah *et al.*, 2000), findings were similar when comparing plasma GABA levels in autistic children who were drug-naïve or without psychotropic medications for a least one month versus those who were on various psychotropics.

Low plasma GABA levels may be a novel familial marker of autism (Fig. 2). In a separate study, we measured plasma GABA in 41 parents of children with and without autism (see box plot of parental plasma GABA levels; median, range, the box length represents the middle 50% of the data). Median GABA levels were 63.7 ng/ml for mothers of autistic children (N = 12), 75.5 ng/ml for fathers of autistic children (N = 7), 120.5 ng/ml for mothers of control children (N = 14) (without autism or any other psychiatric disorder), and 99.6 ng/ml for fathers of control children (N = 8). Plasma GABA levels were significantly lower in parents of autistic children compared to control parents (81.3 ng/ml SD 32.2 versus 112.2 ng/ml SD 38.5; Mann-Whitney test, Z = −2.6, p = 0.008). Plasma GABA levels were significantly different between mothers of autistic children versus control mothers (Mann-Whitney test, Z = −2.4, p = 0.02), but not between fathers of autistic children and control fathers (Mann-Whitney test, Z =−1.1, p = −0.28).

Age was not different between the two groups of fathers and mothers. Information on the menstrual phase was not collected although there is one study showing increased plasma GABA levels from the follicular to the luteal

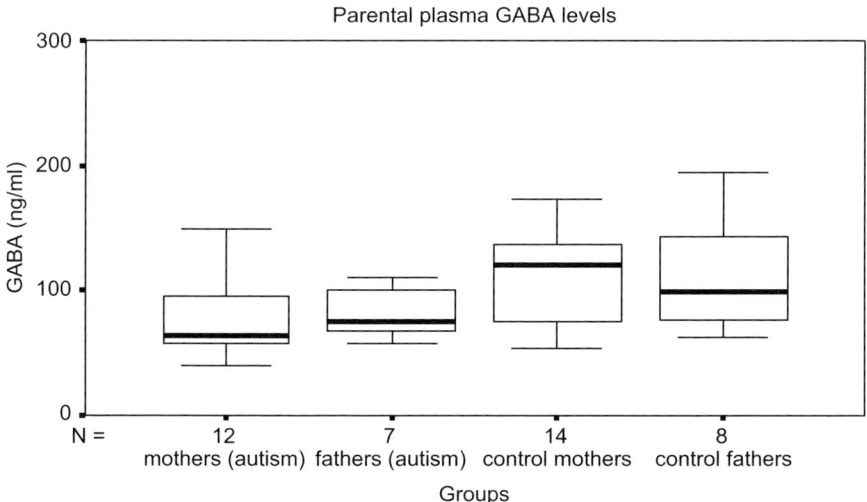

FIG. 2. Plasma GABA levels in parents of children with and without autism (p = 0.008).

phase (Halbreich *et al.*, 1996). None of the mothers indicated pregnancy. In one study, CSF GABA was lower during pregnancy (Altemus *et al.*, 2004). More than half of parents (23 of 41, 56%) were medication-free. Others took a wide variety of medications. Psychotropic medications (including antidepressants, psychostimulants, and antipsychotics) were reported in six parents. About the same proportions of mothers of autistic children and control mothers reported taking medications. The literature on changes of plasma GABA due to medications is limited. A definite increase in plasma GABA has only been shown after treatment with divalproic sodium (Löscher and Schmidt, 1981; Shiah *et al.*, 2000). None of the parents was prescribed divalproic acid or any other anticonvulsant. Medications are therefore unlikely to have changed plasma GABA levels in this pilot study. Overall, plasma GABA levels were lower in parents of autistic children than in control parents. The difference was not statistically significant in fathers possibly owing to the small sample size. Replication in a larger study is warranted.

There are no studies on the familial aggregation of plasma GABA levels in relatives of autistic probands. Twin and family studies have shown that plasma GABA levels are to some extent under genetic control (Petty *et al.*, 1999). In a family study (Bjork *et al.*, 2001), low plasma GABA levels were correlated with measures of aggression in relatives of probands with unipolar depressive illness. These findings support further inquiry in the familial pattern of plasma GABA level as a marker of genetic risk for neuropsychiatric disorders including autism. Previous studies have found low plasma GABA in people with mood disorders and bipolar disorder. Similar findings in parents of autistic children support the theory of DeLong (2004) that there are shared risk factors between types of autism and mood disorder.

There is a discrepancy in our findings between the higher levels of plasma in autistic probands compared to controls, and the lower plasma GABA levels in parents of autistic probands compared to control parents. Current data do not provide an easy explanation, but differences of age and diagnosis between probands and parents are obvious and should be considered. Based on the pilot data in autistic children and their parents, it is hypothesized that plasma GABA levels capture some aspect of central GABA function that is relevant to the pathophysiology and familial risk of autism. These preliminary data suggest that plasma GABA is a promising marker for autism and familial risk of autism.

IV. Insurgent Catatonia

In the past, catatonia was considered a diffuse syndrome, usually associated with schizophrenia. Some have claimed the disappearance of the syndrome due to obscure reasons (Mahendra, 1981). However, empirical findings highlight its increased incidence in psychosis and validity as a syndrome and genetically determined phenotype (Beckmann *et al.*, 1996; Stöber *et al.*, 2000). For example, in recent samples of acute psychiatric inpatients, the prevalence of catatonia was between 7–17% (Fink and Taylor, 2003). Most catatonic patients were diagnosed with mood disorders. Many authors link catatonia to manic-depressive illness (Abrams and Taylor, 1976; Bush *et al.*, 1996b). Twenty-five percent or more of manic patients have enough catatonic features to meet the DSM criteria. More than half of catatonic patients have manic-depressive illness. About 10–15% of patients with catatonia meet the criteria for schizophrenia.

Catatonia is not classified separately in DSM-IV but serves as a specifier of schizophrenia, mood disorder, and psychosis not otherwise specified, *but not autism* (American Psychiatric Association, 1994). There is increasing support for catatonia to be considered as a distinct syndrome in its own right (Taylor and Fink, 2003), reminiscent of the original conceptualization of catatonia by Kahlbaum (1874) and later by Leonhard (1979). Although there are no controlled treatment studies in catatonia satisfying current standards for evaluating therapies, the literature is consistent showing positive effects of anticonvulsant drugs, particularly benzodiazepines and barbiturates, and of electroconvulsive therapy (ECT), regardless of the severity or etiology of catatonia (Caroff *et al.*, 2004; Fricchione *et al.*, 1983; Rohland *et al.*, 1993).

In a recent study (Ungavri *et al.*, 2005), using a narrow definition of catatonia (i.e., the presence of four or more signs/symptoms with at least one having a score 2 or above on the Bush–Francis Catatonia Rating Scale (BFCRS)) (Bush *et al.*, 1996b), 72 subjects (32%) with chronic schizophrenia met the criteria for the catatonia group (mean number of catatonic signs/symptoms = 5.9F2.0; mean sum score of 8.7F3.4 on the BFCRS). The frequency distribution of catatonic signs/symptoms in the catatonic group and in the whole sample was very similar, with mannerisms, grimacing, stereotypes, posturing, and mutism being the most frequent. Catalepsy, mannerisms, posturing, and mutism are the features traditionally associated with catatonic schizophrenia.

In another study (van der Heijden *et al.*, 2005), a large sample of schizophrenics (N = 19,309) was studied. Although the diagnosis of catatonic schizophrenia dropped from 7.8% in 1980–1989 to 1.3% in 1990–2001, a possible underdiagnosis of catatonic schizophrenia was found in an independent sample. In a consecutive sample of patients admitted with psychosis, application of a systematic catatonia rating scale showed that 18% fulfilled criteria for catatonia (van der Heijden *et al.*, 2005).

Together, these studies (Ungavri *et al.*, 2005; van der Heijden *et al.*, 2005) suggest under-diagnosis and non-recognition of catatonia both in patients with schizophrenia and bipolar disorder. The reasons are complex, but may include the historical decision to classify catatonia as a type of schizophrenia, the segregation of severe psychiatrically ill patients in long-term facilities, and the perceived lack of anti-catatonic treatments (Fink and Taylor, 2003). These factors may have suppressed interest in and recognition of catatonia. A study of Wing and Shah (2000) showing that 17% of adolescents and young adults with autism satisfied criteria of catatonia suggests that catatonia may also be underdiagnosed in autism owing to similar or other reasons. This study will be discussed at length in one of the next sections.

V. GABA in Catatonia

> The higher nervous arrangements evolved out of the lower to keep down those lower, just as a government evolved out of a nation as well as directs that nation. If this be the process of evolution, then the reverse process of dissolution is not only "a taking off" of the higher, but is at the same time a "letting go" of the lower. If the governing body of this country were destroyed suddenly, we should have two causes for lamentation: (1) the loss of service of eminent men; and (2) the anarchy of the now uncontrolled people. The loss of the governing body answers to the dissolution in our patient (the exhaustion of the higher two layers of his highest centres); the anarchy anwers to the no longer controlled activity of the next lower level of evolution (third layer).
>
> John Hughlings Jackson (1835–1911)

> Progressive inhibition of tonically inhibited living systems, from cell to society, is coupled with increased variability generation in such a manner that the probability of making an optimally adaptive choice of behavior from among the options available remains constant over a wide range of increasing force parameters.
>
> Eugene Roberts (2000)

The single most important observation leading to the belief that GABA dysfunction plays a role in catatonia is the often dramatic response to treatment with benzodiazepines (i.e., positive modulators of the benzodiazepine/$GABA_A$ receptor complex) (Bush *et al.*, 1996a; Fricchione *et al.*, 1983). Other effective treatments for catatonia (e.g., barbiturates, zolpidem, carbamazepine, and electroconvulsive therapy (ECT)) (Green, 1986; Sanacora *et al.*, 2003), also seem to enhance GABA function. Efficacy of serotonergic agents and antipsychotics in catatonia has been less well documented, but seems less consistent.

If one assumes a central role of GABA dysfunction in catatonia, the scope of GABA functions in the normal brain should allow the expression of catatonia when deficiencies in GABA function develop. Roberts (2000) sees a central role of GABA as neurotransmitter used by neurons that exert tonic inhibition of neural circuits for innate or learned behavioral sequences. In its extreme forms, catatonia is characterized by immobility alternating with purposeless agitation. Both behaviors can be viewed as opposite primitive reflexes in response to overwhelming stress or danger, that are expressed when innate, genetically preprogrammed neuronal circuits are released from tonic inhibition. Following John Hughlings

Jackson's hierarchical concept of dissolution (Jackson, 1958), immobility or hyperactivity are then "positive" symptoms caused by the removal of the influence of higher centers. The neuronal circuitry that is involved in catatonia is not well defined, but probably involves frontal cortex, parietal cortex, basal ganglia, and possibly the cerebellum (Fink and Taylor, 2003).

A few general criteria for any viable GABA theory of catatonia are proposed:

1. The theory has to accommodate findings from GABAergic theories of schizophrenia and affective disorders as catatonia occurs in both disorders.
2. Hypothalamic abnormalities of GABA function should be present and may account for severe autonomic dysfunction in malignant catatonia.
3. Treatments that relieve catatonia should enhance GABA function, directly or indirectly.
4. Genetic studies in catatonia, schizophrenia, and affective disorders should provide support for involvement for genes affecting GABA function, at least in subgroups.

Evidence that GABA dysfunction in catatonia satisfies these criteria is summarized:

1. GABA theories have been formulated for schizophrenia (Roberts, 1972; van Kammen, 1977), psychosis (Kalkman and Loetscher, 2003; Keverne, 1999), and affective disorders (Brambilla *et al.*, 2003; Emrich *et al.*, 1980; Petty, 1995) including bipolar disorder (Petty *et al.*, 1993).

2. GABA is a prominent neurotransmitter in the hypothalamus (Decavel and van den Pol, 1990). Hypothalamic GABAergic mechanisms are considered important for regulation of stress responses by the hypothalamic-pituitary-adrenal (HPA) (Herman and Cullinan, 1997; Engelmann 2004).

3. Most currently used psychotropic medications, including benzodiazepines, antipsychotics (Zink *et al.*, 2004), selective serotonin reuptake inhibitors (Bhagwagar *et al.*, 2004; Sanacora *et al.*, 2002; Tunnicliff *et al.*, 1999), phenelzine (Baker *et al.*, 1991; McManus *et al.*, 1992), and anticonvulsants seem to enhance GABA neurotransmission, albeit through different mechanisms. A few studies suggest a direct role of GABA in ECT. In an iomazenil-SPECT study, increased benzodiazepine receptor uptake in cortical regions (except temporal cortices) was found one week after successful bitemporal ECT (Mervaala *et al.*, 2001). In a MRS study, two-fold increased brain GABA levels were found in depressed patients after a course of ECT (Sanacora *et al.*, 2003). The small number of patients precluded conclusions regarding any correlation between clinical improvement and increased brain GABA level. Previously, it was reported that CSF GABA increased by 50% after ECT (Lipcsey *et al.*, 1986). In another study, cortical glutamate/glutamine levels in the left anterior cingulum of depressed patients normalized after ECT, but only in responders (Pfleiderer *et al.*, 2003). In

non-responders, levels remained low. Limitations in this study's MRS methodology did not allow obtaining separate measurements for GABA because of overlapping resonances of glutamate, glutamine, and GABA. In a study of plasma GABA in depressed patients treated with ECT, plasma GABA levels tended to decrease for about one hour after ECT (Devenand *et al.*, 1995). Limitations of this study include variable storage times known to increase GABA levels in plasma and CSF and PRN administration of chloral hydrate that enhances GABA function similar as barbiturates, during the ECT course.

4. There is some evidence that GABA related genes are involved in affective disorder and schizophrenia. GABRA5 has been associated with bipolar disorder in two studies (Otani *et al.*, 2005; Papadimitriou *et al.*, 1998). In a genome scan of catatonia, a linkage signal in the region 15q11.2-q21.1 (where three GABA-A subunit genes are located) was found (Stöber *et al.*, 2000). There are no (family-based) gene association studies in catatonia available in the literature. Findings from a family-based association study in a sample of children and adolescents with childhood-onset schizophrenia (COS) (n = 72) suggested that the gene encoding GAD(67) may be a common risk factor for schizophrenia (Addington *et al.*, 2005).

Empirical evidence for GABA dysfunction in catatonia comes from one single receptor-imaging study. Findings in a benzodiazepine ligand-binding study of catatonic patients have shown a decreased density of $GABA_A$ receptors in the left sensorimotor cortex in akinetic catatonia (Northoff *et al.*, 1999). Other, more circumstantial, evidence is found in biochemical studies. Cerebrospinal fluid levels of GABA were decreased in 11 patients with Neuroleptic Malignant Syndrome, a condition that may be related to catatonia, compared with GABA CSF levels in eight controls (Nisijima and Ishiguro, 1995). In this study, levels of noradrenalin were increased, but levels of 5-hydroxyindoleacetic acid (5-HIAA), serotonin's main metabolite, were slightly, but not significantly, decreased. Further evidence of impaired GABA function in catatonia must await future studies.

VI. Catatonia in Autism

A. CASE-VIGNETTE (DHOSSCHE, 1998)

John was the full-term product of an uneventful pregnancy. Birth was complicated by cord strangulation and hypoxia. He was described as a peaceful baby. At age 2, John did not speak any words yet, he was overly quiet and passive. Hearing deficits were suspected by his mother. Audiological testing indicated normal hearing.

At age 3, he was evaluated by a child neurologist and child psychiatrist because of problems with speech development and poor social interactions. Psychiatric observation showed avoidant gaze, limited interest in social interactions, limited emphatic contact, inappropriate smiling, and bizarre postures. His vocabulary was limited to a few words. He also used many jargon words and was echolalic. John was greatly fascinated by flickering lights and twirling objects. Neurological investigations were negative, including urinary amino acids, CSF, and pneumoencephalograph. The EEG was abnormal, with mild general slowing. A diagnosis of Autistic Disorder was made. John was placed in special education. His adjustment at school and home improved over time. Social functioning remained poor. Insistence on routines and obsessive preoccupations were constantly present throughout this period. Developmental progress was somewhat jerky. At age 5, his speech improved greatly in a short period. He also showed more interests. Cognitive testing was done at age 11 and showed a total IQ (WISC-R) of 103.

When John was 15, he started to complain about command auditory hallucinations. There was also marked worsening of obsessive-compulsive behaviors, including obsessive slowness. His psychomotor retardation alternated with outbursts of aggression and agitation. School performance deteriorated gradually. He was admitted to the psychiatric hospital.

Initial psychiatric assessment showed a disheveled adolescent with alternating episodes of agitation and severe psychomotor slowing, staring in space, waxy flexibility, posturing, and decreased verbal output. He reported command hallucinations. There was no formal thought disorder. Obsessions were elaborate and had a delusional quality. His mood was constricted. No focal neurological abnormalities were found. CT of the brain and routine laboratory tests were normal. Chromosomal analysis showed 46 XY karyotype without fragile X. Schizophrenia with catatonia and paranoid features, superimposed on AD, was diagnosed.

The patient was treated with several antipsychotic medications but his condition did not improve. John continued to complain about hearing voices. He was aggressive and difficult to manage on the ward. Episodes of catatonic behavior were among the most prominent features throughout the first year of admission. At that point, clozapine was prescribed in monotherapy (400 mg per day). Improvement was seen after a few weeks. John became less aggressive and more social. Catatonic recurrences continued. The frequency of stuporous episodes decreased after lorazepam (2.5 mg three times a day) was added to this medication regimen. A few months later, John was discharged from the hospital. Follow-up shows a young man in his early twenties with autistic symptoms but without catatonic-psychotic recurrences. He attends a day treatment center and lives with his parents. He continues to take clozapine and a small dose of lorazepam (2.5 mg/day).

B. COMMENTS

The clinical presentation of this patient carries some similarity with Neuroleptic Malignant Syndrome (NMS). NMS is an unlikely diagnosis because the onset was before antipsychotic medications were prescribed. During the protracted inpatient course the catatonic syndrome seemed refractory to treatment with various typical antipsychotics. It is possible that catatonia was sustained or exacerbated by this medication regimen.

The marked response to clozapine may indicate superior effect on catatonia (Battegay *et al.*, 1977). On the other hand, there is anecdotal evidence that clozapine improves behavioral problems in autistic children (Zuddas *et al.*, 1996). The positive response to lorazepam suggests specificity of benzodiazepine treatment in catatonic stupor, in accord with controlled studies in general psychiatric patients (Bush *et al.*, 1996a; Ungvari *et al.*, 1994).

The case of John is typical for AD until the onset of psychosis. Hallucinations and delusions in AD are considered rare but have been reported, mostly in high-functioning patients (Kurita, 1999; Petty *et al.*, 1984). Obsessions were difficult to differentiate from delusions. Auditory hallucinations were prominent suggesting a diagnosis of schizophrenia. However, the catatonic symptoms were equally disabling. The progression of stereotypical movements, compulsions, withdrawal into full catatonia is remarkable and has been observed by others (Wing and Shah, 2000; Zaw *et al.*, 1999). In that respect, autistic symptomatology appears as a *form fruste* of full catatonia.

Other case-reports show that catatonia emerges in some people with autism (Ghaziuddin *et al.*, 2005; Realmuto and August, 1991). There seems to be considerable overlap of psychomotor symptoms between the two disorders (e.g., muteness, echolalia, stereotypical movements, and other psychomotor peculiarities). There is one systematic study (Wing and Shah, 2000) of catatonia in autism showing that 17% of a referred sample of adolescents and young adults with autism satisfied modern criteria for catatonia. A semi-structured interview was used to collect information from parents or other caregivers. Patients were diagnosed with catatonia when an exacerbation of certain behavioral features occurred in sufficient degree to interfere with everyday functions of self-care, education, occupation, and leisure. Essential features of catatonia were: increased slowness affecting movements and verbal responses, difficulty in initiating and completing actions, increased reliance on physical or verbal prompting by others, and increased passivity and apparent lack of motivation. Other associated characteristics were reversal of day and night, Parkinsonian features (tremor, eye-rolling, dystonia, odd stiff posture, freezing in postures), excitement and agitation, and increase in repetitive, ritualistic behavior.

Thirty individuals with autism aged 15 or older met criteria for catatonia. Classic autism (AD) was diagnosed in 11 (37%), atypical autism (PDD NOS) in 5

(17%), and Asperger Disorder (AsD) in 14 (47%). All of those with catatonia were aged 15 or older. None of those under age 15 had the full syndrome although isolated catatonic symptoms were often observed. In the majority of cases, catatonic symptoms started between 10 and 19 years of age. Five individuals had brief episodes of slowness and freezing during childhood, before age 10. Obsessive-compulsive and aggressive behavior preceded catatonia in some. Visual hallucinations or paranoid ideas were occasionally reported, but no diagnosis of schizophrenia could be made. Referred patients with catatonia were significantly more likely than patients without catatonia to have had impaired language and passivity in social interaction before the onset of catatonia. Family history and treatments were not recorded.

This study is the only published systematic assessment of catatonic symptoms in autistic people. Although the authors did not use DSM-IV criteria of catatonia, their definition of catatonia includes several DSM-IV core symptoms including severe psychomotor retardation, decreased verbalizations, posturing, and agitated episodes. The high prevalence of catatonic symptoms in autism suggests an intricate, but unaccounted, relation between autistic and catatonic symptoms.

VII. Catatonia in Prader-Willi Syndrome

> During lethargic-refusal states, the patient [with Prader-Willi Syndrome] is practically bed-ridden, refuses any human approach, food or drink, and sinks into a state of self neglect leading to soiling with no evidence of pathophysiologically determined incontinence. These lethargic-refusal states can last for weeks or months, and may resolve spontaneously or with the help of behavioral or pharmacological interventions.
>
> G. Bartolucci and T. Younger (1994)

A. CASE-VIGNETTE (DHOSSCHE AND BOUMAN, 1997)

Mark was the only child of well-educated parents. He was born after 40 weeks of normal gestation. Birth weight was 2461 grams. Apgar score was 7 (after 1'), and nine (after 2'). A pediatric referral was made at 2 months because of hypotonia, feeding problems, hypogonadism, and bilateral retentio testis. EEG, brain CT, and amino acid analysis were normal. Chromosome analysis showed 46 XY configuration. Prader-Willi syndrome (PWS) was diagnosed on clinical grounds. No further cytogenetic studies were done.

Mark started walking when he was 18 months old. He spoke his first words at 14 months and whole sentences at age 3. During the next few years, stigmata of

PWS (e.g., excessive appetite, obesity, short stature, oppositional behavior, and temper tantrums) became clear. Cognitive testing showed scores in the mildly retarded range (IQ 60–65 on WISC-R). He was cheerful as a child. There was no personal or family history of psychosis, epilepsy, or autism.

Mark lived with his parents in a stable family environment. He was well adjusted and he attended a special education school for mentally retarded youngsters. Occasional temper tantrums responded well to behavioral interventions. A few months before the onset of illness, the family moved and Mark started attending a new school, without any obvious adjustment problems.

At age 17, Mark was admitted to a pediatric hospital because of acute onset of catatonia with stupor, staring, incontinence, mutism, rigidity, waxy flexibility, posturing, refusal to eat and drink, and severe disruption of the sleep-wake cycle. Full-blown catatonia developed within one day. A few hours before the onset, Mark had a rare but vehement argument with his mother regarding some school matter. He was sent to his room. When he came back, he was very anxious, talked incoherently about Jesus, and made vague references about visual hallucinations. Within hours, he became catatonic and was brought to the hospital.

Initial treatment consisted of supportive medical care and haloperidol (1 mg three times daily). His condition remained unchanged. He lay in bed, mute, and withdrawn. Neurological examination showed clear consciousness, normal reflexes, intact cranial nerves, drooling, and increased muscle tone. No paresis or sensory abnormalities were found. EEG showed mild, nonspecific slowing in the prefrontal areas. Red and white blood cell count, serum electrolytes, and liver function tests were all within normal limits. The consulting neurologist failed to find evidence for a neurological disorder, and concurred with the diagnosis of stupor of psychiatric origin. He thought that drooling and increased muscle tone were caused by haloperidol.

After 10 days of minimal improvement, a test dose of lorazepam (1 mg) was given orally. Considerable improvement of catatonia occurred within hours. The next day, Mark started walking and eating again. He became more verbal and answered simple questions adequately. He also started making delusional statements that his parents had died, and reported visual hallucinations of his grandparents and Santa Claus. Lorazepam was increased to 4 mg daily over the course of a few days. Stupor gave way to motor restlessness, short attention span, impulsiveness, stereotypies, echolalia, echopraxia, automatic obedience, active resistance to movement, ambitendency, and negativism. Mark continued to make nihilistic-delusional statements for which risperidone was started, in addition to lorazepam, and increased to 6 mg daily. Catatonia and delusions resolved during the next two weeks. Mark was discharged from the hospital and returned back to baseline functioning. Risperidone and lorazepam were discontinued 2 months after discharge. No relapses occurred during five-years follow-up.

B. COMMENTS

PWS is a genetic disorder characterized by hypotonia at birth, small hands and feet, almond shaped eyes, hypogonadism, short stature, and diabetes. Most patients exhibit mild to moderate mental retardation and obesity. Obesity starts in infancy and is accompanied by compulsive eating. The prevalence is estimated at one in 16,000–25,000 live births (Burd *et al.*, 1990; Smith *et al.*, 2003).

This multi-system disorder occurs in all races and both sexes and arises from the lack of expression of genes on the paternally derived chromosome 15q11–13. Candidate genes for PWS in this region are imprinted and silenced on the maternally inherited chromosome. The genetic defect underlying PWS is the absence of expression of one or more genes of paternal origin located on the long arm of chromosome 15 (15q11-13). Several genetic mechanisms have been associated with PWS, mainly paternal 4Mb deletion (in about 60% of cases) and maternal uniparental disomy (UPD) (in about 25% of cases) (Woodage *et al.*, 1994). In a small number of patients (3%), imprinting errors are found because of either a sporadic or inherited microdeletion in the imprinting center. There is a paternal chromosomal translocation in 1% of the cases. Imprinting occurs partly through parent-off-origin allele-specific methylation of CpG residues which is established either during or after gametogenesis and maintained throughout embryogenesis. Alternatively, if the deletion is maternal or there are two paternal copies of chromosome 15q, another syndrome is found (e.g., Angelman Syndrome (AS) characterized by severe mental retardation, attention deficit, inappropriate laughter, ataxia, jerky gait, epilepsy, sleep disturbances, and craniofacial abnormalities) (Angelman, 1965).

Mark satisfied clinical criteria of PWS (Åkefeldt *et al.*, 1991; Holm *et al.*, 1993). Severe neonatal hypotonia, hypogonadism/delayed sexual maturation, obesity, and learning problems are major criteria. All major features were present in Mark. This established the diagnosis with high probability. Unfortunately, no cytogenetic or molecular testing was done.

Psychosis and catatonia started suddenly. There were no prodromal symptoms or premorbid signs of any psychiatric disorder. The episode lasted less than 1 month and there was full return to premorbid level of functioning. These features suggest a diagnosis of Brief Psychotic Disorder, according to DSM-IV (American Psychiatric Association, 1994). The case also satisfies criteria for the DSM-IV catatonic specifier. A clear precipitant of the episode was present, satisfying the specifier "with marked stressors." There was a severe quarrel between Mark and his mother hours before his psychotic break. Possible contributing factors were the recent move and change of school. Although the subsequent course seems out of proportion to the original precipitants, these events may well qualify as very stressful for Mark considering his restricted emotional and cognitive coping.

Catatonic stupor responded quickly to treatment with lorazepam, in accordance with studies in adults (Bush *et al.*, 1996a; Fricchione *et al.*, 1983; Ungvari *et al.*, 1994). A different set of catatonic symptoms appeared after resolution of stupor with lorazepam. It was clear that Mark was delusional and hallucinating once he became verbal again. He responded well to the addition of risperidone. Others have reported positive effects of atypical antipsychotics, including risperidone, in adult catatonia without developmental disorders (Battegay *et al.*, 1977; Cook *et al.*, 1996). More studies are needed to assess the relative and differential efficacy of benzodiazepine monotherapy versus combined regimens (benzodiazepine and atypical antipsychotic) in the treatment of catatonia.

Studies have shown that compulsions (Holm *et al.*, 1993; Wigren and Hansen, 2003) and psychotic disorders (Boer *et al.*, 2002; Clarke, 1993) are associated with PWS. Compulsive behaviors in PWS are autistic-like and include insistence of sameness (Wigren and Hansen, 2005), perseveration, and ordering. Hand washing and checking (i.e., symptoms that are typically found in Obsessive-Compulsive Disorder) are infrequent in PWS. Various psychotic symptoms including paranoia, hallucinations, and bizarre behavior, have been reported in PWS patients, usually in combination with affective symptoms (Clarke, 1993; Whitman and Accardo, 1987). The acute onset and mixture of affective and psychotic symptoms suggest an atypical form of psychosis.

PWS patients with uniparental disomy seem particularly prone to develop psychosis (Boer *et al.*, 2002; Verhoeven *et al.*, 1998; Vogels *et al.*, 2003). This suggests that an abnormal pattern of expression of sex-specific imprinted genes on 15q11–13 is a risk factor for psychosis in PWS. Superior visual recognition memory is another feature in PWS that has been associated with maternal uniparental disomy (Joseph *et al.*, 2001). The genes involved in the development of pychosis and superior recognition memory in PWS are unknown.

The case-vignette is the only case-report in the literature of an adolescent with PWS who developed full-blown catatonia (Dhossche and Bouman, 1997). Catatonic symptoms have been described in a few other cases, but have not been rigorously assessed. For example, Clarke (1993) reported an acutely psychotic PWS patient with significant psychomotor retardation, refusal of food and fluids, and negativistic behavior. Abe and Ohta (1995) described an adolescent with PWS who developed stupor, pallor, and delusional thinking. Bartolucci and Younger (1994) observed, in four of nine youngsters with PWS, discrete episodes of a refusal-lethargy syndrome, characterized by akinesis, refusal of foods and fluids, incontinence, and self-neglect, that occurred independently from psychotic episodes and lasted weeks up to several months. These descriptions suggest the presence of catatonia. Catatonia might have gone unrecognized, as some catatonic symptoms are transient and need to be elicited by trained clinicians. Systematic studies are needed to assess the prevalence of catatonia in PWS.

VIII. GABA in Prader-Willi Syndrome

A few studies have suggested abnormal GABA metabolism in PWS. In one study (Ebert *et al.*, 1997), plasma GABA levels were two to three times higher in people with PWS (and Angelman Syndrome) than controls. This finding was not explained by obesity or level of cognitive impairment. Within the group of subjects with PWS, genetic status (chromosome 15 deletion or disomy) was not related to higher plasma GABA levels. Cerebral GABA$_A$ receptors were studied in PWS (Lucignani *et al.*, 2004). A reduction of [11-C]flumazenil (a ligand of the benzodiazepine binding site associated with the GABA$_A$ receptor) was found in several cortical brain areas in adults with PWS compared to controls. Findings of abnormal GABA metabolism and GABA$_A$ receptors may be related to abnormal expression of GABA$_A$ receptor subunit genes (GABRB3, GABRA5, and GABRG3) located on the PWS/AS chromosomal region (15q11–13) (Lalande *et al.*, 1994; Wagstaff *et al.*, 1991). Some indirect evidence comes from studies showing that treatment with topiramate, an anticonvulsant with GABA-ergic effects, reduces stereotyped and compulsive behaviors in PWS patients (Shapira *et al.*, 2002, 2004; Smathers *et al.*, 2003).

Hypothalamic dysfunction is considered one of the key features of PWS (Swaab, 1997). Abnormal hypothalamic function in PWS may be associated with excessive eating, lack of satiation, growth hormone deficiency, hypogonadism, daytime hypersomnolence, and abnormalities in body temperature control. There are very few neuropathological studies of the hypothalamus in PWS. Swaab *et al.* (1995) found smaller paraventricular nuclei (28% reduction), fewer oxytocin-expressing cells (54% reduction), but normal numbers of vasopressin-expressing cells in the hypothalamus of 5 individuals with PWS. To our knowledge, there are no studies in the literature providing direct evidence of abnormal GABA function in the hypothalamus of individuals with PWS. Given the importance of GABA as neurotransmitter in the hypothalamus (Decavel and van den Pol, 1990), such studies are warranted.

PWS and AS are contiguous gene syndromes (Schmickel, 1986). This type of genomic disorder is characterized by large chromosomal rearrangements with complex phenotypes associated with the dosage effects of multiple unrelated genes. The region on chromosome 15q11-q13 is an example where multiple structural abnormalities, including deletions, duplications, triplications, and supernumerary marker chromosomes have been reported, especially in cases with autism (Lauritsen *et al.*, 1999). Abnormalities of the proximal long arm of chromosome 15 have emerged as the most frequent (up to 3%) cytogenetic abnormality in the autistic population (Bolton *et al.*, 2001; Schroer *et al.*, 1998). Detection is done by testing through FISH studies and additional molecular analyses to determine parental origin of the duplications or abnormal gene segment. So far, all duplications in this region have been maternal in origin (Cook *et al.*, 1997). This suggests that an imprinted gene may be involved in those cases. In cases with inverted duplicated chromosome 15, autistic symptoms are present in varying degrees of severity (Borgatti *et al.*, 2001a). Plasma GABA levels were higher, but not significantly, in cases with this syndrome compared to controls (Borgatti *et al.*, 2001b). The negative finding may be due to the small sample size ($N = 8$) of the study. This suggests that the GABAergic system may be implicated in patients with inverted duplicated chromosome 15 as well as in some patients with idiopathic autism.

Altered brain $GABA_A$ receptor function may account for some of the neuropsychiatric problems associated with PWS as there is increasing evidence that GABA neuronal dysfunction is implicated in various psychiatric disorders including psychosis (Guidotti *et al.*, 2000; van Kammen *et al.*, 1998), mood disorders (Sanacora *et al.*, 1999), anxiety disorders (Goddard *et al.*, 2001), and autism (Blatt *et al.*, 2001; Dhossche *et al.*, 2002). Genetic studies also suggest that polymorphisms of a $GABA_A$ receptor subunit gene (GABRB3) or a gene near this locus increases susceptibility to catatonia (Stöber *et al.*, 2000, 2002), bipolar disorder (Papadimitriou *et al.*, 1998; Otani *et al.*, 2005), and autism (Nurmi *et al.*, 2003; Shao *et al.*, 2003). GABRB3 has been associated with a subtype of autism with high levels of repetitive behaviors and stereotyped patterns (insistence on sameness) (Shao *et al.*, 2003) and with a subtype of autism with savant skills, a characteristic often associated with the development of special skills (O'Conner and Hermelin, 1991). Repetitive or obsessive-compulsive behaviors (Clarke *et al.*, 2002; Dykens *et al.*, 1996), splinter skills (Dykens, 2002), and catatonia (Dhossche and Bouman, 1997) have been associated with PWS. Taken together, these studies suggest that a genetic locus (or loci) in the PWS region (possibly GABRB3 or nearby genes) confer(s) risk for repetitive, ritualistic, and obsessive-compulsive behaviors. Future studies of the association between these psychopathological dimensions and GABAergic parameters are warranted.

IX. Conclusion

There are supportive data for GABA dysfunction in autism, Prader-Willi syndrome, and catatonia. Modern in vivo techniques to measure GABA in the plasma and brain allow further clinical studies to examine common GABAergic mechanisms in these disorders. Such studies may ultimately lead to improved treatments for selected symptoms in these disorders.

Acknowledgments

Research support from the Thrasher Research Fund, Salt Lake City, Utah (to DMD), and US National Institutes of Health (NS44177 to YL) is gratefully acknowledged.

References

Abe, K., and Ohta, M. (1995). Recurrent brief episodes with psychotic features in adolescence: Periodic psychosis of puberty revisited. *Br. J. Psychiat.* **167,** 507–513.

Abrams, R., and Taylor, M. (1976). Catatonia: A prospective clinical study. *Archives General Psychiat.* **33,** 579–581.

Addington, A., Gornick, M., Duckworth, J., Sporn, A., Gogtay, N., Bobb, A., Greenstein, D., Lenane, M., Gochman, P., Baker, N., Balkissoon, R., Vakkalanka, R., Weinberger, D., Rapoport, J., and Straub, R. (2005). GAD1 (2q31.1), which encodes glutamic acid decarboxylase (GAD(67)), is associated with childhood-onset schizophrenia and cortical gray matter volume loss. *Mol. Psychiatry* **10,** 581–588.

Adinoff, B., Kramer, G., and Petty, F. (1995). Levels of gamma-aminobutyric acid in cerebrospinal fluid and plasma during alcohol withdrawal. *Psychiatry Res.* **29,** 137–144.

Åkefeldt, A., Gilberg, C., and Larson, C. (1991). Prader-Willi syndrome in a Swedish rural county: Epidemiological aspects. *Devel. Med. Child Neurol.* **33,** 715–721.

Allot, R. (2001). Autism and the motor theory of language. *In* "The Great Mosaic Eye: Language and Evolution" pp. 93–113.

Altemus, M., Fong, J., Yang, R., Damast, S., Luine, V., and Ferguson, D. (2004). Changes in cerebrospinal fluid neurochemistry during pregnancy. *Biologic. Psychiatry* **56,** 386–392.

American Psychiatric Association (1994). "Diagnostic and Statistical Manual of Mental Disorders," 4th ed. American Psychiatric Association, Washington, D.C.

Angelman, H. (1965). "Puppet" children. A report of three cases. *Dev. Med. Child Neurol.* **7,** 681–688.

Baker, G., Wong, J., Yeung, J., and Coutts, R. (1991). Effects of the antidepressant phenelzine on brain levels of gamma-aminobutyric acid (GABA). *J. Affect. Disorders* **21,** 207–211.

Bakker, S., van der Meulen, E., Buitelaar, J., Sandkuijl, L., Pauls, D., Monsuur, A., van't Slot, R., Minderaa, R., Gunning, W., Pearson, P., and Sinke, R. (2003). A whole-genome scan in 164 Dutch sib pairs with Attention-Deficit/Hyperactivity Disorder: Suggestive evidence for linkage on chromosomes 7p and 15q. *Am. J. Hum. Gene* **72,** 1251–1260.

Bartolucci, G., and Younger, J. (1994). Tentative classification of neuropsychiatric disturbances in Prader-Willi syndrome. *J. Intelle. Disabil. Res.* **38,** 621–629.

Battegay, R., Cotar, B., Fleischauer, J., and Rauchfleisch, U. (1977). Results and side effects of treatment with clozapine. *Comprehensive Psychiatry* **18,** 423–428.

Beckmann, H., Franzek, E., and Stöber, G. (1996). Genetic heterogeneity in catatonic schizophrenia: A family study. *Am. J. Med. Gene. (Neuropsychiatric Genetics)* **67,** 289–300.

Berrettini, W., Nurnberger, J. J., Hare, T., Gershon, E., and Post, R. (1982). Plasma and CSF GABA in affective illness. *British Journal of Psychiatry* **141**, 483–487.

Bhagwagar, Z., Wylezinska, M., Taylor, M., Jezzard, P., Matthews, P., and Cowen, P. (2004). Increased brain GABA concentrations following acute administration of a selective serotonin reuptake inhibitor. *Am. J. Psychiatry* **161**, 368–370.

Bjork, J., Moeller, F., Kramer, G., Kram, M., Suris, A., Rush, A., and Petty, F. (2001). Plasma GABA levels correlate with aggressiveness in relatives of patients with unipolar depressive disorder. *Psychiatric Res.* **101**, 131–136.

Blatt, G., Fitzgerald, C., Guptill, J., Booker, A., Kemper, T., and Bauman, M. (2001). Density and distribution of hippocampal neurotransmitter receptors in autism: An autoradiographic study. *J. Aut. Dev. Disorders* **31**, 537–543.

Boer, H., Holland, A., Whittington, J., Butler, J., Webb, T., and Clarke, D. (2002). Psychotic illness in people with Prader-Willi syndrome due to chromosome 15 maternal uniparental disomy. *Lancet* **12**, 135–136.

Bolton, P., Dennis, N., Browne, C., Thomas, N., Veltman, M., Thompson, R., and Jacobs, P. (2001). The phenotypic manifestations of interstitial duplications of proximal 15q with special reference to the autistic spectrum disorders. *Am. J. Med. Gent. (Neuropsychiatric Genetics)* **105**, 675–685.

Borgatti, R., Piccinelli, P., Passoni, D., Dalpra, L., Miozzo, M., Micheli, R., Gagliardi, C., and Balottin, U. (2001a). Relationship between clinical and genetic features in "inverted duplicated chromosome 15" patients. *Pedia. Neurol.* **24**, 111–116.

Borgatti, R., Piccinelli, P., Passoni, D., Raggi, E., and Ferrarese, C. (2001b). Pervasive developmental disorders and GABAergic system in patients with inverted duplicated chromosome 15. *J. Child Neurol.* **16**, 911–914.

Brambilla, P., Perez, J., Barale, F., Schettini, G., and Soares, J. (2003). GABAergic dysfunction in mood disorders. *Mol. Psychiatry* **8**, 721–737.

Burd, L., Vesely, B., Martsolf, J., and Kerbeshian, J. (1990). Prevalence study of Prader-Willi syndrome in North Dakota. *Am. J. Med. Gent.* **37**, 97–99.

Bush, G., Fink, M., Petrides, G., Dowling, F., and Francis, A. (1996a). Catatonia. II. Treatment with lorazepam and electroconvulsive therapy. *Acta Psychiatr. Scand.* **93**, 137–143.

Bush, G., Fink, M., Petrides, G., Dowling, F., and Francis, A. (1996b). Catatonia: I: Rating scale and standardized examination. *Acta Psychiat. Scandi.* **93**, 129–136.

Caroff, S., Mann, S., Francis, A., and Fricchione, G. (2004). "Catatonia. From Psychopathology to Neurobiology." American Psychiatric Publishing, Inc., Washington, D.C.

Christie, S., Miralles, C., and De Blas, A. (2002). GABAergic innervation organizes synaptic and extrasynaptic GABAA receptor clustering in cultured hippocampal neurons. *J. Neurosci.* **22**, 684–697.

Clarke, D. (1993). Prader-Willi syndrome and psychoses. *Br. J. Psychiatry* **163**, 680–684.

Clarke, D., Boer, H., Whittington, J., Holland, A., Butler, J., and Webb, T. (2002). Prader-Willi syndrome, compulsive and ritualistic behaviours: The first population-based survey. *Br. J. Psychiatry* **180**, 358–362.

Cohen, B. (1999). Elevated levels of plasma and urine gamma-aminobutyric acid: A case study of an autistic child (letter). *Autism* **3**, 437–440.

Cook, E., Olson, K., and Pliskin, N. (1996). Response of organic catatonia to risperidone (letter). *Arch. Gen. Psychiatry* **53**, 82–83.

Cook, E., Lindgren, V., Levental, B., Courchesne, R., Lincoln, A., Shulman, C., Lord, C., and Courchesne, E. (1997). Autism or atypical autism in maternally but not paternally derived proximal 15q duplication. *Am. J. Hum. Gent.* **60**, 928–934.

Cook, E., Courchesne, R., Cox, N., Lord, C., Gonen, D., Guter, S., Lincoln, A., Nix, K., Haas, R., Levental, B., and Courchesne, E. (1998). Linkage-disequilibrium mapping of autistic disorders, with 15q11-13 markers. *Am. J. Hum. Gent.* **62**, 1077–1083.

Courchesne, E., Yeung-Courchesne, R., Press, G., Hesselink, J., and Jernigan, T. (1988). Hypoplasia of cerebellar vermis lobules VI and VII in autism. *New England J. Med.* **318,** 1349–1354.

Courchesne, E., Karns, C., Davis, H., Ziccardi, R., Carper, R., Tigue, Z., Chisum, H., Moses, P., Pierce, K., Lord, C., Lincoln, A., Pizzo, S., Schreibman, L., Haas, R., Akshoomoff, N., and Courchesne, R. (2001). Unusual brain growth patterns in early life in patients with autistic disorder: An MRI study. *Neurology* **24,** 245–254.

Decavel, C., and van den Pol, A. (1990). GABA: A dominant neurotransmitter in the hypothalamus. *J. Com. Neurol.* **302,** 1019–1037.

DeLong, R. (2004). Autism and familial major mood disorders: Are they related? *J. Neuropsychiat. Clin. Neurosci.* **16,** 199–213.

DeLorey, T., Handforth, A., Anagnostaras, S., Homanics, G., Minsassian, B., Asatourian, A., Fanselow, M., Delgado-Escueta, A., Ellison, G., and Olsen, R. (1998). Mice lacking the beta3 subunit of the GABAA receptor have the epilepsy phenotype and many of the behavioral characteristics of Angelman syndrome. *J. Neurosci.* **18,** 8505–8514.

Devenand, D., Shapira, B., Petty, F., Kramer, G., Fitzsimons, L., Lerer, B., and Sackheim, H. (1995). Effects of electroconvulsive therapy on plasma GABA. *Convul. Ther.* **11,** 3–13.

Deykin, E., and MacMahon, B. (1979). The incidence of seizures among children with autistic symptoms. *Am. J. Psychiat.* **136,** 1310–1312.

Dhossche, D., and Bouman, N. (1997). Catatonia in an adolescent with Prader-Willi Syndrome. *Ann. Clin. Psychiat.* **4,** 247–253.

Dhossche, D. (1998). Catatonia in Autistic Disorders (brief report). *J. Autism Dev. Disorders* **28,** 329–331.

Dhossche, D., Applegate, H., Abraham, A., Maertens, P., Bencsath, A., Bland, L., and Martinez, J. (2002). Elevated plasma GABA levels in autistic youngsters: Stimulus for a GABA hypothesis of autism. *Med. Sci. Monitor* PR1–PR6.

Dhossche, D. (2004). Autism as early expression of catatonia. *Med. Sci. Monitor* **10,** RA31–RA39.

Dykens, E., Leckman, J., and Cassidy, S. (1996). Obsessions and compulsions in Prader-Willi syndrome. *J. Child Psychol. Psychiat.* **37,** 995–1002.

Dykens, E. (2002). Are jigsaw puzzle skills "spared" in persons with Prader-Willi syndrome? *J. Child Psychol. Psychiat.* **43,** 343–352.

Earl, C. (1934). The primitive catatonic psychosis of idiocy. *Br. J. Med. Psychol.* **14,** 230–253.

Ebert, M., Schmidt, D., Thompson, T., and Butler, M. (1997). Elevated gamma-aminobutyric acid (GABA) levels in individuals with either Prader-Willi Syndrome or Angelman Syndrome. *J. Neuropsychiat. Clin. Neurosci.* **9,** 75–80.

Emrich, H., von Zerssen, D., Kissling, W., Moller, H., and Windorfer, A. (1980). Effect of sodium valproate on mania. The GABA-hypothesis of affective disorders. *Arch. Psychiatrie und Nervenkrankheiten* **229,** 1–16.

Engelmann, M., Landgraf, R., and Wotjak, C. (2004). The hypothalamic-neurohypophysial system regulates the hypothalamic-pituitary-adrenal axis under stress: An old concept revisited. *Front. Neuroendocrinol.* **25,** 132–149.

Fatemi, S., Halt, A., Stary, J., Kanodia, R., Schulz, S., and Realmuto, G. (2002). Glutamic acid decarboxylase 65 and 67 kDa proteins are reduced in autistic parietal and cerebellar cortices. *Biol. Psychiat.* **52,** 805–810.

Ferkany, J., Smith, L., Seifert, W., Caprioli, R., and Enna, S. (1978). Measurement of GABA in blood. *Life Sci.* **22,** 2121–2128.

Fink, M., and Taylor, M. (2003). "Catatonia. A Clinician's Guide to Diagnosis and Treatment." University Press, Cambridge.

Fricchione, G., Cassem, N., Hooberman, D., and Hobson, D. (1983). Intravenous lorazepam in neuroleptic-induced catatonia. *J. Clin. Psychopharmacol.* **3,** 338–342.

Genazzani, A., Bernardi, F., Monteleone, P., Luisi, S., and Luisi, M. (2000). Neuropeptides, neurotransmitters, neurosteroids, and the onset of puberty. *Ann. NY Acad. Sci.* **9000,** 1–9.

Ghaziuddin, M., Quinlan, P., and Ghaziuddin, N. (2005). Catatonia in autism: a distinct subtype? *J. Intell. Disabil. Res.* **49,** 102–105.

Gillberg, C., and Schaumann, H. (1981). Infantile autism and puberty. *J. Autism Dev. Disorders* **11,** 365–371.

Gillberg, C. (1991a). Outcome in autism and autistic-like symptoms. *J. Am. Acad. Child Adolescent Psychiatry* **30,** 375–382.

Gillberg, C. (1991b). The treatment of epilepsy in autism. *J. Autism Dev. Disorders* **21,** 61–77.

Goddard, A., Mason, G., Almai, A., Rothman, D., Behar, K., Petroff, O., Charney, D., and Krystal, J. (2001). Reductions in occipital cortex GABA levels in panic disorder detected with 1H-magnetic resonance spectroscopy. *Arch. Gen. Psychiat.* **58,** 556–561.

Green, A. (1986). Changes in gamma-aminobutyric acid biochemistry and seizure threshold. *Ann. NY Acad. Sci.* **462,** 105–119.

Grossman, M., Hare, T., Manyam, N., Glaeser, B., and Wood, J. (1980). Stability of GABA levels in CSF under various conditions of storage. *Brain Res.* **182,** 99–106.

Guidotti, A., Auta, J., Davis, J., DiGiorgi Gerevini, V., Dwivedi, Y., Grayson, D., Impagnatiello, F., Pandey, G., Pesold, C., Sharma, R., Uzunov, D., and Costa, E. (2000). Decrease in reelin and glutamic acid decarboxylase67 (GAD67) expression in schizophrenia and bipolar disorder: A postmortem brain study. *Arch. Gen. Psychiat.* **57,** 1061–1069.

Halbreich, U., Petty, F., Yonkers, K., Kramer, G., Rush, A., and Bibi, K. (1996). Low plasma gamma-aminobutyric acid levels during the luteal phase of women with premenstrual dysphoric disorder. *Am. J. Psychiat.* **153,** 718–720.

Herlenius, E., and Lagercrantz, H. (2001). Neurotransmitters and neuromodulators during early human development. *Early Hum. Dev.* **65,** 21–37.

Herman, J., and Cullinan, W. (1997). Neurocircuitry of stress: Central control of the hypothalamo-pituitary-adrenocortical axis. *Trends Neurosci.* **20,** 78–84.

Holm, V., Cassidy, S., Butler, M., Hanchett, J., Greenswag, L., Whitman, B., and Greenberg, F. (1993). Prader-Willi syndrome: Consensus diagnostic criteria. *Pediatrics* **2,** 398–402.

Hussman, J. (2001). Suppressed GABAergic inhibition as a common factor in suspected etiologies of autism (letter). *J. Autism Dev. Disorders* **31,** 247–248.

Jackson, J. (1958). Evolution and dissolution of the nervous system. *In* "Selected Writings of John Hughlings Jackson" (J. Taylor, Ed.), pp. 45–118. Stapes, London.

Jimenez-Jimenez, F., Molina, J., Gomez, P., Vargas, C., de Bustos, F., Benito-Leon, J., Tallon-Barranco, A., Orti-Pareja, M., Gasalla, T., and Arenas, J. (1998). Neurotransmitter amino acids in cerebrospinal fluid of patients with Alzheimer's disease. *J. Neural Transmission* **105,** 269–277.

Joseph, B., Egli, M., Sutcliffe, J., and Thompson, T. (2001). Possible dosage effect of maternally expressed genes on visual recognition memory in Prader-Willi Syndrome. *Am. J. Med. Gent. (Neuropsychiatric Genetics)* **105,** 71–75.

Kahlbaum, K. (1874). "Die Katatonie Oder Das Spannungsirresein." Verlag August Hirshwald, Berlin.

Kakee, A., Takanaga, H., Terasaki, T., Naito, M., Tsuruo, T., and Sugiyama, Y. (2001). Efflux of a suppressive neurotransmitter, GABA, across the blood-brain barrier. *J. Neurochem.* **79,** 110–118.

Kalkman, H., and Loetscher, E. (2003). GAD67: The link between the GABA-deficit hypothesis and the dopaminergic and glutamatergic theories of psychosis. *J. Neural Transmission* **110,** 803–812.

Kanner, L. (1943). Autistic disturbances of affective contact. *Nervous Child* **2,** 217–250.

Kemph, J., Devane, C., Levin, G., Jarecke, R., and Miller, R. (1993). Treatment of aggressive children with clonidine: Results of an open study. *J. Am. Acad. Child Adolescent Psychiat.* **32,** 577–581.

Keverne, E. (1999). GABAergic neurons and the neurobiology of schizophrenia and other psychoses. *Brain Res. Bull.* **48,** 467–473.

Kurita, H. (1999). Delusional disorder in a male adolescent with high-functioning PDDNOS. *J. Autism. Dev. Disorders* **29,** 419–423.

Lalande, M., Sinnett, D., and Glatt, K., *et al.* (1994). Fine mapping of the GABA-A receptor subunit gene cluster in chromosome 15q11q13 and assignment of GABRG3 to this region: Report of the Second International Workshop of Human Chromosome 15 Mapping. *Cytogenetic Cell Genet.* **67,** 17.

Lauritsen, M., Mors, O., Mortensen, P., and Ewald, H. (1999). Infantile autism and associated autosomal chromosome abnormalities: A register-based study and a literature survey. *J. Child Psychol. Psychiat.* **40,** 335–345.

Leonhard, K. (1979). "The Classification of Endogenous Psychoses" (E. Robins, Ed.). Irvington Publications, New York.

Lichtshtein, D. J. D., Ebstein, R., Biederman, J., Rimon, R., and Belmaker, R. (1978). Gamma-aminobutyric acid (GABA) in the CSF of schizophrenic patients before and after neuroleptic treatment. *Br. J. Psychiat.* **132,** 145–148.

Lipcsey, A., Kardos, J., Prinz, G., and Simonyi, M. (1986). Der Effect der electrokonvulsiven Therapie auf den GABA-Spiegel des Liquor cerebrospinalis. *Psychiatrie, Neurologie, und medizinische Psycholologie Leipzig* **38,** 554–555.

Löscher, W., and Schmidt, D. (1981). Plasma GABA levels in neurological patients under treatment with valproic acid. *Life Sci.* **28,** 2383.

Löscher, W., and Frey, H. (1982). Transport of GABA at the blood-CSF interface. *J. Neurochem.* **38,** 1072–1079.

Lucignani, G., Panzacchi, A., Bosio, L., Moresco, R., Ravasi, L., Coppa, I., Chiumello, G., Frey, K., Koeppe, R., and Fazio, F. (2004). GABA-A receptor abnormalities in Prader-Willi syndrome assessed with positron emission tomography and [11-C]flumazenil. *NeuroImage* **22,** 22–28.

Mahendra, B. (1981). Where have all the catatonics gone? (Editorial). *Psycholog. Med.* **11,** 669–671.

McManus, D., Baker, G., Martin, I., Greenshaw, A., and McKenna, K. (1992). Effects of the antidepressant/antipanic drug phenelzine on GABA concentrations and GABA-transaminase activity in rat brain. *Biochem. Pharmacol.* **43,** 2486–2489.

Menold, M., Shao, Y., Wolpert, C., Donnelly, S., Raiford, K., and Martin, E., *et al.* (2001). Association analysis of chromosome 15 gabaa receptor subunit genes in autistic disorder. *J. Neurogenet.* **15,** 245–259.

Mervaala, E., Kononen, M., Fohr, J., Saastamoinen, M., Korhonen, M., Kuikka, J., Viinanmaki, H., Tammi, A., Tiihonen, J., Partanen, J., and Lehtonen, J. (2001). SPECT and neuropsychological performance in severe depression treated with ECT. *J. Affe. Disorders* **66,** 47–58.

Mitushima, D., Hei, D., and Terasawa, E. (1994). Gamma-aminobutyric acid is an inhibitory neurotransmitter restricting the release of luteinizing hormone-releasing hormone before the onset of puberty. *Proc. Nat. Acad. Sci. USA* **91,** 395–399.

Nisijima, K., and Ishiguro, T. (1995). Cerebrospinal fluid levels of monoamine metabolites and gamma-aminobutyric acid in neuroleptic malignant syndrome. *J. Psychiat. Res.* **3,** 233–244.

Northoff, G., Steinke, R., Czcervenka, C., Krause, R., Ulrich, S., Danos, P., Kroph, D., Otto, H., and Bogerts, B. (1999). Decreased density of GABA-A receptors in the left sensorimotor cortex in akinetic catatonia: Investigation of *in vivo* benzodiazepine receptor binding. *J. Neurol. Neurosurg. Psychiat.* **67,** 445–450.

Nurmi, E., Dowd, M., Tadevosyan-Leyfer, O., Haines, J., Folstein, S., and Sutcliffe, J. (2003). Exploratory subsetting of autism families based on savant skills improves evidence of genetic linkage to 15q11–13. *J. Am. Acad. Child Adolescent Psychiat.* **42,** 856–863.

O'Conner, N., and Hermelin, B. (1991). Talents and preoccupations in idiots-savants. *Psychol. Med.* **21,** 959–964.

Odano, I., Anezaki, T., Ohkubo, M., Yonekura, Y., Onishi, Y., Inuzuka, T., Takahashi, M., and Tsuji, S. (1996). Decrease in benzodiazepine receptor binding in a patient with Angelman syndrome detected by iodine-123 iomazenil and single-photon emission tomography. *Eur. J. Nucl. Med.* **23,** 598–604.

Otani, K., Ujike, H., Tanaka, Y., Morita, Y., Katsu, T., Nomura, A., Uchida, N., Hamamura, T., Fujiwara, Y., and Kuroda, S. (2005). The GABA type A receptor alfa 5 subunit gene is associated with bipolar I disorder. *Neurosci. Lett.* **381,** 108–113.

Papadimitriou, G., Dikeos, D., Karadima, G., Avramopoulos, D., Daskalopoulou, E., Vassilopoulos, D., and Stefanis, C. (1998). Association between the GABA(A) receptor alpha 5 subunit gene locus (GABRA5) and bipolar affective disorder. *Am. J. Med. Gene.* **81,** 73–80.

Petroff, O., Rothman, D., Behar, K., and Mattson, R. (1996). Low brain GABA level is associated with poor seizure control. *Ann. Neurol.* **40,** 908–911.

Petty, F., Kramer, G., and Feldman, M. (1987). Is plasma GABA of peripheral origin? *Biol. Psychiat.* **22,** 725–732.

Petty, F., Kramer, G., Gullion, C., and Rush, A. (1992). Low plasma gamma-aminobutyric acid levels in male patients with depression. *Biol. Psychiat.* **32,** 354–363.

Petty, F., Kramer, G., Fulton, M., Moeller, F., and Rush, A. (1993). Low plasma GABA is a trait-like marker for bipolar illness. *Neuropsychopharmacolgy* **9,** 125–132.

Petty, F. (1995). GABA and mood disorders: a brief review and hypothesis. *J. Affect. Disorders* **34,** 275–281.

Petty, F., Fulton, M., Kramer, G., Kram, M., Davis, L., and Rush, A. (1999). Evidence for the segregation of a major gene for human plasma GABA levels. *Mol. Psychiatry* **4,** 587–589.

Petty, L., Ornitz, E., Michelman, J., and Zimmerman, E. (1984). Autistic children who become schizophrenic. *Arch. Gen. Psychiatry* **41,** 129–135.

Pfleiderer, B., Michael, N., Erfurth, A., Ohrmann, P., Hohmann, U., Wolgast, M., Fiebich, M., Arolt, V., and Heindel, W. (2003). Effective electroconvulsive therapy reverses glutamate/glutamine deficit in the left anterior cingulum of unipolar depressed patients. *Psychiatric Research: Neuroimaging* **122,** 185–192.

Rapin, I., and Katzman, R. (1998). Neurobiology of autism. *Ann. Neurol.* **43,** 7–14.

Realmuto, G., and August, G. (1991). Catatonia in autistic disorder: A sign of comorbidity or variable expression. *J. Autism Dev. Disorders* **21,** 517–528.

Roberts, E. (1972). Prospects for research on schizophrenia. An hypotheses suggesting that there is a defect in the GABA system in schizophrenia. *Neurosci. Res. Prog. Bull.* **10,** 468–482.

Roberts, E. (2000). Adventures with GABA. Fifty years on. *In* "GABA in the Nervous System. The View at Fifty Years" (D. Martin and R. Olsen, Eds.), pp. 1–24. Lippincott Williams and Wilkins, Philadephia, PA.

Rohland, B., Carroll, B., and Jacoby, R. (1993). ECT in the treatment of the catatonic syndrome. *J. Affec. Disorders* **29,** 255–261.

Rutter, M. (1983). Cognitive deficits in the pathogenesis of autism. *J. Child Psychol. Psychiat.* **24,** 513–531.

Sanacora, G., Mason, G., Rothman, D., Behar, K., Hyder, F., Petroff, O., Berman, R., Charney, D., and Krystal, J. (1999). Reduced cortical gamma-aminobutyric acid levels in depressed patients determined by 1H-magnetic resonance spectroscopy. *Arch. Gen. Psychiat.* **56,** 1043–1047.

Sanacora, G., Mason, G., Rothman, D., and Krystal, J. (2002). Increased occipital cortex GABA concentrations in depressed patients after therapy with selective serotonin reuptake inhibitors. *Am. J. Psychiat.* **159,** 663–665.

Sanacora, G., Mason, G., Rothman, D., Hyder, F., Ciarcia, J., Ostroff, R., Berman, R., and Krystal, J. (2003). Increased cortical GABA concentrations in depressed patients receiving ECT. *Am. J. Psychiat.* **160,** 577–579.

Schmickel, R. (1986). Contiguous gene syndromes: A component of recognizable syndromes. *J. Pediatri.* **109,** 231–241.

Schmidt, D., and Löscher, W. (1982). Plasma and cerebrospinal fluid gamma-aminobutyric acid in neurological disorders. *J. Neurol. Neurosurg. Psychiat.* **45,** 931–935.

Schroer, R., Phelan, M., Michaelis, R., Crawford, E., Skinner, S., Cuccaro, M., Simensen, R., Bishop, J., Skinner, C., Fender, D., and Stevenson, R. (1998). Autism and maternally derived aberrations of chromosome 15q. *Am. J. Med. Gent.* **76,** 327–336.

Shao, Y., Cuccaro, M., Hauser, E., Raiford, K., Menold, M., Wolpert, C., Ravan, S., Elston, L., Decena, K., Donnelly, S., Abramson, R., Wright, H., De Long, G., Gilbert, J., and Pericak-Vance, M. (2003). Fine mapping of Autistic Disorder to chromosome 15q11-q13 by use of phenotypic subtypes. *Am. J. Hum. Genet.* **72,** 539–548.

Shapira, N., Lessig, M., Murphy, T., Driscoll, D., and Goodman, W. (2002). Topiramate attenuates self-injurious behaviour in Prader-Willi syndrome. *Internat. J. Neuropsychopharmacol.* **5,** 141–145.

Shapira, N., Lessing, M., Lewis, M., Goodman, W., and Driscoll, D. (2004). Effects of topiramate in adults with Prader-Willi syndrome. *Am. J. Ment. Retard.* **4,** 301–309.

Shiah, I., Yatham, L., and Baker, G. (2000). Divalproex sodium increases plasma GABA levels in healthy volunteers. *Internat. Clin. Psychopharmacol.* **15,** 221–225.

Smathers, S., Wilson, J., and Nigro, M. (2003). Topiramate effectiveness in Prader-Willi syndrome. *Pediatric Neurology* **28,** 130–133.

Smith, A., Egan, J., Ridley, G., Haan, E., Montgomery, P., Williams, K., and Elliott, E. (2003). Birth prevalence of Prader-Willi Syndrome in Australia. *Archives of Disease in Childhood* **88,** 263–264.

Soghomonian, J., and Martin, D. (1998). Two isoforms of glutamate decarboxylase: Why? *Trends Pharmacolog. Sci.* **19,** 500–505.

Song, Y., Shenwu, M., Dhossche, D., and Liu, Y.-M. (2005). A capillary liquid chromatographic/tandem mass spectrometry method for the quantification of gamma-aminobutyric acid (GABA) in human plasma and cerebrospinal fluid. *J. Chromatography B* **25,** 295–302.

Stöber, G., Saar, K., Ruschendorf, F., Meyer, J., Nurnberg, G., Jatzke, S., Franzek, E., Reis, A., Lesch, K., Wienker, T., and Beckmann, H. (2000). Splitting schizophrenia: Periodic catatonia susceptibility locus on chromosome 15q15. *Am. J. Hum. Genet.* **67,** 1201–1207.

Stöber, G., Seelow, D., Ruschendorf, F., Ekici, A., Beckmann, H., and Reis, A. (2002). Periodic catatonia: Confirmation of linkage to chromosome 15 and further evidence for genetic heterogeneity. *Hum. Gent.* **111,** 323–330.

Swaab, D., Purba, J., and Hofman, M. (1995). Alterations in the hypothalamic paraventricular nucleus and its oxytocin neurons (putative satiety cells) in Prader-Willi syndrome: A study of five cases. *J. Clin. Endocrinol. Metabolism* **80,** 573–579.

Swaab, D. (1997). Prader-Willi syndrome and the hypothalamus. *Acta Pediatrica (Supplement)* **423,** 50–54.

Taylor, M., and Fink, M. (2003). Catatonia in psychiatric classification: A home of its own. *Am. J. Psychiatry* **160,** 1–9.

Tunnicliff, G., Schindler, N., Crites, G., Goldenberg, R., Yohum, A., and Malatynska, E. (1999). The GABAA receptor complex as a target for fluoxetine action. *Neurochem. Res.* **24,** 1271–1276.

Turner, T. (1989). Schizophrenia and mental handicap: A historical review, with implications for further research. *Psychologi. Medi.* **19,** 301–314.

Ungavri, G., Leung, S., Shing, F., Cheung, H.-K., and Leung, T. (2005). Schizophrenia with prominent catatonic features ("catatonic schizophrenia"). I. Demographic and clinical correlates in the chronic phase. *Prog. Neuro-Psychopharmacol. Biol. Psychiat.* **29,** 27–38.

Ungvari, G., Leung, C., Wong, M., and Lau, J. (1994). Benzodiazepines in the treatment of the catatonic syndrome. *Acta Psychiat. Scand.* **89,** 285–288.

Vaiva, G., Thomas, P., Ducrocq, F., Fontaine, M., Boss, V., Devos, P., Rascle, C., Cottencin, O., Brunet, A., Laffargue, P., and Goudemand, M. (2004). Low posttrauma GABA plasma levels as a predictive factor in the development of acute posttraumatic stress disorder. *Biol. Psychiat.* **55,** 250–254.

van der Heijden, F., Tuinier, S., Arts, N., Hoogendoorn, M., Kahn, R., and Verhoeven, W. (2005). Catatonia: disappeared or under-diagnosed? *Psychopathology* **15,** 3–8.

van Gelder, N., and Elliot, K. (1958). Disposition of gamma-amino-butyric acid administered to mammals. *J. Neurochem.* **3,** 139–143.

van Kammen, D. P. (1977). Gamma-aminobutyric acid (GABA) and the dopamine hypothesis of schizophrenia. *Am. J. Psychiat.* **134,** 138–143.

van Kammen, D. P., Petty, F., Kelley, M., Kramer, G., Barry, E., Yao, J., Gurklis, J., and Peters, J. (1998). GABA and brain abnormalities in schizophrenia. *Psychiatric Research: Neuroimaging Section* **82,** 25–35.

Varju, P., Katarova, Z., Madarasz, E., and Szabo, G. (2001). GABA signaling during development: New data and old questions. *Cell Tissue Res.* **305,** 239–246.

Verhoeven, W., Curfs, L., and Tuinier, S. (1998). Prader-Willi syndrome and cycloid psychoses. *Intelle. Disabil. Res.* **42,** 455–462.

Vogels, A., Matthijs, G., Legius, E., Devriendt, K., and Fryns, J.-P. (2003). Chromosome 15 maternal uniparental disomy and psychosis in Prader-Willi syndrome. *J. Med. Gene.* **40,** 72–73.

Wagstaff, J., Knoll, J., Fleming, J., Kirkness, E., Martin-Gallardo, A., Greenberg, F., Graham, J., Menninger, J., Ward, D., and Venter, J., *et al.* (1991). Localization of the gene encoding the GABAA receptor beta 3 subunit to the Angelman/Prader-Willi region of human chromosome 15. *Am. J. Hum. Gent.* **49,** 330–337.

Whitman, B., and Accardo, P. (1987). Emotional symptoms in Prader-Willi Syndrome patients. *Am. J. Med. Gent.* **28,** 897–905.

Wigren, M., and Hansen, S. (2003). Rituals and compulsivity in Prader-Willi syndrome: Profile and stability. *J. Intelle. Disabil. Res.* **47,** 428–438.

Wigren, M., and Hansen, S. (2005). ADHD symptoms and insistence on sameness in Prader-Willi syndrome. *J. Intelle. Disabil. Res.* **49,** 449–456.

Wing, L., and Shah, A. (2000). Catatonia in autistic spectrum disorders. *Br. J. Psychiat.* **176,** 357–362.

Woodage, T., Deng, Z., Prasad, M., Smart, R., Lindeman, R., Christian, R., Ledbetter, D., Robson, L., Smith, A., and Trent, R. (1994). A variety of genetic mechanisms are associated with the Prader-Willi Syndrome. *Am. J. Med. Gene. (Neuropsychiatric Genetics)* **54,** 219–226.

Zaw, F., Bates, G., Murali, V., and Bentham, P. (1999). Catatonia, autism, and ECT. *Dev. Med. Child Neurol.* **41,** 843–845.

Zink, M., Schmitt, A., May, B., Muller, B., Braus, D., and Henn, F. (2004). Differential effects of long-term treatment with clozapine or haloperidol on GABA transporter expression. *Pharmacopsychiatry* **37,** 171–174.

Zuddas, A., Ledda, M., Fratta, A., Muglia, P., and Cianchetti, C. (1996). Clinical effects of clozapine on Autistic Disorder. *Am. J. Psychiat.* **153.**

ALCOHOL, GABA RECEPTORS, AND NEURODEVELOPMENTAL DISORDERS

Ujjwal K. Rout

Department of Surgery, Division of Pediatric Surgery Research Laboratories
University of Mississippi Medical Center, Jackson, Mississippi 39216, USA

Development of the fetal brain is affected by maternal alcohol consumption during pregnancy. Children exposed to alcohol during gestation suffer from a wide range of physical and neurological damage. Both heavy and moderate maternal drinking is unsafe for the developing fetus. Heavy drinking causes obvious physical and behavioral changes in children, whereas moderate drinking may cause only subtle changes that may remain unnoticed. Neurobehavioral problems in children resulting from both heavy and moderate maternal drinking during pregnancy are long-lasting and are aggravated by stress and the natural process of aging. There are several ways alcohol can disturb the normal process of brain development. These include changes in neurotransmitters, their receptors, and the process of neurotransmission. Recent studies provide strong evidence that γ-aminobutyric acid (GABA) and GABA-receptor systems are affected by alcohol during brain development. GABA and GABA-receptors play significant roles in the proliferation, migration, differentiation, and positioning of different cell types in the developing brain required for the optimum synaptogenesis and balance in excitatory and inhibitory neurotransmission. Studies with rodent brains clearly indicate that alcohol during pregnancy alters these processes by GABA and GABA-receptor dependent mechanisms. This chapter provides an overview on the GABA-receptor system in rodent brains, the role of this system in the normal development of the brain, and disturbances that may cause neurobehavioral disorders with an emphasis on alcohol.

INTERNATIONAL REVIEW OF
NEUROBIOLOGY, VOL. 71
DOI: 10.1016/S0074-7742(05)71010-2

I. Introduction

Maternal alcohol consumption during pregnancy is one of the leading causes of neurobehavioral problems in the Western world. Children exposed to alcohol during gestation suffer from a wide range of physical and neurological abnormalities commonly known as fetal alcohol syndrome (FAS) (Jones et al., 1973; Ulleland, 1972). The characteristics of FAS include growth deficiency, short palpebral fissures, a relatively short nose, indistinct philtrum, thin upper lip, flattened medial midface, small crania, cognitive impairment, intellectual deficiencies, behavioral disturbances, and neurological damage (Sampson et al., 1997).

Studies with human brains that are exposed to alcohol during gestation show that not all parts of the brain are equally sensitive to alcohol's teratogenic effects. Autopsy and magnetic resonance studies combined with quantitative analysis reveal that neuronal migration, basal ganglia (both caudate and lenticular nuclei), cerebellum (anterior vermis; lobule I to V), and corpus callosum (complete to partial agenesis, hypoplasia, and midline abnormalities) are primarily vulnerable to heavy alcohol exposure (Bhatara et al., 2002; Johnson et al., 1996; Jones et al., 1974; Mattson et al., 1992, 1996, 2001; Riley et al., 1995; Swayze et al., 1997). Ocular and auditory abnormalities are also common amongst children with FAS (Church and Kaltenbach, 1997; Stromland and Pinazo-Duran, 1994). Positron emission tomography analyses seem to show low levels of glucose metabolism in the caudate and cerebellum of children with FAS (Roebuck et al., 1998), suggesting zone specific metabolic deregulation induced by prenatal alcohol in human brains.

Moderate alcohol consumption during pregnancy causes subtle impairments in the behavior of offspring that may occur in absence of characteristic facial and severe behavior abnormalities of FAS. This condition is categorized as alcohol related neurodevelopmental disorder (ARND). Cognitive deficits in children with ARND are often not apparent until the child is challenged during the educational years (Conry, 1990; Streissguth et al., 1990) or later in life (Streissguth et al., 1991). Therefore, behavioral consequences of gestational exposure to moderate alcohol may have long lasting effects (Riley, 1990). Signs may diminish as the victim develops compensatory mechanisms to deal with the dysfunction but re-emerge under stressful conditions or with age (Steinhausen and Spohr, 1998). Higher incidence of adult-onset of psychiatric disturbances, including a 44% incidence of major depressive illness and 40% incidence of psychosis is reported in humans exposed to alcohol during gestation (Famy et al., 1998). Animal studies also demonstrate this nature of the problem (Hannigan et al., 1987; Riley, 1990). The combined rate of FAS and ARND is estimated to be at least 9 out of 1000 live births in developed countries (Sampson et al., 1997). Recent studies also document autism in children exposed to alcohol during gestation (Miles et al., 2003; Nanson, 1992).

This Chapter presents a brief overview on γ-aminobutyric acid (GABA) and GABA-receptor systems with empasis on their expression and regulatory function during brain development. Data from rodent models are used to describe the roles of this system in the development of neurobehavioral problems resulting from gestational exposure to alcohol. When possible, information on human studies is provided, but the article is primarily supported by data from rodent models.

A. GABA AND GABA RECEPTORS

GABA is a highly flexible molecule, exists in many low-energy conformations, and binds to different GABA receptors to mediate its actions. In the mature CNS, three classes of GABA receptors are identified (Chebib and Johnston, 1999). Of these, $GABA_A$ and $GABA_C$ types are ionotropic receptors that form ligand-gated Cl^- ion channels (Feigenspan et al., 1993; Sivilotti and Nistri, 1991). The $GABA_B$ type of receptors are metabotropic (G protein-coupled) and are coupled to Ca^{2+} and K^+ channels or adenylate cyclase via $G_{i/o}$ GTP binding proteins (Kerr and Ong, 1992). $GABA_A$ and $GABA_C$ receptors mediate fast neurotransmission, whereas $GABA_B$ receptors mediate slow neurotransmission (Wong et al., 2003).

The $GABA_A$ receptors display extraordinary structural heterogeneity resulting from various subunit compositions. There are eight subunit families, altogether consisting of 21 different subunits (α^{1-6}, β^{1-4}, γ^{1-4}, δ, ε, π, θ, ρ^{1-3}) that may form a hetero-oligomeric $GABA_A$ receptor. However, in the endoplasmic reticulum and Golgi apparatus, a restricted number of subunit combinations assemble together and are properly packaged, processed, and trafficked to the cell surface (Barnes, 2000). Most in vivo $GABA_A$ receptors are formed by the co-assembly of α-, β-, and γ-subunits (Macdonald and Olsen, 1994). Hetero-oligomeric $GABA_A$ receptors are blocked by alkaloid bicuculline and are modulated by barbiturates, benzodiazepines, neuroactive steroids, and alcohol (Chebib and Johnston, 1999; Mehta and Ticku, 1999).

$GABA_C$ receptors are made up of a single type of protein subunit (ρ). The molecular subunits of $GABA_C$ receptors are found in the retina, thalamus, pituitary, and gut. Five different ρ subunits are identified, of which $\rho1-$, $\rho2-$, or $\rho3$-subunits form functional homomeric or pseudoheteromeric receptors (Chebib, 2004). $GABA_C$ receptors are not blocked by bicuculline, nor are they modulated by steroids, barbiturates, or benzodiazepines (Johnston, 1996). GABA is an order of magnitude less potent at $GABA_A$ than $GABA_C$ receptor (Chebib and Johnston, 1999). Muscimol and its analogue, 4,5,6,7-tetrahydroisoxazolo [5,4-c]pyridine-3-ol (THIP), act at both $GABA_A$ and $GABA_C$ receptors. However, at $GABA_A$ receptors, muscimol is an antagonist, whereas THIP is a partial agonist (Krogsgaard-Larsen et al., 1994; Kusama et al., 1993). At $GABA_C$

receptors, muscimol is a partial agonist (Kusama *et al.*, 1993), while THIP is an antagonist (Woodward *et al.*, 1993).

The $GABA_B$ receptors consist of two subunits, $GABA_{B1}$ and $GABA_{B2}$. Different spliced variants of $GABA_{B1}$ receptors are reported and designated as $GABA_{B1a}$, $GABA_{B1b}$, $GABA_{B1c}$, $GABA_{B1d}$, $GABA_{B1e}$, $GABA_{B1f}$, and $GABA_{B1g}$ (Billinton *et al.*, 2001; Bowery *et al.*, 2002; Kawakami *et al.*, 2004). The functional $GABA_B$ receptors require co-assembly of $GABA_{B1}$ and $GABA_{B2}$ (Bowery *et al.*, 2002; Couve *et al.*, 2000). $GABA_B$ receptors are hetero-oligomeric receptors made up of a mixture of subunits, and are selectively activated by baclofen and CCGP27492 and are blocked by phaclofen, the phosphoric analog of baclofen (Chebib and Johnston, 1999).

In adult brains, GABA acts as a major inhibitory neurotransmitter. However, it also mediates excitatory effects in the dorsal root ganglion, CA1 hippocampal pyramidal cells, and cells in layer 5 of cortex in mature brains (Stein and Nicoll, 2003).

B. GABA, GABA RECEPTORS, AND BRAIN DEVELOPMENT

1. *Expression Pattern*

GABA is in abundance and is widespread during embryonic development. It appears with the beginning of neurogenesis when the central nervous system is primarily composed of proliferating neuroepithelium (Behar *et al.*, 1996). In the latter phase, GABAergic neurons become widespread in all the regions of the brain (Zecevic and Milosevic, 1997). The widespread appearance of GABAergic neurons in the cortex and hippocampus during the primary period of neurogenesis is consistent with the notion that GABA acts as a trophic signal during neurogenesis (Barker *et al.*, 1998; Ma and Barker, 1998). Since synapses are not present during neurogenesis in the developing brain, it is considered that GABA is released from precursor cells and acts in a paracrine manner to activate its receptors (Demarque *et al.*, 2002). GABA expression is mostly reduced during the perinatal period as the astrocytes and oligodendrocytes appear and neurons differentiate into transmitting circuits (Zecevic and Milosevic, 1997).

During embryonic development different $GABA_A$ receptor subunits exhibit regional and temporal changes with expression levels of some subunits being more prominent than others. Expression of α_2, α_3, α_4, α_5, β_1, β_2, β_3, γ_1, γ_2, and γ_3 dominates in the embryonic stage, whereas in the adults, expression of α_1, α_2, α_3, β_2, β_3, γ_2, and δ predominates (Ma and Barker, 1995). Expression levels of α_2, γ_1, and γ_3 drop progressively during development, whereas the expression of α_1 subunit increases (Fritschy *et al.*, 1994). Functional significance of subunit switching and change in the expression levels during embryonic development is not completely understood. It is possible that α_4, α_5, and γ_3 subunits form GABA$_A$ receptor/chloride ion channels regulating proliferation, migration, and differentiation of neuronal progenitors rather than synaptogenesis, whereas α_1 subunit may add to form channels for fast GABAergic signals at synapses. The ubiquitous expression of $\beta_{2/3}$ during development possibly allows these subunits to associate with the different α subunits forming distinct receptor subtypes (Fritschy et al., 1994).

The expression pattern of GABA$_B$ subunit mRNA isoforms (GABA$_{B1}$ and GABA$_{B2}$) in developing brain are not coordinately regulated although both transcripts are detected in several structures in developing brains, including the hippocampus, cerebral cortex, neurepithelium, cerebellum, diencephalons, hippocampus, and thalamus (Kim et al., 2003; Li et al., 2004; Martin et al., 2004). The expression of GABA$_{B1}$ subunit during gestation days 11 to 12 is followed by the appearance of GABA$_{B2}$ transcripts about 2 days later. Furthermore, expression of GABA$_{B1a}$ dominates during postnatal development whereas GABA$_{B1b}$ dominates in the adult brains (Fritschy et al., 1999). These differences in the temporal expression patterns of GABA$_B$ receptor isoforms indicate that GABA$_B$ receptor subunit genes are under independent regulation during embryonic development, but the functional significance of this phenomenon is not clear, primarily because information on protein level expression of these subunits are still lacking.

2. Regulation

GABA is a diffusible factor and is released from pioneer and migrating neurons (Rivera et al., 2004), growth cones (Gordon-Weeks et al., 1984), and glia (Barakat and Bordey, 2002). GABA in extracellular space is maintained due to delayed maturation of GABA transporters, that although present by the end of the gestation, remain ineffective until the perinatal period (Yan et al., 1997). Diffused GABA acts in a paracrine manner during brain development via GABA receptors (Wang et al., 2003).

Effects of GABA on immature and mature neurons differ. GABA augments levels of intracellular Ca^{2+}, c-Fos, and BDNF transcripts in a GABA-receptor and Ca^{2+} channel dependent manner in rat hippocampal neurons during maturation (Berninger et al., 1995). This GABA effect is seen only in immature neurons because GABA$_A$ antagonists do not influence expression of c-fos and BDNF messenger RNA levels in mature neurons (Berninger et al., 1995; Obrietan et al., 2002).

GABA is excitatory in immature neurons, which switch primarily to their inhibitory actions in the early postnatal brains (Ben-Ari et al., 1989; Obata et al., 1978). Recent studies suggest that this occurs due to changes in the expression pattern of cation-chloride cotransporters (CCCs) during postnatal brain development (Ben-Ari, 2002; Lu et al., 1999). The immature neurons primarily express Na^+-K^+-$2Cl^-$ co-transporter (NKCC1) that is driven by sodium and potassium gradients and raises intracellular Cl^- concentrations. The K^+-Cl^- co-transporter (KCC2), in contrast, is expressed primarily in mature neurons and couples Cl^- transport to the $K+$ gradient and normally lowers the intracellular Cl^- concentrations. Since $GABA_A$ receptors are permeable to Cl^-, activation of these receptors releases Cl from inside of immature neurons causing depolarization. The reverse happens when $GABA_A$ receptors are activated in adult neurons. It is shown that the GABA switch is delayed by blockade of $GABA_A$ receptors and accelerated by an increase in activation (Ganguly *et al.*, 2001). However, other mechanisms may also regulate this switch, as it may occur in absence of $GABA_A$ receptor activation (Ludwig *et al.*, 2003).

Contrasting effects of GABA on cell proliferation are reported depending upon the type of precursors and the animal studied. GABA inhibits cell cycle progression of precursors in organotypic striatal slices (Nguyen *et al.*, 2003). It increases proliferation of ventricular zone cells, but decreases proliferation of cells in the sub ventricular zone (Haydar *et al.*, 2000). It decreases DNA synthesis in acute slices (LoTurco *et al.*, 1995), but enhances proliferation of cerebellar granule cell precursors (Fiszman *et al.*, 1999). These differences in the outcomes of GABA effects on cell proliferation may be due to additional factors that work in conjunction with GABA and appear in different cell types at specific times of development. Moreover, substitute mechanisms for the GABA mediated proliferation of neuronal precursors may exist as no obvious changes in the brain structure are noticed following GABA deficiency.

In the developing brain, the postmitotic neurons migrate from their site of origin to the final destination, where they make synaptic connection. Neurotransmitters and an N-type calcium channel regulate migration of immature neurons (Komuro and Rakic, 1998). GABA acting on different GABA receptors acts as motility promoting, accelerating, or stop signal (Lujan *et al.*, 2005). In cultured rat brain slices, $GABA_B$ and $GABA_C$–like receptor activation stimulates migration of neurons from the intermediate and ventricular zones respectively, whereas activation of $GABA_A$ receptors arrests migration of neurons as they approach their target destination (Behar *et al.*, 1996, 1998, 2000, 2001). The role of GABA in the tangential migration of neurons is evident by the fact that a blockade of the $GABA_B$ receptors leads to the accumulation of tangentially migrating interncurons in the proliferating zone in organotypic cultures from embryos (Lopez-Bendito *et al.*, 2003). It is also evident that GABA provides boundary information for the migratory neurons in the ventromedial nucleus of the hypothalamus in mice embryos in a $GABA_A$ receptor-mediated mechanism (Dellovade *et al.*, 2001), and activation of $GABA_A$ receptors by GABA inhibits neuronal migration in the anterior sub ventricular zone and rostral migratory stream of juvenile and adult mice (Bolteus and Bordey, 2004). $GABA_A$ receptor agonist muscimol also inhibits migration of luteinizing hormone-releasing hormone neurons in the embryonic olfactory explants (Fueshko *et al.*, 1998).

GABA also modulates neuronal arbour elaboration and differentiation. Activation of the $GABA_A$ receptor promotes neurite outgrowth and maturation of GABAergic interneurons (Barbin *et al.*, 1993; Marty *et al.*, 1996), and $GABA_A$ receptor antagonists reduce the dendritic outgrowth of cultured rat hippocampal neurons (Barbin *et al.*, 1993). Blockage of Ca^{2+}/calmodulin kinase II (CaMKII) or mitogen-activated protein kinase in cerebellar granule cells reduces the differentiating effects of GABA, suggesting that this process requires a Ca^{2+} influx and an activation of Ca^{2+} dependent kinases (Borodinsky *et al.*, 2003; Maric *et al.*, 2001).

C. GABA, GABA Receptors, and Neurodevelopmental Disorders

Almost all GABA in the brain is synthesized by the decarboxylation of glutamate by two forms of glutamic acid decarboxylase (GAD): GAD_{65} and GAD_{67}. These isoforms are the product of two genes (Erlander *et al.*, 1991). Knockout mouse with deleted GAD_{65} and GAD_{67} have only 0.02% of normal GABA levels in the brain. Double mutants do not survive because of cleft palate (Ji *et al.*, 1999). Histological and immunological studies with fetal and newborn mice show no discernible changes in the brain structures, including in cortical lamination, suggesting that the role of GABA in the early brain development is substituted by other molecules such as taurine acting on a glycine receptor/Cl^- channel (Flint *et al.*, 1998) and glutamate acting at N-methyl-d-aspartate (NMDA) receptors (Behar *et al.*, 1999). Ultra structural studies of the brains of mutant mice and global gene expression analysis may elucidate changes that may occur due to loss of GABA during development. Mice deficient in GAD_{65} mRNA and protein demonstrate conditional fear behavior, spontaneous seizure, lowered threshold for seizure inducing drugs, and increased anxiety-like behavior (Asada *et al.*, 1996; Stork *et al.*, 2000, 2003). GAD_{67} null mouse with only 7% GABA in the brain mRNA develops cleft palate and dies the day of birth with no discernible defects in the brain structures.

In adults, altered GABA levels are reported in several neurological disorders. The GABA level is significantly reduced in the cerebrospinal fluid of patients with Huntington's disease and Alzheimer's disease (Enna *et al.*, 1977). In patients with stiff-person syndrome, prominent and significant decrease in the GABA level is observed in the sensorimotor cortex with a smaller decrease in the posterior occipital cortex (Levy *et al.*, 2005; Wong *et al.*, 2004). GABA concentrations are also lower in the cerebrospinal fluid and blood plasma in unipolar depressed patients (Brambilla *et al.*, 2003; Sanacora *et al.*, 2004). Proton magnetic resonance spectroscopy shows decreased GABA concentrations in the occipital cortex of depressed patients. In contrast, higher levels of GABA are reported in the blood and urine of autistic children (Cohen, 2001; Dhossche *et al.*, 2002). Since GABA receptors are also expressed in the peripheral tissues (Erdo and Wolff, 1990) many of the behavioral problems of excess GABA may derive indirectly from the abnormalities that may occur in tissues other than CNS. Relevant to this, the self-stimulatory behaviors in autistic children are assigned to higher GABA levels in peripheral tissues (Cohen, 2001).

Altered GABAergic transmission in adults results in anxiety disorder, schizophrenia, and premenstrual dysphoric disorder (Wong et al., 2003). Gene manipulation of different subunits (α_1, α_2, α_5, α_6, β_2, β_3, γ_{2S}, γ_{2L}, and δ) of the GABA$_A$ receptor indicates that basal behaviors of surviving animals in most mutants do not change significantly. This could be because other subunits may compensate for the loss, as manipulation of specific subunit gene expression is associated with altered expression of other subunits (Nusser et al., 1999; O'Meara et al., 2004; Ramadan et al., 2003). Nevertheless, when challenged by drugs or alcohol, many of the mutants behave differently than the control animals, suggesting that compensatory mechanisms work but the threshold level for tolerance in response to stress is reduced in the mutants (Boehm et al., 2004).

Mice deficient in the α_1 subunit of GABA$_A$ receptors display handling induced tremor (Kralic et al., 2005) and α_2 subunit knock-out mice display a lower basal level of locomotion (Boehm et al., 2004). Studies with α_5 knock-out and knock-in mice suggest that α_5-receptor subunits mediate hippocampal dependent learning processes (Collinson et al., 2002; Crestani et al., 2002). Deletion of β_2 subunit results in higher locomotor activity with no signs of motor dysfunction (Sur et al., 2001), whereas β_3 subunit knock-out mice display a wide range of abnormal behaviors, including hyperactivity, poor motor coordination, learning and memory problems, hypersensitivity to sensory stimuli consistent with Angelman syndrome, and a subtype of autism (Buxbaum et al., 2002; DeLorey et al., 1998; Nurmi et al., 2001). Studies with γ_2 heterozygous mice implicate that a γ_2 dysfunction predisposes animals to anxiety (Crestani et al., 1999) which remain undisturbed in δ subunit knock-out mice (Mihalek et al., 1999).

II. Alcohol and Neurodevelopmental Disorders: Role of GABA and GABA Receptors

Neurodevelopmental disorders are the outcomes of suboptimal neurotransmission resulting from anomalies in the CNS development. This may occur due to abnormal proliferation, differentiation, migration, and apoptosis of various precursors that form the heterogeneous tissues of the brain. Functional changes in the brain are derived from altered membrane fluidity and permeability, functions of enzymes, receptors, and signaling molecules, and changes in gene expression patterns in the developing structures. Alcohol influences this entire repertoire in the developing brain, and therefore is a potential neurobehavioral teratogen.

Rodent models are extensively used to understand the mechanisms of prenatal alcohol induced behavioral problems, and a large number of studies support involvements of GABA and the GABA receptor system in the process. Prenatal alcohol exposure during pregnancy and lactation increases GABA levels in the frontal cortex, olfactory bulbs, anterior colliculus, and amygdala in the rat offspring (Ledig et al., 1988). Fetal rats exposed to alcohol during gestation days 1 to 20 also have higher GABA levels in the brain (Maier et al., 1996). Therefore, many of the toxic effects of prenatal alcohol on the developing brains may derive from changes in the extracellular GABA levels and the paracrine actions it mediates through GABA receptors.

Alcohol, by its ability to alter membrane fluidity (Arienti et al., 1993, 1994; Gutierrez-Ruiz et al., 1995), may affect subunit association and stability of various receptors and ion channels (Lovinger et al., 1989; Messing et al., 1986). In neocortical slices from chick, mice, and rat brains and in rat hippocampal CA1 neurons, the $GABA_A$ receptor mediated membrane current is potentiated by alcohol (Reynolds et al., 1992; Weiner et al., 1994).

Gestational alcohol exposure may disturb GABA receptor function by changing the optimum number and affinity of receptors during development. Extensive studies have been done to examine this possibility. Two doses of alcohol on gestation day 8 alter spatial learning of offspring and increase low affinity GABA receptor numbers in the brain (Minetti et al., 1996). Chronic alcohol exposure during gestation increases $GABA_A$-benzodiazepine receptor number and pharmacology in adult guinea pig cerebral cortex (Bailey et al., 1999). Proportions of GABAergic neurons are altered in layers II/III somatosensory cortex of guinea pigs following gestational exposure to alcohol (Bailey et al., 2004). Chronic exposure to prenatal alcohol increases $GABA_A$ receptor $\beta_{2/3}$-subunit protein expression, reduces the growth of the hippocampus, and causes behavioral dysfunction in young adult offspring of guinea pigs that include changes in spontaneous locomotor activity, cognitive deficits, and impaired spatial learning (Iqbal et al., 2004). Another study also reported that the chronic gestational exposure to alcohol in guinea pigs decreases α_1 and $\beta_{2/3}$-subunit protein expression in the cerebral cortex of adult offspring in association with the increase in locomotor activity (Bailey et al., 2001). The cerebral cortical weight is also reduced.

In adult rats, chronic alcohol exposure decreases binding of benzodiazepine with the $GABA_A$ receptor assemblies derived from α_2, α_3, and γ_3 in the cerebral cortex (Mehta and Ticku, 2005). This alcohol sensitivity of α_2, α_3, and γ_2 subunit assemblies in the brain is a consistent observation (Mhatre and Ticku, 1992; Mhatre et al., 1993; Montpied et al., 1991), and it is suggested that this could be due to changes in the trafficking of the subunits to the cell surface as observed for the α_1 subunit in the cerebral cortex following alcohol exposure (Kumar et al., 2003). All these studies strongly indicate that long-term behavioral effects of gestational alcohol may be due to changes in subunit composition, number, and distribution of GABA receptors in different parts of the brain. There is evidence that the effects of chronic prenatal alcohol on the functions of $GABA_A$ may be different in different brain regions, which would contribute to the complex pattern of cognitive and behavioral dysfunction (Allan *et al.*, 1998).

Exposure to alcohol may change post-translational modification of the $GABA_A$ receptor, affecting allosteric coupling between binding sites and changing receptor function. Intracellular domains of many $GABA_A$ receptors subunits contain a number of consensus sites for serine/threonine and tyrosine protein kinases. Protein kinase A, protein kinase C, tyrosine kinases, cyclic guanosine monophosphate-dependent protein kinase, and calcium/calmodulin dependent protein kinase II modulate $GABA_A$ receptor function by phosphorylating different serine, threonine, and tyrosine residues differently in different subunits of $GABA_A$ receptors (Kumar *et al.*, 2004; Leidenheimer *et al.*, 1991; Stelzer, 1992). For instance, phosphorylation of both serine at the 408 and 409 positions of the β_3 subunit enhances receptor function, whereas phosphorylation of serine at the 408 position reduces channel functions (Moss *et al.*, 1992). Phosphorylation events in association with the differences in subunit composition of receptors may also change receptor function. Phosphorylation of tyrosine residues at the 365 and 367 positions of γ_2 subunit by tyrosine kinase Src with coexpression of α_1 and β_2 causes enhancement of the $GABA_A$ receptor function (Moss *et al.*, 1995).

Several effects of alcohol on the body are due to post translation modification of proteins, such as protein phosphorylation (Mahadev and Vemuri, 1999; Mahadev *et al.*, 2001). Chronic alcohol administration in adult rats alters association of $PKC\gamma$ with the α_1-subunit containing $GABA_A$ receptors in the cerebral cortical membranes (Kumar *et al.*, 2004). Alcohol-induced changes in the phosphorylation of GABA receptor subunit/s in the embryonic brain therefore may change the receptor functions and the developmental events regulated by these receptors.

During the transition of the excitatory function of $GABA_A$ receptors in the developing brain into a predominantly inhibitory function in the mature brain various subunits undergo functional rearrangements to modify the responsiveness of receptors to GABA (Fritschy *et al.*, 1994; Laurie *et al.*, 1992). This process is believed to be regulated by the receptor activity itself because the receptor maturation is vulnerable to disruption by GABA-mimetic agents and antagonists that disturb $GABA_A$ receptor composition, sensitivity, and function (Belhage *et al.*, 1990; Bitran *et al.*, 1991; Elster *et al.*, 1995). Alcohol shares anxiolytic, sedative-hypnotic, and anticonvulsant properties with benzodiazepines and barbiturates (Reynolds *et al.*, 1992; Weiner *et al.*, 1994), and therefore exposure to alcohol during the GABAergic shift may disturb the normal pattern of subunit repertoire and assembly, causing permanent changes in receptor function. This may cause long-lasting behavioral changes.

Synapse formation may be affected by change in the positioning of GABAergic neurons that may result from disruption of neuronal migration induced by gestation alcohol. It is extensively documented that prenatal alcohol alters the migration of neuronal population in rodents (Hirai *et al.*, 1999; Siegenthaler and Miller, 2004). Anatomical studies of human brains exposed to alcohol during gestation also show ectopic neurons (Clarren *et al.*, 1978). Now, there is evidence that prenatal alcohol reduces GABAergic interneurons in layers II/III of the somatosensory cortex of the adult guinea pig (Bailey *et al.*, 2004), and studies with postmortem human brains provide evidence that defects in GABAergic neurotransmission resulting from disturbances in the neuronal migration and defective lamination may cause schizophrenia and bipolar disorders (Benes and Berretta, 2001).

Alcohol during synaptogenesis and a brain growth spurt, a period equivalent to the third trimester in humans, causes massive cell death in rodent models (Olney *et al.*, 2002a,b). Similar incidence occurs when pups are exposed to NMDA antagonists and GABA receptor agonists (Olney *et al.*, 2002c). Based on these observations, a dual mechanism involving blockade of NMDA glutamate receptor and hyperactivation of $GABA_A$ receptors is proposed for apoptotic death of cells. Potentiation of $GABA_A$ receptor activity in hippocampal cells from postnatal days 1–3 rats with neurosteroid $(3\alpha, 5\alpha)$-3-hydroxypegnan-20-one or with benzodiazepines mimics previously mentioned phenomenon, suggesting that alcohol-induced death of brain cells in third trimester is mediated by the activation of $GABA_A$ receptors (Xu *et al.*, 2000). Deletion of large numbers of cells at this time of development in the brain by alcohol may diminish brain size and cause behavioral problems (Farber and Olney, 2003).

Concordance amongst GABA receptor β_3 subunit mutation, cleft palate, autism, and FAS (Buxbaum *et al.*, 2002; Culiat *et al.*, 1993; Gordon, 1993; McCauley *et al.*, 2004; Miles *et al.*, 2003; Nanson, 1992) strongly implies that the $GABA_A$ receptor β_3 subunits may be a common mediator for these disorders.

Several effects of alcohol are mediated by $GABA_B$ receptors in adult rodent brains. The alcohol-induced locomotor stimulation, hyperexcitability, and inhibition of specific cerebral Purkinje neurons are modulated by $GABA_B$ receptors (Humeniuk *et al.*, 1993; Mead and Little, 1995; Yang *et al.*, 2000). Moreover, $GABA_B$ receptors modulate alcohol sensitivity of $GABA_A$ receptor mediated inhibitory postsynaptic currents (Wu *et al.*, 2005). These data hint that neurodevelopmental problems caused by alcohol may also be mediated by $GABA_B$ receptors, but experimental evidence on this possibility is currently missing. Data describing involvement of $GABA_C$ receptors in alcohol effects on the brain is also scarce. Studies examining possible involvement of $GABA_B$ and $GABA_C$ receptors in the alcohol-induced neurodevelopmental disorders are sought to elucidate further the role of GABA and the GABA-receptor system in the process.

III. Conclusions

It is evident that GABA and GABA receptors play a significant role in the development of the brain and disturbance in this system during this time of life may result in long-lasting behavior problems. Suboptimal synapse formation resulting from inadequate number and positioning of cells with changed receptor function in alcohol exposed brains is the primary reason for the neurodevelopmental disorders. So far, most studies are done to examine the involvement of GABA systems in this process using a chronic alcohol exposure paradigm. Unfortunately, this approach does not elucidate the mechanisms of alcohol effect on GABA and the GABA-receptor system because development of the brain is a changing process and alcohol may affect this process differently depending upon the time and dosage of exposure. Examining effects of alcohol at specific times of development will facilitate understanding the progression of pathology better.

Abstaining from alcohol is the best way to avoid the evils of alcohol on the fetus but habits of alcohol drinking have not improved since the outcomes became known. Because alcohol's behavioral effects are derived significantly from altered GABA and GABA receptor function, it is imperative that this information be used to develop drugs against alcohol's effects on the developing brain. It is encouraging to know that $GABA_B$ receptor agonists are now used against spasticity, gastro-oesophageal reflux disease, and addiction (Couve et al., 2000; Vacher and Bettler, 2003) and currently, $GABA_B$ antagonists are on trial to improve cognitive improvements of Alzheimer's disease (Couve et al., 2004). These studies assure feasibility of targeting specific molecule of GABA system in the body. Thus, GS39783 that exerts anxiolytic effects by positive modulation of $GABA_B$ receptors in animal models (Couve et al., 2004) may also be tried to prevent behavior problems in subjects exposed to alcohol during development. Similar studies are required with the $GABA_A$ receptor system because a majority of studies show involvements of $GABA_A$ receptors in the alcohol-induced neuro-developmental disorders. It is promising that compound RY064, which has high a affinity for $GABA_A$ receptor containing α_5 subunit, is being used to antagonize alcohol's neurobehavioral effects in animal models (McKay et al., 2004). This study is yet to be extended in the developmental context.

Vasoactive neuroprotective peptides inhibit prenatal alcohol induced brain malformation in rodent models (Poggi et al., 2003; Spong et al., 2001). Follow up studies examining the effects of these peptides on the GABA system and the behavioral outcomes in offspring are warranted to understand the mechanism better and develop drugs to prevent alcohol's toxic effects on the developing brain. Studies on the effects of other neuroprotective compounds such as RU486, Go6976, secretin, and corticotrophin releasing factor (Ghoumari et al., 2003; Koves et al., 2004; Madtes et al., 2004) on GABA system and behavioral outcomes of gestational alcohol may aid in finding additional ways to reduce occurrence of neurodevelopmental disorders.

Acknowledgment

Dr. Kakooza-Mwesige Angelina is gratefully acknowledged for her suggestions in improving this article.

References

Allan, A. M., Wu, H., Paxton, L. L., and Savage, D. D. (1998). Prenatal ethanol exposure alters the modulation of the gamma-aminobutyric acidA1 receptor-gated chloride ion channel in adult rat offspring. *J. Pharmacol. Exp. Ther.* **284,** 250–257.

Arienti, G., Battistoni, F., Carlini, E., Di Renzo, G. C., Cosmi, E. V., and Reboldi, G. P. (1994). The effect of maternal ethanol intoxication on erythrocyte ghost fluidity in new-born rat pups. *Biochem. Mol. Biol. Int.* **33,** 1–8.

Arienti, G., Di Renzo, G. C., Cosmi, E. V., Carlini, E., and Corazzi, L. (1993). Rat brain microsome fluidity as modified by prenatal ethanol administration. *Neurochem. Res.* **18,** 335–338.

Asada, H., Kawamura, Y., Maruyama, K., Kume, H., Ding, R., Ji, F. Y., Kanbara, N., Kuzume, H., Sanbo, M., Yagi, T., and Obata, K. (1996). Mice lacking the 65 kDa isoform of glutamic acid decarboxylase (GAD65) maintain normal levels of GAD67 and GABA in their brains but are susceptible to seizures. *Biochem. Biophys. Res. Commun.* **229,** 891–895.

Bailey, C. D., Brien, J. F., and Reynolds, J. N. (1999). Altered GABA(A)-benzodiazepine receptor number and pharmacology in the adult guinea pig cerebral cortex after chronic prenatal ethanol exposure. *Alcohol. Clin. Exp. Res.* **23,** 1816–1824.

Bailey, C. D., Brien, J. F., and Reynolds, J. N. (2001). Chronic prenatal ethanol exposure increases GABA(A) receptor subunit protein expression in the adult guinea pig cerebral cortex. *J. Neurosci.* **21,** 4381–4389.

Bailey, C. D., Brien, J. F., and Reynolds, J. N. (2004). Chronic prenatal ethanol exposure alters the proportion of GABAergic neurons in layers II/III of the adult guinea pig somatosensory cortex. *Neurotoxicol. Teratol.* **26,** 59–63.

Barakat, L., and Bordey, A. (2002). GAT-1 and reversible GABA transport in Bergmann glia in slices. *J. Neurophysiol.* **88,** 1407–1419.

Barbin, G., Pollard, H., Gaiarsa, J. L., and Ben-Ari, Y. (1993). Involvement of GABAA receptors in the outgrowth of cultured hippocampal neurons. *Neurosci. Lett.* **152,** 150–154.

Barker, J. L., Behar, T., Li, Y. X., Liu, Q. Y., Ma, W., Maric, D., Maric, I., Schaffner, A. E., Serafini, R., Smith, S. V., Somogyi, R., Vautrin, J. Y., Wen, X. L., and Xian, H. (1998). GABAergic cells and signals in CNS development. *Perspect. Dev. Neurobiol.* **5,** 305–322.

Barnes, E. M., Jr. (2000). Intracellular trafficking of GABA(A) receptors. *Life Sci.* **66,** 1063–1070.

Behar, T. N., Li, Y. X., Tran, H. T., Ma, W., Dunlap, V., Scott, C., and Barker, J. L. (1996). GABA stimulates chemotaxis and chemokinesis of embryonic cortical neurons via calcium-dependent mechanisms. *J. Neurosci.* **16,** 1808–1818.

Behar, T. N., Schaffner, A. E., Scott, C. A., Greene, C. L., and Barker, J. L. (2000). GABA receptor antagonists modulate postmitotic cell migration in slice cultures of embryonic rat cortex. *Cereb. Cortex.* **10,** 899–909.

Behar, T. N., Schaffner, A. E., Scott, C. A., O'Connell, C., and Barker, J. L. (1998). Differential response of cortical plate and ventricular zone cells to GABA as a migration stimulus. *J. Neurosci.* **18,** 6378–6387.

Behar, T. N., Scott, C. A., Greene, C. L., Wen, X., Smith, S. V., Maric, D., Liu, Q. Y., Colton, C. A., and Barker, J. L. (1999). Glutamate acting at NMDA receptors stimulates embryonic cortical neuronal migration. *J. Neurosci.* **19,** 4449–4461.

Behar, T. N., Smith, S. V., Kennedy, R. T., McKenzie, J. M., Maric, I., and Barker, J. L. (2001). GABA(B) receptors mediate motility signals for migrating embryonic cortical cells. *Cereb. Cortex* **11,** 744–753.

Belhage, B., Hansen, G. H., and Schousboe, A. (1990). GABA agonist induced changes in ultrastructure and GABA receptor expression in cerebellar granule cells is linked to hyperpolarization of the neurons. *Int. J. Dev. Neurosci.* **8,** 473–479.

Ben-Ari, Y. (2002). Excitatory actions of gaba during development: The nature of the nurture. *Nat. Rev. Neurosci.* **3,** 728–739.

Ben-Ari, Y., Cherubini, E., Corradetti, R., and Gaiarsa, J. L. (1989). Giant synaptic potentials in immature rat CA3 hippocampal neurones. *J. Physiol.* **416,** 303–325.

Benes, F. M., and Berretta, S. (2001). GABAergic interneurons: Implications for understanding schizophrenia and bipolar disorder. *Neuropsychopharmacology* **25,** 1–27.

Berninger, B., Marty, S., Zafra, F., da Penha Berzaghi, M., Thoenen, H., and Lindholm, D. (1995). GABAergic stimulation switches from enhancing to repressing BDNF expression in rat hippocampal neurons during maturation *in vitro. Development* **121,** 2327–2335.

Bhatara, V. S., Lovrein, F., Kirkeby, J., Swayze, V., 2nd., Unruh, E., and Johnson, V. (2002). Brain function in fetal alcohol syndrome assessed by single photon emission computed tomography. *S. D. J. Med.* **55,** 59–62.

Billinton, A., Ige, A. O., Bolam, J. P., White, J. H., Marshall, F. H., and Emson, P. C. (2001). Advances in the molecular understanding of GABA(B) receptors. *Trends Neurosci.* **24,** 277–282.

Bitran, D., Primus, R. J., and Kellogg, C. K. (1991). Gestational exposure to diazepam increases sensitivity to convulsants that act at the GABA/benzodiazepine receptor complex. *Eur. J. Pharmacol.* **196,** 223–231.

Boehm, S. L., 2nd., Ponomarev, I., Jennings, A. W., Whiting, P. J., Rosahl, T. W., Garrett, E. M., Blednov, Y. A., and Harris, R. A. (2004). gamma-Aminobutyric acid A receptor subunit mutant mice: New perspectives on alcohol actions. *Biochem. Pharmacol.* **68,** 1581–1602.

Bolteus, A. J., and Bordey, A. (2004). GABA release and uptake regulate neuronal precursor migration in the postnatal subventricular zone. *J. Neurosci.* **24,** 7623–7631.

Borodinsky, L. N., O'Leary, D., Neale, J. H., Vicini, S., Coso, O. A., and Fiszman, M. L. (2003). GABA-induced neurite outgrowth of cerebellar granule cells is mediated by GABA(A) receptor activation, calcium influx and CaMKII and erk1/2 pathways. *J. Neurochem.* **84,** 1411–1420.

Bowery, N. G., Bettler, B., Froestl, W., Gallagher, J. P., Marshall, F., Raiteri, M., Bonner, T. I., and Enna, S. J. (2002). International Union of Pharmacology. XXXIII. Mammalian gamma-aminobutyric acid(B) receptors: Structure and function. *Pharmacol. Rev.* **54,** 247–264.

Brambilla, P., Perez, J., Barale, F., Schettini, G., and Soares, J. C. (2003). GABAergic dysfunction in mood disorders. *Mol. Psychiatry* **8,** 715, 721–737.

Buxbaum, J. D., Silverman, J. M., Smith, C. J., Greenberg, D. A., Kilifarski, M., Reichert, J., Cook, E. H., Jr., Fang, Y., Song, C. Y., and Vitale, R. (2002). Association between a GABRB3 polymorphism and autism. *Mol. Psychiatry* **7,** 311–316.

Chebib, M. (2004). GABAC receptor ion channels. *Clin. Exp. Pharmacol. Physiol.* **31,** 800–804.

Chebib, M., and Johnston, G. A. (1999). The 'ABC' of GABA receptors: A brief review. *Clin. Exp. Pharmacol. Physiol.* **26,** 937–940.

Church, M. W., and Kaltenbach, J. A. (1997). Hearing, speech, language, and vestibular disorders in the fetal alcohol syndrome: A literature review. *Alcohol. Clin. Exp. Res.* **21,** 495–512.

Clarren, S. K., Alvord, E. C., Jr., Sumi, S. M., Streissguth, A. P., and Smith, D. W. (1978). Brain malformations related to prenatal exposure to ethanol. *J. Pediatr.* **92,** 64–67.

Cohen, B. I. (2001). GABA-transaminase, the liver and infantile autism. *Med. Hypotheses* **57,** 673–674.

Collinson, N., Kuenzi, F. M., Jarolimek, W., Maubach, K. A., Cothliff, R., Sur, C., Smith, A., Otu, F. M., Howell, O., Atack, J. R., McKernan, R. M., Seabrook, G. R., Dawson, G. R., Whiting, P. J., and Rosahl, T. W. (2002). Enhanced learning and memory and altered GABAergic synaptic transmission in mice lacking the alpha 5 subunit of the GABAA receptor. *J. Neurosci.* **22,** 5572–5580.

Conry, J. (1990). Neuropsychological deficits in fetal alcohol syndrome and fetal alcohol effects. *Alcohol. Clin. Exp. Res.* **14,** 650–655.

Couve, A., Calver, A. R., Fairfax, B., Moss, S. J., and Pangalos, M. N. (2004). Unravelling the unusual signalling properties of the GABA(B) receptor. *Biochem. Pharmacol.* **68,** 1527–1536.

Couve, A., Moss, S. J., and Pangalos, M. N. (2000). GABAB receptors: A new paradigm in G protein signaling. *Mol. Cell Neurosci.* **16,** 296–312.

Crestani, F., Keist, R., Fritschy, J. M., Benke, D., Vogt, K., Prut, L., Bluthmann, H., Mohler, H., and Rudolph, U. (2002). Trace fear conditioning involves hippocampal alpha5 GABA(A) receptors. *Proc. Natl. Acad. Sci. USA* **99,** 8980–8985.

Crestani, F., Lorez, M., Baer, K., Essrich, C., Benke, D., Laurent, J. P., Belzung, C., Fritschy, J. M., Luscher, B., and Mohler, H. (1999). Decreased GABAA-receptor clustering results in enhanced anxiety and a bias for threat cues. *Nat. Neurosci.* **2,** 833–839.

Culiat, C. T., Stubbs, L., Nicholls, R. D., Montgomery, C. S., Russell, L. B., Johnson, D. K., and Rinchik, E. M. (1993). Concordance between isolated cleft palate in mice and alterations within a region including the gene encoding the beta 3 subunit of the type A gamma-aminobutyric acid receptor. *Proc. Natl. Acad. Sci. USA* **90,** 5105–5109.

Dellovade, T. L., Davis, A. M., Ferguson, C., Sieghart, W., Homanics, G. E., and Tobet, S. A. (2001). GABA influences the development of the ventromedial nucleus of the hypothalamus. *J. Neurobiol.* **49,** 264–276.

DeLorey, T. M., Handforth, A., Anagnostaras, S. G., Homanics, G. E., Minassian, B. A., Asatourian, A., Fanselow, M. S., Delgado-Escueta, A., Ellison, G. D., and Olsen, R. W. (1998). Mice lacking the beta3 subunit of the GABAA receptor have the epilepsy phenotype and many of the behavioral characteristics of Angelman syndrome. *J. Neurosci.* **18,** 8505–8514.

Demarque, M., Represa, A., Becq, H., Khalilov, I., Ben-Ari, Y., and Aniksztejn, L. (2002). Paracrine intercellular communication by a Ca2+- and SNARE-independent release of GABA and glutamate prior to synapse formation. *Neuron* **36,** 1051–1061.

Dhossche, D., Applegate, H., Abraham, A., Maertens, P., Bland, L., Bencsath, A., and Martinez, J. (2002). Elevated plasma gamma-aminobutyric acid (GABA) levels in autistic youngsters: Stimulus for a GABA hypothesis of autism. *Med. Sci. Monit.* **8,** PR1–PR6.

Elster, L., Hansen, G. H., Belhage, B., Fritschy, J. M., Mohler, H., and Schousboe, A. (1995). Differential distribution of GABAA receptor subunits in soma and processes of cerebellar granule cells: Effects of maturation and a GABA agonist. *Int. J. Dev. Neurosci.* **13,** 417–428.

Enna, S. J., Stern, L. Z., Wastek, G. J., and Yamamura, H. I. (1977). Cerebrospinal fluid gamma-aminobutyric acid variations in neurological disorders. *Arch. Neurol.* **34,** 683–685.

Erdo, S. L., and Wolff, J. R. (1990). gamma-Aminobutyric acid outside the mammalian brain. *J. Neurochem.* **54,** 363–372.

Erlander, M. G., Tillakaratne, N. J., Feldblum, S., Patel, N., and Tobin, A. J. (1991). Two genes encode distinct glutamate decarboxylases. *Neuron* **7,** 91–100.

Famy, C., Streissguth, A. P., and Unis, A. S. (1998). Mental illness in adults with fetal alcohol syndrome or fetal alcohol effects. *Am. J. Psychiatry* **155,** 552–554.

Farber, N. B., and Olney, J. W. (2003). Drugs of abuse that cause developing neurons to commit suicide. *Brain Res. Dev. Brain Res.* **147,** 37–45.

Feigenspan, A., Wassle, H., and Bormann, J. (1993). Pharmacology of GABA receptor Cl-channels in rat retinal bipolar cells. *Nature* **361,** 159–162.

Fiszman, M. L., Borodinsky, L. N., and Neale, J. H. (1999). GABA induces proliferation of immature cerebellar granule cells grown *in vitro*. *Brain Res. Dev. Brain Res.* **115,** 1–8.

Flint, A. C., Liu, X., and Kriegstein, A. R. (1998). Nonsynaptic glycine receptor activation during early neocortical development. *Neuron* **20,** 43–53.

Fritschy, J. M., Meskenaite, V., Weinmann, O., Honer, M., Benke, D., and Mohler, H. (1999). GABAB-receptor splice variants GB1a and GB1b in rat brain: Developmental regulation, cellular distribution and extrasynaptic localization. *Eur. J. Neurosci.* **11,** 761–768.

Fritschy, J. M., Paysan, J., Enna, A., and Mohler, H. (1994). Switch in the expression of rat GABAA-receptor subtypes during postnatal development: An immunohistochemical study. *J. Neurosci.* **14,** 5302–5324.

Fueshko, S. M., Key, S., and Wray, S. (1998). GABA inhibits migration of luteinizing hormone-releasing hormone neurons in embryonic olfactory explants. *J. Neurosci.* **18,** 2560–2569.

Ganguly, K., Schinder, A. F., Wong, S. T., and Poo, M. (2001). GABA itself promotes the developmental switch of neuronal GABAergic responses from excitation to inhibition. *Cell* **105,** 521–532.

Ghoumari, A. M., Dusart, I., El-Etr, M., Tronche, F., Sotelo, C., Schumacher, M., and Baulieu, E. E. (2003). Mifepristone (RU486) protects Purkinje cells from cell death in organotypic slice cultures of postnatal rat and mouse cerebellum. *Proc. Natl. Acad. Sci. USA* **100,** 7953–7958.

Gordon, A. G. (1993). Alcohol, deafness, epilepsy, and autism. *Alcohol. Clin. Exp. Res.* **17,** 926–928.

Gordon-Weeks, P. R., Lockerbie, R. O., and Pearce, B. R. (1984). Uptake and release of [3H]GABA by growth cones isolated from neonatal rat brain. *Neurosci. Lett.* **52,** 205–210.

Gutierrez-Ruiz, M. C., Gomez, J. L., Souza, V., and Bucio, L. (1995). Chronic and acute ethanol treatment modifies fluidity and composition in plasma membranes of a human hepatic cell line (WRL-68). *Cell Biol. Toxicol.* **11,** 69–78.

Hannigan, J. H., Blanchard, B. A., and Riley, E. P. (1987). Altered grooming responses to stress in rats exposed prenatally to ethanol. *Behav. Neural. Biol.* **47,** 173–185.

Haydar, T. F., Wang, F., Schwartz, M. L., and Rakic, P. (2000). Differential modulation of proliferation in the neocortical ventricular and subventricular zones. *J. Neurosci.* **20,** 5764–5774.

Hirai, K., Yoshioka, H., Kihara, M., Hasegawa, K., Sawada, T., and Fushiki, S. (1999). Effects of ethanol on neuronal migration and neural cell adhesion molecules in the embryonic rat cerebral cortex: A tissue culture study. *Brain Res. Dev. Brain Res.* **118,** 205–210.

Humeniuk, R. E., White, J. M., and Ong, J. (1993). The role of GABAB receptors in mediating the stimulatory effects of ethanol in mice. *Psychopharmacology (Berl)* **111,** 219–224.

Iqbal, U., Dringenberg, H. C., Brien, J. F., and Reynolds, J. N. (2004). Chronic prenatal ethanol exposure alters hippocampal GABA(A) receptors and impairs spatial learning in the guinea pig. *Behav. Brain Res.* **150,** 117–125.

Ji, F., Kanbara, N., and Obata, K. (1999). GABA and histogenesis in fetal and neonatal mouse brain lacking both the isoforms of glutamic acid decarboxylase. *Neurosci Res* **33,** 187–194.

Johnson, V. P., Swayze, V. W., II, Sato, Y., and Andreasen, N. C. (1996). Fetal alcohol syndrome: Craniofacial and central nervous system manifestations. *Am. J. Med. Genet.* **61,** 329–339.

Johnston, G. A. (1996). GABAc receptors. Relatively simple transmitter -gated ion channels? *Trends Pharmacol. Sci.* **17,** 319–323.

Jones, K. L., Smith, D. W., Streissguth, A. P., and Myrianthopoulos, N. C. (1974). Outcome in offspring of chronic alcoholic women. *Lancet* **1,** 1076–1078.

Jones, K. L., Smith, D. W., Ulleland, C. N., and Streissguth, P. (1973). Pattern of malformation in offspring of chronic alcoholic mothers. *Lancet* **1,** 1267–1271.

Kawakami, S., Uezono, Y., Makimoto, N., Enjoji, A., Kaibara, M., Kanematsu, T., and Taniyama, K. (2004). Characterization of GABA(B) receptors involved in inhibition of motility associated with acetylcholine release in the dog small intestine: Possible existence of a heterodimer of GABA (B1) and GABA(B2) subunits. *J. Pharmacol. Sci.* **94,** 368–375.

Kerr, D. I., and Ong, J. (1992). GABA agonists and antagonists. *Med. Res. Rev.* **12,** 593–636.

Kim, M. O., Li, S., Park, M. S., and Hornung, J. P. (2003). Early fetal expression of GABA(B1) and GABA(B2) receptor mRNAs on the development of the rat central nervous system. *Brain Res. Dev. Brain Res.* **143,** 47–55.

Komuro, H., and Rakic, P. (1998). Orchestration of neuronal migration by activity of ion channels, neurotransmitter receptors, and intracellular Ca2+ fluctuations. *J. Neurobiol.* **37,** 110–130.

Koves, K., Kausz, M., Reser, D., Illyes, G., Takacs, J., Heinzlmann, A., Gyenge, E., and Horvath, K. (2004). Secretin and autism: A basic morphological study about the distribution of secretin in the nervous system. *Regul. Pept.* **123,** 209–216.

Kralic, J. E., Criswell, H. E., Osterman, J. L., O'Buckley, T. K., Wilkie, M. E., Matthews, D. B., Hamre, K., Breesc, G. R., Homanics, G. E., and Morrow, A. L. (2005). Genetic essential tremor in gamma-aminobutyric acidA receptor alpha1 subunit knockout mice. *J. Clin. Invest.* **115,** 774–779.

Krogsgaard-Larsen, P., Frolund, B., Jorgensen, F. S., and Schousboe, A. (1994). GABAA receptor agonists, partial agonists, and antagonists. Design and therapeutic prospects. *J. Med. Chem.* **37,** 2489–2505.

Kumar, S., Fleming, R. L., and Morrow, A. L. (2004). Ethanol regulation of gamma-aminobutyric acid A receptors: Genomic and nongenomic mechanisms. *Pharmacol. Ther.* **101,** 211–226.

Kumar, S., Kralic, J. E., O'Buckley, T. K., Grobin, A. C., and Morrow, A. L. (2003). Chronic ethanol consumption enhances internalization of alpha1 subunit-containing GABAA receptors in cerebral cortex. *J. Neurochem.* **86,** 700–708.

Kusama, T., Spivak, C. E., Whiting, P., Dawson, V. L., Schaeffer, J. C., and Uhl, G. R. (1993). Pharmacology of GABA rho 1 and GABA alpha/beta receptors expressed in Xenopus oocytes and COS cells. *Br. J. Pharmacol.* **109,** 200–206.

Laurie, D. J., Seeburg, P. H., and Wisden, W. (1992). The distribution of 13 GABAA receptor subunit mRNAs in the rat brain. II. Olfactory bulb and cerebellum. *J. Neurosci.* **12,** 1063–1076.

Ledig, M., Ciesielski, L., Simler, S., Lorentz, J. G., and Mandel, P. (1988). Effect of pre- and postnatal alcohol consumption on GABA levels of various brain regions in the rat offspring. *Alcohol. Alcohol.* **23,** 63–67.

Leidenheimer, N. J., Browning, M. D., and Harris, R. A. (1991). GABAA receptor phosphorylation: Multiple sites, actions and artifacts. *Trends Pharmacol. Sci.* **12,** 84–87.

Levy, L. M., Levy-Reis, I., Fujii, M., and Dalakas, M. C. (2005). Brain gamma-aminobutyric acid changes in stiff-person syndrome. *Arch. Neurol.* **62,** 970–974.

Li, S., Park, M. S., and Kim, M. O. (2004). Prenatal alteration and distribution of the GABA(B1) and GABA(B2) receptor subunit mRNAs during rat central nervous system development. *Brain Res. Dev. Brain Res.* **150,** 141–150.

Lopez-Bendito, G., Lujan, R., Shigemoto, R., Ganter, P., Paulsen, O., and Molnar, Z. (2003). Blockade of GABA(B) receptors alters the tangential migration of cortical neurons. *Cereb. Cortex* **13,** 932–942.

LoTurco, J. J., Owens, D. F., Heath, M. J., Davis, M. B., and Kriegstein, A. R. (1995). GABA and glutamate depolarize cortical progenitor cells and inhibit DNA synthesis. *Neuron* **15,** 1287–1298.

Lovinger, D. M., White, G., and Weight, F. F. (1989). Ethanol inhibits NMDA-activated ion current in hippocampal neurons. *Science* **243,** 1721–1724.

Lu, J., Karadsheh, M., and Delpire, E. (1999). Developmental regulation of the neuronal-specific isoform of K-Cl cotransporter KCC2 in postnatal rat brains. *J. Neurobiol.* **39,** 558–568.

Ludwig, A., Li, H., Saarma, M., Kaila, K., and Rivera, C. (2003). Developmental up-regulation of KCC2 in the absence of GABAergic and glutamatergic transmission. *Eur. J. Neurosci.* **18,** 3199–3206.

Lujan, R., Shigemoto, R., and Lopez-Bendito, G. (2005). Glutamate and GABA receptor signalling in the developing brain. *Neuroscience* **130,** 567–580.

Ma, W., and Barker, J. L. (1995). Complementary expressions of transcripts encoding GAD67 and GABAA receptor alpha 4, beta 1, and gamma 1 subunits in the proliferative zone of the embryonic rat central nervous system. *J. Neurosci.* **15,** 2547–2560.

Ma, W., and Barker, J. L. (1998). GABA, GAD, and GABA(A) receptor alpha4, beta1, and gamma1 subunits are expressed in the late embryonic and early postnatal neocortical germinal matrix and coincide with gliogenesis. *Microsc. Res. Tech.* **40,** 398–407.

Macdonald, R. L., and Olsen, R. W. (1994). GABAA receptor channels. *Annu. Rev. Neurosci.* **17,** 569–602.

Madtes, P., Jr., Lee, K. H., King, J. S., and Burry, R. W. (2004). Corticotropin releasing factor enhances survival of cultured GABAergic cerebellar neurons after exposure to a neurotoxin. *Brain Res. Dev. Brain Res.* **151,** 119–128.

Mahadev, K., Chetty, C. S., and Vemuri, M. C. (2001). Effect of prenatal and postnatal ethanol exposure on Ca2+/calmodulin-dependent protein kinase II in rat cerebral cortex. *Alcohol.* **23,** 183–188.

Mahadev, K., and Vemuri, M. C. (1999). Effect of pre- and postnatal ethanol exposure on protein tyrosine kinase activity and its endogenous substrates in rat cerebral cortex. *Alcohol.* **17,** 223–229.

Maier, S. E., Chen, W. J., and West, J. R. (1996). Prenatal binge-like alcohol exposure alters neurochemical profiles in fetal rat brain. *Pharmacol. Biochem. Behav.* **55,** 521–529.

Maric, D., Liu, Q. Y., Maric, I., Chaudry, S., Chang, Y. H., Smith, S. V., Sieghart, W., Fritschy, J. M., and Barker, J. L. (2001). GABA expression dominates neuronal lineage progression in the embryonic rat neocortex and facilitates neurite outgrowth via GABA(A) autoreceptor/Cl-channels. *J. Neurosci.* **21,** 2343–2360.

Martin, S. C., Steiger, J. L., Gravielle, M. C., Lyons, H. R., Russek, S. J., and Farb, D. H. (2004). Differential expression of gamma-aminobutyric acid type B receptor subunit mRNAs in the developing nervous system and receptor coupling to adenylyl cyclase in embryonic neurons. *J. Comp. Neurol.* **473,** 16–29.

Marty, S., Berninger, B., Carroll, P., and Thoenen, H. (1996). GABAergic stimulation regulates the phenotype of hippocampal interneurons through the regulation of brain-derived neurotrophic factor. *Neuron* **16,** 565–570.

Mattson, S. N., Riley, E. P., Jernigan, T. L., Ehlers, C. L., Delis, D. C., Jones, K. L., Stern, C., Johnson, K. A., Hesselink, J. R., and Bellugi, U. (1992). Fetal alcohol syndrome: A case report of neuropsychological, MRI and EEG assessment of two children. *Alcohol. Clin. Exp. Res.* **16,** 1001–1003.

Mattson, S. N., Riley, E. P., Sowell, E. R., Jernigan, T. L., Sobel, D. F., and Jones, K. L. (1996). A decrease in the size of the basal ganglia in children with fetal alcohol syndrome. *Alcohol. Clin. Exp. Res.* **20,** 1088–1093.

Mattson, S. N., Schoenfeld, A. M., and Riley, E. P. (2001). Teratogenic effects of alcohol on brain and behavior. *Alcohol. Res. Health* **25,** 185–191.

McCauley, J. L., Olson, L. M., Delahanty, R., Amin, T., Nurmi, E. L., Organ, E. L., Jacobs, M. M., Folstein, S. E., Haines, J. L., and Sutcliffe, J. S. (2004). A linkage disequilibrium map of the 1-Mb 15q12 GABA(A) receptor subunit cluster and association to autism. *Am. J. Med. Genet. B Neuropsychiatr. Genet.* **131,** 51–59.

McKay, P. F., Foster, K. L., Mason, D., Cummings, R., Garcia, M., Williams, L. S., Grey, C., McCane, S., He, X., Cook, J. M., and June, H. L. (2004). A high affinity ligand for GABAA-receptor containing alpha5 subunit antagonizes ethanol's neurobehavioral effects in Long-Evans rats. *Psychopharmacology (Berl)* **172,** 455–462.

Mead, A. J., and Little, H. J. (1995). Do GABAB receptors have a role in causing behavioural hyperexcitability, both during ethanol withdrawal and in naive mice? *Psychopharmacology (Berl)* **117,** 232–239.

Mehta, A. K., and Ticku, M. K. (1999). An update on GABAA receptors. *Brain Res. Brain Res. Rev.* **29,** 196–217.

Mehta, A. K., and Ticku, M. K. (2005). Effect of chronic administration of ethanol on GABAA receptor assemblies derived from alpha2-, alpha3-, beta2- and gamma2-subunits in the rat cerebral cortex. *Brain Res.* **1031,** 134–137.

Messing, R. O., Carpenter, C. L., Diamond, I., and Greenberg, D. A. (1986). Ethanol regulates calcium channels in clonal neural cells. *Proc. Natl. Acad. Sci. USA* **83,** 6213–6215.

Mhatre, M. C., Pena, G., Sieghart, W., and Ticku, M. K. (1993). Antibodies specific for GABAA receptor alpha subunits reveal that chronic alcohol treatment down-regulates alpha-subunit expression in rat brain regions. *J. Neurochem.* **61,** 1620–1625.

Mhatre, M. C., and Ticku, M. K. (1992). Chronic ethanol administration alters gamma-aminobutyric acidA receptor gene expression. *Mol. Pharmacol.* **42,** 415–422.

Mihalek, R. M., Banerjee, P. K., Korpi, E. R., Quinlan, J. J., Firestone, L. L., Mi, Z. P., Lagenaur, C., Tretter, V., Sieghart, W., Anagnostaras, S. G., Sage, J. R., Fanselow, M. S., Guidotti, A., Spigelman, I., Li, Z., DeLorey, T. M., Olsen, R. W., and Homanics, G. E. (1999). Attenuated sensitivity to neuroactive steroids in gamma-aminobutyrate type A receptor delta subunit knockout mice. *Proc. Natl. Acad. Sci. USA* **96,** 12905–12910.

Miles, J. H., Takahashi, T. N., Haber, A., and Hadden, L. (2003). Autism families with a high incidence of alcoholism. *J. Autism. Dev. Disord.* **33,** 403–415.

Minetti, A., Arolfo, M. P., Virgolini, M. B., Brioni, J. D., and Fulginiti, S. (1996). Spatial learning in rats exposed to acute ethanol intoxication on gestational day 8. *Pharmacol. Biochem. Behav.* **53,** 361–367.

Montpied, P., Morrow, A. L., Karanian, J. W., Ginns, E. I., Martin, B. M., and Paul, S. M. (1991). Prolonged ethanol inhalation decreases gamma-aminobutyric acidA receptor alpha subunit mRNAs in the rat cerebral cortex. *Mol. Pharmacol.* **39,** 157–163.

Moss, S. J., Gorrie, G. H., Amato, A., and Smart, T. G. (1995). Modulation of GABAA receptors by tyrosine phosphorylation. *Nature* **377,** 344–348.

Moss, S. J., Smart, T. G., Blackstone, C. D., and Huganir, R. L. (1992). Functional modulation of GABAA receptors by cAMP-dependent protein phosphorylation. *Science* **257,** 661–665.

Nanson, J. L. (1992). Autism in fetal alcohol syndrome: A report of six cases. *Alcohol. Clin. Exp. Res.* **16,** 558–565.

Nguyen, L., Malgrange, B., Breuskin, I., Bettendorff, L., Moonen, G., Belachew, S., and Rigo, J. M. (2003). Autocrine/paracrine activation of the GABA(A) receptor inhibits the proliferation of neurogenic polysialylated neural cell adhesion molecule-positive (PSA-NCAM+) precursor cells from postnatal striatum. *J. Neurosci.* **23,** 3278–3294.

Nurmi, E. L., Bradford, Y., Chen, Y., Hall, J., Arnone, B., Gardiner, M. B., Hutcheson, H. B., Gilbert, J. R., Pericak-Vance, M. A., Copeland-Yates, S. A., Michaelis, R. C., Wassink, T. H., Santangelo, S. L., Sheffield, V. C., Piven, J., Folstein, S. E., Haines, J. L., and Sutcliffe, J. S. (2001). Linkage disequilibrium at the Angelman syndrome gene UBE3A in autism families. *Genomics* **77,** 105–113.

Nusser, Z., Ahmad, Z., Tretter, V., Fuchs, K., Wisden, W., Sieghart, W., and Somogyi, P. (1999). Alterations in the expression of GABAA receptor subunits in cerebellar granule cells after the disruption of the alpha6 subunit gene. *Eur. J. Neurosci.* **11,** 1685–1697.

Obata, K., Oide, M., and Tanaka, H. (1978). Excitatory and inhibitory actions of GABA and glycine on embryonic chick spinal neurons in culture. *Brain Res.* **144,** 179–184.

Obrietan, K., Gao, X. B., and Van Den Pol, A. N. (2002). Excitatory actions of GABA increase BDNF expression via a MAPK-CREB-dependent mechanism—a positive feedback circuit in developing neurons. *J. Neurophysiol.* **88,** 1005–1015.

Olney, J. W., Tenkova, T., Dikranian, K., Qin, Y. Q., Labruyere, J., and Ikonomidou, C. (2002a). Ethanol-induced apoptotic neurodegeneration in the developing C57BL/6 mouse brain. *Brain Res. Dev. Brain Res.* **133,** 115–126.

Olney, J. W., Wozniak, D. F., Farber, N. B., Jevtovic-Todorovic, V., Bittigau, P., and Ikonomidou, C. (2002b). The enigma of fetal alcohol neurotoxicity. *Ann. Med.* **34,** 109–119.

Olney, J. W., Wozniak, D. F., Jevtovic-Todorovic, V., Farber, N. B., Bittigau, P., and Ikonomidou, C. (2002c). Glutamate and GABA receptor dysfunction in the fetal alcohol syndrome. *Neurotox. Res.* **4,** 315–325.

O'Meara, G. F., Newman, R. J., Fradley, R. L., Dawson, G. R., and Reynolds, D. S. (2004). The GABA-A beta3 subunit mediates anaesthesia induced by etomidate. *Neuroreport* **15,** 1653–1656.

Poggi, S. H., Goodwin, K., Hill, J. M., Brenneman, D. E., Tendi, E., Schinelli, S., Abebe, D., and Spong, C. Y. (2003). The role of activity-dependent neuroprotective protein in a mouse model of fetal alcohol syndrome. *Am. J. Obstet. Gynecol.* **189,** 790–793.

Ramadan, E., Fu, Z., Losi, G., Homanics, G. E., Neale, J. H., and Vicini, S. (2003). GABA(A) receptor beta3 subunit deletion decreases alpha2/3 subunits and IPSC duration. *J. Neurophysiol.* **89,** 128–134.

Reynolds, J. N., Prasad, A., and MacDonald, J. F. (1992). Ethanol modulation of GABA receptor-activated Cl-currents in neurons of the chick, rat and mouse central nervous system. *Eur. J. Pharmacol.* **224,** 173–181.

Riley, E. P. (1990). The long-term behavioral effects of prenatal alcohol exposure in rats. *Alcohol. Clin. Exp. Res.* **14,** 670–673.

Riley, E. P., Mattson, S. N., Sowell, E. R., Jernigan, T. L., Sobel, D. F., and Jones, K. L. (1995). Abnormalities of the corpus callosum in children prenatally exposed to alcohol. *Alcohol. Clin. Exp. Res.* **19,** 1198–1202.

Rivera, C., Voipio, J., Thomas-Crusells, J., Li, H., Emri, Z., Sipila, S., Payne, J. A., Minichiello, L., Saarma, M., and Kaila, K. (2004). Mechanism of activity-dependent downregulation of the neuron-specific K-Cl cotransporter KCC2. *J. Neurosci.* **24,** 4683–4691.

Roebuck, T. M., Mattson, S. N., and Riley, E. P. (1998). A review of the neuroanatomical findings in children with fetal alcohol syndrome or prenatal exposure to alcohol. *Alcohol. Clin. Exp. Res.* **22,** 339–344.

Sampson, P. D., Streissguth, A. P., Bookstein, F. L., Little, R. E., Clarren, S. K., Dehaene, P., Hanson, J. W., and Graham, J. M., Jr. (1997). Incidence of fetal alcohol syndrome and prevalence of alcohol-related neurodevelopmental disorder. *Teratology* **56,** 317–326.

Sanacora, G., Gueorguieva, R., Epperson, C. N., Wu, Y. T., Appel, M., Rothman, D. L., Krystal, J. H., and Mason, G. F. (2004). Subtype-specific alterations of gamma-aminobutyric acid and glutamate in patients with major depression. *Arch. Gen. Psychiatry* **61,** 705–713.

Siegenthaler, J. A., and Miller, M. W. (2004). Transforming growth factor beta1 modulates cell migration in rat cortex: Effects of ethanol. *Cereb. Cortex* **14,** 791–802.

Sivilotti, L., and Nistri, A. (1991). GABA receptor mechanisms in the central nervous system. *Prog. Neurobiol.* **36,** 35–92.

Spong, C. Y., Abebe, D. T., Gozes, I., Brenneman, D. E., and Hill, J. M. (2001). Prevention of fetal demise and growth restriction in a mouse model of fetal alcohol syndrome. *J. Pharmacol. Exp. Ther.* **297,** 774–779.

Stein, V., and Nicoll, R. A. (2003). GABA generates excitement. *Neuron* **37,** 375–378.

Steinhausen, H. C., and Spohr, H. L. (1998). Long-term outcome of children with fetal alcohol syndrome: Psychopathology, behavior, and intelligence. *Alcohol. Clin. Exp. Res.* **22,** 334–338.

Stelzer, A. (1992). Intracellular regulation of GABAA-receptor function. *Ion Channels* **3,** 83–136.

Stork, O., Ji, F. Y., Kaneko, K., Stork, S., Yoshinobu, Y., Moriya, T., Shibata, S., and Obata, K. (2000). Postnatal development of a GABA deficit and disturbance of neural functions in mice lacking GAD65. *Brain Res.* **865,** 45–58.

Stork, O., Yamanaka, H., Stork, S., Kume, N., and Obata, K. (2003). Altered conditioned fear behavior in glutamate decarboxylase 65 null mutant mice. *Genes Brain Behav.* **2,** 65–70.

Streissguth, A. P., Aase, J. M., Clarren, S. K., Randels, S. P., LaDue, R. A., and Smith, D. F. (1991). Fetal alcohol syndrome in adolescents and adults. *JAMA* **265,** 1961–1967.

Streissguth, A. P., Barr, H. M., and Sampson, P. D. (1990). Moderate prenatal alcohol exposure: Effects on child IQ and learning problems at age 7 1/2 years. *Alcohol. Clin. Exp. Res.* **14,** 662–669.

Stromland, K., and Pinazo-Duran, M. D. (1994). Optic nerve hypoplasia: Comparative effects in children and rats exposed to alcohol during pregnancy. *Teratology* **50,** 100–111.

Sur, C., Wafford, K. A., Reynolds, D. S., Hadingham, K. L., Bromidge, F., Macaulay, A., Collinson, N., O'Meara, G., Howell, O., Newman, R., Myers, J., Atack, J. R., Dawson, G. R., McKernan,

R. M., Whiting, P. J., and Rosahl, T. W. (2001). Loss of the major GABA(A) receptor subtype in the brain is not lethal in mice. *J. Neurosci.* **21,** 3409–3418.

Swayze, V. W., 2nd., Johnson, V. P., Hanson, J. W., Piven, J., Sato, Y., Giedd, J. N., Mosnik, D., and Andreasen, N. C. (1997). Magnetic resonance imaging of brain anomalies in fetal alcohol syndrome. *Pediatrics* **99,** 232–240.

Ulleland, C. N. (1972). The offspring of alcoholic mothers. *Ann. N. Y. Acad. Sci.* **197,** 167–169.

Vacher, C. M., and Bettler, B. (2003). GABA(B) receptors as potential therapeutic targets. *Curr. Drug Targets CNS Neurol. Disord.* **2,** 248–259.

Wang, D. D., Krueger, D. D., and Bordey, A. (2003). GABA depolarizes neuronal progenitors of the postnatal subventricular zone via GABAA receptor activation. *J. Physiol.* **550,** 785–800.

Weiner, J. L., Zhang, L., and Carlen, P. L. (1994). Potentiation of GABAA-mediated synaptic current by ethanol in hippocampal CA1 neurons: Possible role of protein kinase C. *J. Pharmacol. Exp. Ther.* **268,** 1388–1395.

Wong, C. G., Bottiglieri, T., and Snead, O. C., 3rd. (2003). GABA, gamma-hydroxybutyric acid, and neurological disease. *Ann. Neurol.* **54**(Suppl. 6), S3–S12.

Wong, C. G., Chan, K. F., Gibson, K. M., and Snead, O. C. (2004). Gamma-hydroxybutyric acid: Neurobiology and toxicology of a recreational drug. *Toxicol. Rev.* **23,** 3–20.

Woodward, R. M., Polenzani, L., and Miledi, R. (1993). Characterization of bicuculline/baclofen-insensitive (rho-like) gamma-aminobutyric acid receptors expressed in Xenopus oocytes. II. Pharmacology of gamma-aminobutyric acidA and gamma-aminobutyric acidB receptor agonists and antagonists. *Mol. Pharmacol.* **43,** 609–625.

Wu, P. H., Poelchen, W., and Proctor, W. R. (2005). Differential GABAB receptor modulation of ethanol effects on GABA(A) synaptic activity in hippocampal CA1 neurons. *J. Pharmacol. Exp. Ther.* **312,** 1082–1089.

Xu, W., Cormier, R., Fu, T., Covey, D. F., Isenberg, K. E., Zorumski, C. F., and Mennerick, S. (2000). Slow death of postnatal hippocampal neurons by GABA(A) receptor overactivation. *J. Neurosci.* **20,** 3147–3156.

Yan, X. X., Cariaga, W. A., and Ribak, C. E. (1997). Immunoreactivity for GABA plasma membrane transporter, GAT-1, in the developing rat cerebral cortex: Transient presence in the somata of neocortical and hippocampal neurons. *Brain Res. Dev. Brain Res.* **99,** 1–19.

Yang, X., Criswell, H. E., and Breese, G. R. (2000). Ethanol modulation of gamma-aminobutyric acid (GABA)-mediated inhibition of cerebellar Purkinje neurons: Relationship to GABAb receptor input. *Alcohol. Clin. Exp. Res.* **24,** 682–690.

Zecevic, N., and Milosevic, A. (1997). Initial development of gamma-aminobutyric acid immunoreactivity in the human cerebral cortex. *J. Comp. Neurol.* **380,** 495–506.

EFFECTS OF SECRETIN ON EXTRACELLULAR GABA AND OTHER AMINO ACID CONCENTRATIONS IN THE RAT HIPPOCAMPUS

Hans-Willi Clement, Alexander Pschibul, and Eberhard Schulz

Department of Child and Adolescent Psychiatry, University of Freiburg
Freiburg D-79104, Germany

In 1998, Horvath *et al.* (1998) observed a marked improvement in speech, eye contact, and attention in autistic children 5 weeks after treatment with secretin, during an endoscopic investigation. We investigated the *in vivo* effects of secretin on extracellular amino acids in the rat brain. Studies were carried out on freely moving rats with microdialysis probes in the hippocampus. Amino acids were examined using tandem mass spectroscopy and HPLC/fluorometric detection. Following secretin injection intraperitoneally (8.7 μg/kg), as well as intracerebroventricularly (0.015 μg/0.5 μg/5 μg), considerable increases in microdialysate gamma-aminobutyric acid (GABA) and glutamate levels, as well as a slight increase in microdialysate aspartate levels were observed; other amino acids were not affected. The observed increased microdialysate concentrations of GABA and glutamate following secretin application may contribute to observed effects of secretin in autistic patients.

INTERNATIONAL REVIEW OF
NEUROBIOLOGY, VOL. 71
DOI: 10.1016/S0074-7742(05)71011-4

239

I. Introduction

A. THE REAWAKENED INTEREST IN AN OLD FAMILIAR HORMONE

The story that lead to reawakened scientific interest in secretin 90 years after its discovery (the first substance in which the principle of hormonal regulation was identified) began in 1996 when Parker Beck, a three-and-a-half-year-old boy from New Hampshire (USA), was given an injection with secretin to widen the pancreatic duct in line with an endoscopic examination of the pancreas. He had been suffering from autistic symptoms and chronic strong diarrhea since he was 15 months old. The parents of the child witnessed considerable differences in Parker's receptiveness in the days following the injection with secretin and concluded that it must be related to the injection, and thereby discovered the first indications of a possible efficacy of pharmacotherapy on autistic symptoms with secretin. This was, of course, a small glimmer of hope for several families with autistic children; news of which quickly spread through the Internet.

According to the parents, Parker's chronic diarrhea and sleep disturbances became less directly after the aforementioned examination; the young boy, who had been mute since his 15th month, once again began to take up eye contact and developed speech capacities. These developments continued for a period of 3 months, then came to a close, but did not however develop into any kind of deterioration. In total Parker received three infusions with secretin over the period of a year. In 1997 his parents started to treat their son daily with secretin, which they applied transdermally with the help of dimethyl sulfoxide.

Although the described dramatic effect until now has not been able to be secured and proven in controlled clinical studies and therefore remains a topic of considerable debate, nevertheless scientific interest in the neuromodulatory roll of secretin in the CNS has been reawakened.

B. SECRETIN

The German psychologist Carl F. W. Ludwig (1816–1895) proved in 1851 that secretion from salivary glands is regulated through nerve impulses; around the beginning of the twentieth century, the famous physiologist, Ivan Petrovich Pavlov (1849–1936) investigated the reflex mechanisms that regulate the secretion produced by stomach acids following ingestion. Pavlov developed the well-known theory of the nervous regulation of the salivary and gastrointestinal glands (Henriksen and de Muckadell, 2000).

Taking these studies into account the English doctor and physiologist Ernest Henry Starling (1866–1927) in collaboration with physiologist William Maddoc

Bayliss (1860–1924) undertook an examination into the mechanisms of pancreatic secretion in dogs. In a row of experiments from 1902 to 1904 they showed that the intravenous injection of an extract of duodenal mucosa causes a strong release of bicarbonate and water from the pancreas (Bayliss and Starling, 1902). They postulated that the stimulatory effect of acids in the small intestine on the pancreas secretion is due to the release of a messenger in the upper intestinal mucosa that runs through the circulation system and causes the stimulatory effect on the pancreas. They called this substance "secretin."

In 1905, Bayliss and Starling named this new regulatory principle, where blood is used as the path for transferring information, "hormone." The Greek word "hormao" names a substance that decomposes, stimulates, excites (Henriksen and de Muckadell, 2000).

1. *Isolation and Structure of Secretin*

Secretin was isolated and purified from the porcine intestine for the first time in 1961 by Jorpes and Mutt (1961). They analyzed the sequence of amino acid until 1970 and discovered that secretin is a peptide consisting of 27 amino acids with a molecular weight of around 3000 Daltons (Mutt *et al.*, 1970).

The following chronological order depicts the isolation and sequencing of secretin in pigs (Mutt *et al.*, 1970), cows (Carlquist *et al.*, 1981), humans (Carlquist *et al.*, 1985), in dogs (Shinomura *et al.*, 1987), and the rat (Gossen *et al.*, 1989). Secretin proved to be very much a conservative peptide over the period of the evolution of mammals, where amino acid substitutions only are found in positions 14, 15, and 16.

As claimed by Starling and Bayliss, a precursor-protein with a higher molecular weight was found and identified (Gafvelin *et al.*, 1990).

The relationship of amino acid sequences of secretin to other peptides of the CNS and digestive system showed clearly that it belongs to a family of so-called "brain-gut peptides," to which the following also belong: VIP (vasoactive intestinal peptide), PACAP (pituitary adenylate cyclase activating peptide), GHRH (growth hormone-releasing peptide), PHI (peptide histidine isoleucine) or PHM (peptide histidine methionine), glucagon, GLP-1 (glucagon-like-peptide 1), GLP-2, and GIP (gastric inhibitory peptide) (Ng *et al.*, 2002). These substances, which can be summarized under the formula Sec/Gluc/VIP-super family, all show a structural similarity at the N-terminal end.

The secretin gene of the human was located in the chromosome 11 p 15.5 (Whitmore *et al.*, 2000). Secretin has 25%, or in another context 39%, structural similarity with two other neuropeptides, the hypocretins. These two peptides, that were only found in the dorsal and lateral hypothalamus, are probably not members of the PACAP/Glucagon-superfamily since its identity is found in middle and C-terminal regions, however, not in the N-terminus (de Lecea *et al.*, 1998).

C. THE SECRETIN RECEPTOR

The cloning of a high-affinity secretin receptor was reported first of all by Ishihara *et al.* (1991) who were able to isolate the specific complementary DNA of secretin receptors in the rat out of a cDNA-library of hybrid neuroglial cell line NG 108–15. The nucleotide-sequence analyses showed that it concerned a protein with 449 amino acids and a calculated molecular weight of 49 kDa. Based on these results Ulrich *et al.* were able to isolate exactly the same receptor from a cDNA-library from pancreas cells of the rat (Ulrich *et al.*, 1993).

The secretin receptor contains an N-terminal signal peptide sequence, five possible positions for N-glycosylation, ten cysteine residues that probably lie within the extracellular domain, seven membrane-spanning helices, and three intracellular positions for phosphorylation through PKC; though it belongs to a family of G-protein coupled receptors, interestingly however it shows no significant similarity with other typical receptors of this group, since it discloses the highly conserved sequences of amino acids of the ß-adrenergic family of G-protein coupled receptors, and as a very specific particularization shows a relatively large extracellular domain, which precedes the first transmembrane domain.

With the cloning of the receptors for calcitonin and parathormone and the obvious similarity with the sequence of the secretin receptor it became clear that these receptors belonged to a new branch of the phylogenetic development of the superfamily of G-protein coupled receptors (Ulrich *et al.*, 1998).

Taken together, the receptors of the Sec/Gluc/VIP-family show less than 12% homology with the family of rhodopsin- and ß-adrenoreceptors. The only essential and uniform motifs between the two families are the seven transmembrane helices and the two cysteines that form a disulfide bond between the first and second extracellular domain (Ulrich *et al.*, 1998). The cloning of the human pancreatic secretin receptor showed that it has a 81% uniformity with the pendant of the rat (Jiang , 1995).

The most important functional particularity of the members of the Sec/Gluc/VIP-superfamily lies obviously in their capacity to use two important intracellular signal transduction paths, namely via cAMP as well as through IP_3 (Patel *et al.*, 1995; Sreedharan *et al.*, 1994; Trimble *et al.*, 1987). Secretin creates the synthesis of cAMP by coupling to a G-protein (Fremeau *et al.*, 1986); in addition high concentrations of secretin also stimulate the production of IP_3 with a following intracellular release of Ca^{2+} and the activation of PKC (Trimble *et al.*, 1987).

The strength of specific binding affinity of different ligands on the secretin receptor seems to vary slightly depending on the type of mammal, origin of the organ, and the cell type used. However, radioactive labelled secretin is usually displaced competitively from the receptor in decreasing strength by secretin >

helodermin > PHI (peptide histidine isoleucine) = VIP (Gossen *et al.*, 1990). An opposite succession of the binding affinity was apparent for the VIP-receptor with a decreasing potency in the order VIP > helodermin > GRF > PHI > secretin (Voisin *et al.*, 1991).

Secretin receptors have a relatively high binding affinity for radioactive labelled secretin, where around 50% of the bound 125I-secretin was competitively displaced by 1 nmol/L secretin or by 1 μmol/L VIP (Ulrich *et al.*, 1998). Despite this, the binding behavior shows a certain cross-reactivity between VIP and secretin, which explains the occurrence of high- as well as low-affinity binding sites of both ligands in different organ tissues and the difficulty it caused at the beginning to interpret binding data (Robberecht *et al.*, 1976).

Desensibilization is a ubiquitous phenomenon of receptors of the cell surface and can be achieved through different mechanisms such as decoupling of the receptor from its G-protein, sequestration, and internalization of the receptor or transcriptional down-regulation.

The known mechanism of decoupling of the receptor from its G-protein via phosphorylation of a C-terminal intra cellular domain as a reaction to agonistic stimulation was confirmed for the secretin receptor (Ozcelebi *et al.*, 1995). In addition to this a second, totally independent mechanism of receptor internalization as a reaction to secretin stimulation was reported (Holtmann *et al.*, 1996). VIP-binding to the secretin receptor leads to no evidential desensibilization (Bawab *et al.*, 1991).

D. Effects of Secretin Outside the CNS

Secretin is mainly produced from endocrine cells of the proximal small intestine, the so-called S-cells (Polak *et al.*, 1971). The S-cells belong to the APUD (amine precursor uptake and decarboxylation)-system of the digestion tract and are therefore of neuroectodermal origin. The APUD-system is a peripheral endocrine cell system, whose similarity lies in the production of functional molecules from amine precursors and storage in specific granula. Besides the gastrointestinal tract secretin is expressed in the heart, lungs, kidney, and testicles (Ohta *et al.*, 1992) and in the B-cells of the developing pancreas (Wheeler *et al.*, 1992).

A fundamental function of secretin in the gastrointestinal tract is to stimulate the secretion of water, bicarbonate, and other electrolytes from the epithelium of the ductus pancreaticus as a reaction to the presence of stomach acids and fatty acids in the duodenum (Meyer *et al.*, 1970; Watanabe *et al.*, 1986).

The secretion of bicarbonate is important in the neutralization of the acidic content from the stomach in order to achieve an optimal pH-value for the function of the digestive enzymes of the small intestine. In addition to this the pancreas secretion in combination with the secretion of the duodenal mucosa creates a protective alkaline layer that protects these from ulcerations (Allen *et al.*, 1986). Secretin also increases the stimulatory effect of cholecystokinine on the acinus cells of the pancreas in the production of an enzyme potent secretion (Rausch *et al.*, 1985) and stimulates pancreas growth (Dembinski and Johnson, 1980).

The effect of secretin in the gastrointestinal tract is not limited to the pancreas. It restrains deflation of the stomach (Jin *et al.*, 1994) and the production of stomach acids (You and Chey, 1987). In the gall bladder it promotes the flow of gall by the opening of Cl^- channels (McGill *et al.*, 1994) and by promoting osmotic water transport in the cholangiocytes through increase of aquaporines in their plasma membranes (Marinelli *et al.*, 1997). Furthermore secretin increases the activity and the concentration of epidermal growth factor (EGF) in the secretion of Brunner's glands in the duodenum (Olsen *et al.*, 1994).

On the kidney it functions anti-diuretically through the activation of an adenylate cyclase on the thick ascending limb of the loop of Henle (Charlton *et al.*, 1986).

On the heart, secretin functions in a positively ionotropical manner and once more promotes the activity of an adenylate cyclase (Rice *et al.*, 1999).

E. SECRETIN AS A NEUROPEPTIDE

A functional meaning of secretin as a neuropeptide was increasingly supported in the last few years by anecdotal reports concerning the treatment success in autistic patients, and therefore has been the object of intense speculation. However, even before there were significant and clear indications that secretin had a neuroactive effect in the central and peripheral nervous system.

1. *Central Nervous Expression*

In order to be effective as a physiological neuropeptide, secretin has to be created in specific regions of the nervous system; the expression of mRNA of the secretin precursor peptide in the brain of rats showed at first contradictory results in the data, which on one hand are a clear expression of its existence (Itoh *et al.*, 1991; Whitmore *et al.*, 2000), and on the other hand of its absence (Kopin *et al.*, 1990). Secretin type immunoreactivity in brain tissue of the rat and the pig was already displayed early on (O'Donohue *et al.*, 1981), but could be due to cross-reactivity with other structurally similar peptides. A few years later Fremeau *et al.* (1983) proved the highly specific binding of radioactively labeled secretin to the plasma membrane of cells of different brain regions of the rat. Specific binding was greatest in the cerebellum, intermediate in the cortex, thalamus, striatum, hippocampus, and hypothalamus, and lowest in the midbrain and the pons

(Fremeau, Jr. *et al.*, 1983). Similarly, at first there was no positive proof for the expression of the well-known secretin receptor in human brain tissue (Chow, 1995) or of the rat (Ishihara *et al.*, 1991). Ng, Young, and Chow were able to prove through Northern Blot analysis the presence of secretin receptor transcripts in eight representative brain regions of the rat, namely in cerebellum, cortex, hippocampus, thalamus, hypothalamus, pituitary, brainstem, and striatum, as well as the presence of secretin transcripts in the brainstem and cerebellum (Ng *et al.*, 2002). They also assumed the expression of secretin beneath detection limits in specific neurons of other brain regions. Rindi for example reports the selective expression of marked secretin in the serotoninergic neurons of the dorsal raphe nuclei in mice (Lossi *et al.*, 2004; Rindi *et al.*, 2001).

Studies in rats gave evidence that secretin and its receptor are regulated during postnatal brain development. Tay *et al.* (2004) observed stronger expression of both in weeks 3 to 5 as compared to week 2. The areas of major expression were the cerebellum, the central amygdala, hippocampus, area postrema, and the nucleus tractus solitarius.

2. *Central Nervous Effects*

The oldest description of central nervous effects stems from Fuxe, who pointed out the increase of dopamine turnover in the eminentia mediana, which coincides with a decrease in the prolactin secretion following intracerebroventricular (i.c.v.) injection of secretin; in connection with the observation of further representatives of this peptide family he suspected a hormonal gastrointestinal-hypothalamic-pituitary loop (Fuxe *et al.*, 1979). 1 μg of i.c.v.-injected secretin lead to the stimulation of the hypothalamic tyrosine hydroxylase and a decrease in plasma-prolactin and LH-levels, 5 μg on the other hand increased the concentration of prolactin, but did not change anything in the other measured parameters (Babu and Vijayan, 1983).

Samson indicated secretin-like immunoreactivity in regional brain extracts of the rat, especially in pituitary, hypothalamus, epiphysis, and the septum; apart from this he describes the increase of serum prolactin levels on a high dose application of secretin i.v. (10 μg), as well as a dose-dependent prolactin stimulation *in vitro* in cells of the anterior hemipituitary through synthetic secretin (0.31 μg/ml). However, i.c.v. injection of secretin (0.1–10 μg) led to a restriction of the plasma prolactin level (Samson *et al.*, 1984).

Charlton *et al.* (1983) noticed under i.c.v. application of secretin (5 μg in 5 μL of solution) an increase in defecation, but decreased novel-object approaches and open-field locomotor activity in rats, as well as a change in the respiration rate of anaesthetized animals. None of these effects were able to be shown after i.p. injected secretin. Babarczy *et al.* (1995) described an influence of 0.01 μg i.c.v.-applied secretin on the pain sensitivity of rats through the development of morphine tolerance on one dosage of morphine, but not in regular doses.

VIP and secretin induced the genetic expression of tyrosine hydroxylase in PC12-cells through independently antagonizable, yet not additively effective mechanisms. They were in both cases proteinkinase-A- and adenylate cyclase dependent mechanisms (Roskoski *et al.*, 1989; Wessels-Reiker *et al.*, 1993). The tyrosine hydroxylase is the velocity-defining enzyme in the synthesis of the catecholamines. A modulation of this enzyme was also seen in the upper ganglion of the rat (Ip *et al.*, 1982).

Van Calker *et al.* (1980) described a strong cAMP-stimulation through secretin in cultivated brain cells of the mouse, which mainly consist of glioblasts. VIP showed, in relation to this parameter, a minimal stimulatory potency. Both effects could be inhibited with somatostatin, the first one only with secretin-(5–27).

Further indications of induced cAMP-accumulation in brain slices of these two substances were found by Fremeau (Fremeau *et al.*, 1986).

Very recent electophysiological examinations by means of patch clamp technique on neurons of the nucleus tractus solitarius of the rat have shown that an application of secretin to this region causes a non-selective increase in conductivity of cations causing depolarization of these neurons; this modulatory effect apeared concentration dependent and was not blockable by tetrodotoxin (Yang *et al.*, 2004a).

3. *Interaction of Secretin and GABA in the Cerebellum*

Yung *et al.* (2001) demonstrated the interaction between GABA and secretin and the role of secretin as a direct neuromodulator in the CNS through electrophysiological measurements. First, this group was able to show the expression of secretin and its receptors in the cerebellum of the rat using Northern Blot analysis. In-situ hybridization showed immunoreactivity of secretin, which was confined to the soma and the dendrites of Purkinje cells of the cerebellar cortex, whereas immunoreactivity of the secretin receptor could be shown in Purkinje cells as well as in smaller cells that were positively identified as basket cells. Electrophysiological examinations using whole cell patch clamp technique on Purkinje cells showed at first no change in the resting potential on secretin (3–300 nM), although receptors had been found. After that, the Purkinje cells were stimulated by activation of afferent fibers or interneurons. Resulting was a longlasting (>30 min) increase of the amplitude and frequency of spontaneous and evoked IPSCs after a brief exposition to secretin (3–5 min with 3–300 nM solution), whereas the measured IPSCs through stimulation of the parallel fibers showed no change. The IPSCs were found to be largely mediated by GABA$_A$-receptors. The observed effect of secretin started with a delay of 1–2 min, but lasted for more then 30 min. The first effect backs the theory that secretin increases the probability of vesicular release from presynaptic terminals of GABAergic basket cells in the cerebellum, whilst the aforementioned consideration of this thesis does not necessarily contradict it (Yung *et al.*, 2001). Because secretin receptors had been shown to exist on basket cells as well as on Purkinje cells, the existence of a possible postsynaptic modulation mechanism such as the modulation of $GABA_A$-receptor sensitization was examined through measurement of IPSCs, created by a constant, exogenous utilized amount of GABA. In this instance, unexpectedly, a reduction of the response amplitude under the presence of secretin was observed.

In general this functional knowledge, combined with the detection of secretin in the somatodendritic region of the Purkinje cells and the evidence of receptors on the same cells, as well as in GABAergic basket cells, suggests that secretin may act as a retrograde messenger in the cerebellum of the rat. In summary they observed that secretin facilitates evoked, spontaneous, and miniature IPSCs recorded from cerebellar Purkinje cells. They suggested that secretin is released from the somatodendritic region of Purkinje cells and could serve as a retrograde messenger modulating GABAergic afferent activity.

F. SECRETIN AND AUTISM

In an uncontrolled case study carried out in 1998 Horvath *et al.* (1998) described three patients with autistic symptoms and chronic diarrhea, who not only showed an increased pancreatobilar answer to i.v. applied secretin, but also demonstrated within 5 weeks a marked improvement in their gastrointestinal symptoms, as well as drastic behavioral changes and an increase in their expressive speech capacity. The authors speculated that these clinical observations could be evidence of a connection between gastrointestinal and cerebral function in autistic persons. Horvath noticed that empirical studies were necessary in order to substantiate the connection further and emphasized the importance of placebo control in such studies.

On October 7, 1998, Dateline NBC wrote a report on Parker Beck's dramatic behavior changes, one of the boys described in the Horvath study. The case carried considerable weight for the media, Internet, and amateur press. Word spread immediately of what was in fact the success of an uncontrolled treatment and quickly, worldwide, parents began to demand secretin infusions for their autistic children. It is estimated that thousands of children were given secretin injections in an off-label use following this highly publicized case, which had not been backed by any profound empirical evidence. There were of course those who warned of the dangers and who threatened to obstruct the further therapeutic use of secretin in absence of an appropriation by the FDA (Federal Drug Administration, USA) until studies into the effectiveness of the therapy were carried out.

There was considerable concern about the possibility of a triggering of allergic reactions to multiple injections of porcine secretin, as well as about the danger of an immunization through the application of porcine secretin where an autoimmune activity to the body's endogenous peptide could possibly result (Esch and Carr, 2004). So in the years to follow, numerous randomized double-blind studies attempted to clinically substantiate the treatment effect.

Several studies included persons who had been diagnosed within the spectrum of autistic illnesses ("autistic disorder," "pervasive developmental disorder," or "pervasive developmental disorder not otherwise specified"). In eight of the studies, porcine secretin was used; in two, synthetic human secretin; and in one of the studies biological as well as synthetic porcine secretin was compared to placebo. Ten studies examined a single secretin dose; in one study secretin was injected twice. All studies used different established tests as parameters to determine autistic modes of behavior, including spoken, cognitive, and pro-social capabilities. In addition to the records to be filled in by trained doctors, in some studies the parents' considerations were taken into account using specialized questionnaires. Detailed, study-specific listings of the instruments used, as well as the data relating to the results, have not been included in this work. For a detailed and retrospective overview and discussion of the aforementioned clinical studies we refer to Esch and Carr (2004).

None of the studies were able to reproduce the dramatic treatment success within a larger population under clinically controlled specifications. One critique in reference to some of the studies was that some test persons were not interrupted from other common pharmaco therapies currently in use (Dunn-Geier *et al.*, 2000; Sandler *et al.*, 1999).

A specific effect that is described in several of the studies concerns the fact that within the secretin and the control groups a certain number of children showed improvements in their autistic behavior, which however did not differ by statistical significance. A possible explanation for this phenomenon could possibly be an effect of habituation of the children to the test situation; as well as a strong expectation effect (Coniglio, 2001; Corbett *et al.*, 2001; Dunn-Geier *et al.*, 2000; Sandler *et al.*, 1999; Unis *et al.*, 2002). Lightdale *et al.* (2001) tried to replicate Horvath's results, and treated 20 autistic children with gastrointestinal problems with porcine secretin in an uncontrolled, prospective study. Contrary to the estimations of the doctors in which no significant treatment effect could be recognized, 70% of the parents reported medium to strong changes, and 85% were of the opinion that their child would benefit from a further secretin infusion. In a controlled study that purely had the aim of evaluating the parents' estimation of the behavioral improvements of their autistic children, it became clear that the parents were not able to distinguish between their child having had a secretin or placebo injection (Coplan and Souders, 2003).

Maybe the unusual background of these studies biased parents' assessment. The circumstances under which several of them had wanted to take part in a secretin study are heavily emotionally burdened and laden with expectations of success (Esch and Carr, 2004).

In addition to this there was a further more unusual aspect to be taken into account in these tests, which was that the impulse to begin with these studies was driven strongly by media coverage, and that again and again anecdotal reports were issued, as well as singular case descriptions from clinics about the positive treatment effect of intravenous as well as transdermally applied secretin (Lamson and Plaza, 2001).

Because Horvath had indicated the presence of gastrointestinal co-morbidity in addition to the autistic symptoms in the first description, some of the studies previously described (Kern et al., 2002; Roberts et al., 2001; Unis et al., 2002) included a subgroup analysis of children with gastrointestinal dysfunction. The expectation was to possibly identify a subgroup of secretin responders in what was an extremely heterogenic patient group of autistic persons; yet this did not bring about any significant differences.

Only in Kern's study were there any notable positive effects of a secretin dose in the comparison between autistic children with active chronic diarrhea and those without gastrointestinal problems. The patients with chronic obstipation or chronic diarrhea in a period of remission were not included; the number of those within the group with chronic diarrhea was limited, unfortunately, to only five individuals. These showed, under treatment with secretin, a significant reduction in anxiety, agitation, crying, hyperactivity, non-compliance, lethargy, and social withdrawal, as well as a decrease in stereotypical behavior and a more adequate use of language (Kern et al., 2002).

The question of whether children with autism suffer more regularly from gastrointestinal problems than those of the same age without autism is a highly debated topic. Horvath and Permann described, in a study conducted on twins in 2002, the increased occurrence of gastrointestinal inflammation in autistic persons, fewer digestion enzymes, a decreased sulfation capacity of the liver, and a higher secretory answer to secretin stimulation tests (Horvath and Perman, 2002). D'Eufemia reported a changed intestinal permeability of 9 of 21 (43%) randomly selected autistic patients, yet in none of the 40 control persons (D'Eufemia et al., 1996). Furlano described specific histological changes of the intestinal wall in autistic patients with gastrointestinal symptoms, which coincides with impaired epithelial glycosaminoglycan composition, and which seems to be histologically different to other inflamatory intestinal illnesses (Furlano et al., 2001).

In contradiction to this an analysis of public related data of a research data bank in Great Britain showed the same prevalence of 9% gastrointestinal disorder within the population of autistic as well as non-autistic children (Black et al.,

2002). It is therefore clear that as yet there is no true consensus regarding the connection between the development of autistic and gastrointestinal disorders.

Perhaps it would therefore be more promising to undertake a further clinical study on a larger population of autistic children with active chronic diarrhea to identify a secretin sensitive subgroup. The discussion whether secretin leads to peripheral improvements in the gastrointestinal function or whether it acts directly in the CNS would of course continue.

In the absence of a causal medical therapy for autistic disorders there is an extremely wide spectrum of therapeutic attempts by parents who look for help, which is promoted in Internet forums and by organizations involved. The "autism research institute" has been collecting, amongst other things, data on the experiences of parents who have been involved with interventions of all types since 1967. The published chart (http://www.autismwebsite.com/ari/treatment/form34q.htm) enables an overview of the non-standardized assessments of parents. Remarkably, the top positions in regard to the positive influence on the development of the illness are taken up with some diets, vitamins, and enzyme substitution. With regard to the group of drug therapeutics secretin shows a superior treatment profile to almost all other drugs, except Risperdal.

The company Repligen in Needham, Massachusetts, has attained the US-patent right to the treatment of secretin in autistic patients from Victoria Beck and Bernhard Rimland from the Autism Research Institute, USA. The company planned the production and marketing of a synthetic form of human secretin. The FDA (Food and Drug Administration, USA) agreed in advance to quickly give Repligen the right to use secretin in the treatment of children with autism.

In February 2002, five animal tests were conducted into the toxicity of multiple secretin injections (Adis International, 2002).

Phase-II-studies in the USA showed with regard to 3–4-year-old autistic children, yet not in 5–6-year-olds, an improvement in social interaction in accordance with the Autism Diagnostic Observation Schedule.

A later phase-III-study on 132 children between 2 and 5 years old with six injections over a period of 18 months was not able to give evidence of the superiority of an intravenous secretin treatment over one with placebo (www.repligen.com). Details concerning study design as well as the attempt to identify subgroups of secretin responders have yet not been published. Autistic components are also regularly apparent in other psychiatric illnesses, as for example in schizophrenia. A case study by Alamy describes a temporary, substantial improvement in autistoid symptoms of schizophrenic patients to an adjuvant injection of secretin (Alamy *et al.*, 2004). In order to investigate a possible benefit on social behavior in other psychiatric illnesses further clinical studies are being undertaken by Repligen concerning anxiety disorders and negative symptoms in schizophrenia (www.repligen.com).

G. Neuropathologic Findings in Autism

Through post mortem examinations on autistic brains and *in vivo* examinations of autistic people by MRI, evidence of neuroanatomic changes in the following brain regions were found:

- Hippocampus (Acosta, 2003; Aylward *et al.*, 1999; Raymond *et al.*, 1996; Saitoh *et al.*, 2001)
- Amygdala (Aylward *et al.*, 1999)
- Cerebellum (Acosta , 2003; Fatemi *et al.*, 2002; Purcell *et al.*, 2001)
- Corpus callosum (Egaas *et al.*, 1995)
- Regions of the cortex (Casanova *et al.*, 2002)

Casanova *et al.* (2002) detected changes in the mini-columnal structure of the cortex in autistic persons in post mortem examinations of brain tissue from Area 9 of the prefrontal cortex and the Areas 21 + 22 of the temporal lobe of autistic persons.

Cell columns were increased, but were smaller and less compact in their cell structure and had less peripheral neuropil space (Casanova *et al.*, 2002). In this histological compartment fibers of inhibiting interneurons are evident (Casanova *et al.*, 2003).

By using MRI, Courchesne investigated the growth of the brain in autistic children in comparison to healthy children and discovered an early overly large growth followed by a later abnormally slow growth in volume. From the ages of 2–4 there was a hyperplasia of the grey and white substance of the cerebrum and of the white substance of the cerebellum (Courchesne *et al.*, 2001).

Corresponding results were also attained in a study by Aylward (2002). In older autistic persons a diminished size of the cerebellum (Fatemi *et al.*, 2002) and a smaller total number of Purkinje cells were recorded (Acosta, 2003). In connection with this it is also notable that research in the last few years has lead to evidence that the cerebellum is also involved not only in motoric, but also sensory, cognitive, and affective brain functions (Allen and Courchesne, 2003; Ciesielski and Knight, 1994).

Purcell *et al.* (2001) compared the gene expression in the cerebellum of autistic patients with that of healthy individuals. It was discovered that there was a higher expression of a transporter for excitatory effective amino acids (EAAT1) and for an AMPA-sensitive non-NMDA-Glutamate receptor. The density of these receptors has been proved however to be decreased. The examination of post mortem brain tissue of older autistic persons has the disadvantage that it is not possible to recognize which of the apparent changes are of a primary and which are of a secondary nature.

By using MRI-examinations, Saitoh *et al.* (2001) described the presence of significantly decreased cross-section surfaces of the dentate gyrus (comprising the dentate gyrus and CA4-region) in the hippocampus, however there was no decrease measurable in the totality of the subiculum and the CA1-CA3-regions of autistic persons. Abnormalities of this type are typically to be put down to an increase in the pyramidal cell thickness, reduction in neuron size, and a diminished dendritic tree (Acosta, 2003; Raymond *et al.*, 1996). Using the *in vivo* study it was possible to show that neuropathological changes already were apparent by the time of the clinical manifestation of the symptoms (Saitoh *et al.*, 2001).

Eriksson and colleagues (Eriksson *et al.*, 1998) were able to prove that the neurogenesis of the corn cells in the dentate gyrus is not limited to the pre- and peri-natal life span, but in fact is still evident in adults. The existing neuropathological hippocampal correlates that already prevail whilst the individual is alive and the persistence of autistic symptoms in later life suggest that autism is connected with a non-remitting impairment of hippocampal neurogenesis.

II. Aim

The aim of the current study was to determine the effects of secretin on extracellular amino acid concentrations *in vivo*. To this end, microdialysis was performed in the hippocampus of freely moving rats. Our major interest was in the effects of secretin on glutamate and GABA, given their proposed role in autism (Blatt *et al.*, 2001; IMGSAC, 2001; Muhle *et al.*, 2004b; Purcell *et al.*, 2001). In a row of experiments secretin was applied intraperitoneally; herein it is highly probable that a vagal, as well as a hematogenic transmission into the CNS also took place. Following on from this secretin was injected intracerebroventricularly into different concentrations, in order to compare the found effects with that of a direct neuromodulation.

The hippocampus was chosen as the area for examination, since it is involved in selective attention processes, learning and memory functions, and the creation of motivational states. As a periventricular organ it can be immediately accessed by i.c.v.-applied secretin, and there are indications of the presence of secretin receptors in the hippocampus (Karelson *et al.*, 1995; Ng *et al.*, 2002). Besides this there are signs of a hippocampal dysfunction in autistic patients; neuropathological findings relate, amongst others, to the area dentata (Saitoh *et al.*, 2001).

The aim of the current study was to determine the effects of secretin on extracellular amino acid concentrations *in vivo*. To this end, microdialysis was performed in the hippocampus of freely moving rats. Our major interest was in the effects of secretin on GABA and glutamate, given their proposed role in autism (Blatt *et al.*, 2001; IMGSAC, 2001; Muhle *et al.*, 2004b; Purcell *et al.*, 2001).

III. Materials and Methods

A. ANIMALS AND DRUG TREATMENT

In the first series of experiments 30 clinical units of secretin (Secrelux, Goldham) or 8.7 μg secretin pentahydrochloride/kg of body weight were administered i.p.; control animals received an equal volume of saline solution. In the second set of experiments secretin was applied i.c.v. in 5 μL, infused during 2.5 minutes in three doses: 0.015 μg, 0.5 μg, 5.0 μg.

B. SURGERY AND MICRODIALYSIS PROCEDURE

Microdialysis was performed with CMA/12 microdialysis probes (length: 1.0 mm; o.d.: 0.24 mm, Carnegy Medicine, Sweden). The probe was implanted together with a guide cannula under isoflurane (2%) anesthesia into the left hippocampus, as previously described (Clement et al., 1998). This was done according to coordinates given by Paxinos and Watson (Paxinos and Watson, 1982): A:5.2 mm; L:2.0 mm; V:6.5 mm, and fixed with two screws and dental cement. Probes were each used three times. For i.c.v. application, a guide cannula was implanted in the lateral ventricle according to the following coordinates: A:7.2 mm; L: 1.7 mm; V:4.5 mm from the top of the cortex. Positions of the microdialysis-probe and the cannula were controlled histologically.

Microdialysis experiments were carried out in freely moving rats (CMA/120 freely moving system, Carnegy Medicine, Sweden) directly after implantation of the probe. Ringer's solution (NaCl 140 mM; KCl 3.0 mM; CaCl$_2$ 1.2 mM; MgCl$_2$ 1.0 mM) was used for perfusion at a flow rate of 1.25 μL/min (CMA/100, Carnegy Medicine, Sweden), and 25 μL fractions were collected; outlet tubing volume was 3 μL. In vitro recoveries at room temperature were found to be about 6% for GABA and 9% for aspartate and glutamate. After six fractions had been collected, secretin or saline were applied i.p. and sampling continued for a further two hours. At the conclusion of the experiment, the rat was euthanized with CO$_2$. To verify probe placement, rats were decapitated and the brain removed and frozen. Serial coronal sections (thickness: 20 μm) stained with cresyl violet were used for localization according to Paxinos and Watson (Paxinos and Watson, 1982).

C. ASSAY OF AMINO ACIDS

To assess amino acid concentrations (except GABA, glutamate, and aspartate), 200 μL of a solution of deuterated amino acids (internal standards) in methanol was added to 7 μL of dialysate and the mixture evaporated to dryness at 55 °C under a stream of nitrogen. The residue was heated for 15 min at 65 °C with 120 μL butanolic hydrochloric acid (3 M). After evaporation to dryness, 200 μL acetonitrile/water (1:1 v/v) containing 0.025% formic acid was added and the amino acids estimated as their butylesters by tandem mass spectrometry using a PE Sciex API 365 instrument employing the neutral loss (102 Dalton) mode (Chace *et al.*, 1993). Glutamate, GABA, and aspartate were assayed by HPLC with fluorescence detection (Gerlach *et al.*, 1996), using the precolumn derivatization method with ortho-phthaldialdehyde (OPA) and an automatic system from Kontron Instruments (Neufahrn, Germany) consisting of a 325 pump, a 465 auto-sampler, a SFM 25 fluorescence detector, and a computing integrator equipped with a 450-MT2 data system. Excitation and emission wavelengths of the fluorescence detector were set at 330 nm and 450 nm, respectively. 5 μL of microdialysate was used for derivatization with 30 μL of OPA reagent (Grom Analytik, Herrenberg, Germany) diluted 10 times with 1M borate buffer (pH 10.7). 20 μL of this reaction mixture was injected directly into the HPLC system. Concentrations were calculated from peak height with the aid of external standards.

D. Statistical Analysis

Concentrations were not corrected for in vitro recovery. The mean concentration of the six samples before drug application was arbitrarily defined as baseline control (= 100%). All values are expressed as percentages of control: S.E.M. Microsoft Excel 8.0 and SPSS were employed for statistical analysis by non-parametric repeated measurement, one-way analysis of variance on ranks and paired Student t-test. The statistical level of significance was set at $*p < 0.05$, $**p < 0.01$. Areas under the curve were also calculated and compared by paired Student t-test.

IV. Results

Systemically administered secretin (8.7 μg/kg i.p.) was followed by an increase in extracellular GABA and glutamate levels, persisting up to the end of recording two hours after application, while the saline-treated group showed a slight decrease over time (see Fig. 1A and B). Other amino acids investigated were not affected by secretin application, with the exception of aspartate, showing a slight increase after secretin application.

The i.c.v. application of secretin led to a dose-dependent increase of GABA (see Fig. 2) and glutamate (see Fig. 3), as compared to saline, with the highest effects, $p < 0.01$, induced by the medium dose of 0.5 μg for both neurotransmitters.

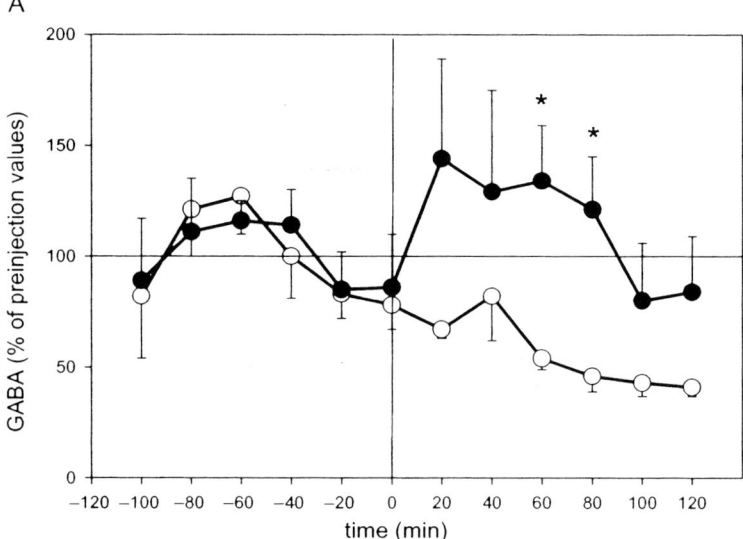

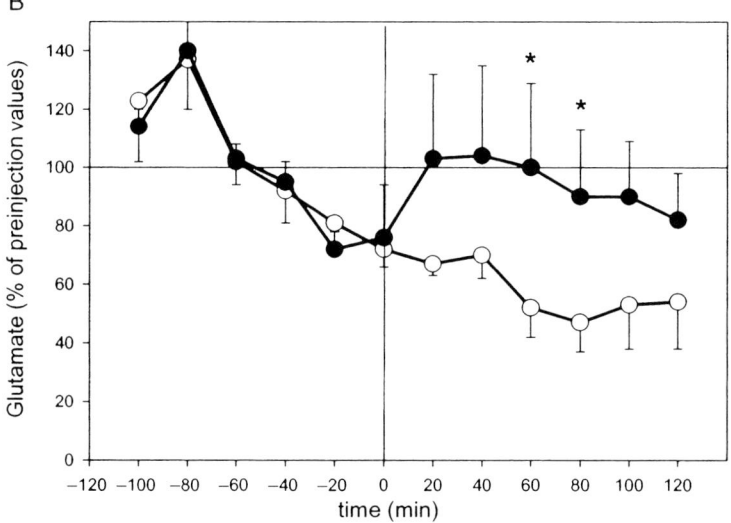

FIG. 1. Time-dependent effect of secretin (8.7 μg/kg i.p., filled circles) and saline (open circles), administered i.p. at 0 min, on extracellular concentration of GABA (A) and glutamate (B) in rat hippocampus. Implantation of microdialysis probes and analysis of the samples by HPLC were performed as detailed in Materials and Methods. Data are presented as mean 1± SEM; n = 5. *p < 0.05 student t-test.

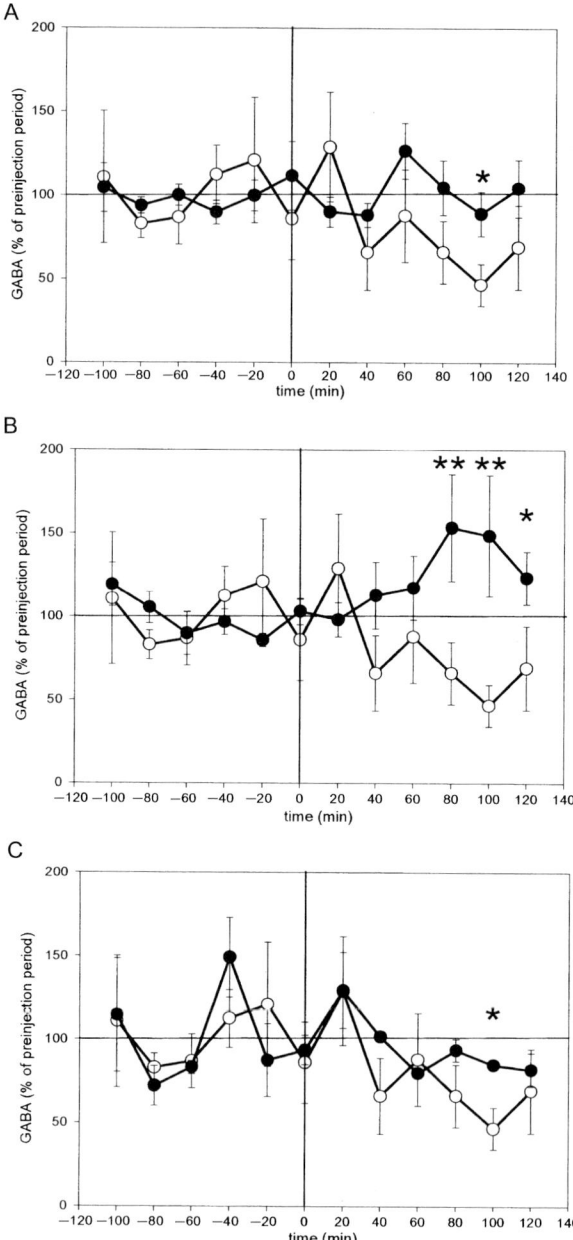

FIG. 2. Time- and dose-dependent effect of secretin (filled circles) applied and saline (open circles), administered i.c.v. in 5 μL at 0 min. A:5.0 μg; B:0.5 μg; C:0.015 μg, on extracellular concentration of GABA in rat hippocampus. Implantation of microdialysis probes and analysis of the samples by HPLC were performed as detailed in Materials and Methods. Data are presented as mean 1 ± SEM; n = 5. *$p < 0.05$, **$p < 0.01$ student t-test.

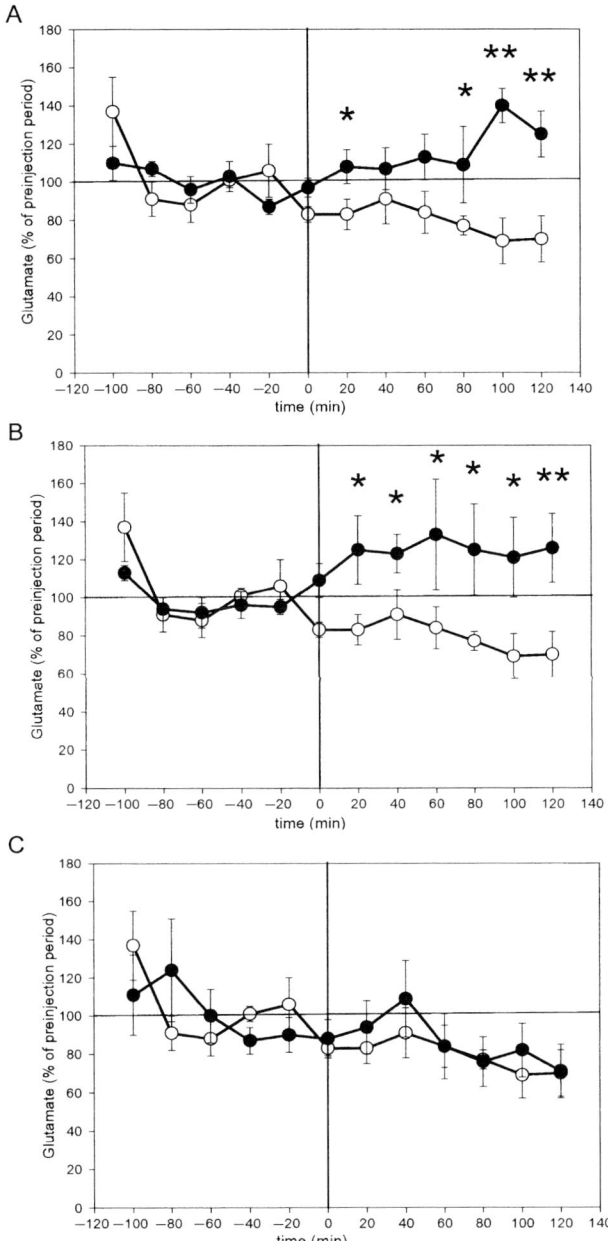

FIG. 3. Time- and dose-dependent effect of secretin (filled circles) applied and saline (open circles), administered i.c.v. in 5 μL at 0 min. A:5.0 μg; B:0.5 μg; C:0.015 μg, on extracellular concentration of glutamate in rat hippocampus. Implantation of microdialysis probes and analysis of the samples by HPLC were performed as detailed in Materials and Methods. Data are presented as mean 1$\pm$ SEM; n = 5. *p < 0.05, **p < 0.01 student t-test.

V. Discussion

More than 150 years after its discovery as a hormone of the gastrointestinal tract secretin has once again become an object of scientific discussion in its role as a neuropeptide. The interest in research into the neuromodular potency of this peptide hormone was reawakened owing to the astounding report issued on the treatment success of secretin on autistic children (Horvath *et al.*, 1998). First indications of the possible use of secretin were already discovered in the 1970s (Fuxe *et al.*, 1979). The effects of secretin on the hippocampus of the rat in the metabolism of amino acids were examined in the present paper. The main focus was placed on glutamate and GABA, which are the two most important brain transmitters in mammals. Secretin was applied intracerebroventiculary in three different dosages, in order to measure the direct neuromodulatory effects of the peptide. The results show an influence of extracellular concentrations of glutamate, aspartate, and GABA in the hippocampus using secretin; the extracellular concentration processes of other measured amino acids, that is, phenylalanine, alanine, methionine, glycine, ornithine, arginine, tyrosine, valine, leucine/isoleucine, and citrullin were not significantly affected.

The stimulatory effect on the three transmitter substances was strongest in a mean dosage of 0.5 μg secretin i.c.v. Curve progressions of glutamate show a prompt and sustained concentration increase up to the end of the experiment; the process of GABA concentration also shows a late but significant elevation. Moreover a weaker effect on the aspartate concentration was identifiable.

Furthermore the high dosage application of secretin (5 μg) demonstrates the described effect on the glutamate/glutamine and also selectively on the glutamate concentration, whereby a noticeably stronger influence in the later phase of the experiment in contrast to the early phase is especially striking. The effect of the GABA concentration can only be assessed as very weak and the influence of aspartate is no longer traceable.

An unusual aspect of the findings of this chapter is the more or less simultaneous rise of glutamate and GABA, since both of these brain transmitters are primarily associated with antagonistic effects on the CNS. Glutamate is the main transmitter of corn cells of the hippocampus, which create the moss fiber system in their totality and as such give excitory impulses of the glutamatergic pyramidal cells of the cornu ammonis, whilst however they also have collaterals to interneurons. Contrary to this most hippocampal interneurons, particularly the basket cells, are involved in GABAergic inhibition of the pyramidal cells, yet there are also other interneurons involved (Freund and Gulyas, 1997).

There are several different ways in which both these main transmitter systems in the hippocampus can interact with one another and thereby influence one another's release. Whilst GABA via GABA$_A$-receptors (predominantly in somatodentric regions) and GABA$_B$-receptors (predominantly located presynaptically) can have an inhibitory effect on the release of glutamate, almost every inhibitory interneuron in the hippocampus conversely gets synaptic input from the main cells. This regulation is rather complex, as it exhibits localization and receptor specific differences. Both synaptic NMDA- and non-NMDA receptors of AMPA type as well as metabotrophic receptors of the somatodentric region have a stimulatory effect on the release of GABA in the hippocampus. There are indications of a glutamatergic inhibition of interneurons on presynaptic Kainate receptors and terminal located metabotrophic glutamate receptors (Vizi and Kiss, 1998).

On the level of the reuptake there are also possibilities for an interaction between GABA and glutamate. Since both are essentially taken up into the synaptic endings by an Na$^+$-dependant, active co-transport, an intracellular strongly rising Na$^+$-concentration uptake induced by the one molecule can promote the inversion of the carrier-transport of the other (Vizi and Kiss, 1998).

Theoretically, in this work, the measured extracellular increase of one of the measured parameters could be the result of an increase of the other, in contradiction to this however there are several microdialysis studies on the hippocampus of the rat, that only describe an independent stimulation of the transmitter system. For example Giovanni *et al.* (2001) examined the influence of locomotor activity associated with investigatory and habitual behavior to new surroundings on the acetylcholine-, glutamate-, and GABA-concentrations in the hippocampus in rats. The experiments showed a lightly positive correlation between locomotor activity and GABA-concentration, yet no significant influence of the glutamate-concentration. The measured extracellular concentrations in microdialysis experiments of functional amino acids mirror the relationship of transmitter release and reuptake in nerve endings and glial structures (Herrera-Marschitz *et al.*, 1996). In regard to glutamate and GABA there are especially controversial discussions taking place about the real origin of these particular microdialytic transmitters, since they are not, as is classically demanded for proof of the synaptic origin of neurotransmitters, able to be inhibited with certainty by TTX-application and Ca^{2+}-depletion of the tissue (for further discussion see Timmerman and Westerink, 1997). It is therefore unclear whether the measured parameters mirror the real neuronal transmission, an influence, or even an inversion of the neural and/or glial reuptakes, or overall changes in the neurotransmitter metabolism.

There are certainly three conceivable ways in which secretin can influence the CNS: auto, that is, paracrine neuro-neuronal, endocrine by manner of blood paths of peripheral organs to the CNS, and lastly over the nervus vagus. The secretin receptor, as well as secretin itself, is functionalized in the CNS by neurons. The distribution of both molecules does not seem to show disregarded type-specific differences: humans and rats demonstrated an extremely similar immuno reactivity to secretin in the CNS, that is in the Purkinje cells of the cerebellar cortex, the central cerebellar nuclei, the pyramidal cells of the motor cortex, and in primary sensory neurons; in addition to this there was evidence of presence in the hippocampus and amygdale of the human and neurons of the auditory system of the rat. Secretin was only traceable in the spinal ganglion of the cat (Koves *et al.*, 2004). Ng and colleagues proved, using Northern Blot, the presence of secretin receptor transcripts in the brain of the rat, in the cerebellum, cortex, hippocampus, thalamus, hypothalamus, hypophysis, brain stem, and striatum. However secretin transcripts themselves were only found in the brain stem and cerebellum (Ng *et al.*, 2002). Transgene mice express secretin in the serotoninergic neurons of the dorsal raphe nuclei (Rindi *et al.*, 2001).

    Even if the expression of the known secretin receptor in the hippocampus of the rat was described, one must be reminded of the fact that secretin shows a substantial cross-reactivity with other members of its peptide family, and can therefore also bind to other receptors, especially to the VIP-receptor. It is known that VIP can strengthen the GABA-release in the hippocampus through presynaptic receptors (Wang *et al.*, 1997). The VIP receptor 1 (VPAC 1) also reacts, besides VIP, to secretin and PACAP with an increase of cAMP (Vaudry *et al.*, 2000). VPAC 1 and VPAC2-receptors are present in the brain of the rat, among other structures in the hippocampus (Joo *et al.*, 2004). The signal path of the VIP- and PACAP-receptors also works, amongst other means, through an activation of the adenylate cyclase (McCulloch *et al.*, 2002). It is therefore also feasible that secretin was also responsible through the VIP-receptor for the increased release of GABA and/or glutamate observed in this study.

    Through the intraventricular application it was confirmed that the effects of secretin took place directly in the CNS. Yung *et al.* (2001) first of all created a hypothetical model based on their electrophysiological measurements on the Purkinje cells in the cerebellum of rats concerning the effects of secretin as a retrograde messenger, which enabled an easier GABA-release from the corn cells using a presynaptic, cAMP-dependent mechanism. This effect is described as relatively long-term (>30 min.). Owing to the fact that secretin receptors were not just found on the corn cells, but also on the Purkinje cells, it would be plausible to suppose, in addition, the existence of an autocrine component in the regulation of this synapse. It is possible that Purkinje cells stabilize their resting potential following a successful depolarization by means of an increase in the inhibitory afferences of the corn cells. Should a similar model be valid in accordance with the hippocampus, for example between basket cells and pyramidal cells, this would explain the GABA-increase measured in this study.

    Electrophysiological examinations conducted by Yang *et al.* (2004a) in the nucleus tractus solitarius of the rat described a different neuro-modulatory effect of secretin, that is, a depolarizing one, not blockable by tetrodotoxin. It is possible that not only secretin of neuronal origin is involved in central nervous regulations, but also secretin of other tissue origins, especially of the gastrointestinal tract. Studies conducted by Banks *et al.* (2002) demonstrate the ability for secretin as well as for secretin analogon ^{131}I-secretin to pass by the blood brain barrier. They investigated the penetration of a radioactive marked ^{131}I-secretin-analogon through the blood brain barrier in 4-week-old mice. The substance passed through the blood brain barrier as an intact molecule and broke through into all brain regions, especially however into the hypothalamus and the cerebrospinal fluid. Banks *et al.* postulated that the secretin analogon passes through the plexus choroideus through a satiable transport process, and passes the vascular blood brain barrier through transmembrane diffusion.

As with most peptides secretin possesses a comparatively short half-life in blood circulation owing to degradation and inactivation. Similar to the amphiphatic molecule VIP and PACAP, which stems from the same protein family, it can however self-assemble into micells in liquid solution and interacts with biomimetic phospholipid membranes, in which the secretin molecules experience a conformation change; it is possible that these physical properties distinctly increase its bioactivity as a pharmacon, yet perhaps already in fact in endogen concentrations (Gandhi *et al.*, 2002; Krishnadas *et al.*, 2003).

The i.c.v. injection of secretin as well as an i.p.-application of secretin led to an increase in concentration of glutamate and GABA in the hippocampus of the rat (Kuntz *et al.*, 2004). The main question whether the measured effect came about through central nervous, vagus mediated, or hormonal type effect is still unanswered. If the effect was due to central neuromodulation, there must have been imitation of central and vagal mediated secretin effects in relation to extracellular transmitter concentrations in the hippocampus; this would only be resolved through an intraperitoneal administration of secretin in vagotomized rats.

Studies by Yang *et al.* (2004) confirm the role of the nervus vagus on i.p. application of secretin through the observation of stronger Fos-expression in different areas of the CNS, even if not in the hippocampus itself. Fos-activation through i.c.v.-application of secretin however seems to have a more extensive central nervous activating spectrum than the administration i.p., and also activates cortical regions (Welch *et al.*, 2003).

The "brain-gut-peptide" secretin seems to be a part of a gastrointestinal-central nervous interaction: as previously mentioned, the central amygdale nucleus area can be activated by secretin in relation to Fos-expression; this is achieved through intraperitoneal administration via vagal afferences (Yang *et al.*, 2004b), through intravenous administration via unclear transmission (Goulet *et al.*, 2001), but, especially remarkable, also directly through i.c.v. administration (Welch *et al.*, 2003). A slightly elevated plasma secretin mirror ($\sim$0.029 μg/kg/h), that did not itself lead to any endocrine activation of the pancreas, was able in experiments on the rat to achieve a significant strengthening of the pancreas secretion through electrical stimulation of the medial amygdale nucleus area (Jo *et al.*, 1994).

It has been proven that the central amygdala nucleus neurons connect to neurons of the vagal dorsal nucleus complex, and in fact to the sensible nucleus tractus solitarius, as well as to the nucleus dorsalis of the nervus vagus; these projections appear to be of a mainly restrictive character and modulate the vagal reflex of the gastrointestinal tract (Lyubashina, 2004; Zhang *et al.*, 2003).

A further indication that the dorsal vagal nucleus complex functions as the deciding relay station in the transmission of vagal stimulation to brain nuclei in the amygdala and hippocampus, is shown in a work completed by Marvel *et al.* (2004). The peripheral application of lipopolysaccharide, an immune activator that induces behavioral symptoms as well as social retreat, was able to be blocked on the level of vagal brain nuclei.

The amygdala is a brain structure whose altered function is seen as a present neuropathological finding in autistic illnesses; its function is considered as essential for the development of a "social brain." In studies using fMRI autistic patients showed, contrary to healthy control persons, no activation of the amygdala in an exercise where parts of the eyes were shown and the patients were to explain the feelings or thoughts being expressed (Baron-Cohen *et al.*, 2000). Since secretin can activate the amygdala, which can itself strengthen the secretinergic pancreas secretion, it is conceivable that a sufficient presence of secretin in the organism, that is, in the gastrointestinal tract, as well as in the CNS, enables for gastrointestinal and amygdala function; should this loop be broken up, exogenic secretin could help to normalize its function.

It is therefore conceivable that secretin could enable positive therapeutic effects only in a subtype of autism since autistic illness is of an extremely heterogenic nature and is associated with a wide spectrum of co-morbidities especially diarrhea, as some authors suggest. Following on from the first descriptions given by Horvath *et al.* (1998), there is also evidence in a study conducted by Kern *et al.* (2002) of the benefit of secretin injections in autistic children with active chronic diarrhea. D'Eufemia described a diarrhea illness observed in autistic people, which coincides with a change in the intestinal permeability (D'Eufemia *et al.*, 1996); there were also diminished amounts of digestive enzymes, sulfation capacity of the liver, and an increased pancreatic secretion in secretin stimulation tests (Horvath and Perman, 2002). Furlano *et al.* (2001) demonstrated histologically in the stomach of autistic children a damaged epithelial glycosaminoglycan composition, and described a lymphocytic colitis, which seemed to be different to other infectious stomach illnesses.

Secretin might in connection with pathophysiological situations of this type, by stimulation of a bicarbonate rich pancreas secretion or by an increase in the amygdala feed back, lead positively to a normalization of gastrointestinal and secondarily or primarily also to neuronal function. There are no descriptions of a lower secretin level in the blood of autistic persons. It is possible that this is a group of children not suffering from gastrointestinal problems to any large extent and would therefore not profit from secretin. In a phase-II-study conducted by Repligen, in an attempt to identify a subgroup of autistic secretin responders, a better response was measured to a secretin injection for a group with calprotectin and chymotrypsin concentrations than for other groups involved in the study (www.repligen.com). There is as yet no published data on the evaluation and study design of the phase-III-study, which followed on from there.

An expression of secretin in the CNS was observed during an embryogenesis in mice (Siu *et al.*, 2005), which is an indication of a possible involvement of secretin in neuronal development. The genes coding for secretin and its receptor are not located in chromosomal areas that have been associated with autism. Several proteins involved in neuronal differentiation and synaptogenesis or in intracellular, not specific neuronal signal transduction path have been associated with autism. Likewise a possible candidate gene for an aetiological participation in autistic child development is a gene for the glutamatergic NMDA receptor (see for an overview IMGSAC, 2001). Furthermore there is evidence of disorders of the glutamatergic system in autism; an abnormally increased availability of the transcription product for the transport protein EAAT1 and for the glutamate receptor AMPA1 has been found, whilst at the same time the actual concentration of AMPA receptors in the cerebellum was abased (Purcell *et al.*, 2001). The function of the complex glutamatergic system of the CNS plays a central role in learning activities like phenomena of LTP and LTD in the hippocampus. Gwag and his colleagues demonstrated that an increase of the endogenous glutamate concentration in the dentate gyrus of rats through blockage of the glial metabolism caused a selective NMDA receptor induced increase of mRNA expression of nerve growth factor (NGF) in the corn cells. For a short time glutamate increase reached double the concentration of the basis level before falling back again within 3 h. This short signal was sufficient to trigger an increase in the transcription rate of the growth factor (Gwag *et al.*, 1997). Autism as an illness, which is associated with disturbed neurogenesis in the amygdala and the hippocampus among other affected brain areas (Aylward *et al.*, 1999), could be positively influenced by such growth stimuli. An increased concentration of serotonin was measured in plasma of autistic people, whereas concentrations of glutamine, glutamate, and GABA were demonstrated to be significantly reduced. Whilst a positive correlation between measured values for serotonin and GABA was found in normal people, these two parameters had a significantly negative correlation in autistic people (Rolf *et al.*, 1993). Dhossche *et al.* (2002) described elevated plasma GABA levels in autistic children aged between 5 and 15 years. Human platelets are classified as an established model, which reflects the monoamine—as well as amino acidic content of neurons.

In the neonatal, still developing hippocampus, GABA still functions as an excitatory brain transmitter, but not however in the adult brain (Ben Ari *et al.*,

1997); yet apart from this the neuronal GABAergic activity is an important signal for the synaptogenesis (Belhage et al., 1998). Hussman (2001) formulated a hypothesis where a reduction of GABAergic inhibition exhibits a possible aetiological factor of autistic symptoms. In a biophysical model of the hippocampal CA3 region described by Wallenstein and Hasselmo (1997), learning and memory ability performance deteriorated as soon as the GABA receptor induced inhibitor was removed, as this caused a deterioration of quality sensory information through concurrent activity of intrinsic and afferent fibers. A decreased GABAergic inhibition could also lead to a glutamatergic hyperexcitation and subsequent damage to vulnerable target neurons, a mechanism considered to be relevant for the onset of diverse neurological illnesses. There is indeed evidence from metabolic studies using magnetic resonance spectroscopy (MRS), which shows an increased degradation of neuronal membranes in the dorsal prefrontal cortex of autistic people (Minshew et al., 1993). In an autoradiographic study by Blatt et al. (2001) eight different receptors from four neurotransmitters (i.e., GABA, serotonin, acetylcholine, and glutamate) were marked with ligands in order to compare the quantity of their hippocampal expression in autistic and normal people. A significant decline in the high affinity binding to $GABA_A$-receptors was detected.

The stimulated release of GABA by secretin in the cerebellum (Ng et al., 2002) may account for an improvement in learning processes in this part of the brain which typically exhibits anomalies in autistic people (Acosta, 2003; Fatemi et al., 2002) and which is essential for the development of higher cognitive functions in child development (Allen and Courchesne, 2003). The long lasting character of modulation likewise affirms the possibility of such an effect.

In summary it seems conceivable that a strengthening of the glutamatergic and the GABAergic transmission could positively influence an existing deficient neuronal development. The connection at this level admittedly remains speculative, as long as a relevant clinically defined subgroup of autistic people cannot be identified, where the application of secretin yields a proven and replicable successful treatment. The Phase III Study by Repligen was prematurely aborted in January 2004, which means that a suitable study design is missing.

Recently a pilot study of 22 people appeared (Sheitman et al., 2004) as well as a further clinical case description (Alamy et al., 2004), where schizophrenic patients with autistic symptoms reacted to a single secretin injection with a clear but very quickly reversible reduction of their clinical symptoms. Whilst people in their immediate environment including their care attendants described these symptom improvements quite impressively, their evaluation is clearly difficult with respect to psychiatric documentation. It is also noteworthy that this deals with schizophrenic patients who in themselves exhibit difficult and therapy resistant characteristics. There are further studies being conducted by Repligen to evaluate the benefit of an application of secretin in schizophrenia and anxiety disorders (www.repligen.com).

Even if the glimmer of hope has dimmed that secretin is the long-awaited "wonder drug" in the treatment of autism, it is equally inappropriate to maintain that the initial changes in behavior described by the parents represent complete misinterpretations. There is now renewed interest in secretin as a neuropeptide. Research into the neuromodulatory potential of endogenous peptides is certainly an important aspect in the development of a more exact knowledge about the creation and therapy of psychiatric illnesses.

References

Acosta, M. T. P. (2003). The neurobiology of autism: New pieces of the puzzle. *Curr. Neurol. & Neurosci. Reports* **3**, 149–156.

Adis International (2002). Secretin - Repligen: SecreFlo. *Drugs in R&D* **3**, 217–219.

Alamy, S. S., Jarskog, L. F., Sheitman, B. B., and Lieberman, J. A. (2004). Secretin in a patient with treatment-resistant schizophrenia and prominent autistic features. *Schizophr. Res.* **66**, 183–186.

Allen, A., Hutton, D. A., Leonard, A. J., Pearson, J. P., and Sellers, L. A. (1986). The role of mucus in the protection of the gastroduodenal mucosa. *Scand. J. Gastroenterol.* **125**, 71–78.

Allen, G., and Courchesne, E. (2003). Differential effects of developmental cerebellar abnormality on cognitive and motor functions in the cerebellum: An fMRI study of autism. *Am. J. Psychiatry* **160**, 262–273.

Aylward, E. H. (2002). Effects of age on brain volume and head circumference in autism. *Neurology* **59**, 175–183.

Aylward, E. H., Minshew, N. J., Goldstein, G., Honeycutt, N. A., Augustine, A. M., Yates, K. O., Barta, P. E., and Pearlson, G. D. (1999). MRI volumes of amygdala and hippocampus in non-mentally retarded autistic adolescents and adults. *Neurology* **53**, 2145–2150.

Babarczy, E., Szabo, G., and Telegdy, G. (1995). Effects of secretin on acute and chronic effects of morphine. *Pharmacol. Biochem. Behav.* **51**, 469–472.

Babu, G. N., and Vijayan, E. (1983). Plasma gonadotropin, prolactin levels and hypothalamic tyrosine hydroxylase activity following intraventricular bombesin and secretin in ovariectomized conscious rats. *Brain Res. Bull.* **11**, 25–29.

Banks, W. A., Goulet, M., Rusche, J. R., Niehoff, M. L., and Boismenu, R. (2002). Differential transport of a secretin analog across the blood-brain and blood-cerebrospinal fluid barriers of the mouse. *J. Pharmacol. Exp. Ther.* **302**, 1062–1069.

Baron-Cohen, S., Ring, H. A., Bullmore, E. T., Wheelwright, S., Ashwin, C., and Williams, S. C. (2000). The amygdala theory of autism. *Neurosci. Biobehav. Rev.* **24**, 355–364.

Bawab, W., Chastre, E., and Gespach, C. (1991). Functional and structural characterization of the secretin receptors in rat gastric glands: Desensitization and glycoprotein nature. *Biosci. Rep.* **11**, 33–42.

Bayliss, W. M., and Starling, E. H. (1902). The mechanism of pancreatic secretion. *J. Physiol. (London)* **28**, 325–353.

Belhage, B., Hansen, G. H., Elster, L., and Schousboe, A. (1998). Effects of gamma-aminobutyric acid (GABA) on synaptogenesis and synaptic function. *Perspect. Dev. Neurobiol.* **5**, 235–246.

Ben Ari, Y., Khazipov, R., Leinekugel, X., Caillard, O., and Gaiarsa, J. L. (1997). GABAA, NMDA and AMPA receptors: A developmentally regulated 'menage a trois'. *Trends Neurosci.* **20**, 523–529.

Black, C., Kaye, J. A., and Jick, H. (2002). Relation of childhood gastrointestinal disorders to autism: Nested case-control study using data from the UK General Practice Research Database. *Br. Med. J.* **325**, 419–421.

Blatt, G. J., Fitzgerald, C. M., Guptill, J. T., Booker, A. B., Kemper, T. L., and Bauman, M. L. (2001). Density and distribution of hippocampal neurotransmitter receptors in autism: An autoradiographic study. *J. Autism Dev. Disord.* **31,** 537–543.

Carlquist, M., Jornvall, H., Forssmann, W. G., Thulin, L., Johansson, C., and Mutt, V. (1985). Human secretin is not identical to the porcine/bovine hormone. *IRCS J. Med. Sci.* **13,** 217–218.

Carlquist, M., Jornwall, H., and Mutt, V. (1981). Isolation and amino acid sequence of bovine secretin. *FEBS Lett.* **127,** 71–74.

Casanova, M. F., Buxhoeveden, D., and Gomez, J. (2003). Disruption in the inhibitory architecture of the cell minicolumn: Implications for autisim. *Neuroscientist* **9,** 496–507.

Casanova, M. F., Buxhoeveden, D. P., Switala, A. E., and Roy, E. (2002). Minicolumnar pathology in autism. *Neurology* **58,** 428–432.

Chace, D. H., Millington, D. S., Terada, N., Kahler, S. G., Roe, C. R., and Hofman, L. F. (1993). Rapid diagnosis of phenylketonuria by quantitative analysis for phenylalanine and tyrosine in neonatal blood spots by tandem mass spectrometry. *Clin. Chem.* **39,** 66–71.

Charlton, C. G., Miller, R. L., Crawley, J. N., Handelmann, G. E., and O'Donohue, T. L. (1983). Secretin modulation of behavioral and physiological functions in the rat. *Peptides* **4,** 739–742.

Charlton, C. G., Quirion, R., Handelmann, G. E., Miller, R. L., Jensen, R. T., Finkel, M. S., and O'Donohue, T. L. (1986). Secretin receptors in the rat kidney: Adenylate cyclase activation and renal effects. *Peptides* **7,** 865–871.

Chow, B. K. (1995). Molecular cloning and functional characterization of a human secretin receptor. *Biochem. Biophys. Res. Comm.* **212,** 204–211.

Ciesielski, K. T., and Knight, J. E. (1994). Cerebellar abnormality in autism: A nonspecific effect of early brain damage? *Acta Neurobiol. Exp. (Warsz.)* **54,** 151–154.

Clement, H. W., Kirsch, M., Hasse, C., Opper, C., Gemsa, D., and Wesemann, W. (1998). Effect of repeated immobilization on serotonin metabolism in different rat brain areas and on serum corticosterone. *J. Neural Transm.* **105,** 1155–1170.

Coniglio, S. J. (2001). A randomized, double-blind, placebo-controlled trial of single-dose intravenous secretin as treatment for children with autism. *J. Pediatr.* **138,** 649–655.

Coplan, J., and Souders, M. C. (2003). Children with autistic spectrum disorders. II: Parents are unable to distinguish secretin from placebo under double-blind conditions. *Arch. Dis. Child* **88,** 737–739.

Corbett, B., Khan, K., Czapansky-Beilman, D., Brady, N., Dropik, P., Goldman, D. Z., Delaney, K., Sharp, H., Mueller, I., Shapiro, E., and Ziegler, R. (2001). A double-blind, placebo-controlled crossover study investigating the effect of porcine secretin in children with autism. *Clin. Pediatr. (Phila)* **40,** 327–331.

Courchesne, E., Karns, C. M., Davis, H. R., Ziccardi, R., Carper, R. A., Tigue, Z. D., Chisum, H. J., Moses, P., Pierce, K., Lord, C., Lincoln, A. J., Pizzo, S., Schreibman, L., Haas, R. H., Akshoomoff, N. A., and Courchesne, R. Y. (2001). Unusual brain growth patterns in early life in patients with autistic disorder: An MRI study. *Neurology* **57,** 245–254.

D'Eufemia, P., Celli, M., Finocchiaro, R., Pacifico, L., Viozzi, L., Zaccagnini, M., Cardi, E., and Giardini, O. (1996). Abnormal intestinal permeability in children with autism. *Acta Paediatr.* **85,** 1076–1079.

de Lecea, L., Kilduff, T. S., Peyron, C., Gao, X., Foye, P. E., Danielson, P. E., Fukuhara, C., Battenberg, E. L., Gautvik, V. T., Bartlett, F. S., Frankel, W. N., van den Pol, A. N., Bloom, F. E., Gautvik, K. M., and Sutcliffe, J. G. (1998). The hypocretins: Hypothalamus-specific peptides with neuroexcitatory activity. *Proc. Natl. Acad. Sci. USA* **95,** 322–327.

Dembinski, A. B., and Johnson, L. R. (1980). Stimulation of pancreatic growth by secretin, caerulein, and pentagastrin. *Endocrinology* **106,** 323–328.

Dhossche, D., Applegate, H., Abraham, A., Maertens, P., Bland, L., Bencsath, A., and Martinez, J. (2002). Elevated plasma gamma-aminobutyric acid (GABA) levels in autistic youngsters: Stimulus for a GABA hypothesis of autism. *Med. Sci. Monit.* **8,** PR1–PR6.

Dunn-Geier, J., Ho, H. H., Auersperg, E., Doyle, D., Eaves, L., Matsuba, C., Orrbine, E., Pham, B., and Whiting, S. (2000). Effect of secretin on children with autism: A randomized controlled trial. *Dev. Med. Child Neurol.* **42,** 796–802.

Egaas, B., Courchesne, E., and Saitoh, O. (1995). Reduced size of corpus callosum in autism. *Arch. Neurol.* **52,** 794–801.

Eriksson, P. S., Perfilieva, E., Bjork-Eriksson, T., Alborn, A. M., Nordborg, C., Peterson, D. A., and Gage, F. H. (1998). Neurogenesis in the adult human hippocampus. *Nat. Med.* **4,** 1313–1317.

Esch, B. E., and Carr, J. E. (2004). Secretin as a treatment for autism: A review of the evidence. *J. Autism. Dev. Disord.* **34,** 543–556.

Fatemi, S. H., Halt, A. R., and Realmuto (2002). Purkinje cell size is reduced in cerebellum of patients with autism. *Cell. Mol. Biol.* **22,** 171–175.

Fremeau, R. T., Korman, L. Y., and Moody, T. W. (1986). Secretin stimulates cyclic AMP formation in the rat brain. *J. Neurochem.* **46,** 1947–1955.

Fremeau, R. T., Jr., Jensen, R. T., Charlton, C. G., Miller, R. L., O'Donohue, T. L., and Moody, T. W. (1983). Secretin: Specific binding to rat brain membranes. *J. Neurosci.* **3,** 1620–1625.

Freund, T. F., and Gulyas, A. I. (1997). Inhibitory control of GABAergic interneurons in the hippocampus. *Can. J. Physiol. Pharmacol.* **75,** 479–487.

Furlano, R. I., Anthony, A., Day, R., Brown, A., McGarvey, L., Thomson, M. A., Davies, S. E., Berelowitz, M., Forbes, A., Wakefield, A. J., Walker-Smith, J. A., and Murch, S. H. (2001). Colonic CD8 and gamma delta T-cell infiltration with epithelial damage in children with autism. *J. Pediatr.* **138,** 366–372.

Fuxe, K., Andersson, K., Hokfelt, T., Mutt, V., Ferland, L., Agnati, L. F., Ganten, D., Said, S., Eneroth, P., and Gustafsson, J. A. (1979). Localization and possible function of peptidergic neurons and their interactions with central catecholamine neurons, and the central actions of gut hormones. *Fed. Proc.* **38,** 2333–2340.

Gafvelin, G., Jornvall, H., and Mutt, V. (1990). Processing of prosecretin: Isolation of a secretin precursor from porcine intestine. *Acad. Sci. USA* **87,** 6781–6785.

Gandhi, S., Rubinstein, I., Tsueshita, T., and Onyuksel, H. (2002). Secretin self-assembles and interacts spontaneously with phospholipids *in vitro*. *Peptides* **23,** 201–204.

Gerlach, M., Gsell, W., Kornhuber, J., Jellinger, K., Krieger, V., Pantucek, F., Vock, R., and Riederer, P. (1996). A post mortem study on neurochemical markers of dopaminergic, GABA-ergic and glutamatergic neurons in basal ganglia-thalamocortical circuits in Parkinson syndrome. *Brain Res.* **741,** 142–152.

Gossen, D., Tastenoy, M., Robberecht, P., and Christophe, J. (1990). Secretin receptors in the neuroglioma hybrid cell line NG108–15. Characterization and regulation of their expression. *Eur. J. Biochem.* **193,** 149–154.

Gossen, D., Vandermeers, A., Vandermeers-Piret, M. C., Rathe, J., Cauvin, A., Robberecht, P., and Christophe, J. (1989). Isolation and primary structure of rat secretin. *Biochem. Biophys. Res. Comm.* **160,** 862–867.

Goulet, M., Blais, V., Shiromani, P., Boismenu, R., Rusche, J., and Rivest, S. (2001). Activation of the rat central nervous system by intravenous secretin infusion. Society for Neuroscience's 31st Annual Meeting, San Diego.

Gwag, B. J., Sessler, F. M., Robine, V., and Springer, J. E. (1997). Endogenous glutamate levels regulate nerve growth factor mRNA expression in the rat dentate gyrus. *Mol. Cells* **7,** 425–430.

Henriksen, J. H., and de Muckadell, O. B. (2000). Secretin, its discovery, and the introduction of the hormone concept. *Scand. J. Clin. Lab. Invest.* **60,** 463–471.

Herrera-Marschitz, M., You, Z. B., Goiny, M., Meana, J. J., Silveira, R., Godukhin, O. V., Chen, Y., Espinoza, S., Pettersson, E., Loidl, C. F., Lubec, G., Andersson, K., Nylander, I., Terenius, L., and Ungerstedt, U. (1996). On the origin of extracellular glutamate levels monitored in the basal ganglia of the rat by *in vivo* microdialysis. *J. Neurochem.* **66,** 1726–1735.

Holtmann, M. H., Roettger, B. F., Pinon, D. I., and Miller, L. J. (1996). Role of receptor phosphorylation in desensitization and internalization of the secretin receptor. *J. Biol. Chem.* **271,** 23566–23571.

Horvath, K., and Perman, J. A. (2002). Autism and gastrointestinal symptoms. *Curr. Gastroenterol. Reports* **4,** 251–258.

Horvath, K., Stefanatos, G., Sokolski, K. N., Wachtel, R., Nabors, L., and Tildon, J. T. (1998). Improved social and language skills after secretin administration in patients with autistic spectrum disorders. *J. Assoc. Acad. Minor. Phys.* **9,** 9–15.

Hussman, J. P. (2001). Suppressed GABAergic inhibition as a common factor in suspected etiologies of autism. *J. Autism Dev. Disord.* **31,** 247–248.

IMGSAC (2001). A genomewide screen for autism: Strong evidence for linkage to chromosomes 2q, 7q, and 16p. *Am. J. Hum. Gen.* **69,** 570–581.

Ip, N. Y., Ho, C. K., and Zigmond, R. E. (1982). Secretin and vasoactive intestinal peptide acutely increase tyrosine 3-monooxygenase in the rat superior cervical ganglion. *Proc. Natl. Acad. Sci. USA* **79,** 7566–7569.

Ishihara, T., Nakamura, S., Kaziro, Y., Takahashi, T., Takahashi, K., and Nagata, S. (1991). Molecular cloning and expression of a cDNA encoding the secretin receptor. *EMBO J.* **10,** 1635–1641.

Itoh, N., Furuya, T., Ozaki, K., Ohta, M., and Kawasaki, T. (1991). The secretin precursor gene. Structure of the coding region and expression in the brain. *J. Biol. Chem.* **266,** 12595–12598.

Jiang, S. (1995). Characterization of a human pancreatic secretin receptor and its expression in human ductal pancreatic adenocarcinomas. *Gastroenterology* **108**(Suppl. 3), A978.

Jin, H. O., Lee, K. Y., Chang, T. M., Chey, W. Y., and Dubois, A. (1994). Secretin: A physiological regulator of gastric emptying and acid output in dogs. *Am. J. Physiol.* **267,** G702–G708.

Jo, Y. H., Yoon, S. H., Hahn, S. J., Rhie, D. J., Sim, S. S., and Kim, M. S. (1994). Effect of medial amygdaloid stimulation on pancreatic exocrine secretion in anesthetized rats. *Pancreas* **9,** 117–122.

Joo, K. M., Chung, Y. H., Kim, M. K., Nam, R. H., Lee, B. L., Lee, K. H., and Cha, C. I. (2004). Distribution of vasoactive intestinal peptide and pituitary adenylate cyclase-activating polypeptide receptors (VPAC1, VPAC2, and PAC1 receptor) in the rat brain. *J. Comp. Neurol.* **476,** 388–413.

Jorpes, J. E., and Mutt, V. (1961). On the biological activity and amino acid composition of secretin. *Acta Chem. Scand.* **151,** 1790–1791.

Karelson, E., Laasik, J., and Sillard, R. (1995). Regulation of adenylate cyclase by galanin, neuropeptide Y, secretin and vasoactive intestinal polypeptide in rat frontal cortex, hippocampus and hypothalamus. *Neuropeptides* **28,** 21–28.

Kern, J. K., Van Miller, S., Evans, P. A., and Trivedi, M. H. (2002). Efficacy of porcine secretin in children with autism and pervasive developmental disorder. *J Autism Dev. Disord.* **32,** 153–160.

Kopin, A. S., Wheeler, M. B., and Leiter, A. B. (1990). Secretin: Structure of the precursor and tissue distribution of the mRNA. *Proc. Natl. Acad. Sci. USA* **87,** 2299–2303.

Koves, K., Kausz, M., Reser, D., Illyes, G., Takacs, J., Heinzlmann, A., Gyenge, E., and Horvath, K. (2004). Secretin and autism: A basic morphological study about the distribution of secretin in the nervous system. *Regul. Pept.* **123,** 209–216.

Krishnadas, A., Onyuksel, H., and Rubinstein, I. (2003). Interactions of VIP, secretin and PACAP (1–38) with phospholipids: A biological paradox revisited. *Curr. Pharm. Des.* **9,** 1005–1012.

Kuntz, A., Clement, H. W., Lehnert, W., van Calker, D., Hennighausen, K., Gerlach, M., and Schulz, E. (2004). Effects of secretin on extracellular amino acid concentrations in rat hippocampus. *J. Neural. Transm.* **111,** 931–939.

Lamson, D. W., and Plaza, S. M. (2001). Transdermal secretin for autism - a case report. *Altern. Med. Rev.* **6,** 311–313.

Lightdale, J. R., Hayer, C., Duer, A., Lind-White, C., Jenkins, S., Siegel, B., Elliott, G. R., and Heyman, M. B. (2001). Effects of intravenous secretin on language and behavior of children with autism and gastrointestinal symptoms: A single-blinded, open-label pilot study. *Pediatrics* **108,** E90.

Lossi, L., Bottarelli, L., Candusso, M. E., Leiter, A. B., Rindi, G., and Merighi, A. (2004). Transient expression of secretin in serotoninergic neurons of mouse brain during development. *Eur. J. Neurosci.* **20,** 3259–3269.

Lyubashina, O. A. (2004). Possible mechanisms of involvement of the amygdaloid complex in the control of gastric motor function. *Neurosci. Behav. Physiol.* **34,** 379–388.

Marinelli, R. A., Pham, L., Agre, P., and LaRusso, N. F. (1997). Secretin promotes osmotic water transport in rat cholangiocytes by increasing aquaporin-1 water channels in plasma membrane. Evidence for a secretin-induced vesicular translocation of aquaporin-1. *J. Biol. Chem.* **272,** 12984–12988.

Marvel, F. A., Chen, C. C., Badr, N., Gaykema, R. P., and Goehler, L. E. (2004). Reversible inactivation of the dorsal vagal complex blocks lipopolysaccharide-induced social withdrawal and c-Fos expression in central autonomic nuclei. *Brain. Behav. Immun.* **18,** 123–134.

McCulloch, D. A., MacKenzie, C. J., Johnson, M. S., Robertson, D. N., Holland, P. J., Ronaldson, E., Lutz, E. M., and Mitchell, R. (2002). Additional signals from VPAC/PAC family receptors. *Biochem. Soc. Trans.* **30,** 441–446.

McGill, J. M., Basavappa, S., Gettys, T. W., and Fitz, J. G. (1994). Secretin activates Cl- channels in bile duct epithelial cells through a cAMP-dependent mechanism. *Am. J. Physiol.* **266,** G731–G736.

Meyer, J. H., Way, L. W., and Grossman, M. I. (1970). Pancreatic response to acidification of various lengths of proximal intestine in the dog. *Am. J. Physiol.* **219,** 971–977.

Minshew, N. J., Goldstein, G., Dombrowski, S. M., Panchalingam, K., and Pettegrew, J. W. (1993). A preliminary 31P MRS study of autism: Evidence for undersynthesis and increased degradation of brain membranes. *Biol. Psychiatry* **33,** 762–773.

Mutt, V., Jorpes, J. E., and Magnusson, S. (1970). Structure of porcine secretin: The amino acid sequence. *Eur. J. Biochem.* **15,** 513–519.

Muhle, R., Trentacoste, S. V., and Rapin, I. (2004b). The genetics of autism. *Pediatrics* **113,** e472–e486.

Ng, S. S., Yung, W. H., and Chow, B. K. (2002). Secretin as a neuropeptide. *Mol. Neurobiol.* **26,** 97–107.

O'Donohue, T. L., Charlton, C. G., Miller, R. L., Boden, G., and Jacobowitz, D. M. (1981). Identification, characterization, and distribution of secretin immunoreactivity in rat and pig brain. *Proc. Natl. Acad. Sci. USA* **78,** 5221–5224.

Ohta, M., Funakoshi, S., Kawasaki, T., and Itoh, N. (1992). Tissue-specific expression of the rat secretin precursor gene. *Biochem. Biophys. Res. Comm.* **183,** 390–395.

Olsen, P. S., Kirkegaard, P., Poulsen, S. S., and Nexo, E. (1994). Effect of secretin and somatostatin on secretion of epidermal growth factor from Brunner's glands in the rat. *Dig. Dis. Sci.* **39,** 2186–2190.

Ozcelebi, F., Holtmann, M. H., Rentsch, R. U., Rao, R., and Miller, L. J. (1995). Agonist-stimulated phosphorylation of the carboxyl-terminal tail of the secretin receptor. *Mol. Pharmacol.* **48,** 818–824.

Patel, D. R., Kong, Y., and Sreedharan, S. P. (1995). Molecular cloning and expression of a human secretin receptor. *Mol. Pharmacol.* **47,** 467–473.

Paxinos, G., and Watson, C. (1982). "The Rat Brain in Stereotaxic Coordinates." Academic Press, New York.

Polak, J. M., Coulling, I., Bloom, S., and Pearse, A. G. (1971). Immunofluorescent localization of secretin and enteroglucagon in human intestinal mucosa. *Scand. J. Gastroenterol.* **6,** 739–744.

Purcell, A. E., Jeon, O. H., Zimmermann, A. W., Blue, M. E., and Pevsner, J. (2001). Postmortem brain abnormalities of the glutamate neurotransmitter system in autism. *Neurology* **57,** 1618–1628.

Rausch, U., Vasiloudes, P., Rudiger, K., and Kern, H. F. (1985). *In vivo* stimulation of rat pancreatic acinar cells by infusion of secretin. II. Changes in individual rates of enzyme and isoenzyme biosynthesis. *Cell Tissue Res.* **242,** 641–644.

Raymond, G. V., Bauman, M. L., and Kemper, T. L. (1996). Hippocampus in autism: A Golgi analysis. *Acta Neuropathol. (Berl)* **91,** 117–119.

Rice, P. J., Lindsay, G. W., Bogan, C. R., and Hancock, J. C. (1999). cAMP and *in vitro* inotropic actions of secretin and VIP in rat papillary muscle. *Peptides* **20,** 519–522.

Rindi, G., Lossi, L., Merighi, A., and Leiter, A. B. (2001). Brain stem serotoninergic neurons express the gastrointestinal hormone secretin during embryonic development and in adult mice: Evidence from normal and transgenic animals. Society for Neuroscience's 31st Annual Meeting, San Diego.

Robberecht, P., Conlon, T. P., and Gardner, J. D. (1976). Interaction of porcine vasoactive intestinal peptide with dispersed pancreatic acinar cells from the guinea pig. Structural requirements for effects of vasoactive intestinal peptide and secretin on cellular adenosine $3':5'$-monophosphate. *J. Biol. Chem.* **251,** 4635–4639.

Roberts, W., Weaver, L., Brian, J., Bryson, S., Emelianova, S., Griffiths, A. M., MacKinnon, B., Yim, C., Wolpin, J., and Koren, G. (2001). Repeated doses of porcine secretin in the treatment of autism: A randomized, placebo-controlled trial. *Pediatrics* **107,** E71.

Rolf, L. H., Haarmann, F. Y., Grotemeyer, K. H., and Kehrer, H. (1993). Serotonin and amino acid content in platelets of autistic children. *Acta Psychiatr. Scand.* **87,** 312–316.

Roskoski, R., Jr., White, L., Knowlton, R., and Roskoski, L. M. (1989). Regulation of tyrosine hydroxylase activity in rat PC12 cells by neuropeptides of the secretin family. *Mol. Pharmacol.* **36,** 925–931.

Saitoh, O., Karns, C. M., and Courchesne, E. (2001). Development of the hippocampal formation from 2 to 42 years: MRI evidence of smaller area dentata in autism. *Brain* **124,** 1317–1324.

Samson, W. K., Lumpkin, M. D., and McCann, S. M. (1984). Presence and possible site of action of secretin in the rat pituitary and hypothalamus. *Life Sci.* **34,** 155–163.

Sandler, A. D., Sutton, K. A., DeWeese, J., Girardi, M. A., Sheppard, V., and Bodfish, J. W. (1999). Lack of benefit of a single dose of synthetic human secretin in the treatment of autism and pervasive developmental disorder. *N. Engl. J. Med.* **341,** 1801–1806.

Sheitman, B. B., Knable, M. B., Jarskog, L. F., Chakos, M., Boyce, L. H., Early, J., and Lieberman, J. A. (2004). Secretin for refractory schizophrenia. *Schizophr. Res.* **66,** 177–181.

Shinomura, Y., Eng, J., and Yalow, R. S. (1987). Dog secretin: Sequence of biological activity. *Life Sci.* **41,** 1243–1248.

Sreedharan, S. P., Patel, D. R., Xia, M., Ichikawa, S., and Goetzl, E. J. (1994). Human vasoactive intestinal peptide1 receptors expressed by stable transfectants couple to two distinct signaling pathways. *Biochem. Biophys. Res. Comm.* **203,** 141–148.

Tay, J., Goulet, M., Rusche, J., and Boismenu, R. (2004). Age-related and regional differences in secretin and secretin receptor mRNA levels in the rat brain. *Neurosci. Lett.* **366,** 176–181.

Timmerman, W., and Westerink, B. H. (1997). Brain microdialysis of GABA and glutamate: What does it signify? *Synapse* **27,** 242–261.

Trimble, E. R., Bruzzone, R., Biden, T. J., Meehan, C. J., Andreu, D., and Merrifield, R. B. (1987). Secretin stimulates cyclic AMP and inositol trisphosphate production in rat pancreatic acinar tissue by two fully independent mechanisms. *Proc. Natl. Acad. Sci. USA* **84,** 3146–3150.

Ulrich, C. D., Holtmann, M., and Miller, L. J. (1998). Secretin and vasoactive intestinal peptide receptors: Members of a unique family of G protein-coupled receptors. *Gastroenterology* **114,** 382–397.

Ulrich, C. D., Pinon, D. I., Hadac, E. M., Holicky, E. L., Chang-Miller, A., Gates, L. K., and Miller, L. J. (1993). Intrinsic photoaffinity labeling of native and recombinant rat pancreatic secretin receptors. *Gastroenterology* **105,** 1534–1543.

Unis, A. S., Munson, J. A., Rogers, S. J., Goldson, E., Osterling, J., Gabriels, R., Abbott, R. D., and Dawson, G. (2002). A randomized, double-blind, placebo-controlled trial of porcine versus synthetic secretin for reducing symptoms of autism. *J. Am. Acad. Child Adolesc. Psychiatry* **41,** 1315–1321.

van Calker, D., Muller, M., and Hamprecht, B. (1980). Regulation by secretin, vasoactive intestinal peptide, and somatostatin of cyclic AMP accumulation in cultured brain cells. *Proc. Natl. Acad. Sci. USA* **77,** 6907–6911.

Vaudry, D., Gonzalez, B. J., Basille, M., Yon, L., Fournier, A., and Vaudry, H. (2000). Pituitary adenylate cyclase-activating polypeptide and its receptors: From structure to functions. *Pharmacol. Rev.* **52,** 269–324.

Vizi, E. S., and Kiss, J. P. (1998). Neurochemistry and pharmacology of the major hippocampal transmitter systems: Synaptic and nonsynaptic interactions. *Hippocampus* **8,** 566–607.

Voisin, T., Couvineau, A., Guijarro, L., and Laburthe, M. (1991). VIP receptors from porcine liver: High yield solubilization in a GTP-insensitive form. *Life Sci.* **48,** 135–141.

Wallenstein, G. V., and Hasselmo, M. E. (1997). GABAergic modulation of hippocampal population activity: Sequence learning, place field development, and the phase precession effect. *J. Neurophysiol.* **78,** 393–408.

Wang, H. L., Li, A., and Wu, T. (1997). Vasoactive intestinal polypeptide enhances the GABAergic synaptic transmission in cultured hippocampal neurons. *Brain Res.* **746,** 294–300.

Watanabe, S., Chey, W. Y., Lee, K. Y., and Chang, T. M. (1986). Secretin is released by digestive products of fat in dogs. *Gastroenterology* **90,** 1008–1017.

Welch, M. G., Keune, J. D., Welch-Horan, T. B., Anwar, N., Anwar, M., and Ruggiero, D. A. (2003). Secretin activates visceral brain regions in the rat including areas abnormal in autism. *Cell. Mol. Biol.* **23,** 817–837.

Wessels-Reiker, M., Basiboina, R., Howlett, A. C., and Strong, R. (1993). Vasoactive intestinal polypeptide-related peptides modulate tyrosine hydroxylase gene expression in PC12 cells through multiple adenylate cyclase-coupled receptors. *J. Neurochem.* **60,** 1018–1029.

Wheeler, M. B., Nishitani, J., Buchan, A. M., Kopin, A. S., Chey, W. Y., Chang, T. M., and Leiter, A. B. (1992). Identification of a transcriptional enhancer important for enteroendocrine and pancreatic islet cell-specific expression of the secretin gene. *Mol. Cell Biol.* **12,** 3531–3539.

Whitmore, T. E., Holloway, J. L., Lofton-Day, C. E., Maurer, M. F., Chen, L., Quinton, T. J., Vincent, J. B., Scherer, S. W., and Lok, S. (2000). Human secretin (SCT): Gene structure, chromosome location, and distribution of mRNA. *Cytogenet. Cell Genet.* **90,** 47–52.

Yang, B., Goulet, M., Boismenu, R., and Ferguson, A. V. (2004a). Secretin depolarizes nucleus tractus solitarius neurons through activation of a nonselective cationic conductance. *Am. J. Physiol. Regul. Integr. Comp. Physiol.* **286,** R927–R934.

Yang, H., Wang, L., Wu, S. V., Tay, J., Goulet, M., Boismenu, R., Czimmer, J., Wang, Y., Wu, S., Ao, Y., and Tache, Y. (2004b). Peripheral secretin-induced Fos expression in the rat brain is largely vagal dependent. *Neuroscience* **128,** 131–141.

You, C. H., and Chey, W. Y. (1987). Secretin is an enterogastrone in humans. *Dig. Dis. Sci.* **32,** 466–471.

Yung, W. H., Leung, P. S., Ng, S. S., Zhang, J., Chan, S. C., and Chow, B. K. (2001). Secretin facilitates GABA transmission in the cerebellum. *J. Neurosci.* **21,** 7063–7068.

Zhang, X., Cui, J., Tan, Z., Jiang, C., and Fogel, R. (2003). The central nucleus of the amygdala modulates gut-related neurons in the dorsal vagal complex in rats. *J. Physiol. (Lond)* **553,** 1005–1018.

PREDICTED ROLE OF SECRETIN AND OXYTOCIN IN THE TREATMENT OF BEHAVIORAL AND DEVELOPMENTAL DISORDERS: IMPLICATIONS FOR AUTISM

Martha G. Welch* and David A. Ruggiero*,†

*Department of Psychiatry, Division of Neuroscience, Columbia University College of
Physicians & Surgeons, New York, New York 10032, USA
†Department of Anatomy & Cell Biology, Columbia University College of Physicians &
Surgeons, New York, New York 10032, USA

The long-term goal of our work is to create a novel treatment for autism and to explain its pathogenesis. Based on a theory that views emotions and emotional behavior as stemming from dysregulations of a unified brain/gut network, we propose a new paradigm for the treatment of mental illness. This chapter reviews evidence that two neuropeptides, secretin and oxytocin, are critical in the conditioning of infant adaptive behavioral patterns and that peptidergic mechanisms are abnormal in developmental disorders such as autism.

Our clinical observations in the treatment of autism justify our laboratory investigations into the role of peptides in the neurological manifestations of visceral diseases encompassing emotional/visceral brain regions abnormal in

INTERNATIONAL REVIEW OF
NEUROBIOLOGY, VOL. 71
DOI: 10.1016/S0074-7742(05)71012-6

273

autism. Importantly, our studies have thus far demonstrated that (1) visceral inflammation activates visceral/emotional brain regions in areas known to be abnormal in autism; (2) secretin, like oxytocin, activates many of the same visceral/emotional brain regions that are dysregulated in chronic cerebral and visceral disorders such as autism; (3) a structural basis for the mechanisms of action of secretin and oxytocin was clarified; (4) secretin as well as oxytocin is synthesized in the hypothalamus and may act on structures involved in the pathophysiology of autism; (5) secretin and oxytocin localize to perivascular and subependymal regions of the paraventricular hypothalamus, suggesting a chemosensory and secretory function.

We believe autism is the result of an adverse cascade of events that stems from one or more genetic/environmental insults. Over time, if uncompensated, the cascade leads to adverse conditioning of stress adaptation networks and results in various interrelated developmental psychological, neurological and immunological pathology, including autism.

Our laboratory is engaged in efforts to translate clinical experience in the treatment of autistic patients into bench findings. We believe that it is possible, regardless of etiology, to treat autism and related developmental disorders by intervening in stress response mechanisms with exogenous administration of peptide combinations.

I. Introduction

To date, there is no comprehensive treatment for the broad range of autistic symptomatology: seizures (Park, 2003); attentional/arousal dysregulation, attentional deficit hyperactive disorder (Booth et al., 2003); obsessive-compulsive disorder (Hollander et al., 2003); stereotypies (Militerni et al., 2002); social isolation (Iqbal, 2002); attachment disorders (Kobayashi et al., 2001; Tinbergen and Tinbergen, 1983); face recognition deficits (Ogai et al., 2003; Schultz et al., 2003); gaze aversions (Richer and Coss, 1976); gastrointestinal disorders (Gershon and D'Autreaux, 2003; Horvath and Perman, 2002; Horvath et al., 1998; Torrente et al., 2002); and altered heart rate variability (Corona et al., 1998; Graveling and Brooke, 1978). Research in the field of autism and related disorders over the last twenty years has produced a large body of knowledge but has yet to produce any significant therapeutic outcomes. The search for even partially ameliorative interventions is a goal for all parents of autistic children.

Current pharmacologic treatments, such as anti-psychotics, mood stabilizers, antidepressants, anticonvulsants, and single peptides treat single symptoms, often with unacceptable side effects and/or limited therapeutic effects (Ansorge et al., 2004; Coniglio et al., 2001; Dunn-Geier et al., 2000; Kern et al., 2002; Lightdale *et al.*, 2001; Owley *et al.*, 2001; Posey and McDougle, 2000; Roberts *et al.*, 2001; Sandler, 1999). Psychotherapeutic measures have also been attempted with limited success (Diggle *et al.*, 2003; Langworthy-Lam *et al.*, 2002). Despite great efforts to understand the etiologies of autism and to devise treatments, autism and many severe behavioral disorders are still generally considered to be idiopathic and incurable.

The current paradigm for treating mental illness was greatly influenced by the work of Walter Cannon and Phillip Bard (Bard, 1928; Cannon, 1927). Their work set aside the visceral theory of emotions of William James and Carl Lange (James, 1884; Lange, 1885), who argued that emotional states of well- or ill-being are the result of visceral sensations (see Fig. 1A). Instead, Cannon and Bard argued that

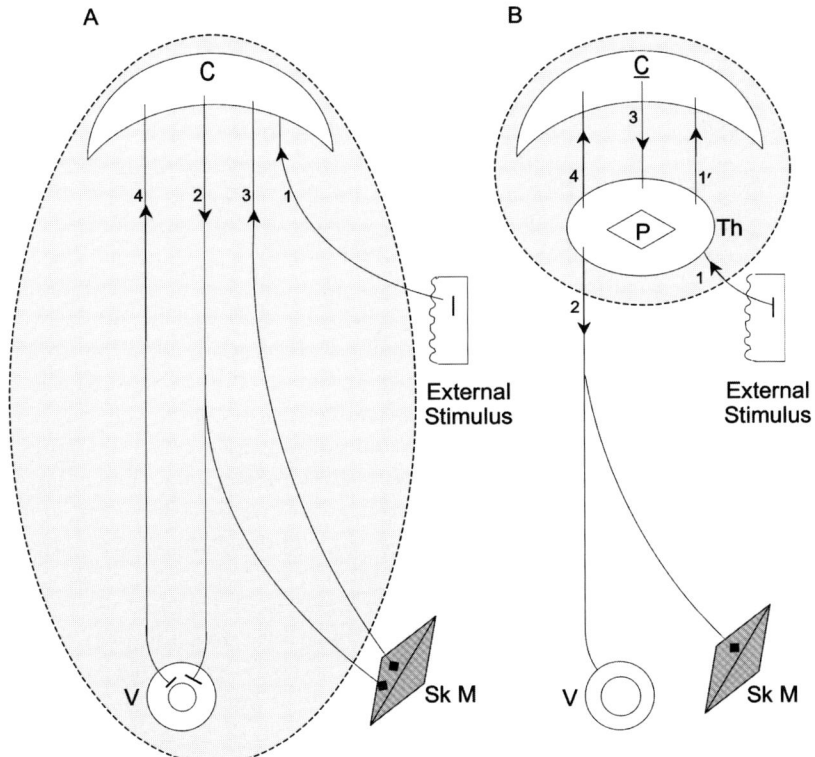

FIG. 1. Modified schematic of (A) James-Lange's visceral theory of emotion (1884–1885). (B) Cannon-Bard's revisionist thalamic theory of emotions (1927–1928) (Lissak and Molnar, 1975). According to James-Lange, emotions are the product of brain and gut, as indicated by bi-directional communication between viscera and brain and dotted line. Canon-Bard believed emotions are generated entirely in the brain, as indicated by one-way communication from brain to viscera and dotted line. **C** = Cortex; **P** = Pattern Generator; **Sk M** = Skeletomuscle; **Th** = Thalamus; **V** = Viscera.

emotional reactions do not stem from the viscera but result rather from behavioral patterns generated by the thalamus/hypothalamus (Fig. 1B). However, there has since been a growing acceptance that the viscera play a role in the generation of emotions (Damasio, 1994; LeDoux, 1998). Nonetheless, the scientific and health care communities still assume to a very large extent that behavior originates in the brain and therefore, in order to affect behavior, one must intervene in the processes of the brain. This chapter will present findings that support a revised theory on the origin and nature of behavior, one that logically calls for a new paradigm in the treatment of developmental and behavioral disorders.

We will argue that rather than originating in the brain, developmental disorders arise from dysregulation of a unified brain/gut system and are the result of a cascade of interrelated psychological, neurological, and immunological reactions to unmodulated stress (see Fig. 2). Further, we will argue that it is possible to ameliorate developmental and behavioral disorders, regardless of etiology, by intervening in stress mechanisms with treatments that target both the brain and periphery simultaneously.

Clinical observations in the psychiatric practice of Welch form the framework for the concepts reviewed in this chapter (Tinbergen and Tinbergen, 1983; Welch, 1983, 1988; Welch and Chaput, 1988; Welch et al., 2004c, 2006). Two seemingly disparate groups of patients, consisting of maternally deprived orphans and autistic children, were treated for two shared symptom complexes: behavioral symptoms such as lack of direct eye contact, indiscriminate approaches toward strangers, inability to respond to normal maternal nurturing, and odd or restricted food preferences, and gastrointestinal (GI) symptoms such as gut motility abnormalities, discomfort, and diarrhea. Welch developed an intervention that employs intense nurturing as a means of conditioning stress adaptation responses. The intervention led to concurrent amelioration of both behavioral and gut symptoms. In many cases following the intervention, direct eye contact between mother and child ensued, the child was able to benefit from normal nurturing, adverse behaviors were dramatically reduced, and GI symptoms abated. At the end of the intervention, mothers who had previously experienced childbirth often described feeling as though they had just given birth. These collective bedside observations led to a theory that the two groups share a common dysregulation of underlying stress mechanisms.

Welch attributed the striking changes observed between mother and child in the therapy to the simultaneous release of natural endogenous peptides, especially the bonding peptide oxytocin (Uvnas-Moberg 1989; Welch et al., 2004c, 2005, 2006). A serendipitous discovery involving secretin (Horvath et al., 1998) provided an additional candidate for the hypothesized peptidergic mechanism, as well as further support for a brain/gut theory of developmental disorders. Secretin, given as a single dose probe of abnormal GI function in three autistic boys, resulted in improved eye contact and verbal communication. Welch reasoned that treatment with continuous exogenous combined secretin/oxytocin peptides

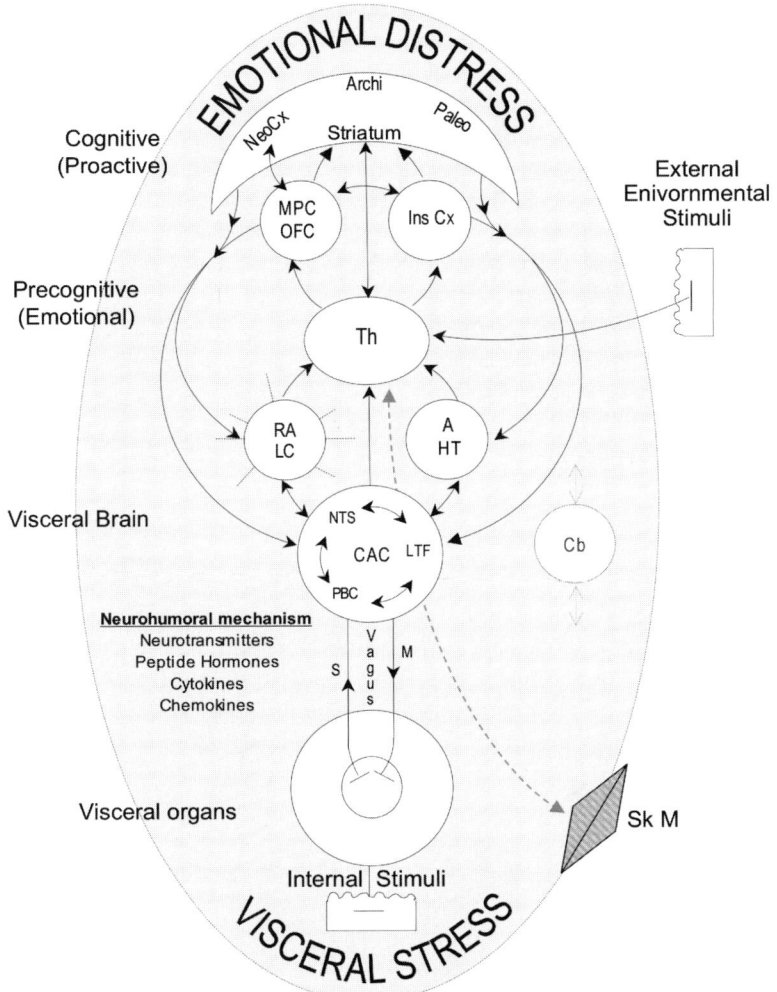

FIG. 2. Abbreviated schematic of the Welch-Ruggiero unified brain/gut theory of emotions. The simplified figure illustrates the fundamental circuits linking the viscera and emotional brain. The structures shown are sites of action of transmitters and peptides in mediating neurohumoral mechanisms sustaining adaptive behaviors accompanying arousal. Shaded area indicates viscera and brain are one system with bi-directional communication between the two, a theory that is in line with James-Lange (Fig. 1A). Most behavioral and pharmacologic therapies continue to target the brain and cognitive processes, assuming as did Cannon-Bard that emotions arise in the brain as a system separate from the viscera (Fig. 1B). **A** = Amygdala; **C** = Cortex; **CAC** = Central Autonomic Core; **Cb** = Cerebellum; **HP** = Hypothalamus; **Ins Cx** = Insular Cortex; **LTF** = Lateral Tegmental Field; **LC** = Locus Ceruleus; **M** = Motor; **MPC** = Medial Prefrontal Cortex; **NTS** = Nucleus of Solitary Tract; **OFC** = Orbital Frontal Cortex; **PBC** = Parabrachial Complex; **RA** = Raphe; **S** = Sensory; **Sk M** = Skeletal Muscle; **Th** = Thalamus; **V** = Viscera.

might replicate the physiological conditions that are elicited in normal reciprocal mother-infant interactions and ameliorate behavioral and GI symptoms in autism (Welch 1983, 1998, 2003b, 2005).

The relationship between brain/gut stress and developmental disorders such as autism is the framework for our scientific investigations. Prefrontal perceptual encoding mechanisms that regulate HPA stress axis and autonomic output to the viscera and immune system are impaired by gestational stress (Berger *et al.*, 2002). It is well established that early environmental stressors can permanently alter perceptual, emotional, intellectual, and social development; in autistic children, all of these are impaired. (Dawson *et al.*, 1998a,b). Autism is associated with high rates of visceral and immune disorders (Gupta *et al.*, 1998, 2000; White *et al.*, 2003) and familial autoimmunity is a risk factor (Szatmari, 1999).

We hypothesize that autism and associated disorders are the result of an adverse cascade of psychoneuroimmunological events that derive from one or more gene/environmental insults or unmitigated stressors. We also hypothesize that early intervention can interrupt the adverse cascade of events, thereby compensating for such insults and averting the further on-going sequelae that lead to severe chronic developmental disorders. Environmental insults can occur in utero or postnatally. Without intervention, the infant's stress profile will result in a failure to activate specific developmental programs, such as glucocorticoid and GABA receptor compositions. The cascade leads to a disruption in the stress-regulatory system of the developing infant, impairing ability to benefit from the caregiver's nurturing or stress-modulation. This disruption results in an interruption of key genetic developmental programs that are normally activated by peptidergic mechanisms. In the face of genetic/environmental stressors, the excess demands on the infant's stress-regulatory mechanisms make peptide modulation critical.

The failure we further hypothesize, persists until the peptide balance is restored. The earlier the silenced or arrested gene programs are activated by successful intervention, the less stress-induced damage will occur. Conversely, the longer the infant is unable to receive stress modulation, the more the infant's adaptation to environmental and emotional challenges is adversely conditioned. In such cases, in the face of unremitting environmental and emotional challenge, the infant adopts various maladaptive defense strategies that result in regressive adverse behaviors and a range of pathology. In the case of autistic children adverse behaviors include stereotypical movements, approach/avoidance behaviors, obsessiveness, compulsiveness, and tantrums.

We will report findings from our laboratory and we will review findings on transmitters and peptides that have been found to be dysregulated in developmental disorders, including peptidergic mechanisms that reset biological systems perturbed pre- and/or postnatally by unmodulated stress. We will examine the role that hypothalamic/gut peptides play in the modulation of these systems, especially as occurs naturally in the process of maternal nurturing. Finally, we will discuss the hypothesis that peptidergic mechanisms may lead to new clinical and pharmacologic therapies, and chart a future course for research and treatment of autism and related disorders.

II. Background

A. GABA, GENES, ENVIRONMENT, AND PEPTIDES

The fact that GABA inhibitory transmitter systems are genetically altered in autism (Lamb *et al.*, 2002; Ma *et al.*, 2005) is complicated by the fact that GABA receptor genes could also be activated or silenced environmentally (Caldji *et al.*, 2003, 2004). These findings may corroborate our clinical observations. Welch observed that autistic infants who were unable to engage in reciprocal bonding behaviors in response to maternal cues showed dramatic physiological change following successful treatment with intense maternal/infant nurturing (Tinbergen and Tinbergen, 1983; Welch, 1983, 1988; Welch and Chaput, 1988; Welch *et al.*, 2004c, 2006). Hofer may have identified the same phenomenon when he referred to "hidden regulators" of the mother/infant interaction (Hofer, 1994). Recent genetic research using animal models may offer a more compelling explanation of the phenomenon in terms of powerful cellular mechanisms underlying maternal/infant interactions.

Environmental effects can determine the activation status of a gene. It is common to think of gene effects as fixed. In fact, genes are activated or silenced continuously. Therefore, when environmental events silence or fail to activate a gene program, the outcome can be as deleterious as a genetic defect or abnormality, such as the $GABA_A$ receptor gene abnormality on Chromosome 15q11 associated with autism (Menold *et al.*, 2001).

GABA gene/environment interactions found in low-nurture rearing environments may be pertinent to children who do not or cannot benefit from normal maternal nurturing, such as in orphans and autistic children with abnormal face recognition and sensory processing. In animals, the level of maternal nurture that the offspring receives determines which gene programs for GABA receptor subunits are activated. The level of maternal care can permanently alter subunit composition of the $GABA_A$ receptor complex in brain regions that regulate responses to stress, including the amygdala. However, cross-fostering animal offspring of low-nurture mothers to high-nurture mothers reverses the subunit composition by selectively producing specific GABA receptor subunits (Caldji *et al.*, 2003). This finding may explain Welch's clinical observations that intense components of maternal nurture can ameliorate behavioral symptoms of low-nurture orphans.

In another experiment, cross-fostering of low-nurture animals to high-nurture mothers raised the numbers of glucocorticoid receptors in areas of hippocampus that determine stress reactivity (Meaney, 2004; Meaney and Szyf, 2005; Weaver *et al.*, 2004). Reduced numbers of glucocorticoid receptors were found in the brains of subjects with depression, bipolar disorder, and schizophrenia (Webster *et al.*, 2002). If this finding occurs in autism as well, it would lend further support to the concept that high nurture interventions might raise the numbers of hippocampal glucocorticoid receptors in autism.

GABA is a major inhibitory neurotransmitter responsible for sensory gating of stress-related information that influences behavioral, endocrine, and autonomic networks. It is implicated in many psychiatric disorders (Kalia, 2005; Lewis *et al.*, 2005; Roy-Byrne, 2005). Secretin and oxytocin facilitate GABA inhibition (Kuntz *et al.*, 2004; Zaninetti and Raggenbass, 2000). Secretin and oxytocin are abnormal in autism (Gershon and D'Autreaux, 2003; Green *et al.*, 2001). In rat hippocampus, an oxytocin agonist facilitates inhibitory transmission by exerting an excitatory action on the soma or dendrites of GABAergic interneurons (Zaninetti and Raggenbass, 2000), and systemic injections of secretin increase concentrations of GABA (Kuntz *et al.*, 2004). Oxytocin and/or secretin deficits in autism could further complicate genetic GABA abnormalities. If so, it is possible that endogenous up-regulation via intense nurture or exogenous secretin and/or oxytocin could compensate for genetic GABA abnormalities in autistic children by triggering key developmental gene programs.

In addition to these abnormal glucocorticoid receptor findings in subjects with severe mental illness, Knable and colleagues found hippocampal abnormalities of reelin and brain-derived neurotrophic factor (BDNF) (Knable *et al.*, 2004). A genetic abnormality of reelin has been identified in autism (Persico *et al.*, 2001), and blood levels of BDNF were abnormal in a cohort of neonates later diagnosed with autism (Nelson *et al.*, 2001). The reelin findings suggest a dysfunction of inhibitory GABAergic interneurons (Knable *et al.*, 2004).

The reason autistic children do not respond to normal maternal cues is poorly understood. However, one possible factor could be central deficits that occur in sensory processing and face recognition. Whatever the reason, we hypothesize that since autistic children are unable to recognize or respond to normal maternal nurturing, they develop the genetic profile of low-nurture animals (with low glucocorticoid and altered GABA receptor composition). If this is the case, it may be possible to mimic the conditions of a high-nurture internal environment in autistic children through early up-regulation of endogenous peptides or early administration of exogenous peptides. In this way, children diagnosed with autistic symptoms at an early age and treated with peptides may be capable of accepting and reciprocating their mothers' nurturing. An important question is the extent to which exogenous peptides might overcome abnormalities of peptides, GABA, glucocorticoid receptors, cytokines, reelin, and other molecules in the stress cascade. If key developmental gene programs are activated, the cascade of adverse stress adaptation conditioning may be averted, thereby preventing the ongoing damage of unremitting stress in autism.

B. SECRETIN'S ROLE IN REGULATING STRESS

The role of secretin in maintaining homeostasis has long been established: anti-stress gastric hormonal action (Bayliss and Starling, 1902), deacidification of the gut (Jin *et al.*, 1994), stimulation of hepatic bile flow (McGill *et al.*, 1994), increase of coronary blood flow (Gunnes *et al.*, 1983, 1985), and increased lipolysis during fasting and muscular exercise (Bell *et al.*, 1984). In an animal, secretin is synthesized by the pancreas and colon (Lopez *et al.*, 1995) and by flora that inhabit the gut (Gauthier *et al.*, 2003).

Secretin's role as a peripheral stress-regulatory hormone and central neuro-modulator of stress-adaptation responses has been suggested by earlier structural and functional studies (Chang *et al.*, 1985; Charleton *et al.*, 1981; Fuxe *et al.*, 1979; Itoh *et al.*, 1991; Mutt *et al.*, 1979; O'Donohue *et al.*, 1981; Samson *et al.*, 1984). Secretin-releasing peptide and secretin are secreted as part of unified, vagally-mediated behavioral and reflex response patterns (Chey and Chang, 2001). These peptides are triggered by stress-related increases in gastrin and gastric acid output in the gut (Li *et al.*, 1998). The prefrontal cortex and subcortical outlets of emotional memory are then conditioned to increase sympathetic output to the GI tract when stressed. The prefrontal perceptual encoding autonomic control mechanism modulates sympathetic/vagal discharges and visceral activity patterns (Ruggiero *et al.*, 1993, 1998). Stress-related increase in hydrochloric acid results in hyper-drive of the dorsal motor vagal output to gastric parietal cells, requiring compensatory increases in secretin cell output. The long-term impact of sustained stress is a dysregulation of finely tuned vagal reflex networks. It is possible that the decreased number of secretin cells (50% fewer) reported in autistic guts (Gershon and D'Autreaux, 2003) may be the result of unremitting stress-induced dysregulation that eventually leads to cell arrest or apoptosis of secretinergic cells. Inadequate amounts of secretin or inability to up-regulate secretin in the face of stress could make the infant vulnerable to GI pathology.

Secretin has been found to peripherally and centrally activate the dorsal vagal complex in the brain via the vagus and spinal nerves (Westlund *et al.*, 1996). The prefrontal cortical perceptual encoding areas (OFC, Ins Cx,) and subcortical outlets of emotional expression (Th, A, HP) (Fig. 2) are impaired by prenatal stress (Fumagalli *et al.*, 2004). This network converges on the dorsal vagal complex, impairing the parasympathetic vagal discharges to the viscera (Ruggiero *et al.*, 1993). Normally, activation of the nucleus of the solitary tract (NTS) within the dorsal vagal complex helps maintain homeostasis through modulation of behavioral, autonomic, and endocrine systems (Williams *et al.*, 2001). This dorsal vagal complex reflex response pattern is viscerally conditioned by components of maternal nurturing, such as breastfeeding and vocalization, which condition the infant to alternate between swallowing and breathing and, later in development, to alternate between talking and listening (Porges, 1995). In their vagal circuit of emotion regulation theory, Porges *et al.* propose that cardiac vagal tone is of such importance that it can serve as an index of emotion regulation (Porges *et al.*, 1994).

It is important to note that Cannon's main criticism of William James's visceral theory of emotions (James, 1884) (Fig. 1A) was based on evidence showing that "total" separation of the viscera from the central nervous system (i.e., total destruction of the sympathetic and spinal sensory roots) does not alter emotional behavior. Cannon offered this as proof that emotions must arise in the brain (Cannon, 1927) (Fig. 1B). Though he was an expert on the autonomic nervous system, a devotee of Darwin, a lifetime friend of Pavlov, and a firm believer in the role of stress adaptation conditioning in the maintenance of homeostasis, Cannon, and until recently most others, overlooked the vagal brain/gut pathway and the profound role that the vagus nerve plays in controlling and conditioning behavior and emotions. We theorize that the vagus nerve serves as a primary pathway between the brain and gut by which peptides such as secretin and oxytocin influence dysregulated stress response patterns such as occur in autism (Fig. 2).

Secretin is secreted in response to breastfeeding (Zabielski *et al.*, 1994). In an animal model, the colostrum content of breast milk during the immediate post-partum period is a greater stimulus to secretin release than the milk itself (Guilloteau *et al.*, 1992). Studies on infantile autism reveal a lower incidence of breastfeeding (Tanoue and Oda, 1989). Further study of the relationship between low or absent breastfeeding and autism could reveal whether replacement of key stress-modulating peptides such as secretin and oxytocin could help offset the deficit.

Secretin levels are elevated in diseases such as cystic fibrosis, hyaline membrane disease (Boccia *et al.*, 2001), and Crohn's disease (Teufel *et al.*, 1986), all of which are associated with GI abnormalities. Interestingly, vasoactive intestinal peptide (VIP), another member of the secretin family, was elevated in neonates later diagnosed as autistic, as compared to children with normal development (Nelson *et al.*, 2001). Human neonates exhibit excessive gastric acid and secretin output, as assayed in two-day olds, whereas secretin expression reaches the mature pattern by the second postnatal week in healthy pre-term infants (Lucas *et al.*, 1980a, b). Sick infants with hyaline membrane disease exhibit sustained up-regulation of secretin secondary to both starvation (Lucas *et al.*, 1980a, b) and respiratory stress analogous to that induced by colchicine (Jones and Gonzalez-Lima, 2001). These data suggest that secretin is up-regulated in the periphery in response to stress.

The central actions of secretin are less clear. Several studies have reported secretin in widespread areas of the central nervous system (Chang *et al.*, 1985; Mutt *et al.*, 1979; O'Donohue *et al.*, 1981), including the hypothalamus (Chang *et al.*, 1985; Charlton *et al.*, 1981; Mutt *et al.*, 1979; Samson *et al.*, 1984). Others reported expression of an mRNA secretin precursor in the brainstem, thalamus, and cerebral cortex, as well as in the hypothalamus (Itoh *et al.*, 1991; Ohta *et al.*, 1992). Several studies suggest that secretin is synthesized endogenously in the central nervous system (Fuxe *et al.*, 1979; Itoh *et al.*, 1991; O'Donohue *et al.*, 1981; Ohta *et al.*, 1992). Earlier studies have localized secretin and its presumptive receptor binding sites to viscerolimbic brain regions involved in central autonomic regulation (Itoh *et al.*, 1991; Nozaki *et al.*, 2002; Ohta *et al.*, 1992). The precise location of the secretinergic cells was not established by any of the above techniques, which lacked the single-cell resolution of our immunocytochemical methods (Welch *et al.*, 2004a).

Abnormalities of the cerebellum could explain a dysfunction of its role in emotional memory, learning of motor skills, and autonomic controls in autism. Cerebellar vermal connections with the hippocampal formation, amygdala, and hypothalamus form an integrated network implicated in adverse conditioning of fear responses (Sacchetti *et al.*, 2005). Immunohistochemical techniques used to localize secretin have shown the highest immunoreactivity in the Purkinje cells of the cerebellum (Koves *et al.*, 2002). Reduced numbers and volume of Purkinje cells have been reported in the cerebellum of autistic patients (Bailey *et al.*, 1998; Bauman and Kemper, 2005). The cerebellum is very important in development of conditioned behaviors. It is the great cerebral ganglion, receiving mental and physical information and transmitting both to the thalamocortical/striatal networks and to the viscera and musculoskeletal systems (Carpenter, 1996). These circuits are highly modified by postnatal experience. Inasmuch as secretin facilitates GABA transmission in the cerebellum (Yung *et al.*, 2001), it is possible that secretin administration could have a beneficial effect on cerebellar neural transmission in autism.

Behavioral changes that follow injection of secretin into the cerebroventricular system in rats include significantly increased defecation, altered respiration, and decreased novel-object approaches and open-field locomotor activity (Charlton *et al.*, 1981). Banks determined that secretin could cross the blood-brain barrier in mice injected with a radiolabeled secretin analogue (Banks *et al.*, 2002). The compound was reported to have entered every brain region, with the fastest uptake in the hypothalamus and the hippocampus, two brain regions that exhibit developmental abnormalities in autistic patients (Bauman and Kemper, 1985; Haznedar *et al.*, 2000; Ogai *et al.*, 2003; Schultz *et al.*, 2003). In another study, secretin was found to bind with specificity and high affinity to receptors in the nucleus of the solitary tract, thalamus, hypothalamus, and cerebral cortex (Nozaki *et al.*, 2002), all of which are sites of pathology in autism.

Secretin may influence behavior by peripheral and central mechanisms that protect against visceral stressors. Both experimental and clinical data previously cited raise the possibility that neuro-psychiatric and functional GI abnormalities in autism could be secondary to hypoxia, a known environmental insult in developmental disorders (Davis *et al.*, 1992). Secretin receptors as well as oxytocin receptors couple to G-proteins. G-proteins stimulate adenylate cyclase, which leads to the production of cyclic adenosine monophosphate (cAMP) (Harmar, 2001). In gut epithelial cells impaired by hypoxia in their ability to generate cAMP, pharmacologic elevation of cAMP normalizes both polymorphonuclear-induced permeability changes and restoration of barrier function (Friedman *et al.*, 1998). Such permeability changes have been found in autistic children (D'Eufemia *et al.*, 1996). Secretin is known to elevate cAMP (Fremeau *et al.*, 1986), and has been reported to decrease intestinal permeability in 13 of 20 autistic children (Horvath and Perman, 2002). We hypothesize that synthesis and secretion of secretin on demand may ameliorate hypoxia and other metabolic challenges of perinatal stress via two known mechanisms of action: as a vasodilator (Gandhi *et al.*, 2002), and as a gastric protective hormone (Bayliss and Starling, 1902). Various studies have illuminated responses to hypoxia by neurotransmitter and neuropeptide systems in the cardiovascular and GI control regions of the medulla oblongata (Iadecola *et al.*, 1993; Ruggiero *et al.*, 1993, 1998; Talman and Kelkar, 1993). Taken together the data suggest that systematic studies should be performed to investigate hypoxic changes in autism.

C. Oxytocin's Role in Regulating Stress

In an experimental model, Uvnas-Moberg showed that oxytocin is an antistress hormone, decreasing blood pressure and offsetting flight-or-fight and corticotropin-releasing-hormone (CRH) and norepinephrine responses to stressors (Uvnas-Moberg, 1997). Oxytocin was administered to rats I.C.V. for a 5-day period, resulting in sedation, diminished blood pressure, decreased corticosterone levels, and increased release of vagally controlled hormones. The fact that these effects were long-lasting, persisted for several weeks, and were not reversed by oxytocin antagonists, indicated to her that "secondary mechanisms" had been activated. While these secondary mechanisms have not been identified, it is possible that they involve the gene program activation in Caldj's experiments cited in the previous section (Caldji *et al.*, 2003; Weaver *et al.*, 2004). In any case, it is clear that oxytocin has a powerful role in regulating stress.

As central oxytocin pathways develop postnatally, they mature in their ability to influence vagally mediated digestive functions. Paraventricular hypothalamic oxytocinergic neurons already present from birth modulate vagal digestive motor functions via projections to the nucleus of the solitary tract and dorsal motor nucleus of the vagus in rats (McCann and Rogers, 1990). The cumulative length of the fibers increases between 23-fold and 94-fold between birth and adulthood (Rinaman, 1998). Oxytocin activates parietal cells that secrete acid into the lumen of the stomach. Secretin is then secreted into the bloodstream in response to acid in the stomach. Oxytocin also stimulates cholecystokinin, with which it is co-localized in the brain. Cholecystokinin stimulates bile excretion, slows gastric emptying, and makes the mother and child sleepy (Uvnas-Moberg, 1989).

Oxytocin affects the gut through regulation of vascular tone (Jankowski *et al.*, 2000). Oxytocin acts on vagal neurons via voltage gated current, which is sodium dependent and modulated by calcium (Raggenbass and Dreifuss, 1992). Oxytocin is delivered by a humoral route and by direct innervation (Sofroniew *et al.*, 1981). The humoral route involves the production of oxytocin in the hypothalamus which projects fibers to the pituitary, from which oxytocin is secreted into the circulation.

Oxytocin release is stimulated by the act of breastfeeding, which in turn stimulates lactation (Hatton *et al.*, 1992; Matthiesen, 2001; Pedersen and Boccia, 2002). Breastfeeding is a major source of vagal stimulation (Uvnas-Moberg, 1989). In infants, vagal function can have profound influences on development. Breastfeeding confers long-term protection against GI inflammation (Barlow *et al.*, 1974). Somatostatin, a polypeptide hormone produced chiefly by the hypothalamus that inhibits the secretion of various other hormones, such as somatotropin, glucagon, insulin, thyrotropin, and gastrin, is inhibited during periods of vagal stimulation (Eriksson *et al.*, 1994). When continuous nursing is absent or low, vagal stimulation is low. Consequently, levels of oxytocin, VIP, prolactin, and other gut hormones such as secretin are decreased and the level of somatostatin is increased. Uvnas-Moberg has demonstrated that somatostatin (1) is secreted by a broad range of tissues, including pancreas, intestinal tract, and regions of the central nervous system; (2) inhibits gut function and is increased in level 10-fold during sickness, resulting in decreased motility, decreased HCl output, and decreased discharge of bile; and (3) blocks nutrient absorption, release of gastrin and cholecystokinin, and their growth-promoting effects (Uvnas-Moberg, 1989). We hypothesize that exogenous secretin/oxytocin peptide therapy could compensate for the effects of increased somatostatin in sick children.

Perhaps most important, suckling by the infant triggers a vagally-induced gut hormone response in the infant as well as in the mother, synchronizing their metabolism and inducing mild sedation or anxiolysis in the nursing pair (Uvnas-Moberg, 1987). The mechanisms by which this interaction occurs are in part related to vagal control of cardiac reactivity found in nursing women. Release of oxytocin influences the physiological state of the mother and her mothering patterns, increasing parasympathetic and decreasing sympathetic tone in the case of normal nurturing (Uvnas-Moberg, 1989). Mothering patterns and the mother's physiological state in turn influence the child's physiological state and behavior (Francis *et al.*, 2002). Indeed, non-nursing mothers experience increased sympathetic and decreased parasympathetic tone (Altemus *et al.*, 2001). Our clinical observation of mothers of adopted children with non-nursing histories reflects this finding. The mothers are anxious and hypervigilant with the children. We also observe that adopted children with non-nursing histories demonstrate symptoms of sympathetic imbalance, including anxiety, hyperactivity, and hypervigilance. Furthermore, after therapy that includes components of intense maternal nurturing, the sympathetic balance (e.g., hypothesized peptidergic balance) of both the adoptive mother and the adopted child is restored. This outcome suggests that it may be possible to intervene in the peptidergic mechanisms of the mother and child to overcome maladaptive stress response patterns, regardless of etiology, in order to restore homeostasis.

Homeostasis is in part maintained by hormonal priming, which normally occurs during pregnancy. Once the mother begins to interact with her offspring, purely endogenous central release of neuropeptides and neurotransmitters decreases her anxiety. In animal mothers, the beneficial effects of hormonal priming cannot continue without mother-pup interaction. Either $GABA_A$ receptor antagonism or oxytocin receptor antagonism in caudal periaqueductal gray increases anxiety to levels typically found in virgin animals (Lonstein, 2005). In fact, without prior hormonal priming, virgin animals induced with maternal behavior by exposure to pups did not show any decrease in anxiety (Ferreira, 2002; Stern and MacKinnon, 1976).

Suckling initiates secretion of oxytocin from the hypothalamus and prolactin from the anterior pituitary. When the vagus is stimulated, such as during suckling, the levels of oxytocin, secretin, prolactin, and other GI hormones are increased. Suckling stimulates the vagus nerve of the mother and infant by way of sensory receptors on the nipple and in the mouth, respectively. The sensory receptors of the breast and mouth transmit via the parallel vagal and spinal pathways converging on the hypothalamus. The descending pathway in both mother and infant is from the vagal motor nucleus to the GI tract, pancreas, and other visceral organs including the heart (Uvnas-Moberg, 1989). The frequency and intensity of suckling stimulates peptidergic mechanisms and regulates cardiorespiratory and GI function in both mother and child, synchronizing their metabolism (Bornstein et al., 2000).

The literature suggests that stressful conditions damage proteins such as peptides or inhibit their processing (Brostrom and Brostrom, 1998). In stressed infants, excess metabolic demands interfere with peptide function (Sanchez-Alvarez et al., 2002). Oxytocin is a chemical antioxidant (Moosman and Behl, 2002). However, when peptides such as oxytocin undergo oxidation while scavenging for stress-induced free radicals, they lose their hormonal function (Ducrocq et al., 1998). Furthermore, as oxytocin undergoes enzymatic conversion by brain stress responsive peptidases, oxytocin action in the brain may be compromised (Burbach et al., 1980). We surmise that if oxytocin is compromised by stress, then it is likely that secretin is subject to similar stress effects.

Impaired peptidergic function due to excess metabolic demand at the cellular level can lead to adverse behavior. Oxytocin affects areas of social recognition and early environmental conditioning of stress adaptation patterns in animal models (Carter, 1998). Abnormally reared male rhesus monkeys with significant social deficits have persistently demonstrated reduced cerebrospinal fluid oxytocin levels and alterations in both CRH and vasopressin, suggesting that abnormal rearing impairs the development of brain systems critical to normal social and emotional competence. The autistic brain has been shown to have abnormal face recognition patterns (Dawson *et al.*, 2005). Autistic-like symptomatology in animals is associated with pathogenic rearing histories (Winslow, 2005). At the same time, there is evidence that restoration of oxytocin levels can ameliorate some of the adverse symptoms. Endogenous up-regulation in humans is concomitant with feelings of trust (Zak *et al.*, 2004), while exogenous up-regulation engenders feelings of trust (Kosfeld *et al.*, 2005). Oxytocin treatment fully restores social recognition in oxytocin mutant mice that fail to recognize familiar conspecifics (Ferguson *et al.*, 2001).

The fact that oxytocin receptors are distributed throughout the cerebellar behavioral and visceral reflex circuits (Vaccari *et al.*, 1998) provides additional insight into the functional neuroanatomical basis for therapeutic efficacy of oxytocin treatment in mediating stress adaptation patterns.

D. SECRETIN AND OXYTOCIN'S INTERACTIONS WITH NEUROTRANSMITTERS

Secretin and oxytocin, acting on the brain and gut, likely subserve similar functions, such as the regulation of blood flow and metabolism (Gandhi *et al.*, 2002; Haraldsen *et al.*, 2002; Jankowski *et al.*, 2000). Oxytocin and its receptors are synthesized in the rat vasculature and may be involved in the regulation of vascular tone (Jankowski *et al.*, 2000). Oxytocin is known to modulate cerebral blood flow via a nitric oxide mechanism (Haraldsen *et al.*, 2002). Nitric oxide is a bi-product of stress, acting as an inhibitory neurotransmitter in the enteric nervous system (Kurjak *et al.*, 1996). Nitric oxide is an important biologic mediator in regulation of GI functions and plays a significant role in secretin-stimulated pancreatic secretion (Jyotheeswaran *et al.*, 2000). Nitric oxide at low concentrations functions as a signal in diverse physiological processes, including blood pressure control, neurotransmission, learning and memory, but at high concentrations nitric oxide functions as a defensive cytotoxin. Nitric oxide's absence or excess alters the cardiac effect of secretin in the same direction (Sitniewska *et al.*, 2000). Furthermore, nitric oxide inhibits oxytocin release (Engelmann and Ludwig, 2004; Kadekaro, 2004) and inhibits onset of maternal behavior in an animal model (Okere *et al.*, 1996).

Nitric oxide findings may have implications for the treatment of autistic children, some of whom have been found to have high plasma nitric oxide levels (Sweeten *et al.*, 2004). Nitric oxide has been implicated in the mechanism of tissue injury of inflammation in gastric mucosa. As stress-induced nitric oxide increases, glutathione levels decrease (Asanuma *et al.*, 2005). Plasma glutathione, another important antioxidant, is lower in autistic children (James *et al.*, 2004).

Decreased nitric oxide production in immune cells is also accompanied by increased IL-10 levels (Chang *et al.*, 2005). IL-10 (interleukin 10) is an anti-inflammatory cytokine that contributes to intestinal homeostasis (Maes *et al.*, 2003). Transgenic mice lacking IL-10 (IL-10-/-) spontaneously develop colitis (Annacker *et al.*, 2003). The interaction of secretin, oxytocin, and nitric oxide may be an indication that a combined secretin and oxytocin mechanism could prove beneficial in treating autism and inflammatory bowel disease (IBD). In duodenum of two autistic children, there were 50% fewer secretin (S) cells and 80% fewer (S) cells co-localizing serotonin (Gershon and D'Autreaux, 2003). Dysregulation of serotonin networks is found in IBD (Coates *et al.*, 2004) and in autistic children with familial hyperserotonemia (Cook, 1990; Leventhal *et al.*, 1990; Yirmiya *et al.*, 2001). Serotonin is proinflammatory to the gut (Linden *et al.*, 2003) and negatively regulates IFN gamma/IL-10 cytokine production ratios (Maes *et al.*, 2003). A recent study demonstrates altered IL-10 function in a population of autistic children with GI symptoms (Ashwood *et al.*, 2004). Findings relating secretin or oxytocin with serotonin and cytokines are important for our investigation of secretin/oxytocin treatment of the IL-10 mutant mouse model of IBD. Preliminary data show that combined secretin/oxytocin infusions ameliorate GI inflammation in the face of an IL-10 deficit (Welch *et al.*, 2003b).

Signaling molecules interact in the regulation of homeostasis and the stress response, notably via the paraventricular hypothalamus (Leong *et al.*, 2002), a site of oxytocinergic and secretinergic neurons (Windle *et al.*, 2004; Welch *et al.*, 2004b). Secretin and oxytocin interact with monoamines and angiotensin, which are abnormal in chronic cerebral and visceral disease (Chirguer *et al.*, 2001; Fuxe *et al.*, 1979; Jezova *et al.*, 2003; Li and Guyenet, 1996; Vacher *et al.*, 2002; Walker *et al.*, 1999). Hypothalamic oxytocin neurons modulate the activity patterns of brainstem structures synthesizing monoamines and regulating behavior, sleep, and arousal (Godino *et al.*, 2005). These interactions in turn subserve behavioral, endocrine, and autonomic regulation in response to homeostatic challenges. Normally, serotonin and norepinephrine control the expression of oxytocin in the endocrine hypothalamus and its secretion into the systemic circulation (Chriguer *et al.*, 2001; Vacher *et al.*, 2002). If secretin and oxytocin are dysregulated, it can be inferred that their role in reducing stress will be compromised.

Social isolation stress is a causal factor in stress ulcers and hippocampal pathology among insubordinate vervets (Uno *et al.*, 1989). The antagonizing of angiotensin II type 1 (AT1) receptors restores modulation of HPA stress axis function (Armando *et al.*, 2001). It also prevents gastric mucosal injury (Bregonzio *et al.*, 2003). Insubordinate vervets and autistic children share non-compliance behaviors and social isolation (Breiner and Beck, 1984). The cingulate/hippo-campal stress adaptation network is also a major site of pathology for both autistic children and vervets (Bauman and Kemper, 1985; Uno *et al.*, 1989). This network is implicated in the adverse conditioning of both GI functions (Gabry *et al.*, 2002; Uno *et al.*, 1989) and behavioral functions (Freeman *et al.*, 1997; Jones and Gonzalez-Lima, 2001). Secretin has been effective in ameliorating both gut and behavioral abnormalities of autistic children (Horvath and Perman, 2002; Hor-vath *et al.*, 1998; Lamson and Plaza, 2001; Kern *et al.*, 2004). Still to be investigated is whether secretin levels are altered by the social isolation of autistic children.

A particular interaction relevant to GI disorders is the relationship between angiotensin and secretin receptors. AT1 receptors are linked to social isolation stress (Armando *et al.*, 2001). AT1 receptors and secretin receptors are co-localized in endocytic vesicles (Walker *et al.*, 1999).

Taken together, these findings suggest that secretin, a vasodilator (Gandhi *et al.*, 2002), could be modulating the effect of AT1, a vasoconstrictor (Helou *et al.*, 2003). This modulation may take place, as Leong suggests, by "cross-talk" (interaction) of peptides at the level of the hypothalamus (Leong *et al.*, 2002) where angiotensin II attenuates GABAergic synaptic inputs (Li and Guyenet, 1996). If secretin is perhaps modulating effects of AT1 receptors via its effect on GABA, this interaction may explain why secretin ameliorates both GI symptoms (Armando *et al.*, 2001; Bregonzio *et al.*, 2003; Horvath and Perman, 2002; Uno *et al.*, 1989) and autistic behavioral symptoms (Horvath *et al.*, 1998; Kern *et al.*, 2004). Secretin and oxytocin have been shown to facilitate GABA inhibition in rat hippocampus (Kuntz *et al.*, 2004; Zaninetti and Raggenbass, 2000). If secretin and oxytocin both modulate GABA inhibition in the paraventricular hypothala-mus, as secretin and oxytocin both do in the hippocampal formation, then a dual anti-stress action of oxytocin and secretin could predict ameliorative effects of combined secretin and oxytocin treatment in both GI and brain pathology at the level of the hippocampus and hypothalamus, as well as in the amygdala, the cerebellum, and in the periphery.

E. Brain/Gut Function

Since the gut and the brain develop from the same part of the human embryo, it is not surprising that the intestinal tract has such a rich nerve supply that it has been referred to as the "abdominal" or "second brain" (Gershon, 1998; Robinson, 1907). The gut shares many of the same types of neurons and chemical transmitters with the brain, to which it is linked through the nucleus of the solitary tract. The nucleus of the solitary tract communicates visceral environmental information patterns to the emotional brain, triggering behavioral, endocrine, and autonomic stress adaptation patterns.

IBD is a GI disorder that activates the anterior olfactory nucleus, the piriform cortex, and the amygdala, providing specific support for a brain/gut connection in autism (Welch et al., 2005). These brain areas are important in social recognition and early environmental conditioning of neonatal adaptive behaviors (Ferguson, et al., 2001; Haxby et al., 2002), both of which are deficient in autism (Dawson et al., 1998; Ogai et al., 2003; Pelphrey et al., 2002; Winslow and Insel, 2002). A single injection of oxytocin has rescued the social recognition deficit of oxytocin mutant mice (Ferguson et al., 2001).

There is evidence from both animal studies and human research that various peptides can ameliorate GI disorders. Oxytocin is known to protect infants against enterocolitis (Barlow et al., 1974). Oxytocin also improves the antioxidative state of the colonic tissue and ameliorates oxidative colonic injury (Iseri et al., 2005). VIP, a member of the secretin family of peptides, successfully treats IBD in an animal model (Abad et al., 2003). In human research, epidermal growth factor enemas ameliorate IBD in humans (Sinha et al., 2003). The actions of secretin may be additive and/or as important as oxytocin in brain/gut regulation. Secretin has long been known as a deacidification hormone that protects the gut (Bayliss and Starling, 1902). Gastroenterologist Horvath reported that systemic administration of secretin, given as a probe of abnormal GI function, resulted in improved behavior in autistic boys with abnormalities of gut function and decreased intestinal permeability in another group (Horvath and Perman, 2002; Horvath et al., 1998).

Recent research points to the connection between behavioral and gut disorders. Early adverse events are associated with IBD (Ringel and Drossman, 2001). There is evidence of brain inflammation and gut pathology in autism (Vargas et al., 2005; White, 2003). Visceral inflammation is a potent dysregulator of cognitive and emotional brain regions/states.

Both secretin and oxytocin have vasodilator functions (Gandhi et al., 2002; Haraldsen et al., 2002; Jankowski et al., 2000). Secretin vasodilates the gut vasculature, as does VIP, another member of the secretin family of peptides (Furness et al., 2004; Naruse et al., 1998). The neuroprotective actions of VIP involve its ability to vasodilate cerebral blood vessels in the face of high stress (Dalsgaard et al., 2003). Our detection of perivascular secretin- and oxytocin-like immunoreactivity (Fig. 3) provides evidence of their role in the regulation of cerebral blood flow similar to that played by VIP (Welch et al., 2004b). These findings suggest that secretin and oxytocin influence cerebral perfusion and that both confer neuroprotection against hypoxia through peripheral visceral and cerebral vasodilation. The facts that cerebral perfusion deficits exist in autism

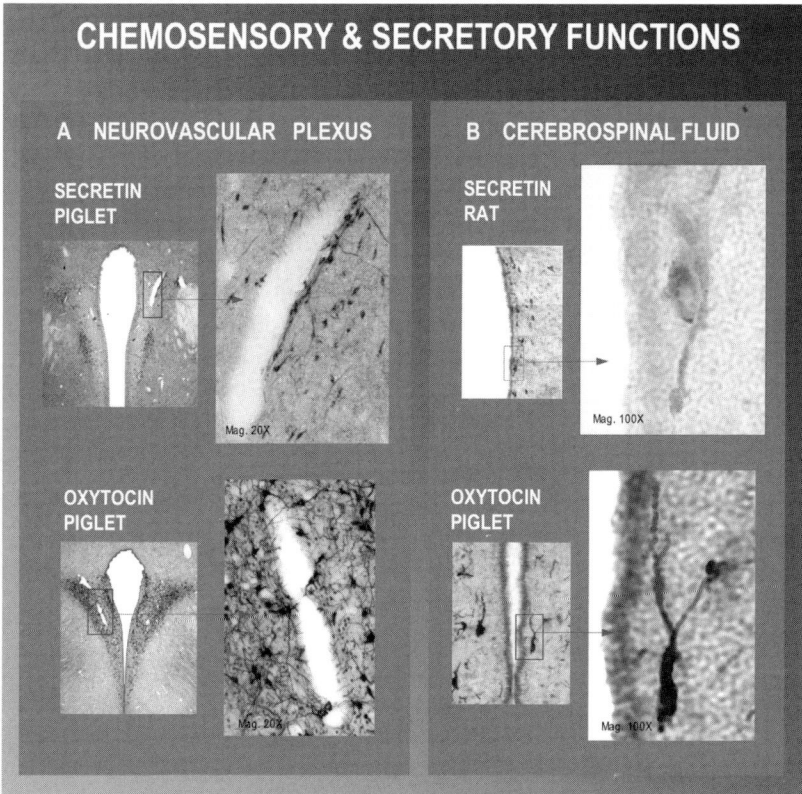

FIG. 3. Our studies showing secretin and oxytocin exchange across (A) the neurovascular plexus and (B) the CSF/ependymal interface of the third ventricle. Theoretically, neuropeptides may be secreted into or extracted from the CSF and blood stream. Secretin/oxytocin neurons may act as baro- and chemosensors, responding on demand to elevations of messenger molecules, stress hormones, and neurotransmitters. Secretin and oxytocin may be secreted into or extracted from the blood/CSF, where interactions may occur with classical neurotransmitter systems, such as GABA, serotonin, and dopamine, and with other stress regulatory peptides, such as corticotropin releasing hormone, vasopressin, and angiotensin (Welch *et al.*, 2004b).

(Haznedar *et al.*, 2000; Ohnishi *et al.*, 2000) and that oxytocin and secretin may be dysregulated in autism (Gershon and D'Autreaux, 2003; Green *et al.*, 2001) suggest that the two peptides may be connected in function and may share an oxidative stress mechanism.

To the best of our knowledge, no study has compared the cerebrovascular and brain activity patterns of autistic children and patients diagnosed with IBD. If the primary site of pathology in autism is an inflammatory/autoimmune condition involving the gut, then we would expect comparable patterns of brain activation, altered transmitter/peptide receptor binding sites, and concomitant dysregulation of peripheral and cerebral blood flows in patients with either IBD or autism. It has been suggested that the neurologic complications of IBD involve an immune-mediated inflammatory process of cerebral vasculature (Dietrich and Erbguth, 2003; Lossos *et al.*, 1995). Both autism and IBD are associated with signs of immune dysregulation: major histocompatibility complex expression (Gupta, 2000; Matri *et al.*, 2003), familial predisposition to each (Szatmari, 1999; Cho, 2003), peptide dysregulations (Gershon and D'Autreaux, 2003; Green *et al.*, 2001; Kimura *et al.*, 1994; Kulman *et al.*, 2000; Teufel *et al.*, 1986), and altered cytokines (Singh, 1996; Gupta *et al.*, 1998; Dohi *et al.*, 2000).

Visceral stress is a dysregulator of cytokine/peptide interactions (Licinio *et al.*, 1999; Tannenbaum *et al.*, 2002). Such interactions are governed in complex ways in the brain via neural and humoral signaling pathways along the HPA axis and are critical to the neuro-immune-endocrine system (Haddad *et al.*, 2002). It is also known that chronic visceral stress is a potent dysregulator of cognitive/emotional brain regions (Traub *et al.*, 1996). It is conceivable, we think probable, that chronic inflammation compromises the brain's capacity to sense, synthesize, and react normally to inflammatory signals, thereby resulting in a range of neurologic and psychiatric pathology. If so, it would follow that visceral and emotional dysregulation should be viewed as inseparable and treated as a single dysregulation.

III. Findings

A. Central Hypotheses of Our Work

Our central hypotheses are as follows. The brain and body form a single physiological unit (Fig. 2). Neuropeptides and their transmitters act through neural-humoral pathways to influence the brain/gut simultaneously. Release or inhibition of neuropeptides determines physiologically regulated or dysregulated states and results in positive or negative stress adaptation conditioning. Physiologically dysregulated states may be ameliorated by neuropeptide therapy.

B. Secretin Activates Visceral Brain Regions That Are Abnormal in Autism

The aim of this study was to determine whether central networks are involved in the presumptive behavioral and autonomic regulatory actions of secretin (Welch *et al.*, 2003a), a gut hormone that has been reported to have ameliorative effects in autistic children.

Central neural responses monitored by regional c-*fos* gene expression were examined in response to intracerebroventricular secretin injection in awake, Sprague-Dawley male rats. Tissue sections were incubated in an antibody to the c-*fos* gene product, Fos, and processed immunohistochemically.

Secretin-infused rats showed altered numbers of Fos-immunoreactive nuclei, mainly in visceral and limbic areas of the brain. Secretin induced c-*fos* protein expression in the dorsal vagal complex, the area postrema, and its sub-postremal region of transition with the nucleus of the solitary tract and the commissural, medial parvicellular, and periventricular subnuclei. Secretin activated cells localized to the nucleus of the solitary tract projection fields: lateral reticular formation, locus ceruleus, ventral periaqueductal gray, and paraventricular thalamic nucleus, corresponding to the non-discriminative, stress-reactive, visceral thalamus. Based on these results and prior findings (Ruggiero *et al.*, 1998), we hypothesize that the visceral thalamus/cortical/striatal stress axis processes the drives (tensions) motivating expression of adaptive maternal/infant bonding behavior. In the hypothalamus, the predominant labeling mapped to the paraventricular hypothalamic nucleus, mainly its periventricular region and magnocellular subdivision. Secretin induced c-*fos* in the medial and central amygdala and the lateral septal complex. Dramatic secretin-activation of ependymal and subependymal nuclei lining the third ventricle contrasted with the absence of immunoreactivity in the age-matched controls.

Specific areas of the cerebral hemispheres were heavily labeled in the secretin-treated rats as compared to controls. The nuclear immunoreaction product was most heavily concentrated along the medial bank of the PFC, the orbitofrontal cortex, the anterior olfactory nucleus, and the piriform cortex. Relative to the Fos expression in untreated controls, secretin attenuated Fos immunoreactivity in the dorsal periaqueductal gray, the intralaminar thalamus, the lateral amygdala, the medial parvicellular hypothalamus, the somatosensory and association areas of the parietal cortex, and the motor cortex.

Significantly, secretin altered the activity of structures involved in behavioral conditioning of stress adaptation and visceral reflex reactions. Additionally, activation of third ventricular ependymal and subependymal cells provides a possible cellular mechanism for the behavioral regulatory actions. Secretin's behavioral effects in autistic children may involve these cellular mechanisms (Welch *et al.*, 2003a, 2004a, 2005).

Some of the brain areas that were either activated or attenuated in the secretin-treated rats overlap with areas known to be abnormal in the brains of autistic patients. Among the important areas where c-*fos* was decreased in the brains of rats treated with secretin was the medial parvicellular hypothalamus, which synthesizes the stress hormone releasing factor CRH.

Our findings in the rat suggest that secretin exerts a regulatory action on the brain, including stress adaptation networks and brain areas that influence the gut.

This new observation supports the idea that previously unsuccessful trials with secretin either included patients whose intestinal function was not appropriate for secretin treatment (Kern *et al.*, 2004), that the dosages and scheduling needed to be changed, or both.

Our study is the first systematic analysis of the actions of secretin in the brain of the laboratory rat using c-*fos* activation as a monitor of altered metabolism in various regions of interest. These areas are of interest because of their role in regulating behaviors related to stress adaptation and optimal function of visceral and immune organ systems. Data in this study predict that secretin may activate dysregulated behavioral and visceral regulatory circuits in autistic children.

C. Secretin Distribution and Specificity Support a Central Stress Neuroregulatory Role Applicable in Autism

We sought to determine whether secretin is synthesized centrally, specifically by the HPA axis, and to discuss secretin's possible neuroregulatory role in autism (Welch *et al.*, 2004a). Previous biochemical and radioimmunoassays demonstrated secretin immunoreactivity and high-affinity secretin receptor binding sites in forebrain regions (Charlton *et al.*, 1981; Mutt *et al.*, 1979). In our previous study, we demonstrated that secretin activated the same forebrain regions.

This study provided the first direct immunocytochemical demonstration of secretin immunoreactivity in the forebrain. It was the first to establish with single cell resolution that secretin is synthesized in the forebrain, specifically by the HPA stress axis. The specificity of our findings was demonstrated by preadsorption control data and by the fact that secretin and other members of its peptide family, PACAP, glucagon, and VIP, had different distributional patterns. For example, VIP staining cells were prominent in the suprachiasmatic nucleus which was virtually devoid of secretin.

Secretinergic neurons were heavily labeled in colchicine-treated rats, as compared to the untreated group which showed undetectable or light labeling. Secretin immunoreactivity was cytoplasmic and restricted to neurons of the anterior and middle regions of the hypothalamus and adjoining periventricular gray. Presumptive secretinergic neurons were concentrated in precise loci within the paraventricular/supraoptic and intercalated regions of the hypothalamus. Secretinergic cells were heavily concentrated and intensely stained in magnocellular, parvicellular, and periventricular divisions of the paraventricular nucleus.

Cells were localized to specific regions of the paraventricular and supraoptic hypothalamic nuclei, and to the ependyma and subependymal zone of the

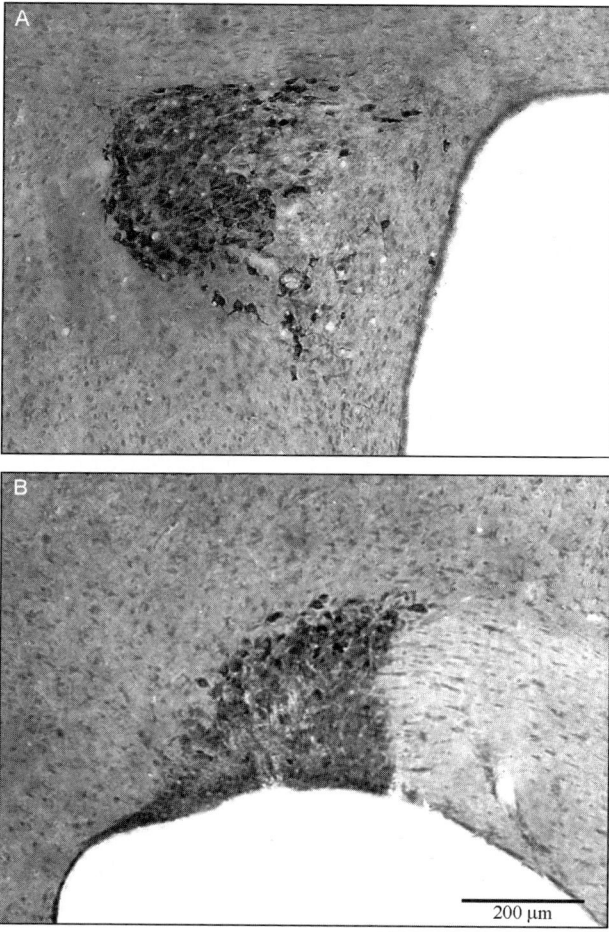

200 μm

FIG. 4. Presumptive secretinergic neurons in the hypothalamic (A) paraventricular nucleus, and (B) supraoptic nucleus of a colchicine treated rat. This is the first direct immunocytochemical demonstration of secretin immunoreactivity in the forebrain. Control studies support specificity of secretin immunoreactivity. Dramatic differences in topographic distribution and density of secretinergic neurons from the distribution patterns of other members of the secretin/VIP/glucagon/PACAP family extend evidence of the existence of a secretinergic brain/gut stress-regulatory system (Welch *et al.*, 2004a).

third ventricle (see Fig. 4A). Secretinergic cells in the supraoptic nucleus were concentrated dorsally and extended medially, arching over the optic tract (Fig. 4B). Small numbers of cells were scattered among a heavily labeled neuropil in the ventral region of the supraoptic nucleus. A related mapping study established an anatomical basis for potential interactions of secretin, oxytocin, and CRH in stress-responsive brain regions activated by visceral inflammation (Ruggiero et al., 2005). The mapping study provided anatomical evidence that brain regions immunoreactive for S, OT, and/or CRH overlap. These regions are activated by stress and are common sites of pathology in cerebral and visceral diseases, including autism. Together with prior findings the results of the mapping study support the hypothesis that treatment with combined S and OT will counteract the stress effects of CRH in chronic visceral diseases with concomitant neurological manifestations.

The findings of our secretin distribution study offer evidence that the hypothalamus, like the gut, is capable of synthesizing secretin. The wide spectrum of behavioral, endocrine, and autonomic visceral effects of systemic peptide administration is consistent with this concept. A neuroregulatory relationship between the peripheral and central stress response systems is suggested, as is a dual peripheral/central role for secretin in conditioning both stress adaptation systems.

Secretin levels were up-regulated by colchicine, an exemplar of homeostatic stressors, compared to the low expression in untreated animals. Thus, secretin expression by brain and gut secretin cells is likely stress-related and, as suggested by the distribution patterns, may interact with other neuropeptides in conditioning stress-adaptation.

The fact that central colchicine up-regulated hypothalamic secretin expression suggests that secretin may be synthesized on demand in response to stress, a possible mechanism that may underlie secretin's role in autism. Secretinergic cells were found in comparable hypothalamic regions of the newborn pigs not treated with colchicine (Welch et al., 2004b). This finding is consistent with evidence in human neonates expressing transient up-regulation of secretin in response to birth stressors (Lucas et al., 1980a, b). Secretin is low constitutively and up-regulated by a physiological stressor in adult rats (Welch et al., 2004a). Therefore, it is conceivable that secretin cells are depleted by stress in the same way that epinephrine cells are depleted in neurodegenerative diseases (Burke et al., 2004). The fact that Gershon demonstrated 50% fewer S cells in the intestines of two autistic children than in controls (Gershon and D'Autreaux, 2003) suggests that autistic children with fewer secretin cells would not be able to up-regulate secretin in response to GI stressors, making them vulnerable to unremitting stress.

High-power optics revealed secretinergic periventricular neuronal processes contributing to subependymal perivascular fiber plexuses, which may serve as central chemosensors (see Fig. 3). Arrays of cells were diagonally organized in the nucleus intercalatus. In general, peripheral and central chemosensory receptors are activated by changes in chemical composition of the internal milieu, including a vast latticework of extracellular spaces referred to as Virchow-Robin spaces, which are abnormal in autism (Taber et al., 2004). Chemosensors are activated by cerebral hypoxica/ischemic insults, hypercapnia, and chemical stress factors triggering visceral/emotional stress reactions. Pathological physiological states result in altered permeabilities of cerebrospinal fluid (CSF) and blood brain barrier functions (Mark *et al.*, 2004). Secretin as well as oxytocin may be up-regulated on demand and released by terminals of hypothalamic, pituitary, bulbar, and spinal projections in response to such alterations in the chemical composition of this milieu. Furthermore, interactions of these peptides may occur with classical neurotransmitter systems, such as GABA, serotonin, and dopamine, as well as other stress regulatory peptides, such as CRH, vasopressin and angiotensin. Our findings extend previous studies characterizing central chemosensors that synthesize monoamines and peptides (Ruggiero *et al.*, 1985).

D. Brain Effects of Chronic IBD in Areas Abnormal in Autism and Treatment by Single Neuropeptides Secretin and Oxytocin

Recent research points to the connection between behavioral and gut disorders. In this study we sought to determine the extent to which chronic GI inflammation alters the functional activity of specific regions of the visceral and emotional brain in a rat model of acquired inflammatory bowel disease (IBD). Concomitantly, we tested the hypothesis that continuous infusion of stress regulatory peptides secretin or oxytocin will resolve gut inflammation and secondary neurological manifestations (Welch *et al.*, 2005).

Early adverse events are associated with IBD. In animal models, maternal deprivation and social isolation predispose to gastric erosion and brain pathology, as previously discussed. There is controversy over the effectiveness of secretin or oxytocin in the treatment of autism. The neurobiological basis of the efficacy of combined secretin/oxytocin treatment is supported by our preliminary findings in progress. This combined continuous neuropeptide therapy may ameliorate brain/gut dysregulation in autism.

IBD was induced in male Sprague-Dawley rats (n=10) with trinitrobenzene sulfonic acid (TNBS) vs. controls (n=11). IBD was characterized by moderate/severe infiltration of inflammatory cells. Secretin or oxytocin or equivolume saline was administered I.V. by Alzet pump for 20 days after disease onset.

IBD saline-, IBD secretin, and IBD oxytocin-treated animals had diarrhea and exhibited clearcut signs of sickness behaviors, lethargy, and lack of exploratory behavior. Qualitative analysis of the gut of TNBS treated animals showed substantial inflammatory infiltrates in the submucosa of the IBD colons (see Fig. 5A), whereas the guts of control animals demonstrated no inflammation (Fig. 5B). Qualitative analysis of the forebrain revealed c-*fos* induction in the rats with IBD: paleo, archi, meso, insular, and orbitofrontal cortices and subcortical regions

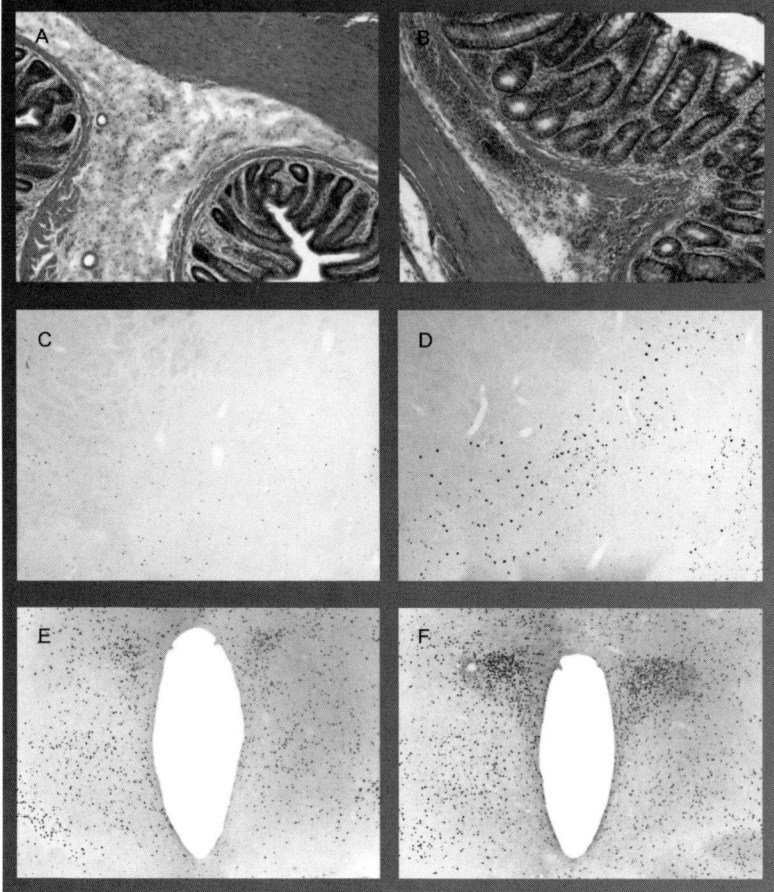

FIG. 5. Evidence showing brain and gut areas activated by visceral inflammation in a rat. Brain areas overlap with those often abnormal in autism, suggesting that inflammatory bowel disease (IBD) could be a model for testing treatments of autism. (A) Colon of control. (B) Colon showing infiltrates from TNBS-induced colitis. Concomitant cerebral metabolic activity patterns were compared by examining the regional distribution of c-*fos* gene expression. Colitis-induced changes localized to structures that are sensitive to stress and to secretin infusion by I.V. (C) Central amygdaloid nucleus of a control rat. (D) Colitis-induced stress reaction of central amygdala. (E) Hypothalamus of control rat. (F) Colitis-induced stress reaction of hypothalamus (Welch *et al.*, 2005).

were analyzed in healthy controls. Cortical and subcortical areas in control rats exhibited low constitutive c-*fos* expression (Fig. 5C and E). The IBD group exhibited robust c-*fos* activation of precognitive networks, specifically piriform, endopiriform/insular, central and medial amygdala (Fig. 5D), and paraventricular hypothalamic nucleus (Fig. 5F), midline intralaminar thalamus, and habenula. The brain activation patterns of IBD animals treated with secretin or oxytocin et al., alone did not differ significantly from that of the untreated group of IBD animals.

We concluded from these results that visceral stress is processed centrally. In this study, experimental animals with IBD demonstrated activation of brain regions that control stress response patterns. Brain and gut areas affected in this study of acquired chronic visceral inflammation also overlap with those regions affected in autism. It is interesting to note that our IBD experiment provides evidence that contradicts one of Walter Cannon's five criticisms of the James' visceral theory of emotions (James, 1884) (Fig. 1A), namely that artificial induction of visceral changes typical of strong emotions does not produce corresponding emotions (Cannon, 1927) (Fig. 1B). As noted in our study, emotional brain regions were robustly activated by artificially induced visceral inflammation.

The overlap of brain areas concomitantly activated by visceral inflammation and those abnormal in autism suggests that IBD could be a model for testing treatments of autism. The affected areas express receptors for stress-regulatory peptides including secretin and oxytocin, which modulate the actions of classical transmitters (Gould and Zingg, 2003; Tay et al., 2004). Maternal nurture behaviors such as breastfeeding and holding stimulate the release of brain/gut peptides, including secretin and oxytocin (Lucas et al., 1980a, b; Matthiesen et al., 2001). Deserving of study is whether combinations of peptides will be effective in resolving visceral inflammation and its neurological manifestations.

E. Visceral Inflammation Model of Autism

Although a direct pathophysiological link between autism and GI disorders has not been established, there is considerable evidence that visceral inflammation is co-morbid in autism (Ashwood et al., 2004; Goldberg, 2004; Horvath et al., 1999; White, 2003).

To date, it is not known whether the gut and/or the brain are primary sites of pathology in childhood developmental disorders. We presume, based upon clinical observations, that many chronic visceral and cerebral diseases share a common underlying peptidergic dysregulation, regardless of their sites of pathology. If this hypothesis is correct, one would expect peptidergic dysregulation to have an adverse impact on the body's ability to maintain homeostasis in the face of inflammation and to result in multiple brain/gut co-morbidities.

Chronic visceral disease generates pathological visceral activity patterns that transmit to emotional/visceral brain networks. The fact that neurological manifestations occurred in brain regions abnormal in autism in both an induced (TNBS colitis) model and a genetic (IL-10-/- spontaneous colitis) model indicates that primary visceral pathology can potentially cause secondary pathological brain activity. (Welch *et al.*, 2003b, 2005).

Previous uses of animal models are relevant to the study of autism and related pathologies. In an animal model of hyperserotonemia, there were two important findings: oxytocin is low in the paraventricular hypothalamus and calcitonin gene related peptide (CGRP), a gut peptide belonging to the secretin family, is elevated in the amygdala (Whitaker-Azmitia, 2005). In addition to hyperserotonemia in autism, two other parallel findings are reported in humans: serum oxytocin is low in autism (Green *et al.*, 2001) and serum CGRP is elevated in neonates later diagnosed with autism (Nelson *et al.*, 2001).

In another animal model, autoimmune disease-sensitive SJL/J mice exposed to thimerosal, an ethyl mercury-containing preservative used in vaccines, showed behavioral changes and hippocampal neurodegenerative changes (Hornig *et al.*, 2004). Open to question is whether thimerosal vaccines are a risk factor for autism, given the high rates of autoimmune disease in autistic patients (Szatmari, 1999). Recent findings indicate that innate neuroimmune reactions play a central pathogenic role in inflammatory or neurodegenerative changes in some autistic patients (Vargas *et al.*, 2005).

A third animal model may pertain to autism and be useful in testing brain/gut pathology. In prenatally-stressed rats, brain-derived neurotrophic factor (BDNF) is reduced in the prefrontal cortex and striatum into adulthood, implying that adverse life events during gestation may interfere with the expression and function of this neurotrophin throughout development (Fumagalli *et al.*, 2004). BDNF, a molecular determinant of synaptic plasticity and cellular homeostasis, is important in central and visceral neurodevelopment and neuroprotection (El Shamy and Ernfors, 1997; Husson *et al.*, 2005). BDNF is abnormal in neonatal blood samples of infants later diagnosed with autism (Nelson *et al.*, 2001). Along with altered BDNF expression, post-mortem and neuroimaging studies demonstrate defective sensory as well as cerebellar/thalamic/cortical/striatal networks in autistic brains (Bauman and Kemper, 1985, 2005).

Sapolsky's vervet's (Uno *et al.*, 1989) are an excellent animal model of both gut and hippocampal pathology. Autistic children and vervets share non-compliance behaviors, social isolation, hippocampal deficits, and GI abnormalities, as cited in the previous section II. D. Long-term alterations of visceral sensitivity and gut mucosal integrity are found in animal models of maternal deprivation, one of the most profound forms of social isolation (Barreau *et al.*, 2004). Social isolation stress exacerbates both GI disorders and autistic spectrum disorders. Low nurture or social isolation predispose to gastric erosion and brain pathology in experimental models in areas abnormal in autism (Ackerman *et al.*, 1978; Andersen and

Teicher, 2004; Meaney *et al.*, 1988; Uno *et al.*, 1989). Socially insubordinate vervets and autistic children share behavioral, brain, and GI abnormalities (Breiner and Beck, 1984; Bauman *et al.*, 1985; Lightdale *et al.*, 2001; Uno *et al.*, 1989; White, 2003).

Using animals with gut inflammation to study central manifestations could be useful in learning more about the cellular response to homeostatic challenge. In particular, animal models of low/high-nurture offspring become more vulnerable to hippocampal neuronal loss via cell death, or apoptosis, when they receive low nurture (Weaver *et al.*, 2002). Offspring are protected against apoptosis when they receive high nurture, suggesting that high level nurture may protect against apoptosis. Forrester posited that when homeostasis is not restored, inflammatory responses persist and lead to apoptosis. Further, he states that this tissue response is universal, since all cells derive from the same blastocyst. Cell differentiation that results in different organ systems manifests itself in separate clinical disorders, obscuring the common underlying disease process (Forrester, 2004). This concept could explain the wide range of symptoms present in autism. Using a visceral inflammation animal model could help determine the extent to which cell arrest or cell death occurs in the brain as a result or the peripheral inflammation.

Such a model could also be useful in determining the extent to which neuropeptides will prevent or intervene in the process of cell arrest or apoptosis peripherally and/or centrally. We believe that apoptosis is the end stage of the cascade of adverse events that stems from unmodulated stress (Welch *et al.*, 2003a, 2004b, c, 2005). Whereas peripheral and central inflammation can stem from a wide variety of causes, in the case of developmental disorders we believe that peptidergic mechanisms may protect against cell arrest and apoptosis. Our clinical experience shows that it is possible to overcome the negative developmental effects of unmodulated stress via early intervention with intense maternal nurturing therapy. Development, including normal speech, cognition, and improved social interaction, proceeded rapidly following the therapy in some cases (Welch *et al.*, 2006). It is possible that some developmental cell groups had not suffered apoptosis, but rather were in a stage of cell arrest. The cells could have been activated by mechanisms of components of maternal nurture (Welch, 1987, 1988; Welch and Chaput, 1988; Welch *et al.*, 2005). A visceral inflammation animal model could be useful in testing whether this is in fact the case and whether change occurs simultaneously in the gut and the brain as we predict.

While the causal relationship between gut disturbance and autism remains in question, the use of an animal model of IBD presents a novel approach to understanding the multifactoral causes and comorbidities associated with complex developmental disorders.

IV. Future Directions

A. NEUROPEPTIDE DOSING IN FUTURE CLINICAL TRIALS

The amount, delivery methods and schedule of dosing are all factors that can have profound influence on the outcomes of clinical trials involving secretin and other peptides. We do not believe that a single peptide given as a single dose will be effective in re-establishing homeostasis. Rather, we believe that combinations of two or more peptides administered as a continuous infusion will more closely mimic the natural physiologic stress-adaptation pattern elicited by maternal nurturing and will ameliorate visceral inflammation and central activation.

The idea of single dosage in clinical trials of secretin seems to have followed from Horvath's single injection of secretin (Horvath *et al.*, 1998). Ferguson's finding that restoration of social recognition in oxytocin mutant mice can be achieved with one injection may also have encouraged a single dose protocol (Ferguson *et al.*, 2001). It is unreasonable to expect that effects from a single dose will have lasting effects. Indeed, Kuntz showed that the effects of secretin administration on glutamate and GABA diminish in just 2 hours (Kuntz *et al.*, 2004). It is more likely that long-range effects on stress adaptation response patterns will require continuous administration of peptides for days, weeks, or months. This fact should be kept in mind when assessing the results of studies that use single peptides or short-term administration of peptides. Whether multiple doses or continuous dosing would prolong the effects in humans needs further investigation.

B. LONG-TERM GOALS OF OUR RESEARCH EFFORTS

We seek to demonstrate that combined secretin/oxytocin neuropeptide therapy is effective in ameliorating chronic developmental disorders of the brain/gut stress axis. Studies in progress focus on the brain, behavior, and circulatory/GI immune functions consequent to visceral inflammation before and after secretin/oxytocin treatment. The long-term effects of this combined peptide therapy may be demonstrated by reversal of predicted alteration of stress transmitters and peptides, as well as receptor binding sites abnormal in chronic mental and visceral metabolic disorders.

Clinical trials will assess whether combined secretin/oxytocin administration is effective in resolving visceral inflammation, autism, and autism with GI symptoms. Comparative neuroimaging studies will be designed to determine brain activation patterns of patients with IBD, patients with autism, and patients with co-morbid autism and symptoms of visceral inflammation.

Though some genetic and epigenetic risk factors for autism and inflammation have been established, we seek to determine a marker expressed by brain and gut that can identify a precursor stage of the diseases.

V. Conclusion

Our laboratory is engaged in efforts to translate bedside observations and experience into bench findings that will lead to a new paradigm in the treatment of developmental and behavioral disorders.

Our studies have demonstrated that secretin is synthesized in the hypothalamus, and that secretin activates visceral/emotional brain regions. We have also established that inflammatory bowel disease activates visceral/emotional brain regions in areas known to be abnormal in autism. Secretin and oxytocin may be secreted into or extracted from the blood/CSF, where interactions may occur with classical neurotransmitter systems, such as GABA, serotonin, and dopamine, and with other stress regulatory peptides such as CRH, vasopressin, and angiotensin.

The literature provides ample evidence that autistic children exhibit classic stress-induced inflammatory symptoms in brain and gut areas that are acted upon by secretin and oxytocin. Secretin and oxytocin are both important in determining stress response patterns and to maintaining homeostasis. The actions of secretin and oxytocin in the development and conditioning of stress adaptation networks via maternal nurturing are well documented. It follows that therapies intervening in the peptide mechanisms underlying maternal nurturing will be most effective in treating developmental disorders.

We believe developmental and behavioral disorders such as autism are the end result of an adverse cascade of events that stems from one or more genetic/environmental insults. Such insults can arrest the activation of key developmental programs. Over time, if uncompensated, the cascade can lead to maladaptive stress response patterns and to various interrelated psychological, neurological, and immunological pathology, including autism. Our clinical observations and the evidence presented in this chapter support the idea that it may be possible, regardless of etiology, to treat autism and other developmental and behavioral disorders effectively by intervening in stress mechanisms with exogenous combined secretin/oxytocin peptide treatment.

Acknowledgments

We wish to acknowledge Nobel Laureate Niko Tinbergen's participation and support in the development of the family-based intense nurturing intervention as a means of conditioning stress adaptation responses of autistic children and his encouragement to pursue the underlying basic science. We are especially indebted to Robert Ludwig for photomicrography and diagrams, for synthesis and articulation of concepts, and for overall editorial contributions. We again owe much to Sue Ann Power for critical reading and editing.

References

Abad, C., Martinez, C., Juarranz, M. G., Arranz, A., Leceta, J., Delgado, M., and Gomariz, R. P. (2003). Therapeutic effects of vasoactive intestinal peptide in the trinitrobenzene sulfonic acid mice model of Crohn's disease. *Gastroenterology* **124,** 961–971.

Ackerman, S. H., Hofer, M. A., and Weiner, H. (1978). Predisposition to experimental ulcers. *Gastroenterology* **75,** 930–931.

Altemus, M., Redwine, L. S., Leong, Y. M., Frye, C. A., Porges, S. W., and Carter, C. S. (2001). Responses to laboratory psychosocial stress in postpartum women. *Psychosom. Med.* **63,** 814–821.

Andersen, SL, and Teicher, M. H. (2004). Delayed effects of early stress on hippocampal development. *Neuropsychopharmacology* **29,** 1988–1993.

Annacker, O., Asseman, C., Read, S., and Powrie, F. (2003). Interleukin-10 in the regulation of T cell-induced colitis. *J. Autoimmun.* **20,** 277–279.

Ansorge, M. S., Zhou, M., Lira, A., Hen, R., and Gingrich, J. A. (2004). Early-life blockade of the 5-HT transporter alters emotional behavior in adult mice. *Science* **306**(5697), 879–881.

Armando, I., Carranza, A., Nishimura, Y., Hoe, K. L., Barontini, M., Terron, J. A., Falcon-Neri, A., Ito, T., Juorio, A. V., and Saavedra, J. M. (2001). Peripheral administration of an angiotensin II AT(1) receptor antagonist decreases the hypothalamic-pituitary-adrenal response to isolation stress. *Endocrinology* **142,** 3880–3889.

Asanuma, K., Iijima, K., Sugata, H., Ohara, S., Shimosegawa, T., and Yoshimura, T. (2005). Diffusion of cytotoxic concentrations of nitric oxide generated luminally at the gastro-oesophageal junction of rats. *Gut* **28,** 1–20.

Ashwood, P., Anthony, A., Torrente, F., and Wakefield, A. J. (2004). Spontaneous mucosal lymphocyte cytokine profiles in children with autism and gastrointestinal symptoms: Mucosal immune activation and reduced counter regulatory interleukin-10. *J. Clin. Immunol.* **24,** 664–673.

Bailey, A., Luthert, P., Dean, A., Harding, B., Janota, I., Montgomery, M., Rutter, M., and Lantos, P. (1998). A clinicopathological study of autism. *Brain* **121,** 889–905.

Banks, W. A., Goulet, M., Rusche, J. R., Niehoff, M. L., and Boismenu, R. (2002). Differential transport of a secretin analog across the blood-brain and blood-cerebrospinal fluid barriers of the mouse. *J. Pharmacol. Exp. Ther.* **302,** 1062–1069.

Bard, P. (1928). A diencephalic mechanism for the expression of rage with special reference to the sympathetic nervous system. *Am. J. Physiol. Acad. Sci. Hung.* **26,** 149–155.

Barlow, B., Santulli, T. V., Heird, W. C., Pitt, J., Blanc, W. A., and Schullinger, J. N. (1974). An experimental study of acute neonatal enterocolitis–the importance of breast milk. *J. Pediatr. Surg.* **9,** 587–595.

Barreau, F., Cartier, C., Ferrier, L., Fioramonti, J., and Bueno, L. (2004). Nerve growth factor mediates alterations of colonic sensitivity and mucosal barrier induced by neonatal stress in rats. *Gastroenterology* **127,** 524–534.

Bauman, M., and Kemper, T. L. (1985). Histoanatomic observations of the brain in early infantile autism. *Neurology* **35,** 866–874.

Bauman, ML, and Kemper, T. L. (2005). Neuroanatomic observations of the brain in autism: A review and future directions. *Int. J. Dev. Neurosci.* **23,** 183–187.

Bayliss, WM, and Starling, E. H. (1902). The mechanism of pancreatic secretion. *J. Physiol. (London)* **28,** 325–353.

Bell, P. M., Henry, R. W., Buchanan, K. D., and Alberti, K. G. (1984). The effect of starvation on the gastro-entero-pancreatic hormonal and metabolic responses to exercise. (GEP hormones in starvation and exercise). *Diabete Metab.* **10,** 194–198.

Berger, M. A., Barros, V. G., Sarchi, M. I., Tarazi, F. I., and Antonelli, M. C. (2002). Long-term effects of prenatal stress on dopamine and glutamate receptors in adult rat brain. *Neurochem Res.* **15,** 488–496.

Boccia, D., Stolfi, I., Lana, S., and Moro, M. L. (2001). Nosocomial necrotising enterocolitis outbreaks: Epidemiology and control measures. *Eur. J. Pediatr.* **160,** 385–391.

Booth, R., Charlton, R., Hughes, C., and Happe, F. (2003). Disentangling weak coherence and executive dysfunction: Planning drawing in autism and attention-deficit/hyperactivity disorder. *Philos. Trans. R Soc. Lond. B Biol. Sci.* **358,** 387–392.

Bornstein, MH, and Suess, P. E. (2000). Child and mother cardiac vagal tone: Continuity, stability, and concordance across the first 5 years. *Dev. Psychol.* **36,** 54–65.

Bregonzio, C., Armando, I., Ando, H., Jezova, M., Baiardi, G., and Saavedra, J. M. (2003). Anti-inflammatory effects of angiotensin II AT1 receptor antagonism prevent stress-induced gastric injury. *Am. J. Physiol. Gastrointest. Liver Physiol.* **285,** G414–G423.

Breiner, J, and Beck, S. (1984). Parents as change agents in the management of their developmentally delayed children's noncompliant behaviors: A critical review. *Appl. Res. Ment. Retard.* **5,** 259–278.

Brostrom, CO, and Brostrom, MA. (1998). Regulation of translational initiation during cellular responses to stress. *Prog. Nucleic Acid Res. Mol. Biol.* **58,** 79–125.

Burbach, J. P., De Kloet, E. R., and De Wied, D. (1980). Oxytocin biotransformation in the rat limbic brain: Characterization of peptidase activities and significance in the formation of oxytocin fragments. *Brain Res.* **202,** 401–414.

Burke, W. J., Li, S. W., Chung, H. D., Ruggiero, D. A., Kristal, B. S., Johnson, E. M., Lampe, P., Kumar, V. B., Franko, M., Williams, E. A., and Zahm, D. S. (2004). Neurotoxicity of MAO metabolites of catecholamine neurotransmitters: Role in neurodegenerative diseases. *Neurotoxicology* **25,** 101–115.

Caldji, C., Diorio, J., and Meaney, M. J. (2003). Variations in maternal care alter GABA(A) receptor subunit expression in brain regions associated with fear. *Neuropsychopharmacology* **28,** 1950–1959.

Caldji, C., Diorio, J., Anisman, H., and Meaney, M. J. (2004). Maternal behavior regulates benzodiazepine/GABAA receptor subunit expression in brain regions associated with fear in BALB/c and C57BL/6 mice. *Neuropsychopharmacology* **29,** 1344–1352.

Cannon, WB (1927). The James-Lange theory of emotions: A critical examination and an alteration. *Amer. J. Psychol.* **39,** 106–124.

Carpenter, M. B. (1996). "Human Neuroanatomy." Williams and Wilkins, Media, PA.

Carter, C. S. (1998). Neuroendocrine perspectives on social attachment and love. *Psychoneuroendocrinology* **23,** 779–818.

Chang, H. P., Huang, S. Y., and Chen, Y. H. (2005). Modulation of cytokine secretion by garlic oil derivatives is associated with suppressed nitric oxide production in stimulated macrophages. *J. Agric. Food Chem.* **53,** 2530–2534.

Chang, T. M., Berger-Ornstein, L., and Chey, W. Y. (1985). Presence of biologically and immunologically active secretin-like substance in the mammalian brain. *Peptides* **6,** 193–198.

Charlton, C. G., O'Donohue, T. L., Miller, R. L., and Jacobowitz, D. M. (1981). Secretin immunoreactivity in rat and pig brain. *Peptides* **2**(Suppl. 1), 45–49.

Chey, WY, and Chang, T. (2001). Neural hormonal regulation of exocrine pancreatic secretion. *Pancreatology* **1,** 320–335.

Cho, J. H. (2003). Significant role of genetics in IBD: The NOD2 gene. *Rev. Gastroenterol Disord.* **3,** S18–S22.

Chriguer, R. S., Rocha, M. J., Antunes-Rodrigues, J., and Franci, C. R. (2001). Hypothalamic atrial natriuretic peptide and secretion of oxytocin. *Brain Res.* **889,** 239–242.

Coates, M. D., Mahoney, C. R., Linden, D. R., Sampson, J. E., Chen, J., Blaszyk, H., Crowell, M. D., Sharkey, K. A., Gershon, M. D., Mawe, G. M., and Moses, P. L. (2004). Molecular defects in mucosal serotonin content and decreased serotonin reuptake transporter in ulcerative colitis and irritable bowel syndrome. *Gastroenterology* **126,** 1657–1664.

Cook, E. H. (1990). Autism: Review of neurochemical investigation. *Synapse.* **6,** 292–308.

Coniglio, S. J., Lewis, J. D., Lang, C., Burns, T. G., Subhani-Siddique, R., Weintraub, A., Schub, H., and Holden, E. W. (2001). A randomized, double-blind, placebo-controlled trial of single-dose intravenous secretin as treatment for children with autism. *J. Pediatr.* **138,** 649–655.

Corona, R., Dissanayake, C., Arbelle, S., Wellington, P., and Sigman, M. (1998). Is affect aversive to young children with autism? Behavioral and cardiac responses to experimenter distress. *Child Dev.* **69,** 1494–1502.

Dalsgaard, T., Hannibal, J., Fahrenkrug, J., Larsen, C. R., and Ottesen, B. (2003). VIP and PACAP display different vasodilatory effects in rabbit coronary and cerebral arteries. *Regul. Pept.* **110,** 179–188.

Damasio, A. (1994). "Descartes Error: Emotion, Reason, and the Human Brain." Grosset/Putnam, New York.

Davis, E., Fennoy, I., Laraque, D., Kanem, N., Brown, G., and Mitchell, J. (1992). Autism and developmental abnormalities in children with perinatal cocaine exposure. *J. Natl. Med. Assoc.* **84,** 315–319.

Dawson, G., Meltzoff, A. N., Osterling, J., Rinaldi, J., and Brown, E. (1998a). Children with autism fail to orient to naturally occurring social stimuli. *J. Autism. Dev. Disord.* **28,** 479–485.

Dawson, G., Meltzoff, A. N., Osterling, J., and Rinaldi, J. (1998b). Neuropsychological correlates of early symptoms of autism. *Child Dev.* **69,** 1276–1285.

Dawson, G., Webb, S. J., and McPartland, J. (2005). Understanding the nature of face processing impairment in autism: Insights from behavioral and electrophysiological studies. *Dev. Neuropsychol.* **27,** 403–424.

D'Eufemia, P., Celli, M., Finocchiaro, R., Pacifico, L., Viozzi, L., Zaccagnini, M., Cardi, E., and Giardini, O. (1996). Abnormal intestinal permeability in children with autism. *Acta Paediatr.* **85,** 1076–1079.

Dietrich, W., and Erbguth, F. (2003). [Neurological complications of inflammatory intestinal diseases.]. *Fortschr. Neurol. Psychiatr.* **71,** 406–414.

Diggle, T., McConachie, H. R., and Randle, V. R. (2003). Parent-mediated early intervention for young children with autism spectrum disorder. *Cochrane. Database. Syst. Rev.* **1,** CD003496.

Dohi, T., Fujihashi, K., Kiyono, H., Elson, C. O., and McGhee, J. R. (2000). Mice deficient in Th1- and Th2-type cytokines develop distinct forms of hapten-induced colitis. *Gastroenterology* **119,** 724–733.

Ducrocq, C., Dendane, M., Laprevote, O., Serani, L., Das, B. C., Bouchemal-Chibani, N., Doan, B. T., Gillet, B., Karim, A., Carayon, A., and Payen, D. (1998). Chemical modifications of the vasoconstrictor peptide angiotensin II by nitrogen oxides (NO, HNO2, HOONO)–evaluation by mass spectrometry. *Eur. J. Biochem.* **253,** 146–153.

Dunn-Geier, J., Ho, H. H., Auersperg, E., Doyle, D., Eaves, L., Matsuba, C., Orrbine, E., Pham, B., and Whiting, S. (2000). Effect of secretin on children with autism: A randomized controlled trial. *Dev. Med. Child Neurol.* **42,** 796–802.

El Shamy, WM, and Ernfors, P. (1997). Brain-derived neurotrophic factor, neurotrophin-3, and neurotrophin-4 complement and cooperate with each other sequentially during visceral neuron development. *J. Neurosci.* **17,** 8667–8675.

Engelmann, M., and Ludwig, M. (2004). The activity of the hypothalamo-neurohypophysial system in response to acute stressor exposure: Neuroendocrine and electrophysiological observations. *Stress* **7**, 91–96.

Eriksson, M., Bjorkstrand, E., Smedh, U., Alster, P., Matthiesen, A. S., and Uvnas-Moberg, K. (1994). Role of vagal nerve activity during suckling. Effects on plasma levels of oxytocin, prolactin, VIP, somatostatin, insulin, glucagon, glucose and of milk secretion in lactating rats. *Acta. Physiol. Scand.* **151**, 453–459.

Ferguson, J. N., Aldag, J. M., Insel, T. R., and Young, L. J. (2001). Oxytocin in the medial amygdala is essential for social recognition in the mouse. *J. Neurosci.* **21**, 8278–8285.

Ferreira, A., Pereira, M., Agrati, D., Uriarte, N., and Fernandez-Guasti, A. (2002). Role of maternal behavior on aggression, fear and anxiety. *Physiol. Behav.* **77**, 197–204.

Forrester, J. S. (2004). Common ancestors: Chronic progressive diseases have the same pathogenesis. *Clin. Cardiol.* **27**, 186–190.

Francis, D. D., Young, L. J., Meaney, M. J., and Insel, T. R. (2002). Naturally occurring differences in maternal care are associated with the expression of oxytocin and vasopressin (V1a) receptors: Gender differences. *J. Neuroendocrinol.* **14**, 349–353.

Freeman, J. H., Jr., Weible, A., Rossi, J., and Gabriel, M. (1997). Lesions of the entorhinal cortex disrupt behavioral and neuronal responses to context change during extinction of discriminative avoidance behavior. *Exp. Brain Res.* **115**, 445–457.

Fremeau, R. T., Jr., Korman, L. Y., and Moody, T. W. (1986). Secretin stimulates cyclic AMP formation in the rat brain. *J. Neurochem.* **46**, 1947–1955.

Friedman, G. B., Taylor, C. T., Parkos, C. A., and Colgan, S. P. (1998). Epithelial permeability induced by neutrophil transmigration is potentiated by hypoxia: Role of intracellular cAMP. *J. Cell Physiol.* **176**, 76–84.

Fumagalli, F., Bedogni, F., Perez, J., Racagni, G., and Riva, M. A. (2004). Corticostriatal brain-derived neurotrophic factor dysregulation in adult rats following prenatal stress. *Eur. J. Neurosci.* **20**, 1348–1354.

Furness, J. B., Jones, C., Nurgali, K., and Clerc, N. (2004). Intrinsic primary afferent neurons and nerve circuits within the intestine. *Prog. Neurobiol.* **72**, 143–164.

Fuxe, K., Andersson, K., Hokfelt, T., Mutt, V., Ferland, L., Agnati, L. F., Ganten, D., Said, S., Eneroth, P., and Gustafsson, J. A. (1979). Localization and possible function of peptidergic neurons and their interactions with central catecholamine neurons, and the central actions of gut hormones. *Fed. Proc.* **38**, 2333–2340.

Gabry, K. E., Chrousos, G. P., Rice, K. C., Mostafa, R. M., Sternberg, E., Negrao, A. B., Webster, E. L., McCann, S. M., and Gold, P. W. (2002). Marked suppression of gastric ulcerogenesis and intestinal responses to stress by a novel class of drugs. *Mol. Psychiatry* **7**, 474–483.

Gandhi, S., Tsueshita, T., Onyuksel, H., Chandiwala, R., and Rubinstein, I. (2002). Interactions of human secretin with sterically stabilized phospholipid micelles amplify peptide-induced vasodilation *in vivo*. *Peptides* **23**, 1433–1439.

Gauthier, A., Puente, J. L., and Finlay, B. B. (2003). Secretin of the enteropathogenic *Escherichia coli* type III secretion system requires components of the type III apparatus for assembly and localization. *Infect. Immun.* **71**, 3310–3319.

Gershon, M. D. (1998). "The Second Brain." Harper Collins, New York.

Gershon, M. D., and D'Autreaux. (2003). Personal communication.

Godino, A., Giusti-Paiva, A., Antunes-Rodrigues, J., and Vivas, L. (2005). Neurochemical brain groups activated after an isotonic blood volume expansion in rats. *Neuroscience* **133**, 493–505.

Goldberg, E. A. (2004). The link between gastroenterology and autism. *Gastroenterol Nurs.* **27**, 16–19.

Gould, B. R., and Zingg, H. H. (2003). Mapping oxytocin receptor gene expression in the mouse brain and mammary gland using an oxytocin receptor-LacZ reporter mouse. *Neuroscience* **122**, 155–167.

Graveling, R. A., and Brooke, J. D. (1978). Hormonal and cardiac response of autistic children to changes in environmental stimulation. *J. Autism. Child Schizophr.* **8**, 441–455.

Green, L., Fein, D., Modahl, C., Feinstein, C., Waterhouse, L., and Morris, M. (2001). Oxytocin and autistic disorder: Alterations in peptide forms. *Biol. Psychiatry* **50**, 609–613.

Guilloteau, P., Chayvialle, J. A., Toullec, R., Grongnet, J. F., and Bernard, C. (1992). Early-life patterns of plasma gut regulatory peptide levels in calves: Effects of the first meals. *Biol. Neonate.* **61**, 103–109.

Gunnes, P., Smiseth, O. A., Lygren, I., and Jorde, R. (1985). Effects of secretin infusion on myocardial performance and metabolism in the dog. *J. Cardiovasc. Pharmacol.* **7**, 1183–1187.

Gunnes, P., Waldum, H. L., Rasmussen, K., Ostensen, H., and Burhol, P. G. (1983). Cardiovascular effects of secretin infusion in man. *Scand J. Clin. Lab. Invest.* **43**, 637–642.

Gupta, S., Aggarwal, S., Rashanravan, B., and Lee, T. (1998). Th1- and Th2-like cytokines in CD4+ and CD8+ T cells in autism. *J. Neuroimmunol.* **85**, 106–109.

Gupta, S. (2000). Immunological treatments for autism. *J. Autism. Dev. Disord.* **30**, 475–479.

Haddad, J. J., Saade, N. E., and Safieh-Garabedian, B. (2002). Cytokines and neuro-immune-endocrine interactions: A role for the hypothalamic-pituitary-adrenal revolving axis. *J. Neuroimmunol.* **133**, 1–19.

Haraldsen, L., Soderstrom-Lauritzsen, V., and Nilsson, G. E. (2002). Oxytocin stimulates cerebral blood flow in rainbow trout (Oncorhynchus mykiss) through a nitric oxide dependent mechanism. *Brain Res.* **929**, 10–14.

Harmar, A. J. (2001). Family-B G-protein-coupled receptors. *Genome Biol.* **2**, Reviews 3013.

Hatton, G. I., Modney, B. K., and Salm, A. K. (1992). Increases in dendritic bundling and dye coupling of supraoptic neurons after the induction of maternal behavior. *Ann NY Acad. Sci.* **12**, 142–155.

Haxby, J. V., Hoffman, E. A., and Gobbini, M. I. (2002). Human neural systems for face recognition and social communication. *Biol. Psychiatry* **51**, 59–67.

Helou, C. M., Imbert-Teboul, M., Doucet, A., Rajerison, R., Chollet, C., Alhenc-Gelas, F., and Marchetti, J. (2003). Angiotensin receptor subtypes in thin and muscular juxtamedullary efferent arterioles of rat kidney. *Am. J. Physiol. Renal. Physiol.* **285**, F507–F514.

Hofer, M. A. (1994). Hidden regulators in attachment, separation, and loss. *Monogr. Soc. Res. Child. Dev.* **59**, 192–207.

Hollander, E., Novotny, S., Hanratty, M., Yaffe, R., De Caria, C. M., Aronowitz, B. R., and Mosovich, S. (2003). Oxytocin infusion reduces repetitive behaviors in adults with autistic and Asperger's disorders. *Neuropsychopharmacology* **28**, 193–198.

Hornig, M., Chian, D., and Lipkin, W. I. (2004). Neurotoxic effects of postnatal thimerosal are mouse strain dependent. *Mol. Psychiatry* **9**, 833–845.

Horvath, K., Papadimitriou, J. C., Rabsztyn, A., Drachenberg, C., and Tildon, J. T. (1999). Gastrointestinal abnormalities in children with autistic disorder. *J. Pediatr.* **135**, 559–563.

Horvath, K, and Perman, J. A. (2002). Autism and gastrointestinal symptoms. *Curr. Gastroenterol Rep.* **4**, 251–258.

Horvath, K., Stefanatos, G., Sokolski, K. N., Wachtel, R., Nabors, L., and Tildon, J. T. (1998). Improved social and language skills after secretin administration in patients with autistic spectrum disorders. *J. Assoc. Acad. Minor. Phys.* **9**, 9–15.

Husson, I., Rangon, C. M., Lelievre, V., Bemelmans, A. P., Sachs, P., Mallet, J., Kosofsky, B. E., and Gressens, P. (2005). BDNF-induced white matter neuroprotection and stage-dependent neuronal survival following a neonatal excitotoxic challenge. *Cereb. Cortex* **15**, 250–261.

Iadecola, C., Faris, P. L., Hartman, B. K., and Xu, X. (1993). Localization of NADPH diaphorase in neurons of the rostral ventral medulla: Possible role of nitric oxide in central autonomic regulation and oxygen chemoreception. *Brain Res.* **603**, 173–179.

Iqbal, Z. (2002). Ethical issues involved in the implementation of a differential reinforcement of inappropriate behaviour programme for the treatment of social isolation and ritualistic behaviour in an individual with intellectual disabilities. *J. Intellect. Disabil. Res.* **46,** 82–93.

Iseri, S. O., Sener, G., Saglam, B., Gedik, N., Ercan, F., and Yegen, B. C. (2005). Oxytocin ameliorates oxidative colonic inflammation by a neutrophil-dependent mechanism. *Peptides* **26,** 483–491.

Itoh, N., Furuya, T., Ozaki, K., Ohta, M., and Kawasaki, T. (1991). The secretin precursor gene. Structure of the coding region and expression in the brain. *J. Biol. Chem.* **266,** 12595–12598.

James, S. J., Cutler, P., Melnyk, S., Jernigan, S., Janak, L., Gaylor, D. W., and Neubrander, J. A. (2004). Metabolic biomarkers of increased oxidative stress and impaired methylation capacity in children with autism. *Am. J. Clin. Nutr.* **80,** 1611–1617.

James, W. (1884). What is an emotion? *Mind* **9,** 188–205.

Jankowski, M., Wang, D., Hajjar, F., Mukaddam-Daher, S., McCann, S. M., and Gutkowska, J. (2000). Oxytocin and its receptors are synthesized in the rat vasculature. *Proc. Natl. Acad. Sci. USA* **97,** 6207–6211.

Jezova, M., Armando, I., Bregonzio, C., Yu, Z. X., Qian, S., Ferrans, V. J., Imboden, H., and Saavedra, J. M. (2003). Angiotensin II AT(1) and AT(2) receptors contribute to maintain basal adrenomedullary norepinephrine synthesis and tyrosine hydroxylase transcription. *Endocrinology* **144,** 2092–2101.

Jin, H. O., Lee, K. Y., Chang, T. M., Chey, W. Y., and Dubois, A. (1994). Secretin: A physiological regulator of gastric emptying and acid output in dogs. *Am. J. Physiol.* **267,** 702–708.

Jones, D., and Gonzalez-Lima, F. (2001). Mapping Pavlovian conditioning effects on the brain: Blocking, contiguity, and excitatory effects. *J. Neurophysiol.* **86,** 809–823.

Jyotheeswaran, S., Li, P., Chang, T. M., and Chey, W. Y. (2000). Endogenous nitric oxide mediates pancreatic exocrine secretion stimulated by secretin and cholecystokinin in rats. *Pancreas* **20,** 401–407.

Kadekaro, M. (2004). Nitric oxide modulation of the hypothalamo-neurohypophyseal system. *Braz. J. Med. Biol. Res.* **37,** 441–450.

Kalia, M. (2005). Neurobiological basis of depression: An update. *Metabolism* **54**(5 Suppl 2), 24–27.

Kern, J. K., Espinoza, E., and Trivedi, M. H. (2004). The effectiveness of secretin in the management of autism. *Expert. Opin. Pharmacother.* **5,** 379–387.

Kern, J. K., Van Miller, S., Evans, P. A., and Trivedi, M. H. (2002). Efficacy of porcine secretin in children with autism and pervasive developmental disorder. *J. Autism. Dev. Disord.* **32,** 153–160.

Kimura, M., Masuda, T., Hiwatashi, N., Toyota, T., and Nagura, H. (1994). Changes in neuropeptide-containing nerves in human colonic mucosa with inflammatory bowel disease. *Pathol Int.* **44,** 624–634.

Knable, M. B., Barci, B. M., Webster, M. J., Meador-Woodruff, J., and Torrey, E. F. (2004). Stanley Neuropathology Consortium. Molecular abnormalities of the hippocampus in severe psychiatric illness: Postmortem findings from the Stanley Neuropathology Consortium. *Mol. Psychiatry* **9,** 609–620.

Kobayashi, R., Takenoshita, Y., Kobayashi, H., Kamijo, A., Funaba, K., and Takarabe, M. (2001). Early intervention for infants with autistic spectrum disorders in Japan. *Pediatr Int.* **43,** 202–208.

Kosfeld, M., Markus, H., Zak, P. J., Fischbacher, U., and Fehr, E. (2005). Oxytocin increases trust in humans. *Nature* **435,** 673–676.

Koves, K., Kausz, M., Reser, D., and Horvath, K. (2002). What may be the anatomical basis that secretin can improve the mental functions in autism? *Regul. Pept.* **109,** 167–172.

Kulman, G., Lissoni, P., Rovelli, F., Roselli, M. G., Brivio, F., and Sequeri, P. (2000). Evidence of pineal endocrine hypofunction in autistic children. *Neuro. Endocrinol. Lett.* **21,** 31–34.

Kuntz, A., Clement, H. W., Lehnert, W., Van Calker, D., Hennighausen, K., Gerlach, M., and Schulz, E. (2004). Effects of secretin on extracellular amino acid concentrations in rat. *J. Neural. Transm.* **111,** 931–939.

Kurjak, M., Schusdziarra, V., and Allescher, H. D. (1996). Presynaptic modulation by VIP, secretin and isoproterenol of somatostatin release from enriched enteric synaptosomes: Role of cAMP. *Eur. J. Pharmacol.* **24,** 165–173.

Lamb, J. A., Parr, J. R., Bailey, A. J., and Monaco, A. P. (2002). Autism: In search of susceptibility genes. *Neuromolecular Med.* **2,** 11–28.

Lamson, DW, and Plaza, S. M. (2001). Transdermal secretin for autism – a case report. *Altern. Med. Rev.* **6,** 311–313.

Lange, C. (1885). "Om Sindsbehavgelser." Copenhagen.

Langworthy-Lam, K. S., Aman, M. G., and Van Bourgondien, M. E. (2002). Prevalence and patterns of use of psychoactive medicines in individuals with autism in the Autism Society of North Carolina. *J. Child Adolesc. Psychopharmacol.* **12,** 311–321.

LeDoux, J. (1998). "The Emotional Brain: The Emotional Underpinnings of Emotional Life." Simon & Schuster, New York.

Leong, D. S., Terron, J. A., Falcon-Neri, A., Armando, I., Ito, T., Johren, O., Tonelli, L. H., Hoe, K. L., and Saavedra, J. M. (2002). Restraint stress modulates brain, pituitary and adrenal expression of angiotensin II AT(1A), AT(1B) and AT(2) receptors. *Neuroendocrinology* **75,** 227–240.

Leventhal, B. L., Cook, E. H., Jr., Morford, M., Ravitz, A., and Freedman, D. X. (1990). Relationships of whole blood serotonin and plasma norepinephrine within families. *J. Autism. Dev. Disord.* **20,** 499–511.

Lewis, D. A., Hashimoto, T., and Volk, D. W. (2005). Cortical inhibitory neurons and schizophrenia. *Nat. Rev. Neurosci.* **6,** 312–324.

Li, P., Chang, T. M., and Chey, W. Y. (1998). Secretin inhibits gastric acid secretion via a vagal afferent pathway in rats. *Am. J. Physiol.* **275,** G22–28.

Li, YW, and Guyenet, P. G. (1996). Angiotensin II decreases a resting K+ conductance in rat bulbospinal neurons of the C1 area. *Circ. Res.* **78,** 274–282.

Licinio, J., and Wong, M. L. (1999). The role of inflammatory mediators in the biology of major depression: Central nervous system cytokines modulate the biological substrate of depressive symptoms, regulate stress-responsive systems, and contribute to neurotoxicity and neuroprotection. *Mol. Psychiatry.* **4,** 317–327.

Lightdale, J. R., Hayer, C., Duer, A., Lind-White, C., Jenkins, S., Siegel, B., Elliott, G. R., and Heyman, M. B. (2001). Effects of intravenous secretin on language and behavior of children with autism and gastrointestinal symptoms: A single-blinded, open-label pilot study. *Pediatrics* **108,** E90.

Linden, D. R., Chen, J. X., Gershon, M. D., Sharkey, K. A., and Mawe, G. M. (2003). Serotonin availability is increased in mucosa of guinea pigs with TNBS-induced colitis. *Am. J. Physiol. Gastrointest. Liver Physiol.* **285,** G207–G216.

Lissak, K., and Molnar, P. (1975). Emotional behavior: The control of the emotions and their expression. *In* "The life and contributions of Walter Bradford Cannon" (C. M. Brooks, K. Koizumi, and J. O. Pinkston, Eds.), pp. 115–145. State University of New York Press, Albany.

Lonstein, J. S. (2005). Reduced anxiety in postpartum rats requires recent physical interactions with pups, but is independent of suckling and peripheral sources of hormones. *Horm. Behav.* **47,** 241–255.

Lopez, M. J., Upchurch, B. H., Rindi, G., and Leiter, A. B. (1995). Studies in transgenic mice reveal potential relationships between secretin-producing cells and other endocrine cell types. *J. Biol. Chem.* **270,** 885–891.

Lossos, A., River, Y., Eliakim, A., and Steiner, I. (1995). Neurologic aspects of inflammatory bowel disease. *Neurology.* **45,** 416–421.

Lucas, A., Adrian, T. E., Bloom, S. R., and Aynsley-Green, A. (1980a). Plasma secretin in neonates. *Acta Paediatr. Scand.* **69,** 205–210.

Lucas, A., Bloom, S. R., and Aynsley-Green, A. (1980b). Development of gut hormone responses to feeding in neonates. *Arch. Dis. Child.* **55,** 678–682.

Ma, D. Q., Whitehead, P. L., Menold, M. M., Martin, E. R., Ashley-Koch, A. E., Mei, H., Ritchie, M. D., Delong, G. R., Abramson, R. K., Wright, H. H., Cuccaro, M. L., Hussman, J. P., Gilbert, J. R., and Pericak-Vance, M. A. (2005). Identification of significant association and gene-gene interaction of GABA receptor subunit genes in autism. *Am. J. Hum. Genet.* **77,** 377–388.

Maes, M., Kenis, G., and Kubera, M. (2003). In humans, corticotropin releasing hormone antagonizes some of the negative immunoregulatory effects of serotonin. *Neuro. Endocrinol. Lett.* **24,** 420–424.

Mark, K. S., Burroughs, A. R., Brown, R. C., Huber, J. D., and Davis, T. P. (2004). Nitric oxide mediates hypoxia-induced changes in paracellular permeability of cerebral microvasculature. *Am. J. Physiol. Heart Circ. Physiol.* **286,** H174–H180.

Matri, S., Boubaker, J., Hamzaoui, S., Bardi, R., Ayed, K., and Filali, A. (2003). [The role of major histocompatibility complex genes in the pathogenesis of chronic inflammatory bowel diseases.]. *Tunis. Med.* **81,** 289–294.

Matthiesen, A. S., Ransjo-Arvidson, A. B., Nissen, E., and Uvnas-Moberg, K. (2001). Postpartum maternal oxytocin release by newborns: Effects of infant hand massage and sucking. *Birth* **28,** 13–19.

McCann, M. J., and Rogers, R. C. (1990). Oxytocin excites gastric-related neurones in rat dorsal vagal complex. *J. Physiol.* **428,** 95–108.

McGill, J. M., Basavappa, S., Gettys, T. W., and Fitz, J. G. (1994). Secretin activates Cl⁻ channels in bile duct epithelial cells through a cAMP-dependent mechanism. *Am. J. Physiol.* **266,** G731–G736.

Meaney, M. J. (2004). Environmental 'programming' of individual differences in defensive and reproductive behaviors through maternal effects on chromatin structure and gene expression. Special Lecture. Society for Neuroscience, 34th Annual Meeting, San Diego.

Meaney, M. J., and Szyf, M. (2005). Maternal care as a model for experience-dependent chromatin plasticity? *Trends Neurosci.* **28,** 456–463.

Meaney, M. J., Viau, V., Aitken, D. H., and Bhatnagar, S. (1988). Stress-induced occupancy and translocation of hippocampal glucocorticoid receptors. *Brain Res.* **445,** 198–203.

Menold, M. M., Shao, Y., Wolpert, C. M., Donnelly, S. L., Raiford, K. L., Martin, E. R., Ravan, S. A., Abramson, R. K., Wright, H. H., Delong, G. R., Cuccaro, M. L., Pericak-Vance, M. A., and Gilbert, J. R. (2001). Association analysis of chromosome 15 gabaa receptor subunit genes in autistic disorder. *J. Neurogenet.* **15,** 245–259.

Militerni, R., Bravaccio, C., Falco, C., Fico, C., and Palermo, M. T. (2002). Repetitive behaviors in autistic disorder. *Eur. Child Adolesc. Psychiatry.* **11,** 210–218.

Mutt, V., Carlquist, M., and Tatemoto, K. (1979). Secretin-like bioactivity in extracts of porcine brain. *Life Sci.* **25,** 1703–1707.

Naruse, S., Ito, O., Kitagawa, M., Ishiguro, H., Nakajima, M., and Hayakawa, T. (1998). Effects of PACAP/VIP/secretin on pancreatic and gastrointestinal blood flow in conscious dogs. *Ann. NY Acad. Sci.* **865,** 463–465.

Nelson, K. B., Grether, J. K., Croen, L. A., Dambrosia, J. M., Dickens, B. F., Jelliffe, L. L., Hansen, R. L., and Phillips, T. M. (2001). Neuropeptides and neurotrophins in neonatal blood of children with autism or mental retardation. *Ann. Neurol.* **49,** 597–606.

Nozaki, S., Nakata, R., Mizuma, H., Nishimura, N., Watanabe, Y., Kohashi, R., and Watanabe, Y. (2002). In vitro autoradiographic localization of (125)I-secretin receptor binding sites in rat brain. *Biochem. Biophys. Res. Commun.* **292,** 133–137.

O'Donohue, T. L., Charlton, C. G., Miller, R. L., Boden, G., and Jacobowitz, D. M. (1981). Identification, characterization, and distribution of secretin immunoreactivity in rat and pig brain. *Proc. Natl. Acad. Sci. USA* **78,** 5221–5224.

Ogai, M., Matsumoto, H., Suzuki, K., Ozawa, F., Fukuda, R., Uchiyama, I., Suckling, J., Isoda, H., Mori, N., and Takei, N. (2003). fMRI study of recognition of facial expressions in high-functioning autistic patients. *Neuroreport* **14,** 559–563.

Ohnishi, T., Matsuda, H., Hashimoto, T., Kunihiro, T., Nishikawa, M., Uema, T., and Sasaki, M. (2000). Abnormal regional cerebral blood flow in childhood autism. *Brain.* **123,** 1838–1844.

Ohta, M., Funakoshi, S., Kawasaki, T., and Itoh, N. (1992). Tissue-specific expression of the rat secretin precursor gene. *Biochem. Biophys. Res. Commun.* **183,** 390–395.

Okere, C. O., Higuchi, T., Kaba, H., Russell, J. A., Okutani, F., Takahashi, S., and Murata, T. (1996). Nitric oxide prolongs parturition and inhibits maternal behavior in rats. *Neuroreport.* **7,** 1695–1699.

Owley, T., McMahon, W., Cook, E. H., Laulhere, T., South, M., Mays, L. Z., Shernoff, E. S., Lainhart, J., Modahl, C. B., Corsello, C., Ozonoff, S., Risi, S., Lord, C., Leventhal, B. L., and Filipek, P. A. (2001). Multisite, double-blind, placebo-controlled trial of porcine secretin in autism. *J. Am. Acad. Child Adolesc. Psychiatry* **40,** 1293–1299.

Park, Y. D. (2003). The effects of vagus nerve stimulation therapy on patients with intractable seizures and either Landau-Kleffner syndrome or autism. *Epilepsy Behav. Jun* **4**(3), 286–290.

Pedersen, C. A., and Boccia, M. L. (2002). Oxytocin links mothering received, mothering bestowed and adult stress responses. *Stress.* **5,** 259–267.

Pelphrey, K. A., Sasson, N. J., Reznick, J. S., Paul, G., Goldman, B. D., and Piven, J. (2002). Visual scanning of faces in autism. *J. Autism. Dev. Disord.* **32,** 249–261.

Persico, A. M., D'Agruma, L., Maiorano, N., Totaro, A., Militerni, R., Bravaccio, C., Wassink, T. H., Schneider, C., Melmed, R., Trillo, S., Montecchi, F., Palermo, M., Pascucci, T., Puglisi-Allegra, S., Reichelt, K. L., Conciatori, M., Marino, R., Quattrocchi, C. C., Baldi, A., Zelante, L., Gasparini, P., and Keller, F; Collaborative Linkage Study of Autism. (2001). Reelin gene alleles and haplotypes as a factor predisposing to autistic disorder. *Mol. Psychiatry* **6,** 150–159.

Porges, S. W., Doussard-Roosevelt, J. A., and Maiti, A. K. (1994). Vagal tone and the physiological regulation of emotion. *Monogr. Soc. Res. Child Dev.* **59,** 167–186.

Porges, S. W. (1995). Orienting in a defensive world: Mammalian modifications of our evolutionary heritage. A polyvagal theory. *Psychophysiology* **32,** 301–318.

Posey, D. J., and McDougle, C. J. (2000). The pharmacotherapy of target symptoms associated with autistic disorder and other pervasive developmental disorders. *Harv. Rev. Psychiatry* **8,** 45–63.

Raggenbass, M., and Dreifuss, J. J. (1992). Mechanism of action of oxytocin in rat vagal neurones: Induction of a sustained sodium-dependent current. *J. Physiol.* **457,** 131–142.

Richer, J. M., and Coss, R. G. (1976). Gaze aversion in autistic and normal children. *Acta. Psychiatr. Scand.* **53,** 193–210.

Rinaman, L. (1998). Oxytocinergic inputs to the nucleus of the solitary tract and dorsal motor nucleus of the vagus in neonatal rats. *J. Comp. Neurol.* **399,** 101–109.

Ringel, Y., and Drossman, D. A. (2001). Psychosocial aspects of Crohn's disease. *Surg. Clin. North. Am.* **81,** 231–252.

Roberts, W., Weaver, L., Brian, J., Bryson, S., Emelianova, S., Griffiths, A. M., Mac Kinnon, B., Yim, C., Wolpin, J., and Koren, G. (2001). Repeated doses of porcine secretin in the treatment of autism: A randomized, placebo-controlled trial. *Pediatrics* **107,** E71.

Robinson, B. (1907). "The Abdominal and Pelvic Brain." Frank S. Betz, Hammond, Indiana.

Rogers, I. M., Davidson, D. C., Lawrence, J., and Buchanan, K. D. (1975). Neonatal secretion of secretin. *Arch. Dis. Child.* **50,** 120–122.

Roy-Byrne, P. P. (2005). The GABA-benzodiazepine receptor complex: Structure, function, and role in anxiety. *J. Clin. Psychiatry.* **66**(Suppl 2), 14–20.

Ruggiero, D. A., Anwar, S., Kim, J., and Glickstein, S. B. (1998). Visceral afferent pathways to the thalamus and olfactory tubercle: Behavioral implications. *Brain Res.* **799,** 159–171.

Ruggiero, D. A., Pickel, V. M., Milner, T. A., Anwar, M., Otake, K., Mtui, E.P, and Park, D. (1993). Viscerosensory processing in nucleus tractus solitarii: Structural and neurochemical substrates. *In* "Nucleus of the Solitary Tract" (R. A. Barraco, Ed.), pp. 3–34. CRC Press, Boca Raton.

Ruggiero, D. A., Ross, C. A., Anwar, M., Park, D. H., Joh, T. H., and Reis, D. J. (1985). Distribution of neurons containing phenylethanolamine N-methyltransferase in medulla and hypothalamus of rat. *J. Comp. Neurol.* **239,** 127–154.

Ruggiero, D. A., Gootman, P. M., and Sica, A. (1998). Presence of a non-NMDA glutamate receptor subtype in the sympathetic nervous system of neonatal swine. *J. Auton. Nerv. Syst.* **10,** 101–108.

Ruggiero, D. A., Welch-Horan, T. B., Anwar, M., Anwar, N., Ludwig, R. J., Jafri, F., and Welch, M. G. (2005). Central sites of secretin-, oxytocin- and CRH-immunoreactivity provide anatomical basis for clinically effective interaction in the treatment of autism and IBD. Abstracts 2005 Society for Neuroscience Abstracts, Washington, DC.

Sacchetti, B., Scelfo, B., and Strata, P. (2005). The cerebellum: Synaptic changes and fear conditioning. *Neuroscientist* **11,** 217–227.

Samson, W. K., Lumpkin, M. D., and McCann, S. M. (1984). Presence and possible site of action of secretin in the rat pituitary and hypothalamus. *Life Sci.* **34,** 155–163.

Sanchez-Alvarez, R., Almeida, A., and Medina, J. M. (2002). Oxidative stress in preterm rat brain is due to mitochondrial dysfunction. *Pediatr. Res.* **51,** 34–39.

Sandler, A. D. (1999). Assessment of an itinerant medical evaluation program for school dysfunction. *J. Sch. Health.* **69,** 140–144.

Schultz, R. T., Grelotti, D. J., Klin, A., Kleinman, J., Van der Gaag, C., Marois, R., and Skudlarski, P. (2003). The role of the fusiform face area in social cognition: Implications for the pathobiology of autism. *Philos. Trans. R Soc. Lond. B Biol. Sci.* **358,** 415–427.

Singh, V. K. (1996). Plasma increase of interleukin-12 and interferon-gamma. Pathological significance in autism. *J. Neuroimmunol.* **66,** 143–145.

Sinha, A., Nightingale, J., West, K. P., Berlanga-Acosta, J., and Playford, R. J. (2003). Epidermal growth factor enemas with oral mesalamine for mild-to-moderate left-sided ulcerative colitis or proctitis. *N. Engl. J. Med.* **349,** 350–357.

Sitniewska, E. M., Wisniewska, R. J., and Wisniewski, K. (2000). Influence of nitric oxide on the cardiovascular action of secretin in intact rats. Part B. Does nitric oxide influence the effect of secretin on isolated heart function? *Pol. J. Pharmacol.* **52,** 375–381.

Sofroniew, M. V., Weindl, A., Schrell, U., and Wetzstein, R. (1981). Immunohistochemistry of vasopressin, oxytocin and neurophysin in the hypothalamus and extrahypothalamic regions of the human and primate brain. *Acta Histochem. Suppl.* **24,** 79–95.

Stern, J. M., and Mackinnon, D. A. (1976). Postpartum, hormonal, and nonhormonal induction of maternal behavior in rats: Effects on T-maze retrieval of pups. *Horm. Behav.* **7,** 305–316.

Sweeten, T. L., Posey, D. J., Shankar, S., and McDougle, C. J. (2004). High nitric oxide production in autistic disorder: A possible role for interferon-gamma. *Biol. Psychiatry* **55,** 434–437.

Szatmari, P. (1999). Heterogeneity and the genetics of autism. *J. Psychiatry. Neurosci.* **24,** 159–165.

Taber, K. H., Shaw, J. B., Loveland, K. A., Pearson, D. A., Lane, D. M., and Hayman, L. A. (2004). Accentuated Virchow-Robin spaces in the centrum semiovale in children with autistic disorder. *J. Comput. Assist. Tomogr.* **28,** 263–268.

Talman, W. T., and Kelkar, P. (1993). Neural control of the heart. Central and peripheral. *Neurol. Clin.* **11,** 239–256.

Tannenbaum, B., Tannenbaum, G. S., Sudom, K., and Anisman, H. (2002). Neurochemical and behavioral alterations elicited by a chronic intermittent stressor regimen: Implications for allostatic load. *Brain Res.* **25,** 82–92.

Tanoue, Y., and Oda, S. (1989). Weaning time of children with infantile autism. *J. Autism. Dev. Disord.* **19,** 425–434.

Tay, J., Goulet, M., Rusche, J., and Boismenu, R. (2004). Age-related and regional differences in secretin and secretin receptor mRNA levels in the rat brain. *Neurosci. Lett.* **366,** 176–181.

Teufel, M., Luik, G., and Niessen, K. H. (1986). [Gastrin, secretin, VIP and motilin in children with mucoviscidosis and Crohn disease.]. *Monatsschr Kinderheilkd.* **134,** 132–137.

Tinbergen, N, and Tinbergen, E. A. (1983). "Autistic Children – New Hope for a Cure." George, Allen and Unwin, London and Boston.

Torrente, F., Ashwood, P., Day, R., Machado, N., Furlano, R. I., Anthony, A., Davies, S. E., Wakefield, A. J., Thomson, M. A., Walker-Smith, J. A., and Murch, S. H. (2002). Small intestinal enteropathy with epithelial IgG and complement deposition in children with regressive autism. *Mol. Psychiatry* **7,** 375–382.

Traub, R. J., Silva, E., Gebhart, G. F., and Solodkin, A. (1996). Noxious colorectal distention induced-c-Fos protein in limbic brain structures in the rat. *Neurosci. Lett.* **13,** 165–168.

Uno, H., Tarara, R., Else, J. G., Suleman, M. A., and Sapolsky, R. M. (1989). Hippocampal damage associated with prolonged and fatal stress in primates. *J. Neurosci.* **9,** 1705–1711.

Uvnas-Moberg, K. (1989). The gastrointestinal tract in growth and reproduction. *Sci. Am.* **261,** 78–83.

Uvnas-Moberg, K. (1997). Oxytocin linked antistress effects—the relaxation and growth response. *Acta Physiol. Scand.* **640**(Suppl.), 38–42.

Uvnas-Moberg, K., Widstrom, A. M., Marchini, G., and Winberg, J. (1987). Release of GI hormones in mother and infant by sensory stimulation. *Acta Paediatr. Scand.* **76,** 851–860.

Vaccari, C., Lolait, S. J., and Ostrowski, N. L. (1998). Comparative distribution of vasopressin VIb and oxytocin receptor messenger ribonucleic acids in brain. *Endocrinology.* **139,** 5015–5033.

Vacher, C. M., Fretier, P., Creminon, C., Calas, A., and Hardin-Pouzet, H. (2002). Activation by serotonin and noradrenaline of vasopressin and oxytocin expression in the mouse paraventricular and supraoptic nuclei. *J. Neurosci.* **22,** 1513–1522.

Vargas, D. L., Nascimbene, C., Krishnan, C., Zimmerman, A. W., and Pardo, C. A. (2005). Neuroglial activation and neuroinflammation in the brain of patients with autism. *Ann. Neurol.* **57,** 67–81.

Walker, J. K., Premont, R. T., Barak, L. S., Caron, M. G., and Shetzline, M. A. (1999). Properties of secretin receptor internalization differ from those of the beta(2)-adrenergic receptor. *J. Biol. Chem.* **274,** 31515–31523.

Weaver, I. C., Grant, R. J., and Meaney, M. J. (2002). Maternal behavior regulates long-term hippocampal expression of BAX and apoptosis in the offspring. *J. Neurochem.* **82,** 998–1002.

Weaver, I. C., Cervoni, N., Champagne, F. A., D'Alessio, A. C., Sharma, S., Seckl, J. R., Dymov, S., Szyf, M., and Meaney, M. J. (2004). Epigenetic programming by maternal behavior. *Nat. Neurosci.* **7,** 847–854.

Webster, M. J., Knable, M. B., O'Grady, J., Orthmann, J., and Weickert, C. S. (2002). Regional specificity of brain glucocorticoid receptor mRNA alterations in subjects with schizophrenia and mood disorders. *Mol. Psychiatry* **7,** 924, 985–994.

Welch, M. G. (1983). Retrieval from autism through mother-child holding. *In* "Autistic Children – New Hope for a Cure" (N. Tinbergen and E. A., Eds.). George, Allen and Unwin, London and Boston.

Welch, M. G. (1987). Toward prevention of developmental disorders. *Pa. Med.* **90,** 47–52.

Welch, M. G. (1988). "Holding Time." Simon and Schuster, New York.

Welch, M. G., and Chaput, P. (1988). Mother-child holding therapy and autism. *Pa. Med.* **91,** 33–38.

Welch, M. G., Keune, J. D., Welch-Horan, T. B., Anwar, M., Anwar, N., and Ruggiero, D. A. (2003a). Secretin activates visceral brain regions in rat including areas abnormal in autism. *Cell. Mol. Neurobiol.* **23,** 817–837.

Welch, M. G., Welch-Horan, T. B., Keune, J. D., Anwar, N., Anwar, M., Ludwig, R. J., and Ruggiero, D. A. (2003b). Neurohormonal resolution of genetic and acquired IBD and secondary brain activation in areas abnormal in autism. Prog. # 318.5 2003 Abstracts. Society for Neuroscience Abstracts, Washington, DC.

Welch, M. G., Welch-Horan, T. B., Anwar, M., Keune, J. D., Anwar, N., Ludwig, R. J., and Ruggiero, D. A. (2004a). Secretin: Hypothalamic distribution and hypothesized neuroregulatory role in autism. *Cell. Mol. Neurobiol.* **24,** 219–241.

Welch, M. G., Welch-Horan, T. B., Anwar, M., Keune, J. D., Anwar, N., Ludwig, R. J., Sica, A. L., and Ruggiero, D. A. (2004b). Brain/gut peptides in developmental and inflammatory processes. FASEB Abstract 833.19. 2004 Abstracts. Federation of Affiliated Societies for Experimental Biology, Washington, DC.

Welch, M. G., Welch-Horan, T. B., Anwar, M., Ludwig, R. J., Power, S. A., and Ruggiero, D. A. (2004c). Behavioral anatomy of intensive maternal nurturing in childhood disorders. Prog. # 801.18 2004 Abstracts. Society for Neuroscience Abstracts, Washington, DC.

Welch, M. G., Welch-Horan, T. B., Anwar, M., Anwar, N., Ludwig, R. J., and Ruggiero, D. A. (2005). Brain effects of chronic IBD in areas abnormal in autism and treatment by single neuropeptides secretin and oxytocin. *J. Mol. Neurosci.* **25,** 259–274.

Welch, M. G., Northrup, R. S., Welch-Horan, T. B., Ludwig, R. J., Austin, C. L., and Jacobson, J. S. (2006). Outcomes of prolonged parent-child embrace therapy among 102 children with behavioral disorders. *Complement. Ther. Clin. Pract.* (In Press.).

Westlund, K. N., and Craig, A. D. (1996). Association of spinal lamina I projections with brainstem catecholamine neurons in the monkey. *Exp. Brain Res.* **110,** 151–162.

Whitaker-Azmitia, P. M. (2005). Behavioral and cellular consequences of increasing serotonergic activity during brain development: A role in autism? *Int. J. Dev. Neurosci.* **23,** 75–83.

White, J. F. (2003). Intestinal pathophysiology in autism. *Exp. Biol. Med. (Maywood)* **228,** 639–649.

Williams, G., Bing, C., Cai, X. J., Harrold, J. A., King, P. J., and Liu, X. H. (2001). The hypothalamus and the control of energy homeostasis: Different circuits, different purposes. *Physiol. Behav.* **74,** 683–701.

Windle, R. J., Kershaw, Y. M., Shanks, N., Wood, S. A., Lightman, S. L., and Ingram, C. D. (2004). Oxytocin attenuates stress-induced c-fos mRNA expression in specific forebrain regions associated with modulation of hypothalamo-pituitary-adrenal activity. *J. Neurosci.* **24,** 2974–2982.

Winslow, J. T. (2005). Neuropeptides and non-human primate social deficits associated with pathogenic rearing experience. *Int. J. Dev. Neurosci.* **23,** 245–251.

Winslow, J. T., and Insel, T. R. (2002). Neuroendocrine basis of social recognition. *Curr. Opin. Neurobiol.* **14,** 248–253.

Yirmiya, N., Pilowsky, T., Nemanov, L., Arbelle, S., Feinsilver, T., Fried, I., and Ebstein, R. P. (2001). Evidence for an association with the serotonin transporter promoter region polymorphism and autism. *Am. J. Med. Genet.* **105,** 381–386.

Yung, W. H., Leung, P. S., Ng, S. S., Zhang, J., Chan, S. C., and Chow, B. K. (2001). Secretin facilitates GABA transmission in the cerebellum. *J. Neurosci.* **21,** 7063–7068.

Zabielski, R., Terui, Y., Onaga, T., Mineo, H., and Kato, S. (1994). Plasma secretin fluctuates in phase with periodic pancreatic secretion and the duodenal migrating myoelectric complex in calves. *Res. Vet. Sci.* **56,** 332–337.

Zak, P. J., Kurzban, R., and Matzner, W. T. (2004). The neurobiology of trust. *Ann. NY. Acad. Sci.* **1032,** 224–227.

Zaninetti, M., and Raggenbass, M. (2000). Oxytocin receptor agonists enhance inhibitory synaptic transmission in the rat hippocampus by activating interneurons in stratum pyramidale. *Eur. J. Neurosci.* **12,** 3975–3984.

IMMUNOLOGICAL FINDINGS IN AUTISM

Hari Har Parshad Cohly* and Asit Panja[†]

*Department of Biology, Jackson State University
Jackson, Mississippi 39217, USA
[†]Department of Medicine, Division of Gastroenterology, University of Medicine and Dentistry of New Jersey-Robert Wood Johnson Medical School, New Brunswick, New Jersey 08903, USA

Autism is a disorder of neurobiological origin characterized by impairment of contact and communications. Typical symptoms of autism include extreme withdrawal and an abnormal absorption in fantasy, accompanied by delusion, hallucination, and an inability to communicate verbally or to otherwise relate to people. The cause of autism remains unknown. However, there are several factors including infectious, neurological, metabolic, environmental, and immunologic origin that have been thought to be involved in the disease development process of autism. The cellular entities playing a role in the pathologic processes in the autistic brain are the neurons, glial cells, endothelial cells, microglial cells, and astrocytes with blood brain barrier permeability playing an important role for the trafficking of the immune cells and mediators. In this chapter

INTERNATIONAL REVIEW OF
NEUROBIOLOGY, VOL. 71
DOI: 10.1016/S0074-7742(05)71013-8

317

immunologic findings on autism are discussed. Particular emphasis is made on the aspects of immunological dysfunctions and inflammation as the two important immunological principles contributing to the diseases process in autism.

I. Introduction

The initial identification of autism dates back to 1943, when Dr. Kanner first observed a syndrome of abnormal neurological development and impaired social interactions, restricted stereotyped interests, and abnormalities in verbal and nonverbal behavior among several children. Kanner called this stereotypic behavioral disorder autism. Since then, the incidence rate in autism has increased and this disorder is currently one of the major pediatric health concerns in the United States. In 1997, the Centers for Disease Control and Prevention (1999) estimated that a broad definition of autism or autistic spectrum disorders (ASD) may be present in as many as one out of every 500 children. Studies in neuroimaging (Minshew et al., 1993), anatomy, and cytotechnology (Bailey et al., 1998a; Bauman and Kemper, 1994), and epidemiologic (Gillberg, 1990) findings suggest that ASD results from a variety of quantitative and qualitative abnormalities in brain structure. Some molecular, genetic, and cellular characteristics have been identified in cell types including the neurons, glial cells, endothelial cells, microglial cells, and astrocytes of the central nervous system.

Symptomatic manifestations of autism occur within the first 5 years of life and persist into adulthood. The neuropathological abnormalities in this disease have been largely confined to the cerebellum and medial temporal structures. Thus, their possible involvement in autistic development has been the subject of much interest. Several investigators reported cerebellar abnormality in autistic samples (Bauman and Kemper, 1994; Courchesne et al., 1988, 1994; Ritvo et al., 1986). However, some of these studies are debatable and need further confirmation (Bailey and Cox, 1996; Bailey et al., 1998b). Furthermore, evidence for a decrease in cerebellar cell size with no differences in Purkinje cell densities between the normal and autistic children has been reported in literature (Fatemi et al., 2002). Studies of Carper and Courchesne (2000), and Bailey and Cox (1998) have demonstrated that the degree of frontal lobe abnormality correlated with the degree of cerebellar abnormality. The frontal lobe appears to have an excess of neural tissue while the cerebellum has too little neuronal cells in autistic patients.

Even though the causes of autism remain debatable, some scientific findings provide further clues. Large/small brain size and volume, asymmetry in the right hemisphere, attention to details, overlooking the whole along with clumsy behavior, and chronic inflammation in the central nervous system (CNS) are hallmarks of autism. Studies by Courchesne and his colleagues have shown that newborns who later develop autism have a smaller head size at birth but their head size grows rapidly between 1–2 months and 6–24 (Courchesne et al., 2003). In addition, studies by Herbert and colleagues (2003) demonstrated that there is asymmetrical development of the brain's white matter in autistic children. The brain of children with autism seems to grow normally until age 9 months followed by a rapid period of white matter growth between the period 9–24 months (Courchesne et al., 2003). Thus, in autism, there is asymmetrical-brain maldevelopment and potential abnormality/ies either how the brain is processing information or in the ability of the corpus callosum to network the two sides together where the right hemisphere is especially affected. In addition, studies from Just and his colleagues (2004) illustrated an alteration in brain circuitry causes the inability of autistic patients to utilize the right hemisphere of their brain that normally processes structures to recall the alphabet. However, autistic patients have good ability to appreciate details but little or no ability at conceiving the whole picture. This suggests an overconnectivity of local brain networks while long-range brain wiring are under-connected. Moreover, Teitelbaum and his colleagues (2004) showed that due to skewed brain wiring autistic subjects are clumsy and therefore use unusual strategies for locomotion. In conjunction with these reports is the finding of Goldberg (2000), who demonstrated that the parts of the cerebellum that govern the ability to restore balance operate normally in autistic children. Finally, Vargas et al. (2005) reported that the brain tissue of people with autism shows signs of chronic inflammation in the same areas that show excessive growth. The inflammation appears to last a lifetime with a characteristic increase in the number of astroglial cells. The brain areas that show hyperproliferation in white matter also show inflammation. There is also evidence for activated microglia in the spinal fluid (Vargas et al., 2005). Thus, in autistic inflammation there is involvement of astroglial and microglial cells in the absence of lymphocyte infiltration or immunoglobulin deposition in the CNS. There is also increased production of pro-inflammatory and anti-inflammatory cytokines such as MCP-1 and TGFß-1 by neuroglia (Vargas et al., 2005). All of these findings support a potential role for dysregulated immunoregulatory process and neuroinflammation in the CNS of patients with autism.

II. Immune Dysfunction in Autism

Substantial evidence suggests that the immune system plays an important role in the pathogenesis of autism (Bock et al., 2002; Gupta, 2000; Wakefield et al., 1998). While the exact mechanism of immune dysfunction in autistic patients remains undefined, two general possibilities have been outlined. First, there might be a defect in immune regulation that causes hyper- or hypo-activation of the cellular components of the nervous system. This causes a homeostatic imbalance among the immunoregulatory factors in the brain and/or other affected organs such as the gastrointestinal tract. Second, an alternative mechanism of autistic development has been viewed as autoimmune reaction directed toward a specific target molecule in the brain.

A. Involvement of Neuronal Major Histocompatibility Complex (MHC) in Autism

Class I and Class II major histocompatibility complex were originally thought to be specific to immune cells, but are also expressed by various other cell types in the brain. In fact, certain allelic products of these genes have been thought to be associated with autism (Daniels et al., 1995; Warren et al., 1991, 1992), including the null allele of the C4B gene (located in the class III region of the MHC), the extended haplotype B44-S30- DR4 (the 44 allele of the HLA-B region, the S allele of the BF gene, the 3 allele of C4A, and the null allele of C4B and the DR4 allele) (Daniels et al., 1995; Warren et al., 1992, 1996). It has also been reported that the third hypervariable region (HVR-3) of certain DRb1 alleles has a very strong association with autism (Warren et al., 1996). These observations provide evidence that MHC genes may be involved in autism.

Furthermore, accumulating evidence indicates that neuronal MHC Class I does not simply function in an immune capacity, but is also crucial for normal brain development, neuronal differentiation, synaptic plasticity, and even behavior. The observation that MHC exists not only in injured brain neurons (Neumann et al., 1995, 1997; Wong et al., 1984, 1985), but also in normal uninfected neurons (i.e., in vivo) opens up the possibility of it being involved in normal development as well as in diseases as in autism contributing to the pathophysiology of autistic development (Boulanger and Shatz, 2004). Class I molecules are expressed also by neurons that undergo activity-dependent, long-term structural and synaptic modifications including axonal branching and dendritic growth. In the adult hippocampus, MHC is required for normal long-term potentiation (LTP) and long-term depression (LTD), and is thought to be crucial to learning and memory building process (Boulanger, 2001). Thus, there is no room for immune molecules to reduce neural links (Helmuth, 2000).

Experimental studies with mutant mice genetically deficient in class I MHC or for a class I MHC receptor component, CD3 zeta showed an incomplete refinement of connections between retina and central targets during development (Huh, 2000). In the hippocampus of adult mutants, N-methyl-D-aspartate receptor-dependent long-term potentiation (LTP) is enhanced, long-term depression (LTD) was absent, and specific class I MHC mRNAs were expressed by distinct mosaics of neurons. These results demonstrated an important role for Class I molecules in the activity-dependent remodeling and plasticity of connections in the developing and mature mammalian central nervous system (CNS). These results clearly show that MHC-I molecules are required for proper development in the CNS (Huh, 2000). Since the pattern of MHC expression is very diverse, it could be reasoned that the expression of MHC is directly related to the neuronal activity (Huh, 2000).

B. Impaired Cell-Mediated Immunity

Cell-mediated immunity is impaired in autism. This includes changes in the numbers and functions of macrophages, T cells, B cells, and natural killer cell activity (Gupta 2000; Warren *et al.*, 1986, 1987). In autistic patients who suffered from frequent gastrointestinal symptoms, Wakefield and colleagues demonstrated that CD3(+) cells were significantly increased in affected children compared with developmentally normal non-inflamed control groups ($p < 0.01$) reaching levels similar to inflamed controls (Ashwood *et al.*, 2003).

C. T-Cell Polarity in Autism

Helper T-lymphocytes have been shown to differentiate into two mutually regulatory subsets. Th1-like (IL-2, IFN-γ) mediates classical cell-mediated immune responses such as delayed-type hypersensitivity. Th2-like (IL-4, IL-6, and IL-10) cells promote humoral immune responses, in particular the production of IgE and IgG4 (human) or IgG1 (rodents). Over-activity of either cell type can result in a tissue-damaging autoimmune disease. A number of human diseases including asthma and some kidney diseases are thought to be caused by a Th-2 type autoimmune response. A shift occurs from T helper 1(Th1) to T helper 2 (Th2) T cells in autism as evidenced by a decrease in the production of inteleukin-2 (IL-2) and gamma interferon(IFN-γ), but there is an increase in the production of IL-4 (37). An imbalance of Th1/Th2 subsets of CD4$^+$/CD8$^+$ T cells towards Th2 may play a role in the pathogenesis of autism involving an autoimmune phenomenon (discussed later) (van Gent *et al.*, 1997).

D. Impaired Humoral Immunity

A number of studies have documented abnormal humoral responses in autistic individuals. Warren and collaborators (1997) found decreased serum IgA in 8 of 40 (20%) individuals with autism. Gupta (2000) has demonstrated that the immune system within autistics shows a tendency for upper respiratory tract infections, increased allergy and increased gut yeast infection, and the presence of parasites in some cases. He also found a link with serum immunoglobulin in autistics. These children had increased levels of IgM and IgE and low levels of IgA and IgG1 along with low antibody response to protein antigens (Gupta, 2000). It is also apparent that there is a low response from Th1 as its levels decrease and Th2 increases. It is assumed that if Th1 drops then the gut is open to viral infection and fungal overgrowth. Therefore, low IgA leads to poor gut protection. This results in lymphatic hyperplasia, due to altered self-antigens that may lead to auto-immunity. However, elevated levels of interleukin-12 and interferon-gamma (IFN-γ) are found in autism (Singh, 1996). It has been postulated by Singh that abnormal production of interleukin-12 (IL-12), a critical Th1 promoting cytokine, may be a compensatory mechanism in the body of autistic patients that leads to defective cell-mediated immunity and augmented humoral responses.

E. Brain-Specific Antibodies

Several antibodies reacting to brain tissue have been reported. Approximately 58% (19 of 33) sera of autistic children (less than or equal to 10 years of age) are found to be positive for anti-myelin basic protein (MBP) in autistic patients (Singh, 1993).

Autistic children, but not normal children, had antibodies to caudate nucleus (49% positive sera) implying that autoimmune reaction to caudate nucleus of the brain region may cause neurological impairments in autistic children (Singh and Rivas, 2004). The occurrence of autoreactivities to brain tissue in autistic patients may represent the immune system's neuroprotective response to a previous brain injury that may have occurred during neurodevelopment (Silva et al., 2004). Antibodies against Purkinje cells and gliadin peptides were also observed in autistic patients (Vojdani et al., 2004). There was further evidence that serum IgG anti-nuclear autoantibodies and IgM anti-brain endothelial cells antibodies were found in the sera of autistic patients (Connolly et al., 1999). All these reports strengthen the hypothesis that there is an antibody response which cross-reacts with some component in the brain causing dysfunction of the affected area.

III. Role of Viral Infections in Autistic Development

Given the immunopathogenic features of autism, the development process of this disease is likely to include infection. In fact, it has been shown in neonatal rat infection with Borna disease virus, a neurotropic noncytolytic RNA virus, is associated with marked alterations in the cerebellum, along with reductions in granule and Purkinje cell numbers. In this infectious model, neurons are lost predominantly by apoptosis, by an increase in mRNA levels for pro-apoptotic products (Fas, caspase-1), a decrease in mRNA levels for the anti-apoptotic bcl-x and in situ labeling of fragmented DNA (Hornig et al., 1999). The inflammatory infiltrates that accompany this infection are observed transiently in frontal cortex. Glial activation (microgliosis > astrocytosis) is prominent throughout the brain and persists for several weeks in concert with increased levels of proinflammatory cytokine mRNAs (interleukins 1alpha, 1beta, and 6 and tumor necrosis factor alpha) and progressive hippocampal and cerebellar damage (Hornig et al., 1999).

Maternal exposure to a sublethal intranasal administration of human influenza virus (H1N1) in C57BL/6 mice in Day 9 corresponding to about the second trimester in humans has a very significant effect in the brain development. Prenatal exposure of pregnant mice with H1N1 virus has both short-term and long-lasting deleterious effects on developing brain structure in the progeny. This was evidenced by altered pyramidal and nonpyramidal cell density values, atrophy of pyramidal cells despite normal cell proliferation rate, and final enlargement of brain (Fatemi et al., 2002).

A. Association of Measles Virus with Inflammatory Process in Autism

Maternal infection is a risk factor for many neurodevelopmental disorders, including autism (Ciaranello and Ciaranello, 1995; Patterson, 2002; Pletnikov et al., 2002).

It was reported that 43% of mothers with an autistic child experienced upper respiratory tract, influenza-like, urinary, or vaginal infections during pregnancy compared to only 26% of control mothers (Comi et al., 1999). Studies show that, in rats, maternal exposure to infection alters proinflammatory cytokine levels in the fetal environment, including the brain. It has been proposed that these changes may have a significant impact on the developing brain (Giralt et al., 2002; Urakubo et al., 2001). These observations suggest certain cases of autism may be a sequela of pathogenic infections, especially those of a viral origin (Ciaranello and Ciaranello, 1995; Hornig et al., 2002; Pletnikov et al., 2002).

The target sites for measles virus (MV) are similar to the sites affected by autism. These include the cerebellum, the hippocampus, amygdala, cingulate gyrus, hypothalamus, and the frontal and temporal lobes of the cerebral cortex. Although the route of infection by MV is respiratory, and despite its widespread dissemination to the skin, the intestinal tract and the nervous system are the organs affected. The virus has a strong predilection for lymphoid tissues in the early as well as late stages of the disease.

Human CD46 and CDw150 serve as two receptors for MV induced immunosuppression. CD46 molecule, a member of the complement regulatory cascade of proteins (Dorig et al., 1993; Manchester et al., 1994; Naniche et al., 1993) is ubiquitously expressed on all nucleated cells (McQuaid and Cosby, 2002). CDw150 (signaling lymphocyte activation molecule, or SLAM) is a T-cell costimulatory molecule and is expressed only on immature thymocytes, activated and memory T cells, B cells, activated monocytes, and dendritic cells (Cocks et al., 1995; McQuaid and Cosby, 2002; Minagawa et al., 2001; Punnonen et al., 1997; Sidorenko and Clark, 1993). These two receptors induce marked host immune suppression. Although monocytes express CD46, they are considerably resistant to MV. Once monocytes differentiate into immature myeloid dendritic cells (iDCs) (GM-CSF + IL-4-treated), the cells become susceptible to MV (Murabayashi et al., 2002). DCs that matured via stimulation of their Toll-like receptors (TLRs) 2 and/or 4 exhibited an approximately fivefold increase in CDw150 at the protein level, resulting in higher levels of MV amplification in mixed culture of lymphocytes than in iDCs without TLR2/4 stimuli (Murabayashi et al., 2002).

Measles stimulates maturation of antigen-presenting cells in skin, gut, and lungs. Measles also induces IL-6 from fibroblasts and interferon β and colony-stimulating activity from granulocytes and monocytes such as granulocyte-macrophage colony stimulating factor (GM-CSF) (Van Damme et al., 1989). This consequently requires regulation of the immune system during future infections (Murabayashi et al., 2002). Aberrant expression of TGFβ1 can stimulate inflammatory and fibrotic tissue formation and high intracellular TGFβ1 may induce over-expression of CD46 receptors, a portal for measles virus entry (Pasch et al., 1999).

At the cellular level, MV causes cell cycle cessation, especially during the G0/G1 phase where major decisions regarding the cell's fate are determined (Schrag et al., 1999). If the CD46 receptor is unavailable (Dorig et al., 1993), then growth factor receptors (e.g., IGF-1 and epidermal growth factor (EGF) receptors) are used for viral entry (Schneider et al., 2000). Immunologically, MV was found to be capable of suppressing immune responses (McChesney and Oldstone, 1989; Tsujimura et al., 1998). Recent studies have suggested that MV infects and alters functions of T cells (Fugier-Vivier et al., 1997; Hahm et al., 2004; Niewiesk et al., 2000) and antigen-presenting cells (APC) (Grosjean et al., 1997;

Schnorr *et al.*, 1997; Servet-Delprat *et al.*, 2000). This infection skews the T-cell response to a Th2 phenotype (Griffin and Ward, 1993). MV generates type I interferon (IFN) that acts via a signal transducer and an activator of a transcription (STAT) 2-dependent, but STAT1-independent, pathway (Hahm *et al.*, 2005). Thus, it is possible that the MV contributes in autism by suppressing immune function.

IV. Role of Environmental Factors in Autistic Development

Occupational and/or environmental exposure to mercury is believed to harm human health possibly through modulation of immune homeostasis (Lawrence and McCabe, 2002). Several studies have demonstrated that imbalances in immune regulation by metals can lead to inadequate or excessive production of inflammatory cytokines (Gilmore, 2003; Croonenberghs *et al.*, 2002; Safieh-Garabedian *et al.*, 2004). Alternatively, metals can lead to inappropriate activation of lymphoid subsets involved in acquired immunity to specific antigens. Some resultant pathologies may include chronic inflammatory processes and autoimmune diseases. Metals may change the response repertoire by direct and indirect means by influencing expression of new antigens, new peptides, and/or may change antigen presentation by modifying the antigen-presenting complex (Lawrence and McCabe, 2002).

A. Mercury Link with Autism

Exposure to methyl mercury (MeHg) in high doses has profound effects on the CNS and can be fatal. Neuropathological studies indicate that the occipital cortex and cerebellum are most affected. Prenatal exposure studies from Japan and Iraq demonstrated diffused CNS damage with disruption of cellular migration (Bernard *et al.*, 2000; Choi, 1989). It has been hypothesized that postnatal exposure to thimerosal, a mercurial preservative added to the vaccines, may be associated with autism and learning/speech disorders. However, no direct test of this association has yet been reported (Tager-Flusberg *et al.*, 2000). No human studies as yet document any adverse effects of prenatal or early postnatal exposure to elemental mercury or mercury vapor (Davidson *et al.*, 2004).

A study by Vahter *et al.* (2000) examined the different species of mercury in the blood of pregnant women. They found high correlations between inorganic mercury levels in blood and urine during early pregnancy, a significant correlation between cord and maternal blood, and decreased mercury levels during lactation—presumably the result of excretion in milk. The fetal brain is especially susceptible to damage from exposure to organic mercury.

Astrocytic swelling, excitatory amino acid (EAA) release and uptake inhibition, as well as EAA transporter expression inhibition, are known sequelae of MeHg exposure. The presence of Hg causes an inability of astrocytes to maintain control of the proper milieu of the extracellular fluid and, in turn leads to neuronal demise (Shanker et al., 2003).

Heavy metals have been shown to exert immunotoxic effects on humoral immunity as well. IgG3 production is most sensitive to inhibition by mercuric chloride ($HgCl_2$) followed by IgG1 and IgG2b and then IgM and IgG2a. $HgCl_2$ exerts early, inhibitory effects on B-cell activation. This is manifested by the inhibition of RNA, DNA, and antibody synthesis. (Daum et al., 1993). Metals by binding to SH radicals in proteins and other such groups, can cause autoimmunity by modifying proteins which via T cells activate B cells that target the altered proteins fibrillarin, a 34-kDa protein component of many small nucleolar ribonucleoprotein particles inducing autoimmunity. They also cause aberrant MHC II expression on altered target cells (Hu et al., 1997a, b; Hultman et al., 1994; Pollard et al., 1997). Thimerosal (ethyl mercury) in individuals with pre-disposing HLA molecules bind to CD26 or CD69 and induce antibodies against these molecules (Vojdani et al., 2003). Furthermore, the CD95/Fas apoptotic signaling pathway that is of critical importance in regulating peripheral tolerance, is disrupted by low and environmentally relevant concentrations of Hg^{2+} (McCabe et al., 2005).

V. Inflammatory Mediators in Autism

Inflammation has an important repairing function, but, in CNS, frequently is the cause of damage. Usually neuroinflammation has the tendency to succumb to damage, which would explain the CNS pathology associated with autism (Chavarria and Alcocer-Varela, 2004).

The various components involved in the inflammatory response in the CNS include the participation of different cellular types of the immune system (macrophages, mast cells, T and B lymphocytes, dendritic cells), resident cells of the CNS (microglia, astrocytes, neurons), adhesion molecules, complement proteins, cytokines, and chemokines among other proteic components. Chemotaxis plays an important role in the recruitment of cells to the CNS. The lymphocyte recruitment implies the presence of chemokines and chemokine receptors, the expression of adhesion molecules, the interaction between lymphocytes and the bloodbrain barrier (BBB) endothelium, and their passage through the BBB to arrive at the site of inflammation (Little et al., 2002). Recent studies by Vargas *et al.*, 2005) demonstrated that the two cells involved in inflammation are microglia and astroglia, which are essential for many neuronal functions. The presence of activated cells in their samples suggests that there is a chronic and a sustained neurological damage in autism. This activation could lead to abnormal function of neurons and synapses. Absence of any lymphocytes (T cells), plasma cells/antibody in the brain parenchyma suggests only the activation of astrocytes and microglial cells. These two cells are the classical players of innate immune response and thus demonstrated that an adaptive immunity is not playing an active role in autism. However, one cannot rule out the possibility of adaptive immunity playing a prominent role early in the prenatal or postnatal development phase of autism.

A. Proinflammatory Cytokines and Chemokines in Autism

As previously discussed, several lines of evidence indicate the immune system can influence the normal activities of the CNS. Cells of the immune system are present in the CNS where they show increased chemical activities under conditions of inflammation and disease. Cytokines and chemokines, secreted by either immune or non-immune cells, play critical roles in many chronic and acute inflammatory conditions. Therefore, it is likely that mediators of immune cell-CNS interactions under normal conditions and in diseased states are skewed. Indeed, studies from several laboratories provide evidence for an altered/unique cytokine profile in autism. Two pro-inflammatory chemokines, macrophage chemoattractant protein–1 (MCP-1), thymus, activation-regulated chemokine (TARC), and an anti-inflammatory and modulatory cytokine, TGF-ß1, were consistently elevated in the brain regions studied (Vargas *et al.*, 2005). MCP-1, a chemokine involved in monocyte and T-cell activation for trafficking into areas of tissue injury, was elevated both in brain parenchyma and CSF in cytokine protein array studies (Vargas *et al.*, 2005). In that study, immunochemistry revealed that astrocytes had infiltrated the cerebellum and cerebral cortex. The increased expression of MCP-1 in autism implies that it is linked to microglial activation and perhaps also to the recruitment of additional macrophages and microglia to areas of the cerebellum (Vargas *et al.*, 2005). There is evidence that MCP-1 may serve a signaling function in the damaged CNS that is distinct from its role in proinflammatory events (Little *et al.*, 2002). Its role in autism is not clear but its presence signifies inflammatory insult (Perrin *et al.*, 2005) or neuronal survival and protection (Uicker *et al.*, 2005). The observation that human fetal glial cells and their progenitors express specific receptors for chemokines and can be stimulated to produce MCP-1 as well as proliferate in response to chemokines, supports a role for these cytokines as regulatory factors during ontogeny (Rezaie and Male, 1999). MCP-1 expression in the cerebellum during prenatal development suggests an association with maturation of Purkinje cells (Meng et al., 1999). Like MHC-class II expression in microglia during CNS modeling, MCP-1 elevation in the brain of autistic patients may reflect persistent fetal patterns of brain development.

Other cytokines with pro-inflammatory and anti-inflammatory effect were also increased in the brain of patients with autism (Vargas et al., 2005). An example of anti-inflammatory cytokines is TGF-ß1, a key anti-inflammatory cytokine involved in tissue remodeling following injury. Upregulating extracellular matrix proteins accomplish this. It can suppress specific immune responses by inhibiting T-cell proliferation and maturation while downregulating MHC class II expression especially in the brain stem (Johns et al., 1992). In the study of Vargas et al. (2005), immunocytochemical studies, TGF-ß1 was localized mostly within reactive astrocytes and neurons in the cerebellum. Purkinje cells that exhibited microscopic features of degeneration showed marked reactivity for TGF-ß1. The elevation of this cytokine in autism may reflect a compensation mechanism to diminish neuroinflammation or remodel and repair injured tissue.

The prominent inflammatory cytokine profile was repeated in cerebrospinal fluid (CSF) as well in patients with autism (Vargas et al., 2005). The marked increase of MCP-1 in CSF is indicative of pro-inflammatory pathway activation in the brain of autistic patients. This may be associated with activation of microglial cells as seen in brain parenchyma studies. These studies indicate that cytokine activation plays an important role in immune mediated processes and that their presence in the CSF in autistic patients may reflect an ongoing stage of inflammatory reactions. These reactions are associated with neuroglial activation and/or neuronal injury. The persistent elevation of cytokines in CSF also might reflect a neurodevelopmental arrest, as some of the cytokines are normally elevated during phases of neurodevelopment. Elevated levels of IL-6 are also observed in diseases associated with developmental disorders. IL-6 and cilliary neurotrophic factor (CNTF), a neuronal growth factor, share the same intracellular receptor. This suggests that IL-6 may influence the nervous system via pathways normally used by growth factors. Relatively little is known about the effect of IL-6 or chemokines on CNS neurons, the transduction mechanism linked to IL-6 or chemokine receptors, the pathways involved in IL-6, or chemokine induced neuropathology (Nelson et al., 2004). In a small sample, Gupta and his colleagues (1998) found that tumor necrosis factor-alpha (TNF-α), another potent proinflammatory cytokine, was significantly increased in autistic populations. This finding was further corroborated in a study of Jyonouchi and collaborators (2001) who tested 71 autistic children aged 2–14 years and compared them with healthy siblings and other controls. In this study, innate immune responsiveness showed that in 59 of 71 (83.1%) autistic patients, lipopolysaccharide (lps) activated peripheral blood mononuclear cells (PBMCs) produced levels of TNF-α, IL-1β, and/or IL-6 that were greater than 2 SD above the control mean (CM) values. Without stimulus, the basal level of proinflammatory/counter-regulatory cytokines was high in autistic patients. With stimulants phytohemagglutinin (PHA), tetanus, IL-12p70, and IL-18 of adaptive immunity PBMCs from 47.9 to 60% of autistic patients produced greater than 2 SD above the CM values of TNF-α depending on stimulants. The investigators concluded that a majority of the autistic children in their group exhibited excessive or poorly regulated innate immune responses especially involving increased TNF-α (Jyonouchi et al., 2001). It should also be noted that although NO has been known to exert neuroprotective effects at low to moderate concentrations, NO becomes neurotoxic as the concentration increases. Excessive NO production can cause oxidative stress to neurons, ultimately impairing neuronal function and resulting in neuronal cell death (Abbott and Nahm, 2004). Indeed, plasma NO is high in some children with autism. This elevation may be related to IFN-γ activity (Sweeten et al., 2004).

Taken together these studies suggest that several chemokines/cytokines and/or inflammatory mediators are involved in the pathogenesis of autism. However, their exact cellular source or mechanism of actions remains to be the subject of further investigations.

VI. Involvement of Toll-Like Receptors (TLRs) in Autism

The inflammatory signaling cascades leading to c-fos activation in glial cells have shown that activation by LPS in glial cells occurs via the serum response element (SRE) or cyclic AMP/calcium response element (CRE) in an independent manner, and involves the Elk1 or CREB/ATF-1 transcription factors. Elk1-mediated transactivation was dependent on p38 mitogen-activated protein kinase (MAPK), suggesting a crucial role of these factors in mediating inflammatory responses in the CNS (Simi et al., 2005).

Additionally, Ozato et al. (2002) described the response of cell-surface toll-like receptors (TLRs) upon binding to microbial pathogens. There are at least 10 TLRs that recognize ligands from bacteria, viruses, yeast, and nucleic acids from viruses as well. There is a high binding specificity of the different TLRs for each microbial structure referred to as pathogen-associated molecular patterns (PAMPs) (Ozato et al., 2002). The best studied is TLR4 that binds LPS from gram-negative bacteria. The ligation of LPS to cell surface TLR4 initiates a signal cascade that results in the activation of intracellular nuclear factor kappa beta (NFκB) and the transcription of numerous genes involved in immune responses. This signaling pathway appears to be common to all the TLRs whether the PAMPs originate from bacteria, virus, or yeast.

The central nervous system exhibits a similar immune reaction to pathogenic infection. There is a broad expression of TLRs in human brain astrocytes, oligodendrocytes, and microglia (Bsibsi *et al.*, 2002). Astrocytes and oligodendrocytes express mRNA for TLR2 that recognizes fungal, gram-positive, mycobacterial components and TLR3 that recognize, double-stranded RNA. Microglia cells express mRNA for a wide range of TLR family members (TLR2, TLR3, TLR4, TLR5, TLR6, TLR7, TLR8, and TLR9) much like other cells of the monocytic lineage (Bsibsi *et al.*, 2002). The binding of LPS to TLR on microglia cells (brain macrophage) leads to the innate expression of cytokines, chemokines, extracellular matrix proteins, proteolytic enzymes, and complement proteins in the brain parenchyma (Aloisi, 2001; Nguyen *et al.*, 2002). It is also well established that glial cells participate in innate immune responses in human CNS (Nguyen *et al.*, 2002). The sharing of the TLR receptors between the astrocyte and microglia is another example where the neurology is communicating with immunology using common molecules. Microglial cells are the resident macrophage-like population in the CNS. Microglial cells remain quiescent until injury or infection activates the cells to perform effector inflammatory and antigen presenting cell (APC) functions. Mouse microglial cells express mRNA for all of the recently identified TLRs, TLR1–9. Furthermore, stimulation of quiescent microglia with various TLR agonists, including LPS (TLR4), peptidoglycan (TLR2), polyinosinic-polycytidylic acid (TLR3), and CpG DNA (TLR9) activated the cells to up-regulate unique patterns of innate and effector immune cytokines and chemokines at the mRNA and protein levels. In addition, TLR stimulation activated up-regulation of MHC class II and costimulatory molecules, enabling the microglia to efficiently present myelin Ags to CD4+ T cells. Thus, microglia appear to be a unique and important component of both the innate and adaptive immune response, providing the CNS with a means to rapidly and efficiently respond to a wide variety of pathogens (Olson and Miller, 2004).

VII. Autoimmunity in Autism

Inflammation has been linked with autoimmune insult. Aberrant innate immune response against endotoxin and immune reactivity to dietary proteins may be associated with apparent dietary product associated gastrointestinal inflammation in autistic children (Jyonouchi *et al.*, 2002, 2005). Another piece of information is the virus-induced autoimmune response to developing brain myelin that may impair anatomical development of neural pathways in autistic children (Singh *et al.*, 1993). The consequent anatomical changes of such autoimmune reactions could impair the nerve-impulse transmission and ultimately lead to life-long disturbances of higher mental functions (such as learning, memory, communication, social interaction, etc.) that are seen in autistic populations.

A. MATERNAL ANTIBODIES CAN TRIGGER THE ATTACK IN AUTISM

Conceptually, it is possible that IgG from the mother can pass through the placental barrier and can react with antigenic proteins expressed on cell surface of lymphoid and/or neuronal tissues of the fetus and result in neuronal cell death. Since antigens expressed on lymphocytes are found on cells of the central nervous system and, perhaps, on other tissues of the developing embryo, it has been suggested that aberrant maternal immunity may be associated with the development of autism (Warren et al., 1990). In fact, there is evidence in the literature supporting the importance of maternal antibodies in autism. Dalton and his colleagues (2003) have shown that serum antibodies that bind to rodent Purkinje cells and other neurons were detectable in a mother of three children: the first normal, the second with autism, and the third with a severe specific language disorder. The same serum when injected into pregnant mice during gestation produced altered exploration and motor coordination and changes in cerebellar magnetic resonance spectroscopy in the mouse offspring.

B. MMR VACCINATION MAY INCREASE RISK VIA AN AUTOIMMUNE MECHANISM

As previously mentioned, antibodies from autistic patients against MBP and neuron-axon filament protein (NAFP) cross-reacts with anti-measles antibody and human herpes-6 antibody (Singh et al., 1998). This observation supports the hypothesis that a virus-induced autoimmune response may play a causal role in autism (Singh, 2000). Seventy-five of 125 (60%) autistic sera specifically detected measles hemagglutinin (HA) protein of measles-mumps-rubella (MMR) and over 90% of MMR antibody-positive autistic sera were also positive for MBP autoantibodies, suggesting a strong association between MMR and CNS autoimmunity in autism (Singh et al., 2002). In another study, Singh and Jensen (2003) showed that there were elevated levels of measles antibodies in autistic children with no reaction to mumps or rubella.

There is some evidence that autism arises shortly after immunization with measles-mumps-rubella (MMR) and/or diphtheria-pertussis-tetanus (DPT) vaccines (Megson, 2000). Antibody levels to three vaccines, MMR, DPT, and DT (diphtheria-tetanus), were measured and it was found that the level of MMR antibodies was significantly higher in autistic children as compared to normal children (Singh et al., 2002). There was a very high degree of specificity for MMR antibodies, particularly for measles (Singh et al., 2002). The same result was also found when monovalent measles vaccine was used instead of the trivalent MMR vaccine, furthermore pointing to a problem of only the measles subunit (Singh and Jensen, 2003). A high positive correlation (90% or greater) between the MMR antibody and the MBP autoantibody (Singh *et al.*, 2002) was detected. The deduction drawn from these studies is that the measles subunit of the MMR vaccine triggers an autoimmune reaction in a significant number of autistic children (Singh, 2000; Singh and Jensen, 2003; Singh *et al.*, 2002). MMR vaccine seems to induce interferon-gamma (IFNγ) only in breast-fed infants after primary measles immunization, a Th-1 cellular response. These results imply that the feeding pattern of infants can have a long-term effect on the immune modulation beyond weaning (Pabst *et al.*, 1991).

C. POTENTIAL LINKAGE OF ENVIRONMENTAL FACTORS WITH AUTOIMMUNE EVENTS IN AUTISM

Autoantibodies (primarily IgG) to neuronal cytoskeletal proteins, neurofilaments (NFs), MBP, were prevalent in male workers exposed to mercury. These findings were confirmed in rats and mice. There were significant correlations between IgG titers and subclinical deficits in sensorimotor function. Thus, peripheral autoantibodies to neuronal proteins are predictive of neurotoxicity, since histopathological findings were associated with disease damage. There was also evidence of astrogliosis (indicative of neuronal CNS damage) and the presence of IgG concentrated along the blood brain barrier (El-Fawal *et al.*, 1999). Autoimmune response to mercury has also been shown by the transient presence of antinuclear antibodies (ANA) and antinucleolar antibodies (ANolA) (Fagala and Wigg, 1992; Hu *et al.*, 1997; Nielsen and Hultman, 1999).

In an interesting study with newborns and thimerosol, autoimmune disease-sensitive mice were compared to strains resistant to autoimmunity. Mice were injected solely with thimerosal, a thimerosal-vaccine combination, or a saline solution. The comparative study showed growth delay, reduced locomotion, exaggerated response to novelty, and densely packed, hyperchromic hippocampal neurons with altered glutamate receptors and transporters in autoimmune mice. These animal studies implicate that impaired immunity might put some children at risk of developing autism after being exposed to thimerosal (Chian and Lipkin, 2004).

VIII. Summary

The immunopathogenesis of autism is presented schematically in Fig. 1. Two main immune dysfunctions in autism are immune regulation involving

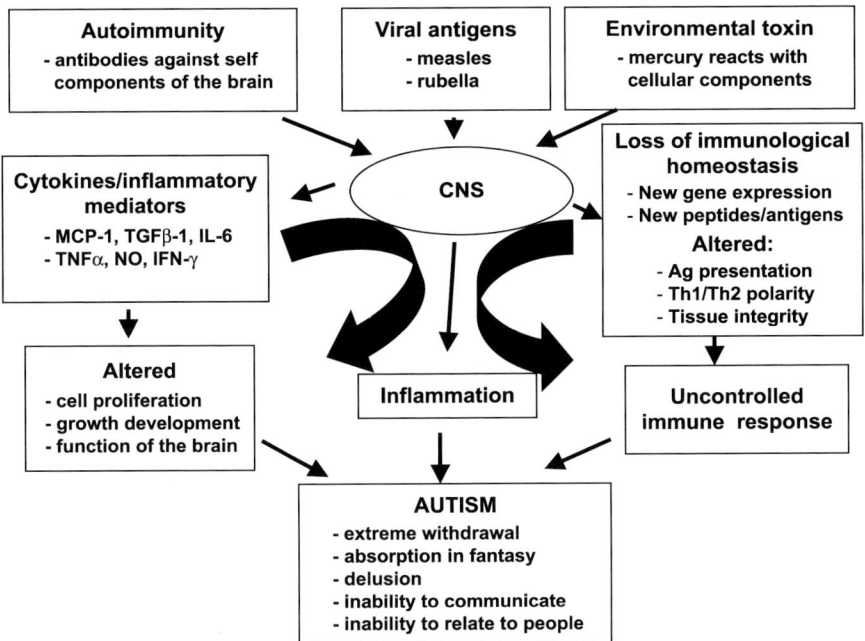

FIG. 1. Schematic presentation of immunopathogenesis of autism.

pro-inflammatory cytokines and autoimmunity. Mercury and an infectious agent like the measles virus are currently two main candidate environmental triggers for immune dysfunction in autism. Genetically immune dysfunction in autism involves the MHC region, as this is an immunologic gene cluster whose gene products are Class I, II, and III molecules. Class I and II molecules are associated with antigen presentation.

The antigen in virus infection initiated by the virus particle itself while the cytokine production and inflammatory mediators are due to the response to the putative antigen in question. The cell-mediated immunity is impaired as evidenced by low numbers of CD4 cells and a concomitant T-cell polarity with an imbalance of Th1/Th2 subsets toward Th2. Impaired humoral immunity on the other hand is evidenced by decreased IgA causing poor gut protection. Studies showing elevated brain specific antibodies in autism support an autoimmune mechanism. Viruses may initiate the process but the subsequent activation of cytokines is the damaging factor associated with autism. Virus specific antibodies associated with measles virus have been demonstrated in autistic subjects.

Environmental exposure to mercury is believed to harm human health possibly through modulation of immune homeostasis. A mercury link with the immune system has been postulated due to the involvement of postnatal exposure to thimerosal, a preservative added in the MMR vaccines. The occupational hazard exposure to mercury causes edema in astrocytes and, at the molecular level, the CD95/Fas apoptotic signaling pathway is disrupted by Hg^{2+}. Inflammatory mediators in autism usually involve activation of astrocytes and microglial cells. Proinflammatory chemokines (MCP-1 and TARC), and an anti-inflammatory and modulatory cytokine, TGF-ß1, are consistently elevated in autistic brains. In measles virus infection, it has been postulated that there is immune suppression by inhibiting T-cell proliferation and maturation and downregulation MHC class II expression. Cytokine alteration of TNF-α is increased in autistic populations. Toll-like-receptors are also involved in autistic development. High NO levels are associated with autism. Maternal antibodies may trigger autism as a mechanism of autoimmunity. MMR vaccination may increase risk for autism via an autoimmune mechanism in autism. MMR antibodies are significantly higher in autistic children as compared to normal children, supporting a role of MMR in autism. Autoantibodies (IgG isotype) to neuron-axon filament protein (NAFP) and glial fibrillary acidic protein (GFAP) are significantly increased in autistic patients (Singh *et al.*, 1997). Increase in Th2 may explain the increased autoimmunity, such as the findings of antibodies to MBP and neuronal axonal filaments in the brain. There is further evidence that there are other participants in the autoimmune phenomenon. (Kozlovskaia *et al.*, 2000). The possibility of its involvement in autism cannot be ruled out. Further investigations at immunological, cellular, molecular, and genetic levels will allow researchers to continue to unravel the immunopathogenic mechanisms' associated with autistic processes in the developing brain. This may open up new avenues for prevention and/or cure of this devastating neurodevelopmental disorder.

References

Abbott, L. C., and Nahm, S. S. (2004). Neuronal nitric oxide synthase expression in cerebellar mutant mice. *Cerebellum* **3,** 141–151.

Aloisi, F. (2001). Immune function of microglia. *Glia* **36,** 165–179.

Ashwood, P., Anthony, A., Pellicer, A. A., Torrente, F., Walker-Smith, J. A., and Wakefield, A. J. (2003). Intestinal lymphocyte populations in children with regressive autism: Evidence for extensive mucosal immunopathology. *J. Clin. Immunol.* **23,** 504–517.

Bailey, A., and Cox, T. (1996). Neuroimaging in child and developmental psychiatry. *In* "Brain Imaging in Psychiatry" (S. Lewis and J. P. Higgins, Eds.), vol. 96. pp. 301–315. Blackwell Scientific, Oxford.

Bailey, A., Luthert, P., Dean, A., Harding, B., Janota, I., Montgomery, M., Rutter, M., and Lantos, P. (1998b). A clinicopathological study of autism. *Brain* **121,** 889–905.

Bailey, A., Palferman, S., Heavey, L., and Le Couteur, A. (1998a). Autism: The phenotype in relatives. *J. Autism Dev. Disord.* **28,** 369–392.

Bauman, M. L., and Kemper, T. L. (1994). Neuroanatomic observations of the brain in autism. *In* "The Neurobiology of Autism" (M. L. Bauman and T. L. Kemper, Eds.), pp. 119–145. Johns Hopkins University Press, Baltimore.

Bernard, S. L., Ewen, J. R., Barlow, C. H., Kelly, J. J., McKinney, S., Frazer, D. A., and Glenny, R. W. (2000). High spatial resolution measurements of organ blood flow in small laboratory animals. *Am. J. Physiol. Heart Circ. Physiol.* **279,** H2043–H2052.

Bock, K. A. (2002). Integrative approach to autism spectrum disorders. *In* DAN! (Defeat Autism Now!) Spring 2002 Conference Practitioner Training, (B. Rimland, Ed.). San Diego, CA.

Boulanger, L. M., Huh, G. S., and Shatz, C. J. (2001). Neuronal plasticity and cellular immunity: Shared molecular mechanisms. *Curr. Opin. Neurobiol.* **11,** 568–578.

Boulanger, L. M., and Shatz, C. J. (2004). Immune signaling in neural development, synaptic plasticity and disease. *Nat. Rev. Neurosci.* **5,** 521–553.

Bsibsi, M., Ravid, R., Gveric, D., and Van Noort, J. M. (2002). Broad expression of toll-like receptors in the human central nervous system. *J. Neuropath. Exp. Neu.* **61,** 1013–1021.

Carper, R. A., and Courchesne, E. (2000). Inverse correlation between frontal lobe and cerebellum sizes in children with autism. *Brain* **123,** 836–844.

Chavarria, A., and Alcocer-Varela, J. (2004). Is damage in central nervous system due to inflammation? *Autoimmun Rev.* **3**(4), 251–260.

Chian, M. D., and Lipkin, W. I. (2004). Neurotoxic effects of postnatal thimerosal are mouse strain-dependent. *Mol. Psy.* **9,** 833–845.

Choi, B. H. (1989). The effects of methylmercury on the developing brain. *Prog. Neurobiol.* **32,** 467–470.

Ciaranello, A. L., and Ciaranello, R. D. (1995). The neurobiology of infantile autism. *Annu. Rev. Neurosci.* **18,** 101–128.

Cocks, B. G., Chang, C. C., Carballido, J. M., Yssel, H., de Vries, J. E., and Aversa, G. (1995). A novel receptor involved in T-cell activation. *Nature* **376,** 260–263.

Comi, A. M., Zimmerman, A. W., Frye, V. H., Law, P. A., and Peeden, J. N. (1999). Familial clustering of autoimmune disorders and evaluation of medical risk factors in autism. *J. Child Neurol.* **14,** 388–394.

Connolly, A. M., Chez, M. G., Pestronk, A., Arnold, S. T., Mehta, S., and Deuel, R. K. (1999). Serum autoantibodies to brain in Landau-Kleffner variant, autism, and other neurological disorders. *J. Pediatrics* **134,** 607–613.

Courchesne, E., Carper, R., and Akshoomoff, N. (2003). Evidence of brain overgrowth in the first year of life in autism. *JAMA* **290,** 337–344.

Courchesne, E., Yeung-Courchesne, R., Press, G. A., Hesselink, J. R., and Jernigan, T. L. (1988). Hypoplasia of cerebellar vermal lobules VI and VII in autism. *N. Engl. J. Med.* **318,** 1349–1354.

Courchesne, E., Townsend, J., Akshoomoff, N. A., Saitoh, O., Yeung-Courchesne, R., Lincoln, A. J., James, H. E., Haas, R. H., Schreibman, L., and Lau, L. (1994). Impairment in shifting attention in autistic and cerebellar patients. *Behav. Neurosci.* **108,** 848–865.

Croonenberghs, J., Bosmans, E., Deboutte, D., Kenis, G., and Maes, M. (2002). Activation of the inflammatory response system in autism. *Neuropsychobiology* **45,** 1–6.

Dalton, P., Deacon, R., Blamire, A., Pike, M., McKinlay, I., Stein, J., Styles, P., and Vincent, A. (2003). Maternal neuronal antibodies associated with autism and a language disorder. *Ann. Neurol.* **53,** 533–537.

Daniels, W. W., Warren, R. P., Odell, J. D., Maciulis, A., Burger, R. A., Warren, W. L., and Torres, A. R. (1995). Increased frequency of the extended or ancestral haplotype B44-SC30-DR4 in autism. *Neuropsychobiology* **32,** 120–123.

Daum, J. R., Shepherd, D. M., and Noelle, R. J. (1993). Immunotoxicology of cadmium and mercury on B-lymphocytes–I. Effects on lymphocyte function. *Int. J. Immunopharmacol.* **15,** 383–394.

Davidson, P. W., Myers, G. J., and Weiss, B. (2004). Mercury exposure and child development outcomes. *Pediatrics* **113**(4 Suppl.), 1023–1029.

Dorig, R. E., Marcil, A., Chopra, A., and Richardson, C. D. (1993). The human CD46 molecule is a receptor for measles virus (Edmonston strain). *Cell* **75,** 295–305.

El-Fawal, H. A., Waterman, S. J., De Feo, A., and Shamy, M. Y. (1999). Neuroimmunotoxicology: Humoral assessment of neurotoxicity and autoimmune mechanisms. *Environ. Health Perspect.* **107** (Suppl. 5), 767–775.

Fagala, G. E., and Wigg, C. L. (1992). Psychiatric manifestations of mercury poisoning. *J. Am. Acad. Child Adolesc. Psychiatry* **31,** 306–311.

Fatemi, S. H., Earle, J., Kanodia, R., Kist, D., Emamian, E. S., Patterson, P. H., Shi, L., and Sidwell, R. (2002). Prenatal viral infection leads to pyramidal cell atrophy and macrocephaly in adulthood: Implications for genesis of autism and schizophrenia. *Cell Mol. Neurobiol.* **22,** 25–33.

Fugier-Vivier, I., Servet-Delprat, C., Rivailler, P., Rissoan, M. C., Liu, Y. J., and Rabourdin-Combe, C. (1997). Measles virus suppresses cell-mediated immunity by interfering with the survival and functions of dendritic and T cells. *J. Exp. Med.* **186,** 813–823.

Gillberg, C. (1990). Autism and pervasive developmental disorders. *J. Child Psychol. Psychiatry* **31,** 99–119.

Gilmore, J. H., Jarskog, L. F., and Vadlamudi, S. (2003). Maternal infection regulates BDNF and NGF expression in fetal and neonatal brain and maternal-fetal unit of the rat. *J. Neuroimmunol.* **138,** 123–127.

Giralt, M., Penkowa, M., Hernandez, J., Molinero, A., Carrasco, J., Lago, N., Camats, J., Campbell, I. L., and Hidalgo, J. (2002). Metallothionein-1+2 deficiency increases brain pathology in transgenic mice with astrocyte-targeted expression of interleukin 6. *Neurobiology Dis.* **9,** 319–338.

Goldberg, M. C., Landa, R., Lasker, A., Cooper, L., and Zee, D. S. (2000). Evidence of normal cerebellar control of the vestibulo-ocular reflex (VOR) in children with high-functioning autism. *J. Autism Dev. Disord.* **30,** 519–524.

Griffin, D. E., and Ward, B. J. (1993). Differential CD4 T cell activation in measles. *J. Infect. Dis.* **168,** 275–281.

Grosjean, I., Caux, C., Bella, C., Berger, I., Wild, F., Banchereau, J., and Kaiserlian, D. (1997). Measles virus infects human dendritic cells and blocks their allostimulatory properties for CD4+ T cells. *J. Exp. Med.* **186,** 801–812.

Gupta, S. (2000). Immunological treatments for autism. *J. Autism Dev. Disord.* **30,** 475–479.

Gupta, S. T., Lee, T., and Aggarval, S. (1998). Alterations in Th1 and Th2 subsets of CD4+ and CD8+ T-cells in autism. *J. Neuroimmunol.* **14,** 499–504.

Hahm, B., Arbour, N., Naniche, D., Homann, D., Manchester, M., and Oldstone, M. B. (2004). Measles virus infects and suppresses proliferation of T lymphocytes from transgenic mice bearing human signaling lymphocytic activation molecule. *J. Virol.* **77,** 3505–3515.

Hahm, B., Trifilo, M. J., Zuniga, E. I., and Oldstone, M. B. (2005). Viruses evade the immune system through type I interferon-mediated STAT2-dependent, but STAT1-independent, signaling. *Immunity* **22,** 247–257.

Helmuth, L. (2000). Neuroscience. Immune molecules prune neural links. *Science* **290,** 2051.

Herbert, M. R., Ziegler, D. A., Deutsch, C. K., O'Brien, L. M., Lange, N., Bakardjiev, A., Hodgson, J., Adrien, K. T., Steele, S., Makris, N., Kennedy, D., Harris, G. J., and Caviness, V. S., Jr. (2003). Dissociations of cerebral cortex, subcortical and cerebral white matter volumes in autistic boys. *Brain* **126**(Pt 5), 1182–1192.

Hornig, M., Mervis, R., Hoffman, K., and Lipkin, W. I. (2002). Infectious and immune factors in neurodevelopmental damage. *Mol. Psychiatry* **7**(Suppl. 2), S34–S35.

Hornig, M., Weissenbock, H., Horscroft, N., and Lipkin, W. I. (1999). An infection-based model of neurodevelopmental damage. *Proc. Natl. Acad. Sci. USA* **96,** 12102–12107.

Hu, H., Abedi-Valugerdi, M., and Moller, G. (1997a). Pretreatment of lymphocytes with mercury *in vitro* induces a response in T cells from genetically determined low-responders and a shift of the interleukin profile. *Immunology* **90,** 198–204.

Hu, H., Moller, G., and Abedi-Valugerdi, M. (1997b). Major histocompatibility complex class II antigens are required for both cytokine production and proliferation induced by mercuric chloride *in vitro*. *J. Autoimmun.* **10,** 441–446.

Huh, G. S., Boulanger, L. M., Du, H., Riquelme, P. A., Brotz, T. M., and Shatz, C. J. (2000). Functional requirement for class I MHC in CNS development and plasticity. *Science* **290,** 2155–2159.

Hultman, P., Johansson, U., and Turley, S. J. (1994). Adverse immunological effects and autoimmunity induced by dental amalgam in mice. *FASEB J.* **8,** 1183–1190.

Johns, L. D., Babcock, G., Green, D., Freedman, M., Sriram, S., and Ransohoff, R. M. (1992). Transforming growth factor-beta 1 differentially regulates proliferation and MHC class-II antigen expression in forebrain and brainstem astrocyte primary cultures. *Brain Res.* **585**(2), 229–236.

Just, M. A., Cherkassky, V. L., Keller, T. A., and Minshew, N. J. (2004). Cortical activation and synchronization during sentence comprehension in high-functioning autism: Evidence of underconnectivity. *Brain* **127**(Pt 8), 1811–1821.

Jyonouchi, H., Sun, S., and Itokazu, N. (2002). Innate immunity associated with inflammatory responses and cytokine production against common dietary proteins in patients with autism spectrum disorder. *Neuropsychobiology* **46,** 76–84.

Jyonouchi, H., Sun, S., and Le, H. (2001). Proinflammatory and regulatory cytokine production associated with innate and adaptive immune responses in children with autism spectrum disorders and developmental regression. *J. Neuroimmunol.* **120,** 170–179.

Jyonouchi, H., Geng, L., Ruby, A., and Zimmerman-Bier, B. (2005). Dysregulated innate immune responses in young children with autism spectrum disorders: Their relationship to gastrointestinal symptoms and dietary intervention. *Neuropsychobiology* **51,** 77–85.

Kozlovskaia, G. V., Kliushnik, T. P., Goriunova, A. V., Turkova, I. L., Kalinina, M. A., and Sergienko, N. S. (2000). Nerve growth factor auto-antibodies in children with various forms of mental dysontogenesis and in schizophrenia high risk group. *Zh. Nevrol. Psikhiatr. Im. S. S. Korsakova* **100,** 50–52.

Lawrence, D. A., and McCabe, M. J., Jr. (2002). Immunomodulation by metals. *Int. Immunopharmacol.* **2,** 293–302.

Little, A. R., Benkovic, S. A., Miller, D. B., and O'Callaghan, J. P. (2002). Chemically induced neuronal damage and gliosis: Enhanced expression of the proinflammatory chemokine, monocyte chemoattractant protein (MCP)-1, without a corresponding increase in proinflammatory cytokines(1). *Neuroscience* **115,** 307–320.

Manchester, M., Liszewski, M. K., Atkinson, J. P., and Oldstone, M. B. (1994). Multiple isoforms of CD46 (membrane cofactor protein) serve as receptors for measles virus. *Proc. Natl. Acad. Sci. USA* **91,** 2161–2165.

McCabe, M. J., Jr., Eckles, K. G., Langdon, M., Clarkson, T. W., Whitekus, M. J., and Rosenspire, A. J. (2005). Attenuation of CD95-induced apoptosis by inorganic mercury: Caspase-3 is not a direct target of low levels of Hg2+. *Toxicol. Lett.* **155,** 161–170.

McChesney, M. B., and Oldstone, M. B. A. (1989). Virus-induced immunosuppression: Infections with measles virus and human immunodeficiency virus. *Adv. Immunol.* **45,** 335–380.

McQuaid, S., and Cosby, S. L. (2002). An immunohistochemical study of the distribution of the measles virus receptors, CD46 and SLAM, in normal human tissues and subacute sclerosing panencephalitis. *Lab. Invest.* **82,** 403–409.

Megson, M. N. (2000). Is autism a G-alpha protein defect reversible with natural vitamin A? *Med. Hypotheses* **54,** 979–983.

Meng, S. Z., Oka, A., and Takashima, S. (1999). Developmental expression of monocyte chemoattractant protein-1 in the human cerebellum and brainstem. *Brain Dev.* **21,** 30–35.

Minagawa, H., Tanaka, K., Ono, N., Tatsuo, H., and Yanagi, Y. (2001). Induction of the measles virus receptor SLAM (CD150) on monocytes. *J. Gen. Virol.* **82**(Pt 12), 2913–2917.

Minshew, N. J., Goldstein, G., Dombrowski, S. M., Panchalingam, K., and Pettegrew, J. W. (1993). A preliminary ^{31}P MRS study of autism: Evidence for undersynthesis and increased degradation of brain membranes. *Biological Psychiatry* **33,** 762–773.

Murabayashi, N., Kurita-Taniguchi, M., Ayata, M., Matsumoto, M., Ogura, H., and Seya, T. (2002). Susceptibility of human dendritic cells (DCs) to measles virus (MV) depends on their activation stages in conjunction with the level of CDw150: Role of Toll stimulators in DC maturation and MV amplification. *Microbes Infect.* **4,** 785–794.

Naniche, D., Varior-Krishnan, G., Cervoni, F., Wild, T. F., Rossi, B., Rabourdin-Combe, C., and Gerlier, D. (1993). Human membrane cofactor protein (CD46) acts as a cellular receptor for measles virus. *J. Virol.* **67,** 6025–6032.

Nelson, T. E., Netzeband, J. G., and Gruol, D. L. (2004). Chronic interleukin 6 exposure alters metabotropic glutamate receptor-mediated calcium signaling in cerebellar Purkinje neurons. *Eur. J. Neurosci.* **20,** 2387–2400.

Neumann, H., Cavalie, A., Jenne, D. E., and Wekerle, H. (1995). Induction of MHC class I genes in neurons. *Science* **269,** 549–552.

Neumann, H., Schmidt, H., Cavalie, A., Jenne, D., and Wekerle, H. (1997). *J. Exp. Med.* **185,** 305–316.

Nguyen, M. D., Mushynski, W. E., and Julien, J. P. (2002). Cycling at the interface between neurodevelopment and neurodegeneration. *Cell Death Differ.* **9,** 1294–1306.

Nielsen, J. B., and Hultman, P. (1999). Experimental studies on genetically determined susceptibility to mercury-induced autoimmune response. *Ren. Fail.* **21,** 343–348.

Niewiesk, S., Gotzelmann, M., and ter Meulen, V. (2000). Selective *in vivo* suppression of T lymphocyte responses in experimental measles virus infection. *Proc. Natl. Acad. Sci. USA* **97,** 4251–4255.

Olson, J. K., and Miller, S. D. (2004). Microglia initiate central nervous system innate and adaptive immune responses through multiple TLRs. *J. Immunol.* **173,** 3916–3924.

Ozato, K., Tsujimura, H., and Tamura, T. (2002). Toll-like receptor signaling and regulation of cytokine gene expression in the immune system. *BioTechniques* **33,** S66–S75.

Pabst, H. F., Spady, D. W., Carson, M. M., Stelfox, H. T., Beeler, J. A., and Krezolek, M. P. (1991). The role of complement in inflammation and phagocytosis. *Immunol. Today* **12,** 322–326.

Pasch, M. C., Bos, J. D., Daha, M. R., and Asghar, S. S. (1999). Transforming growth factor-beta isoforms regulate the surface expression of membrane cofactor protein (CD46) and CD59 on human keratinocytes [corrected]. *Eur. J. Immunol.* **29,** 100–108.

Patterson, P. H. (2002). Maternal infection: Window on neuroimmune interactions in fetal brain development and mental illness. *Curr. Opin. Neurobiol.* **12,** 115–118.

Perrin, F. E., Lacroix, S., Aviles-Trigueros, M., and David, S. (2005). Involvement of monocyte chemoattractant protein-1, macrophage inflammatory protein-1alpha and interleukin-1beta in Wallerian degeneration. *Brain* **128**(Pt 4), 854–866.

Pletnikov, M. V., Moran, T. H., and Carbone, K. M. (2002). Borna disease virus infection of the neonatal rat: developmental brain injury model of autism spectrum disorders. *Fro. Biosci.* **7,** d593–d607.

Pollard, K. M., Lee, D. K., and Casiano, C. A. (1997). The autoimmunity-inducing xenobiotic mercury interacts with the autoantigen fibrillarin and modifies its molecular structure and antigenic properties. *J. Immunol.* **158,** 3421–3428.

Punnonen, J., Cocks, B. G., Carballido, J. M., Bennett, B., Peterson, D., Aversa, G., and de Vries, J. E. (1997). Soluble and membrane-bound forms of signaling lymphocytic activation molecule

(SLAM) induce proliferation and Ig synthesis by activated human B lymphocytes. *Exp. Med.* **185,** 993–1004.

Rezaie, P., and Male, D. (1999). Colonisation of the developing human brain and spinal cord by microglia: A review. *Microsc. Res. Tech.* **45,** 359–382.

Ritvo, E. R., Freeman, B. J., Scheibel, A. B., Duong, T., Robinson, H., Guthrie, D., and Ritvo, A. (1986). Lower Purkinje cell counts in the cerebella of four autistic subjects: Initial findings of the UCLA-NSAC Autopsy Research Report. *Am. J. Psychiatry* **143,** 862–866.

Safieh-Garabedian, B., Haddad, J. J., and Saade, N. E. (2004). Cytokines in the central nervous system: Targets for therapeutic intervention. *Neurol. Disord.* **3,** 271–280.

Schneider, U., Bullough, F., Vongpunsawad, S., Russell, S. J., and Cattaneo, R. (2000). Recombinant measles viruses efficiently entering cells through targeted receptors. *J. Virol.* **74,** 9928–9936.

Schnorr, J. J., Xanthakos, S., Keikavoussi, P., Kampgen, E., ter Meulen, V., and Schneider-Schaulies, S. (1997). Induction of maturation of human blood dendritic cell precursors by measles virus is associated with immunosuppression. *Proc. Natl. Acad. Sci. USA* **94,** 5326–5331.

Schrag, S. J., Rota, P. A., and Bellini, W. J. (1999). Spontaneous mutation rate of measles virus: Direct estimation based on mutations conferring monoclonal antibody resistance. *J. Viral.* **73,** 51–54.

Servet-Delprat, C., Vidalain, P. O., Bausinger, H., Manie, S., Le Deist, F., Azocar, O., Hanau, D., Fischer, A., and Rabourdin-Combe, C. (2000). Measles virus induces abnormal differentiation of CD40L-activated human dendritic cells. *J. Immunol.* **164,** 1753–1760.

Shanker, G., Syversen, T., and Aschner, M. (2003). Astrocyte-mediated methylmercury neurotoxicity. *Biol. Trace Elem. Res.* **95,** 1–10.

Sidorenko, S. P., and Clark, E. A. (1993). Characterization of a cell surface glycoprotein IPO-3, expressed on activated human B and T lymphocytes. *J. Immunol.* **151,** 4614–4624.

Silva, S. C., Correia, C., Fesel, C., Barreto, M., Coutinho, A. M., Marques, C., Miguel, T. S., Ataide, A., Bento, C., Borges, L., Oliveira, G., and Vicente, A. M. (2004). Autoantibody repertoires to brain tissue in autism nuclear families. *J. Neuroimmunol.* **152,** 176–182.

Simi, A., Edling, Y., Ingelman-Sundberg, M., and Tindberg, N. (2005). Activation of c-fos by lipopolysaccharide in glial cells via p38 mitogen-activated protein kinase-dependent activation of serum or cyclic AMP/calcium response element. *J. Neurochem.* **92,** 915–924.

Singh, B. (2000). Stimulation of the developing immune system can prevent autoimmunity. *J. Autoimmun.* **14,** 15–22.

Singh, V. K. (1996). Plasma increase of interleukin-12 and interferon-gamma. Pathological significance in autism. *J. Neuroimmunol.* **66,** 143–145.

Singh, V. K., and Jensen, R. L. (2003). Elevated levels of measles antibodies in children with autism. *Ped. Neu.* **28,** 292–294.

Singh, V. K., and Rivas, W. H. (2004). Prevalence of serum antibodies to caudate nucleus in autistic children. *Neurosci. Lett.* **355,** 53–56.

Singh, V. K., Lin, S. X., Newell, E., and Nelson, C. (2002). Abnormal measles-mumps-rubella antibodies and CNS autoimmunity in children with autism. *J. Biomedical Sci.* **9,** 359–364.

Singh, V. K., Lin, S. X., and Yang, V. C. (1998). Serological association of measles virus and human herpesvirus-6 with brain autoantibodies in autism. *Clin. Immunol. Immunopathol.* **89,** 105–108.

Singh, V. K., Warren, R., Averett, R., and Ghaziuddin, M. (1997). Circulating autoantibodies to neuronal and glial filament proteins in autism. *Pediatr. Neurol.* **17,** 88–90.

Singh, V. K., Warren, R. P., Odell, J. D., Warren, W. L., and Cole, P. (1993). Antibodies to myelin basic protein in children with autistic behavior. *Brain Behav. Immun.* **7,** 97–103.

Sweeten, T. L., Posey, D. J., Shankar, S., and McDougle, C. J. (2004). High nitric oxide production in autistic disorder: A possible role for interferon-gamma. *Biol. Psychiatry* **55,** 434–437.

Tager-Flusberg, H., Joseph, R., and Folstein, S. (2000). Current directions in research on autism. *Ment. Retard. Dev. Dis. Res. Rev.* **7,** 21–29.

Teitelbaum, O., Benton, T., Shah, P. K., Prince, A., Kelly, J. L., and Teitelbaum, P. (2004). Eshkol-Wachman movement notation in diagnosis: The early detection of Asperger's syndrome. *Proc. Natl. Acad. Sci. USA* **101,** 11909–11914. Epub 2004 Jul 28.

Tsujimura, A., Shida, K., Kitamura, M., Nomura, M., Takeda, J., Tanaka, H., Matsumoto, M., Matsumiya, K., Okuyama, A., Nishimune, Y., Okabe, M., and Seya, T. (1998). Molecular cloning of a murine homologue of membrane cofactor protein (CD46): Preferential expression in testicular germ cells. *Biochem. J.* **330,** 163–168.

Uicker, W. C., Doyle, H. A., McCracken, J. P., Langlois, M., and Buchanan, K. L. (2005). Cytokine and chemokine expression in the central nervous system associated with protective cell-mediated immunity against Cryptococcus neoformans. *Med. Mycol.* **43,** 27–38.

Urakubo, A., Jarskog, L. F., Lieberman, J. A., and Gilmore, J. H. (2001). Prenatal exposure to maternal infection alters cytokine expression in the placenta, amniotic fluid and fetal brain. *Schizophrenia Res.* **47,** 27–36.

Van Damme, J., Schaafsma, M. R., Fibbe, W. E., Falkenburg, J. H., Opdenakker, G., and Billiau, A. (1989). Simultaneous production of interleukin 6, interferon-beta and colony-stimulating activity by fibroblasts after viral and bacterial infection. *Eur. J. Immunol.* **19,** 163–168.

van Gent, T., Heijnen, C. J., and Treffers, P. D. A. (1997). Autism and the immune system. *J. Child. Psychol. Psychiatry* **38,** 337–349.

Vahter, M., Akesson, A., Lind, B., Bjors, U., Schutz, A., and Berglund, M. (2000). Longitudinal study of methylmercury and inorganic mercury in blood and urine of pregnant and lactating women, as well as in umbilical cord blood. *Environ. Res.* **84,** 186–194.

Vargas, D. L., Nascimbene, C., Krishnan, C., Zimmerman, A. W., and Pardo, C. A. (2005). Neuroglial activation and neuroinflammation in the brain of patients with autism. *Ann. Neurol.* **57,** 67–81.

Vojdani, A., O'Bryan, T., Green, J. A., Mccandless, J., Woeller, K. N., Vojdani, E., Nourian, A. A., and Cooper, E. L. (2004). Immune response to dietary proteins, gliadin and cerebellar peptides in children with autism. *Nutr. Neurosci.* **7,** 151–161.

Vojdani, A., Pangborn, J. B., Vojdani, E., and Cooper, E. L. (2003). Infections, toxic chemicals and dietary peptides binding to lymphocyte receptors and tissue enzymes are major instigators of autoimmunity in autism. *Int. J. Immunopathol. Pharmacol.* **16,** 189–199.

Wakefield, A. J., Murch, S. H., Anthony, A., Linnell, J., Casson, D. M., Malik, M., Berlowitz, M., Dillon, A. P., Thompson, M. A., Harvey, P., Valentine, A., Davies, S. E., and Walker-Smith, J. A. (1998). Ileal-lymphoid-nodular hyperplasia, non-specific colitis, and pervasive developmental disorder in children. *Lancet* **35,** 637–641.

Warren, R. P., Foster, A., and Margaretten, N. C. (1987). Reduced natural killer cell activity in autism. *J. Am. Acad. Child Adolesc. Psychiatry* **26,** 333–335.

Warren, R. P., Margaretten, N. C., Pace, N. C., and Foster, A. (1986). Immune abnormalities in patients with autism. *J. Autism. Dev. Disord.* **16,** 189–197.

Warren, R. P., Odell, J. D., Warren, W. L., Burger, R. A., Maciulis, A., Daniels, W. W., and Torres, A. R. (1996). Strong association of the third hypervariable region of HLA-DR beta 1 with autism. *J. Neuroimmunol.* **67,** 97–102.

Warren, R. P., Odell, J. D., Warren, W. L., Burger, R. A., Maciulis, A., Daniels, W. W., and Torres, R. A. (1997). Brief report: Immunoglobulin A deficiency in a subset of autistic subjects. *J. Autism Dev. Disord.* **27,** 187–192.

Warren, R. P., Singh, V. K., Cole, P., Odell, J. D., Pingree, C. B., Warren, W. L., and White, E. (1991). Increased frequency of the null allele at the complement C4b locus in autism. *Clin. Exp. Immunol.* **83,** 438–440.

Warren, R. P., Cole, P., Odell, J. D., Pingree, C. B., Warren, W. L., White, E., Yonk, J., and Singh, V. K. (1990). Detection of maternal antibodies in infantile autism. *J. Am. Acad. Child Adolesc. Psychiatry* **29,** 873–877.

Warren, R. P., Singh, V. K., Cole, P., Odell, J. D., Pingree, C. B., Warren, W. L., De Witt, C. W., and McCullough, M. (1992). Possible association of the extended MHC haplotype B44-SC30-DR4 with autism. *Immunogenetics* **36,** 203–207.

Wong, G. H., Bartlett, P. F., Clark-Lewis, I., Battye, F., and Schrader, J. W. (1984). Inducible expression of H-2 and Ia antigens on brain cells. *Nature* **310,** 688–691.

Wong, G. H., Bartlett, P. F., Clark-Lewis, I., McKimm-Breschkin, J. L., and Schrader, J. W. (1985). Interferon-gamma induces the expression of H-2 and Ia antigens on brain cells. *J. Neuroimmunol.* **7,** 255–278.

CORRELATES OF PSYCHOMOTOR SYMPTOMS IN AUTISM

Laura Stoppelbein,* Sara Sytsma-Jordan,[†] and Leilani Greening*

*Department of Psychiatry, University of Mississippi Medical Center
Jackson, Mississippi 39216, USA
[†]Department of Psychology, University of Southern Mississippi
Hattiesburg, Mississippi 39406, USA

Stereotypical behaviors are defined as repetitive motor or vocal responses that serve no obvious adaptive function. The current diagnostic classification system, the DSM-IV-TR, includes the presence of stereotypical behaviors or interest in its criteria for autism. Research suggests that as many as 85% of children with autism exhibit repetitive behaviors or mannerisms. However, stereotypical behaviors are not specific to autism and are associated with other disorders such as Tourette's syndrome, schizophrenia, and mental retardation. Although the DSM-IV-TR criteria for stereotypical behaviors, as outlined in the diagnostic criteria for autistic disorder, focuses on motor symptoms that tend to occur in excess (e.g., twirling, spinning, head-banging), a broader conceptualization of the types of motor abnormalities observed in individuals with autism has been proposed more recently. Stereotyped patterns of behavior include not only excessive atypical movement but also the loss of typical movement (e.g., catatonia) in this broader definition. Support for this definition is evidenced by both clinical observations and empirical research. Research examining the overlap between catatonia and other stereotypic behaviors among individuals with autism suggest that the greatest risk for catatonic behaviors occurs in adolescence and may be precipitated by stressful events. Assessment tools for autism often include some measure of stereotyped behaviors and interest, but the presence of stereotypy is not in and of itself a pathognomonic sign of autism. Focusing primarily on the presence of classic stereotypical behaviors in diagnoses may subsequently lead to overidentifying autism in very young or mentally retarded individuals.

INTERNATIONAL REVIEW OF
NEUROBIOLOGY, VOL. 71
DOI: 10.1016/S0074-7742(05)71014-X

343

A number of theories have been proposed over the years to explain the function and etiology of stereotypical behaviors. Lovaas and his colleagues, for example, proposed that the sensory and perceptual stimuli created through repetitive behaviors may be self stimulating. Others suggest that stereotypical behaviors are maintained by socially mediated positive and negative reinforcers; whereas biological theories focus on dysfunctions in the serotonin, opioid, and dopaminergic systems in the brain.

I. Introduction

Autism was first recognized as a pervasive developmental disorder (PDD) in the mid-twentieth century when Leo Kanner, M.D. noted 11 children engaging in atypical behaviors that were not consistent with other identified psychiatric conditions such as schizophrenia. Kanner observed three areas of impairment in these children including abnormal language, insistence on sameness, and social isolation, and is credited with labeling this constellation of symptoms as infantile autism. Wing and Gould (1979) noted similar symptoms approximately 30 years later and proposed the "triad of impairment," which conceptualized autism as a disorder characterized by impairment in reciprocal social interaction, verbal and non-verbal communication, and a restricted repertoire of activities or interests. The current diagnostic criteria for autism reflect these general categories and include qualitative impairment in social interaction (a minimum of 2 symptoms), communication (a minimum of 1 symptom), and a restricted repertoire and repetitive stereotyped pattern of behaviors, interests, or activities (a minimum of 1 symptom) with a total of 6 symptoms needed to meet full criteria [American Psychiatric Association (APA), 2000]. The current chapter focuses on symptoms characteristic of the latter category with an emphasis on the types of motor abnormalities manifested in children with autism, the function of these behaviors, and the underlying biological mechanisms to explain the occurrence of these behaviors.

Symptoms encompassing this latter category, including a restricted and stereotyped range of activities and interests, have typically been conceptualized as behaviors that are observable and atypical. The APA lists a cluster of symptoms representative of this category in the *Diagnostic and Statistical Manual of Mental Disorders*, 4th edition text revision (DSM-IV-TR; APA, 2000). The first symptom listed is a preoccupation with one or more stereotyped and restricted patterns of interest that is abnormal either in intensity or focus. This may be manifested in seeking out and retaining vast amounts of information about one topic such as geography or presidents or collecting information about people's birthdates. It is important to note that up to 75% of children with autism also have comorbid mental retardation and that this symptom tends to be more common in high functioning children with autism who do not score in the range of mental retardation on measures of intellectual functioning. A second symptom of this category is the compulsive adherence to nonfunctional routines and rituals. Children with autism may have a need to follow a particular route to school each day or complete daily tasks in a particular order. If they are prevented from engaging in their routine they exhibit distress which can escalate to physical resistance. A third symptom characteristic of this category as listed in the DSM-IV-TR is a preoccupation with parts of objects at the exclusion of recognizing and utilizing objects in a functional manner. A child with autism may be satisfied spinning the wheels of a toy car repetitively instead of rolling it on the ground or racing it with another car. Or a child may be preoccupied with and focus on the spinning blades of a ceiling fan or repetitively pressing buttons on a remote control rather than using the objects in a functional manner.

II. Stereotypic Behaviors

A fourth symptom characteristic of the restricted behavioral category is stereotyped or repetitive motor mannerisms. Stereotypic behavior has been defined as repetitive motor or vocal responses that serve no obvious adaptive function (LaGrow and Repp, 1984). Others define stereotypies as motor behaviors that are repetitive, topographically invariant, often rhythmical, and without purpose (Powell *et al.*, 1999). Regardless of the definitions used, they all share a common focus on the atypical nature and the repetitive quality of the behaviors. These behaviors have been recognized as a fundamental feature of Autism Spectrum Disorders (ASD) (APA, 2000; Lewis and Bodfish, 1998).

Empirical research has revealed that as many as 85% of children with autism exhibit repetitive behaviors or mannerisms (Volkmar *et al.*, 1986). Based on parental reports of a large sample of children with autism, the most common stereotypies observed are rocking (65%), toe-walking (57%), arm, hand, or finger flapping (52%), and whirling (50%). Repetitive behaviors can also include self-injurious behaviors such as head-banging, hand or finger-biting, hitting self with a closed or opened fist, hair-pulling, or scratching. These behaviors are often the focus of treatment for children diagnosed with autism because of the risk for physical harm.

In addition to reports of repetitive motor behaviors among children with developmental delays, these behaviors are commonly seen in normally developing children, occurring in up to 15–20% of pediatric patients (Matthews *et al.*, 2001). Stereotyped movements such as those observed in individuals with developmental delays have been reported to occur in as many as 3.5% of typically developing young children (DeLissovoy, 1961). However, in typical children, these tend to be transient developmental phenomena, that usually disappear by age 5. Hence, stereotypical movements are not unique to autism. They are also observed in a number of different neurological and psychiatric disorders including Tourette's syndrome, Huntington's disease, Parkinson's disease, schizophrenia, and obsessive-compulsive disorder. Children with cognitive impairments, such as mental retardation (Lewis and Bodfish, 1998), and sensory impairments including visual and hearing impairments (Bachara and Phelan, 1980; Troster *et al.*, 1991) may also exhibit stereotypic behaviors. These symptoms, therefore, may simply be associated with severe cognitive impairment and, therefore, are not unique diagnostic features of autism (Volkmar and Lord, 1998). Matson and his colleagues (1996) refute this assertion with data revealing a significant difference in the number and intensity of stereotypic behaviors among adults with a dual diagnosis of autism and severe or profound mental retardation (75%) versus adults with severe or profound mental retardation *without* autism (7%). Similarly, Campbell and her colleagues (1990) found that all 224 autistic children included in their study exhibited some form of stereotypic behaviors. The most common types were lower body extremity movement (28%) and object stereotypy (25%). Between 15–20% also displayed upper extremity movement, hand flapping, or body rocking, and 12% engaged in head tilting behaviors. Children with more severe manifestations of autism tended to exhibit more severe stereotypical behaviors. These behaviors are often the focus of clinical attention because stereotypical movements can interfere with the acquisition of new behaviors and the application of previously learned skills (Epstein *et al.*, 1974; Koegel and Covert, 1972; Morrison and Rosales-Ruiz, 1997). They may also yield prognostic utility as they have been observed to lead to more severe self-injurious behaviors (Guess and Carr, 1991; Schroeder *et al.*, 1990).

A. Differential Diagnosis with Stereotypic Movement Disorder

Stereotypical movements are also characteristics of Stereotypic Movement Disorder (SMD), a disorder characterized by repetitive, nonfunctional motor behaviors that are severe enough to warrant treatment and interfere with daily activities or are of sufficient intensity to cause injury (APA, 2000). In the latter case, SMD can be coded as "with self-injurious behavior" per DSM-IV-TR. However, SMD is not diagnosed if the stereotypy is due to a compulsion, tic, PDD, or hairpulling. SMD is most commonly seen in individuals with mental retardation, and although it is seen in 2–3% of children and adolescents in community settings, it is much more common in adults with severe and profound mental retardation (MR) residing in institutional settings, where the prevalence is approximately 25% (APA, 2000; Rojahn *et al.*, 1998). SMD may develop following a stressful or painful event and tends to peak in adolescence, followed by a gradual decline, although in severe/profound MR adults it may persist for years. In its most severe forms, SMD can be associated with self injury, including self biting, head banging, and other physical forms of self harm, which can be severe enough to cause physical damage (e.g., retinal detachment or blindness) if untreated. Self-injurious behavior (SIB) is sometimes observed among individuals with medical conditions including Lesch-Nyhan and Cornelia de Lange syndromes but can be present in the absence of a medical condition and warrant symptom-specific treatment because of the risk for physical injury.

B. Function of Stereotypic Behaviors in Autism

Hypotheses about the function of stereotypic behaviors in individuals with autism range from self-stimulation to communication. Lovaas *et al.* (1987) have suggested that the sensory and perceptual stimuli created through behaviors such as hand flapping or twirling are rewarding. Support for the stimulating benefits include the circumstances in which stereotypies occur, which is primarily when the individual is in an under-stimulating or over-stimulating environment. Hutt and Hutt (1968) observed that children with autism would increase their stereotypic behavior by more than 25% when their environment changed from an empty familiar room to a familiar room with an adult sitting quietly in the corner.

In a landmark study, Iwata *et al.* (1982) introduced functional analysis, an assessment methodology for testing hypotheses regarding contingencies for maintaining self injurious behavior. Since then, this methodology has been expanded and applied to numerous problem behaviors, including stereotypy (Durand and Carr, 1987; Repp *et al.*, 1988). Three main functional hypotheses for stereotypic behavior have emerged from this literature: sensory self stimulation, positive reinforcement, and negative reinforcement (Iwata *et al.*, 1994; Repp *et al.*, 1988). The sensory self-stimulation hypothesis is consistent with prior reports that stereotypy is maintained by access to reinforcing sensory and perceptual stimulation that may be a by-product of the stereotypic behavior itself. For example, repetitive eye poking may be maintained by access to visual stimulation, and repetitive behaviors such as rocking and spinning may be maintained by vestibular stimulation. This hypothesis suggests that stereotypy is *not* socially mediated. Rather, it appears that stereotypy is maintained by either positively or negatively reinforcing sensory consequences including access to pleasant visual, auditory, or vestibular stimuli or by the removal of aversive sensory stimulation, such as pain. Some researchers have subsequently suggested the term *automatic reinforcement* to explain maintenance of behaviors through non-social sensory mechanisms (Iwata *et al.*, 1994).

Behavioral functional analyses have also revealed that stereotypy is maintained by socially-mediated consequences (Durand and Carr, 1987; Kennedy *et al.*, 2000). In the case of positive reinforcement, this may occur in the form of delivery of social attention or access to preferred tangible items such as food or toys contingent upon the occurrence of stereotypic behaviors. The behaviors may be negatively reinforced by response-contingent escape from difficult task demands (Iwata *et al.*, 1994). Studies have found empirical support for the influence of social consequences in the maintenance of stereotypic behaviors among individuals with autism spectrum disorders (Durand and Carr, 1987; Mace and Belfiore, 1990) as well as successful application of treatments based on these functional causes (Repp *et al.*, 1988). Durand and Carr (1987) found that individuals with autism and other PDDs would engage in higher rates of rocking and hand flapping when presented with difficult tasks. These researchers showed that they could increase the stereotypics by making removal of the difficult task contingent upon the occurrence of these behaviors. When participants were taught to say "Help me" and provided with assistance in high demand situations, the rocking and hand flapping declined. Dawson and her colleagues (1998) have also observed that individuals with autism tend to engage in stereotypic, self-injurious, or aggressive behaviors primarily for sensory stimulation or to escape from social or work situations.

Others have suggested that stereotypic behaviors may serve as substitute behaviors in the absence of a preferred activity, or as a form of communication to gain access to or to remove particular types of environmental stimuli (Carr and Durand, 1985). Kennedy and his colleagues (2000) noted in their study of five children that stereotypical behaviors such as hand flapping, body rocking, and object manipulation occurred across a broad range of social settings including when demands were made of them and attention was given, when there were no demands or attention, and when the child was given a preferred toy or object. When taught an alternative and functional behavior such as a behavioral sign for "break" and "more," the children's stereotypic behaviors decreased and they exhibited more functional behaviors (Kennedy *et al.*, 2000).

C. Associated Motor Abnormalities and Catatonia in Autism

Although the DSM-IV-TR (APA, 2000) criteria for autism address motoric symptoms that occur in excess, a broader conceptualization that encompasses the absence of behavioral responses may better explain the motor abnormalities observed in autistic individuals. Leary and Hill (1996) proposed that the types of motor abnormalities classified under the third category of criterion A in the DSM-IV-TR (APA, 2000)—restricted repetitive and stereotyped patterns of behavior—might be more appropriately conceptualized as movement disturbances that "involve both the loss of typical movement and excessive atypical movement" (p. 40). This definition would include difficulties with the dynamics of movement such as starting, executing, or switching movements. Such difficulties affect very simple movements such as head nodding to more complex movements such as coordinating verbal and non-verbal communication. Support for this broader definition includes the wide range of abnormal motor movements and paucity of movement in individuals diagnosed with autistic disorder. Clinical observations of children with autism have revealed catatonic behaviors including immobility, extreme negativism, mutism, and peculiarities of voluntary move-ments can often co-occur with symptoms of echolalia and bursts of hyperactivity (Ghaziuddin *et al.*, 2005). Dhossche (2004) describes three different case vignettes of individuals with PDDs who exhibit catatonic behaviors; one of which was an autistic adolescent male who was demonstrating aggressive outbursts and agi-tated behaviors along with such catatonic-type symptoms as psychomotor retar-dation, staring, waxy flexibility, and posturing. Ghaziuddin *et al.* (2005) describes a case vignette of an adolescent boy with autism who began exhibiting catatonic behaviors during adolescence. These behaviors included an excessive slowing of movements, progressive mutism, and a gradual loss of independent self-care activities. In the only large-scale empirical study to date examining catatonia in autism, Wing and Shah (2000) found that 17% of referrals (15 years of age or older) to a tertiary center for autism had severe exacerbation of catatonic features. These patients were more likely to have had impaired language and passivity in social interactions prior to the onset of the catatonic-like behaviors compared to patients without catatonic behaviors. Although there is limited research, the current literature suggests that individuals with autism are most at risk for developing catatonic symptoms during pre-adolescence and adoles-cence (Wing and Shah, 2000), possibly due to the stressors associated with adolescence that may affect the biological systems implicated in the development of stereotypic movements (e.g., dopamine).

Inspection of the diagnostic criteria for "catatonic features specifier" as set forth by the DSM-IV-TR (APA, 2000) reveals significant overlap in the criteria with psychomotor disturbances observed in some individuals with autism (e.g., posturing, stereotypies, alterations in the level of activity). Although this overlap is supported empirically (Bush *et al.*, 1996; Joseph, 1992; Rutter *et al.*, 1978, 1985; Wing, 1996), further investigation is warranted to understand the commonalities between these two disorders. Pending further research, it should not be assumed that all patients with autism who demonstrate behaviors consistent with catatonia warrant a comorbid diagnosis of catatonia. Rather, catatonic symptoms could be associated with other comorbid conditions such as depression or anxiety or due to medications as in the case of neuroleptic malignant syndrome (Takaoka and Takata, 2003). Nevertheless, patients who begin to show exacerbation of these symptoms or a significant decline in movement should be examined for catatonia (Ghaziuddin *et al.*, 2005).

D. Etiology of Stereotypies, and Associated Motor Disturbances in Autism

A number of biological theories have been proposed to explain the development of autism and more specifically the development of stereotypical behaviors in autism. Hypotheses include structural and neurochemical abnormalities as etiological explanations. Research examining these neuroanatomical explanations have typically relied on the findings from Magnetic Resonance Imaging (MRI) and Positron Emission Tomography (PET). These findings suggest that there may be some abnormalities in the cerebellum and the neuronal systems that are directly influenced by the cerebellum including those that regulate attention, sensory modulation, autonomic activity, and behavior initiation (Courchesne *et al.*, 1988).

Of greater interest are biochemical findings for the etiology of autism and the implication of these findings for understanding the biological basis for stereotypical behaviors. As with many other psychiatric conditions, serotonin (5-HT) and the 5-HT system have been implicated as one of the biological processes that may contribute to the development of autism. Serotonin is involved in many aspects of human behavior including sleep, pain, motor function, appetite, and others (Volkmar and Anderson, 1989). Research examining 5-HT has consistently revealed that up to 50% of individuals with autism are hyperserotonemic (Geller *et al.*, 1982). The mechanism through which 5-HT influences the symptoms of autism, however, is still unknown. Findings specifically related to repetitive and stereotypic behaviors suggest that decreases in 5-HT levels are related to *decreased* stereotypies. Consistent with this hypothesis, Curzon (1990) found that rats administered acute treatment with agents that increase 5-HT activity were observed to engage in behaviors that resemble stereotypies (Curzon, 1990). However, contrary to this hypothesis, others have reported that when the reuptake of 5-HT is inhibited the rate of stereotyped behaviors tends to decrease (Powell *et al.*, 1997). Similarly, McDougle *et al.* (1996) observed an exacerbation of stereotypic behavior among a sample of adults with autism following a reduction of 5-HT through the depletion of its precursor, tryptophan. The effects of central versus peripheral 5-HT may differentially influence the manifestation of stereotypic behaviors.

Other neurochemicals, such as opioids, have also been implicated, specifically in the development of self-injurious and stereotyped behaviors. The opioid hypothesis suggests that when individuals engage in SIB, the brain releases neurochemicals such as endorphins that block pain and produce mild euphoria. Thus, individuals engage in self-injury and other forms of stereotypy to experience euphoria and escape/avoid painful stimulation. Although this may seem paradoxical, it is believed that continued self-injury actually blocks the painful stimulation that it would ordinarily produce, thus contributing to the maintenance of this behavior. Evidence to support this hypothesis includes both animal and human studies. Research with rodents, for example, has revealed that self-injurious behaviors and stereotypies increase when opiate agonists are administered; whereas spontaneous stereotypies in farm animals were reduced or inhibited through opiate antagonists (Dantzer, 1986; Iwamoto and Way, 1977). Similar findings have been observed in human studies. Both Campbell *et al.* (1993) and Rojahn *et al.* (1998) reported a decrease in SIB following the administration of opiate antagonists (naltrexone). However, the declines observed by Campbell *et al.* were not statistically significantly.

More recent research has investigated the interaction between opiates, 5-HT, and the dopaminergic system to explain autism, and more specifically stereotypies and SIB. Serotonin receptors interact with the dopamine (DA) system and there is empirical evidence to suggest that stereotypical animal behaviors, such as head weaving, are decreased by lesions in the nigrostriatal and mesolimbic DA pathways and when DA blockers, such as haloperidol, are administered (Lewis and Bodfish, 1998). Additionally, there is evidence that the interaction between opioid and DA systems is an important mediator of abnormal or stereotyped behaviors (Lewis and Bodfish, 1998). For example, terminal fields that receive substantial amounts of DA innervation also contain large amounts of opioid peptides and receptors (Angulo and McEwan, 1994). The DA supersensitivity hypothesis proposes that repetitive and self-injurious behaviors result from low levels of DA in postsynaptic cells of the basal ganglia, resulting in supersensitivity of the post-synaptic receptors. Thus, the presence of small amounts of DA can be activating. Some support for this hypothesis can be seen in animal models of self-injurious and stereotypic behaviors in which these behaviors are induced following the administration of dopamine agonists such as L-Dopa (Lewis and Baumeister, 1982). Depriving animals of sensory stimulation and restricting their interactions with the environment in controlled laboratory experiments have been found to prevent DA innervation and subsequently produce spontaneous stereotypies (Martin *et al.*, 1991). Suomi and Harlow (1971) observed similar behaviors in non-human primates who had experienced early social deprivation. Interestingly, stereotypies were a long-term consequence of social deprivation in this research. It is hypothesized that early social deprivation or restricted environmental interaction results in a loss of dopaminergic innervation of important brain regions that results in DA receptor supersensitivity (Lewis and Bodfish, 1998).

Non-biological explanations for stereotypical behaviors include behavioral theories such as the reinforcement or communication-based theories. According to the reinforcement theory, there are four categories of reinforcement that maintain behavior. These categories are determined by whether the stimulus for the behavioral response is internal or external to the individual and whether a reinforcing stimulus is presented (positive) to precipitate the response or an aversive stimulus is removed (negative) to cease the response. Positive external reinforcement is hypothesized to maintain a response such as a stereotypy with social attention, whereas positive internal reinforcement maintains repetitive behaviors through the production of pleasant internal sensory consequences (e.g., visual stimulation, vestibular stimulation, endorphin release). In contrast, negative external reinforcement is thought to maintain stereotypical behavior through the escape or avoidance of unpleasant conditions (e.g., task demands), whereas negative internal reinforcement maintains stereotypic or self-injurious behavior through the reduction or cessation of painful stimulation (e.g., otitis media) (Rojahn *et al.*, 1998).

A communication-based theory for stereotypical behavior asserts that these behaviors are functionally equivalent to attempts to communicate in nonverbal individuals. In the absence of language skills, these individuals are thought to have learned maladaptive behaviors as a means of communication. A child may engage in head banging, for example, to escape from task demands or to obtain social attention. That is, head banging may be functionally equivalent to saying "I want a break" or "I want attention" (Rojahn *et al.*, 1998). Studies have shown that efforts to train communicative alternatives (e.g., pointing to a sign that says "break") are effective in reducing self-injury in MR individuals (Carr and Durand, 1985).

To date, much of the literature on the etiology of movement disturbances in autism has focused on understanding the excesses of behavior (e.g., repetitive behaviors). With more recent attention given to a broader conceptualization of movement disorders in individuals with autism (Leary and Hill, 1996), researchers are beginning to investigate etiological explanations for the overlap between catatonia and autism. One neurochemical—gamma-aminobutyric acid (GABA)—has received some attention lately. Catatonia tends to respond well to treatment with benzodiazepines and other effective treatments for catatonia are known to enhance GABA functioning (e.g., barbiturates, ECT). Recently a GABA hypothesis has been proposed to explain autism (Dhossche *et al.*, 2002). While GABA is known to have an excitatory trophic role affecting neuronal wiring, organization, and plasticity of neuronal networks in early development, recent findings indicate that the effects of abnormal trophic GABA functioning in early development are consistent with the brain abnormalities reported so far in autistic individuals. At the present time, there is empirical evidence supporting overlapping etiologies for catatonia and autism including genetic studies implicating the long arm of chromosome 15 in both autism and catatonia (Borgatti *et al.*, 2001; Lauritsen *et al.*, 1999); however, further empirical investigations are needed to test hypotheses about the role of GABA in autism and catatonia before drawing conclusions.

E. Assessment of Stereotypy

Some of the common assessment measures for autism are the Autism Diag-nostic Observation Schedule (ADOS; Lord *et al.*, 1989), the Autism Diagnostic Interview-Revised (ADI-R; Lord *et al.*, 1994), the Childhood Autism Rating Scale (CARS; Schopler *et al.*, 1988), and the Autism Behavior Checklist (ABC; Krug *et al.*, 1980, 1993). These measures also include items pertaining to stereotypic movements. The ADI-R is a structured interview that assesses func-tioning in three domains mirroring the diagnostic criteria for autism: Language and Communication, Reciprocal Social Interactions, and Restricted, Repetitive, and Stereotyped Behaviors and Interests. Items in the third domain in-clude repetitive motor movements and self-injury. The CARS incorporates historical interview information from the parent and direct observation by a professional who rates the child's behavior in 15 domains. An item pertaining to "body use" measures the severity of stereotyped behaviors such as repetitive movements, rocking, spinning, and self-injury and an item pertaining to "object use" addresses inappropriate use of objects, which may include stereo-typed use. The ABC is a 57-item parent or teacher rating scale that is included as part of the Autism Screening Instrument for Educational Planning-2. One of the five subscales, "Body and Object Use," contains several items measuring stereotypic motor movements, such as whirling, rocking, spinning, and flapping, as well as stereotyped use of objects (e.g., spinning or banging objects).

Research on differential diagnosis of autism has consistently found that very young children (under age 3) with mental retardation cannot be reliably distin-guished from children with autism (Rutter and Schopler, 1987; Wing and Gould, 1979). This is because both groups exhibit language delays and social impair-ments, and have a great deal of overlap in restricted, repetitive, and stereotyped behaviors (Vig and Jedrysek, 1999; Wing and Gould, 1979). For instance, young children with mental retardation, adults with severe or profound mental retarda-tion, as well as children with autism have all been shown to engage in hand stereotypies at varying rates (Vig and Jedrysek, 1999). Not surprisingly, some autism assessment measures, including the CARS and ABC, have been noted to have limitations in accurately diagnosing autism in children under the age of 3 (New York State Department of Health, 1999). The ADI-R and CARS have also overidentified very young (under 2 years) and mentally retarded children (mental age < 18 months) (DiLavore *et al.*, 1995). Overidentification is further complicated by the fact that mental retardation can co-occur in 70–80% of children with autism.

III. Conclusion

Stereotypic behaviors and movement disorders are observed in individuals with autism and the DSM-IV recognizes this symptom as a diagnostic feature of autistic disorder. However, behavioral abnormalities of this nature are common early in development, and among children with mental retardation, other PDDs, and neurological conditions, thus limiting its specificity to autism. Diagnosing children with autism should, therefore, not be predicated on observations of stereotypic movements. Nevertheless, the occurrence of these behaviors and the interference they cause in the child's social and cognitive development has led to a body of research on the cause and treatment of stereotypic behaviors in individuals with autism. Etiological explanations for these abnormal movements are largely based on biological and neurochemical theories but may be reinforced, managed, and maintained by social and sensory consequences. More recently, movement disorders in autism have been expanded to include catatonia—a paucity of movement. Catatonic symptoms tend to appear in adolescence and may be precipitated by stressful events when observed in individuals diagnosed with autism. A different neurochemical process (e.g., GABA), may explain the lack of motor movements; however, further research is warranted before drawing definitive conclusions. Continued research will hopefully shed further light on the role of these movement disorders in autism.

References

American Psychiatric Association (2000). "Diagnostic and Statistical Manual of Mental Disorders-text revision, 4th edition, text revision." American Psychiatric Association. Washington, D.C.

Angulo, J. A., and McEwen, B. S. (1994). Molecular aspects of neuropeptide regulation and functioning in the corpus striatum and nucleus accumbens. *Brain Res. Rev.* **19,** 1–28.

Bachara, G. H., and Phelan, W. J. (1980). Rhythmic movement in deaf children. *Perceptual and Motor Skills* **50,** 933–934.

Borgatti, R., Piccinelli, P., Passoni, D., Dalpra, L., Miozzo, M., Micheli, R., Gagliardi, C., and Balottin, U. (2001). Relationship between clinical and genetic features in "inverted duplicated chromosome 15" patients. *Ped. Neu.* **24,** 111–116.

Bush, G., Fink, M., Petrides, G., Dowling, F., and Francis, A. (1996). Catatonia: I. Rating scale and standardized examinations. *Acta Psychiatrica Scand.* **93,** 129–136.

Campbell, M., Locascio, J., Choroco, M. C., Spencer, E. K., Malone, R. P., Kafantaris, V., and Overall, J. E. (1990). Stereotypies and tardive dyskinesia: Abnormal movements in autistic children. *Psychopharmacology Bull.* **26,** 260–266.

Campbell, M., Anderson, L. T., Small, A. M., Adams, P., Gonzales, N. M., and Ernst, M. (1993). Naltrexone in autistic children: Behavioral symptoms and attentional learning. *J. Am. Acad. Child Adoles. Psychiatry* **32,** 1283–1289.

Carr, E. G., and Durand, V. M. (1985). Reducing behavior problems through functional communication training. *J. Appl. Behav. Anal.* **18,** 111–126.

Courchesne, E., Yeung-Courchesne, R., Press, G. A., Hesselink, J. R., and Jernigan, T. L. (1988). Hypoplasia of cerebellar vermal lobules VI and VII in autism. *N. Engl. J. Med.* **318,** 1349–1354.

Curzon, G. (1990). Stereotyped and other motor responses to 5-hydroxytryptamine receptor activation. *In* "Neurobiology of Stereotyped Behaviour" (S. J. Cooper and C. T. Dourish, Eds.), pp. 142–168. Oxford University Press, Oxford.

Dantzer, R. (1986). Behavioral, physiological and foundational aspects of stereotyped behavior: A review and re-interpretation. *J. Ani. Sci.* **62,** 1776–1786.

Dawson, J. E., Matson, J. L., and Cherry, K. E. (1998). An analysis of maladaptive behaviors in persons with autism, PDD-NOS, and mental retardation. *Res. Dev. Disabil.* **19,** 439–448.

DeLissovoy, V. (1961). Head-banging in early childhood: A study of incidence. *J. Ped.* **58,** 803–805.

Dhossche, D. M., Applegate, H., Abraham, A., Maertens, P., Bland, L., Bencsath, A., and Martinez, J. (2002). Elevated plasma gamma-aminobutyric acid (GABA) levels in autistic youngsters: Stimulus for a GABA hypothesis of autism. *Med. Sci. Monit.* **8,** PR1–PR6.

Dhossche, D. M. (2004). Autism as early expression of catatonia. *Med. Sci. Monit.* **10,** 31–39.

DiLavore, P. C., Lord, C., and Rutter, M. (1995). The pre-linguistic autism diagnostic observation schedule. *J. Autism Dev. Disord.* **25,** 355–379.

Durand, V. M., and Carr, E. G. (1987). Social influences on self-stimulatory behavior: Analysis and treatment application. *J. Appl. Behav. Anal.* **20,** 119–132.

Epstein, L. H., Duke, L. A., Sajwaj, T. E., Sorrell, S., and Rimmer, B. (1974). Generality and side effects of overcorrection. *J. Appl. Behav. Anal.* **7,** 385–390.

Geller, E., Ritvo, E. R., Freeman, B. J., and Yuwiler, A. (1982). Preliminary observations on the effect of fenfluramine on blood serotonin and symptoms in three autistic boys. *N. Engl. J. Med.* **307,** 165–167.

Ghaziuddin, M., Quinlan, P., and Ghaziuddin, N. (2005). Catatonia in autism: A distinct subtype? *J. Intellect. Disabil. Res.* **49,** 102–105.

Guess, D., and Carr, E. (1991). Emergence and maintenance of stereotypy and self-injury. *Am. J. Men. Retard.* **96,** 299–329.

Hutt, S. J., and Hutt, C. (1968). Stereotypy, arousal and autism. *Hum. Dev.* **11,** 277–286.

Iwamoto, E. T., and Way, E. L. (1977). Circling behavior and stereotypy induced by intranigral opiate microinjection. *J. Pharmacol. Exp. Ther.* **20,** 347–359.

Iwata, B. A., Dorsey, M. F., Slifer, K. J., Bauman, K. E., and Richman, G. S. (1982). Toward a functional analysis of self-injury. *Anal. Inter. Dev. Disabil.* **2,** 3–20.

Iwata, B. A., Pace, G. M., Dorsey, M. F., Zarcone, J. R., Vollmer, T. R., Smith, R. G., Rodgers, T. A., Lerman, D. C., Shore, B. A., Mazaleski, J. L., Goh, H. L., Cowdery, G. E., Kalsher, M. J., McCosh, K. C., and Willis, K. D. (1994). The functions of self-injurious behavior: An experimental-epidemiological analysis. *J. Appl. Behav. Anal.* **27**(4), 215–240.

Joseph, A. B. (1992). Catatonia. *In* "Movement Disorders in Neurology and Neuropsychiatry" (A. B. Joseph and R. R. Young, Eds.), pp. 335–342. Blackwell, Oxford.

Kennedy, C. H., Meyer, K. A., Knowles, T., and Shukla, S. (2000). Analyzing the multiple functions of stereotypical behavior for students with autism: Implications for assessment and treatment. *J. Appl. Behav. Anal.* **33,** 559–571.

Koegel, R. L., and Covert, A. (1972). The relationship of self-stimulation to learning in autistic children. *J. Appl. Behav. Anal.* **5,** 381–387.

Krug, D. A., Arick, J. R., and Almond, P. J. (1980). Behavior checklist for identifying severely handicapped individuals with high levels of autistic behavior. *J. Child Psychol. Psychiatry* **21,** 221–229.

Krug, D. A., Arick, J. R., and Almond, P. J. (1993). "Autism Screening Instrument for Educational Planning, 2nd edition." Pro-Ed, Austin, TX.

LaGrow, S. J., and Repp, A. C. (1984). Stereotypic responding: A review of intervention research. *Am. J. Men. Deficiency* **88,** 595–609.

Lauritsen, M., Mors, O., Mortensen, P., and Ewald, H. (1999). Infantile autism and associated autosomal chromosome abnormalities: A register-based study and a literature survey. *J. Child Psychol. Psychiatry* **40,** 335–345.

Leary, M. R., and Hill, D. A. (1996). Moving on: Autism and movement disturbance. *Mental Retardation* **34,** 39–53.

Lewis, M. H., and Baumeister, A. A. (1982). Stereotyped mannerisms in mentally retarded persons: Animal models and theoretical anlayses. *In* "International Review of Research in Mental Retardation," (N. R. Ellis, Ed.), Vol. 11, pp. 123–161. Academic Press, New York.

Lewis, M. H., and Bodfish, J. W. (1998). Repetitive behavior disorders in autism. *Men. Retarda. Dev. Disrod.* **4,** 80–89.

Lord, C., Rutter, M., Goode, S., Heemsbergen, J., Jordan, H., Mawhood, L., and Schopler, E. (1989). Autism diagnostic observation schedule: A standardized observation of communicative and social behavior. *J. Autism Dev. Disord.* **19,** 185–212.

Lord, C., Rutter, M., and Le Couteur, A. (1994). Autism diagnostic interview revised: A revised version of a diagnostic interview for caregivers of individuals with possible pervasive developmental disorders. *J. Autism Dev. Disord.* **24,** 659–685.

Lovaas, O. I., Newsom, C., and Hickman, C. (1987). Self-stimulatory behavior and perceptual reinforcement. *J. Appl. Behav. Anal.* **20,** 45–68.

Mace, F. C., and Belfiore, P. (1990). Behavioral momentum in the treatment of escape-motivated stereotypy. *J. Appl. Behav. Anal.* **23,** 507–514.

Martin, L. J., Spicer, D. M., Lewis, M. H., Gluck, S. P., and Cork, L. C. (1991). Social deprivation of infant rhesus monkeys alters the chemoarchitecture of the brain. I. Subcortical regions. *J. Neurosci.* **11,** 3344–3358.

McDougle, C., Naylor, S., Cohen, D., Aghajanian, G., Heninger, G., and Price, L. (1996). Effects of tryptophan depletion in drug-free adults with autistic disorder. *Arch. Gen. Psychiatry* **53,** 993–1000.

Matson, J. L., Baglio, C. S., Smiroldo, B. B., Hamilton, M., Packlowskyj, T., Williams, D., and Kirkpatrick-Sanchez, S. (1996). Characteristics of autism as assessed by the diagnostic assessment for the severely handicapped –II (DASH-II). *Res. Dev. Disabil.* **17,** 135–143.

Matthews, L. H., Matthews, J. R., and Leibowitz, J. M. (2001). Tics, stereotypic movements, and habits. *In* "Handbook of Clinical Child Psychology, 3rd edition" (C. E. Walker and M. C. Roberts, Eds.), pp. 338–358. John Wiley and Sons, New York.

Morrison, K., and Rosales-Ruiz, J. (1997). The effect of object preference on task performance and stereotypy in a child with autism. *Res. Dev. Disabil.* **18,** 127–137.

New York State Department of Health (1999). "Clinical Practice Guideline: Report of the Recommendations. Autism/Pervasive Developmental Disorders, Assessment and Intervention for Young Children (Age 0–3 years)." New York State Department of Health. Albany, NY.

Powell, S., Wechsler, E., Newman, H., and Lewis, M. H. (1997). Behavioral and biochemical effects of chronic fluoxetine treatment in an animal model of repetitive behavior disorder. *Society for Neuroscience Abstracts* **23,** 802.

Powell, S. B., Newman, H. A., Pendergast, J. F., and Lewis, M. H. (1999). A rodent model of spontaneous stereotypy: Initial characterization of developmental, environmental, and neurobiological factors. *Physiology and Behavior* **66,** 355–363.

Repp, A. C., Felce, D., and Barton, L. E. (1988). Basing the treatment of stereotypic and self-injurious behaviors on hypotheses of their causes. *J. Appl. Behav. Anal.* **21,** 281–289.

Rojahn, J., Tasse, M. J., and Morin, D. (1998). Self-injurious behavior and stereotypies. *In* "Handbook of Child Psychopathology, 3rd edition." (T. H. Ollendick and M. Hersen, Eds.), pp. 307–336. Plenum, New York.

Rutter, M. (1978). Diagnosis and definition. *In* "Autism: A Reappraisal of Concepts and Treatments, 2nd edition" (M. Rutter and E. Schopeler, Eds.), pp. 545–566. Blackwell, Oxford.

Rutter, M. (1985). Infantile autism and other pervasive developmental disorders. *In* "Child and Adolescent Psychiatry: Modern Approaches, 2nd edition" (M. Rutter and I. Hersov, Eds.), pp. 545–566. Blackwell, Oxford.

Rutter, M., and Schopler, E. (1987). Autism and pervasive developmental disorders: Concepts and diagnostic issues. *J. Autism Dev. Disord.* **17,** 159–186.

Schopler, E., Reichler, R. J., and Renner, B. R. (1988). "The Childhood Autism Rating Scale." Western Psychological Services, Los Angeles.

Schroeder, S. R., Rojahn, J., Mulick, J. A., and Schroeder, C. S. (1990). Self-injurious behavior. *In* "Handbook of behavior modification with the mentally retarded, 2nd Edition" (J. L. Matson, Ed.), pp. 141–180. Plenum, New York.

Suomi, S. J., and Harlow, H. F. (1971). Abnormal social behavior in young monkeys. *In* "The Exceptional Infant" (J. Hellmuth, Ed.), Vol. 2, pp. 483–529. Bruner-Mazel, New York.

Takaoka, K., and Takata, T. (2003). Catatonia in childhood and adolescence. *Psychiatry Clin. Neurosci.* **57,** 129–137.

Troster, H., Brambring, M., and Beelman, A. (1991). Prevalence and situational causes of stereotyped behaviors in blind infants and preschoolers. *J. Abnormal Child Psychology* **19,** 569–590.

Vig, S., and Jedrysek, E. (1999). Autistic features in young children with significant cognitive impairment: Autism or mental retardation? *J. Autism Dev. Disord.* **29**(3), 235–248.

Volkmar, F. R., and Anderson, G. M. (1989). Neurochemical perspectives on infantile autism. *In* "Autism: Nature, Diagnosis, and Treatment" (G. Dawson, Ed.), pp. 208–224. Guilford Press, New York.

Volkmar, F. R., Cohen, D. J., and Paul, R. (1986). An evaluation of DSM-III criteria for infantile autism. *J. Am. Acad. Child Adolesc. Psychiatry* **25,** 190–197.

Volkmar, F. R., and Lord, C. (1998). Diagnosis and definition of autism and other pervasive developmental disorders. *In* "Autism and Pervasive Developmental Disorders" (F. R. Volkmar, Ed.), pp. 1–31. Cambridge University Press, Cambridge, UK.

Wing, L. (1996). "The Autistic Spectrum: A Guide for Parents and Professionals." Constable, London.

Wing, L., and Gould, J. (1979). Severe impairments of social interaction and associated abnormalities in children: Epidemiology and classification. *J. Autism Dev. Disord.* **9,** 11–29.

Wing, L., and Shah, A. (2000). Catatonia in autistic spectrum disorders. *Br. J. Psychiatry* **176,** 357–362.

GABRB3 GENE DEFICIENT MICE: A POTENTIAL MODEL OF AUTISM SPECTRUM DISORDER

Timothy M. DeLorey

Molecular Research Institute, Mountain View, California 94043, USA

Human chromosome 15q11–13 is associated with the neurodevelopmental disorders autism spectrum disorder, Angelman syndrome, and Prader-Willi syndrome. A number of genes have been identified within this region including a cluster of γ-amino butyric acid type A receptor subunit genes, GABRB3, GABRA5, and GABRG3 (encoding the β_3, α_5, and γ_3 subunits, respectively). Numerous studies have demonstrated the importance of the GABAergic system in neurodevelopment; therefore the presence of a group of $GABA_A$ receptor genes within this locus is intriguing. The β_3 subunit is widely expressed during the late embryonic to early postnatal period of brain development. A deficiency in the β_3 subunit during this critical period would be expected to negatively impact the temporal ordering of neurogenesis and synaptogenesis. This would have subsequent ramifications in the maturation of circuits involved in supporting complex behaviors, motor skills, and cognition. As expected, mice deficient in the

INTERNATIONAL REVIEW OF
NEUROBIOLOGY, VOL. 71
DOI: 10.1016/S0074-7742(05)71015-1

359

gabrb3 gene exhibit a wide assortment of neurochemical, electrophysiological, and behavioral abnormalities, many overlapping with traits typically observed in autism spectrum disorder and Angelman syndrome. These findings suggest a potential involvement of the GABRB3 gene in the etiology of these neuro-developmental disorders. The gabrb3 gene deficient mouse has proved to be a valuable model in the critical examination of the interconnection be-tween development, pathology, and behavior as they relate to disorders of neurodevelopment.

I. Introduction

Autism spectrum disorder (ASD) results from a disruption in the rigid sequence of developmental events necessary for proper brain formation and function. Although the nature of this disruption has yet to be determined, there is a general consensus that idiopathic ASD is an oligomeric, multifactorial disorder (Pickles *et al.*, 1995) involving multiple genes (Risch *et al.*, 1999) with environmental factors likely contributing to the pathogenics of the wide con-tinuum of observed phenotypic traits (Hornig and Lipkin, 2001; London, 2000). The high prevalence rate of ASD, 1 in 166 births (Fombonne, 2003), has prompted an urgency to develop animal models in advance of a comprehensive understanding of the molecular underpinnings associated with this disorder. However, establishing an acceptable animal model for any complex human disorder is a difficult task, as one cannot expect to fully replicate the human be-havioral phenotype in a lower mammal. Therefore, the investigator must first ascertain that the animal model is valid for the intended disorder. A suitable animal model should possess (1) face validity, characteristics associated with the human disorder being investigated; (2) construct validity, similarities in the underlying etiology believed to be associated with the human disorder; and (3) predictive validity, use of the animal model to predict the potential outcome of a treatment regime applied to humans with the disorder. Consequently, Crawley (2004) has elegantly outlined a set of behavioral tasks, relevant to autistic-like behaviors, to serve as a guideline in which to establish the face validity of any proposed rodent model of ASD. The fundamental core symptoms of ASD center around three behavioral domains: (1) inappropriate social interactions; (2) poor communica-tion skills; and (3) restrictive, repetitive (stereotypical) behaviors (Kanner, 1943; Rutter and Schopler, 1987). A broader, more variable ASD behavioral phenotype would likewise include anxiety (Gillott *et al.*, 2001), cognitive impairment (Vig and Jedrysek, 1999), epilepsy (Ballaban-Gil and Tuchman, 2000), motor deficits (Ghaziuddin and Butler, 1998), aggression (Cox and Schopler, 1993), sleep disturbances (Harvey and Kennedy, 2002; Limoges *et al.*, 2005), idiosyncratic responses to sensory stimuli (Ayres and Tickle, 1980), attentional deficits (Allen and Courchesne, 2001), and hyperactivity (Gillberg et al., 1996; Teitelbaum et al., 1998).

Extensive evidence favors a role for GABAergic mechanisms in the etiology of ASD. As other chapters in this book address this issue, only a general overview will be given here in order to be succinct. $GABA_A$ receptors are the most ubiquitous neurotransmitter receptor in the mammalian nervous system. These heterooligomeric GABA-gated chloride channels are constructed from eight classes of subunits, some with multiple variants (α_{1-6}, β_{1-4}, γ_{1-3}, δ, ρ_{1-2}, π, ε, θ) (Olsen and DeLorey, 1999; Simon et al., 2004). The various $GABA_A$ receptors isoforms display differing GABA sensitivities, distinct pharmacologies, and exhibit unique regional and temporal distribution patterns within the central nervous system. Interestingly, $GABA_A$ receptors are excitatory during development and primarily inhibitory in adulthood. During development, $GABA_A$ receptors, along with AMPA/Kainate glutamate receptors, mediate Ca^{2+}-dependent signal transduction pathways (Owens et al., 1996) capable of influencing many brain developmental processes including proliferation, synaptogenesis, and circuit formation. Approximately 50% of all synapses in the developing cortex are GABA-releasing synapses, declining to 15% in the adult (De Felipe et al., 1997).

A number of ASD susceptibility loci have been identified giving rise to the view that defects at several loci can result in overlapping phenotypes. Association studies suggest one of these ASD susceptibility loci is located within the chromosome 15q11–13 region (Bass et al., 2000; Buxbaum et al., 2002; Cook et al., 1998; Martin et al., 2000). Moreover, individuals with a maternal duplication of the 15q11–13 region are often diagnosed as having ASD (Martinsson et al., 1996; Rineer et al., 1998; Schroer et al., 1998). Interestingly, maternal deficiency of the same 15q11–13 region results in the neurodevelopmental disorder Angelman syndrome (AS), a distinct syndrome, but with substantial phenotypic overlap with ASD (Steffenburg et al., 1996). The core features of AS are severe mental retardation, epilepsy, ataxia, impaired language, and a happy demeanor (Williams et al., 1995). The prevalence rate for AS has been reported to be as high as 1 in 10,000 live births (Petersen et al., 1995). Interestingly, a deficiency of the paternal allele of 15q11–13 causes Prader-Willi syndrome (PWS), which exhibits a clinically distinct phenotype (mild mental retardation, obsessive-compulsive features, obesity, and hypogonadism) from that of AS (Cassidy and Morris, 2002). AS and PWS stem from a poorly understood phenomenon called genomic imprinting, involving a select group of genes expressed in the offspring from the chromosomal allele of only one parent instead of both, as is the usual situation for most genes. The above observations imply that the correct parental chromosomal gene(s) dosage from the 15q11–13 region is essential for normal brain development.

Progress in identifying ASD candidate gene(s) in the 15q11–13 region has been hampered by the high rate of recombination that occurs within this region, potentially masking linkage. However, several genes in the 15q11–13 region have been proposed as ASD candidate genes. These include the UBE3A (encoding a ubiquitin ligase), ATP10C (encoding a putative aminophospho-lipid translocase), and a cluster of GABA$_A$ receptor subunit genes, GABRB3, GABRA5, and GABRG3, (encoding the β_3, α_5, and γ_3 subunits respectively) (see Fig. 1). To date, evidence is lacking in support of the ATP10C gene's involvement in either ASD or AS. Whereas, a mutation in the UBE3A gene alone is capable of causing AS (Kishino et al., 1997; Matsuura et al., 1997). However, in most AS individuals the entire maternal 15q11–13 allele is deleted, resulting in a more robust epileptic phenotype than observed in AS individuals with just a mutation in the UBE3A gene (Minassian et al., 1998). This has led to the suggestion that deficiencies in at least two genes, UBE3A and GABRB3, are likely required in order to manifest the more severe epileptic phenotype (Kishino et al., 1997; Matsumoto et al., 1992; Minassian et al., 1998). In postmortem brain tissue taken from ASD cases, a reduction in both the UBE3A and GABRB3 gene products has been reported (Samaco et al., 2005). However, association studies have reported mixed results for the UBE3A gene association in ASD (Nurmi et al., 2001; Veenstra-VanderWeele et al., 1999). Whereas, a strong association between the GABRB3 gene and specific subsets of ASD individuals has been reported (Nurmi et al., 2003; Shao et al., 2003). Further support for a GABAergic mechanism in the etiology of ASD comes from a case in which an individual with a rare autosomal-recessive disorder that prevented the proper synthesis of GABA, was likewise diagnosed with ASD, seizures and severe mental retardation (Burd et al., 2000). Collectively, the previous observations and those presented in other chapters of this book provide a compelling argument that a developmental disruption in GABAergic mechanisms would adversely influence the sequence of events necessary to construct a properly functioning neural network.

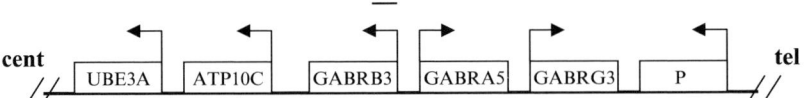

FIG. 1. Arrangement of several of the genes located within the human chromosomal region 15q11–13 (this gene arrangement is conserved on the mouse chromosome band 7B4). The chromosome 15q11–13 region is approximately 4 Mb in size. The maternal allele of this region is deleted in the majority of Angelman syndrome cases, whereas, the maternal duplication of this region often results in the individual being diagnosed with autism spectrum disorder. Association studies have also identified this region as being an autism spectrum disorder susceptibility locus. The black bar above the GABRB3 gene represents the 2.8 Kb region that was disrupted in producing the gabrb3 null mice (Homanics et al., 1997). Cent: centromere, tel: telomere.

The following review examines a mouse line engineered to lack the expression of the gabrb3 gene (syntenic to the GABRB3 gene in human) (Homanics *et al.*, 1997). These mice exhibit decreased GABA$_A$ receptor density and high neonatal mortality rates, with survivors exhibiting a variety of abnormal neurochemical, electrophysiological, and behavioral features. Where appropriate, observations made in regards to these gabrb3 null mice will be correlated and contrasted with those observed in ASD and/or AS (see Table I). This mouse model has provided valuable insight into neurodevelopment and its disorders, in particular ASD and AS. In addition to the gabrb3 gene deficient mouse line, gene deficient mouse lines have been created for other genes found in the 15q11–13 region including the ube3a and gabra5 genes (Crestani *et al.*, 2002; Jiang *et al.*, 1998). In addition, GABA associated genes from other ASD susceptibility regions have likewise been disrupted. These include the Dlx1 and Dlx2 genes, important to the development of telencephalic GABAergic neurons (Anderson *et al.*, 1997; Marin and Rubenstein, 2001) and the reelin gene, encoding a protein important to the function of cortical GABAergic neurons (Alcantara *et al.*, 1998). Collectively, these mouse models will each contribute to a better overall understanding of the role of GABAergic mechanisms in ASD.

II. Molecular Characteristics of GABRB3 Null Mice

Gabrb3 deficient mice were produced by the targeted disruption of exons 1–3 of the gabrb3 gene by homologous recombination in 129/SvJ mouse embryonic stem cells and bred on a hybrid background (129/SvJ X C57Bl/6J) (Homanics *et al.*, 1997). This disruption was verified by northern blot indicating the complete absence of gabrb3 mRNA in homozygous null mice and a 30% reduction of gabrb3 mRNA in heterozygous mice. About 90% of the gabrb3 null mice die within 24 hrs of birth (Homanics *et al.*, 1997), similar to what has been observed in closely associated radiation-induced mouse mutants, p^{cp} mice, that lack the gabrb3 gene as well as the gabra5, gabrg3, and pink-eyed dilution (p) genes, however, sparing the ube3a gene (Lyon *et al.*, 1992; Nakatsu *et al.*, 1993).

III. Morphology

The high mortality rates exhibited by homozygous gabrb3 null mice and p^{cp} mutant mice are likely due to the high incidence of cleft palate, common in these mouse mutants. Cleft palate occurs in about 57% of the gabrb3 null mice (Homanics *et al.*, 1997) and 95% of the p^{cp} mutant mice (Nakatsu *et al.*, 1993).

TABLE I

Qualitative Comparisons Between Autism, Angelman Syndrome, and gabrb3 Null Mice

Characteristic	Autism	Angelman syndrome	gabrb3 null mice	Reference
Impaired social interactions	+	+	poor maternal care	Gillberg et al., 1996; Penner et al., 1993; Homanics et al., 1997
Repetitive, stereotypical behavior	+	+	+	Barthelemy et al., 1997; Summers et al., 1995; DeLorey et al., 1998
Poor motor coordination	+	+	+	Barthelemy et al., 1997; Williams et al., 1995; Homanics et al., 1997
Hyperactivity	+	+	+	Barthelemy et al., 1997; Summers et al., 1995; DeLorey et al., 1998
Tactile hyperresponsivity	+	hyperreflexive	+	Ayres and Tickle, 1980; Viani et al., 1995; Ugarte et al., 2000
Sensitivity to temperature	+	?	+	Harrison and Hare, 2004; Ugarte et al., 2000
Cognitive impairment	+	+	+	Vig and Jedrysek, 1999; Williams et al., 1995; DeLorey et al., 1998
Craniofacial dysmorphism	+	+	+	Wolpert et al., 2000; Williams et al., 1995; Homanics et al., 1997
Reduced benzodiazepine binding				
–Hippocampus	+	+	+	Blatt et al., 2001; Holopainen et al., 2001; Sinkkonen et al., 2003
–Cerebellum	+	+	newborn	Pfund et al., 2001; Holopainen et al., 2001; Homanics et al., 1997
Sleep disturbances	+	+	+	Barthelemy et al., 1997; Williams et al., 1995; Wisor et al., 2002
Epilepsy				
–Age dependent evolution	+	+	+	Rossi et al., 1995; Matsumoto et al., 1992; DeLorey et al., 1998
–Myoclonic jerks	+	+	+	Matsumoto et al., 1992; Guerrini et al., 1996; DeLorey et al., 1998
–Multiple seizure types	+	+	+	Carod et al., 1995; Minassian et al., 1998; DeLorey et al., 1998
–Worsens with carbamazepine	one case	+	+	Monji et al., 2004; Guerrini et al., 2003; DeLorey et al., 1998
–Ethosuximide is beneficial	?	+	+	Sugiura et al., 2001; DeLorey et al., 1998
–Slow wave EEG abnormalities	+	+	+	Hughes and Melyn, 2005; Minassian et al., 1998; DeLorey et al., 1998
–Paroxysmal EEG abnormalities	+	+	+	Rossi et al., 1995; Kumada et al., 2005; DeLorey et al., 1998

+ effect reported, ? not tested.

The higher penetrance of cleft palate in p^{cp} mice, compared to gabrb3 null mice, is likely due to either a difference in genetic backgrounds of the two strains or to the additional loss of genes in the p^{cp} mouse. A significant proportion of the 43% of gabrb3 null mice born without an obvious cleft palate likewise die within the first day and are described as having neonatal feeding difficulties. Survivors are runted until weaning, achieving normal body size by adulthood, and are fertile but have somewhat reduced lifespans. Cleft palates also arise in mice deficient in the gad67 gene (encoding gamma-amino decarboxylase that synthesizes GABA) as well as mice prenatally exposed to drugs that alter GABA signaling (Condie et al., 1997; Jurand and Martin, 1994). Additional studies have likewise implicated GABAergic mechanisms in normal palate development (Culiat et al., 1995). The gross morphology of the brain is normal in gabrb3 null mice (Homanics et al., 1997) and p^{cp} mutant mice (Nakatsu et al., 1993). However, detailed assessments of brain morphology has not been performed in these mouse mutants.

Although cleft palates are rare in ASD and AS, craniofacial dysmorphic features such as protruding jaws, wide-spaced teeth, large mouths, and feeding difficulties in infancy are characteristic of AS (Williams et al., 1995) whereas, high-arched palates are reported in 40% of ASD cases associated with the inverted duplication of chromosome 15q11–13 (Mann et al., 2004; Wolpert et al., 2000).

IV. Neurochemistry

Whole brain homogenates from newborn gabrb3 null mice exhibited a 50% reduction in both [^{3}H]muscimol binding to the GABA binding site and [^{3}H]Ro15-4513 to the benzodiazepine binding site on the GABA$_A$ receptor, as compared to whole brain homogenates taken from wildtype littermates (Homanics et al., 1997). The same study also reported a 50% reduction in [^{3}H]Ro15-4513 binding in whole brain homogenates taken from adult gabrb3 knockout mice compared to adult wildtype mice. Brain slice autoradiography from 1 day old newborn mice provided a more insightful assessment of [^{3}H]Ro15-4513 binding in various brain regions including a significant reduction in binding to hippocampus (40%), thalamus (49%), frontoparietal cortex (30%), caudate/putamen (25%), and olfactory bulb (59%) (Homanics et al., 1997). In a separate study in which the above brain regions were assessed in adult gabrb3 knockout mice, and compared to wildtype littermates, a significant reduction in [^{3}H]Ro15-4513 binding was observed in hippocampus (45%), cerebral cortex (32%), and caudate/putamen (32%) with non-significant reductions in thalamus (28%) and the granule and molecular layers of the cerebellum (11% and 5%, respectively) (Sinkkonen et al., 2003). The consequence of this reduction in GABA$_A$ receptor binding density in regards to electrophysiological function and behavior will be addressed in the following sections.

Neurochemical assessments of ASD and AS brains, likewise, provide support for a role of GABA$_A$ receptors in these disorders. For example, single photon emission computer tomography revealed a 60% reduction in the binding of the benzodiazepine radioligand [^{123}I]iomazenil in the cerebellum of an AS patient, with a deletion of the maternal allele of chromosome 15q11–13, compared to three normal control subjects (Odano et al., 1996). Additionally, a Positron Emission Tomography (PET) study also detected significant bilateral reductions in binding of the benzodiazepine radioligand [^{11}C]flumazenil in cerebellum (34–43% reduction) and hippocampus (18–22% reduction) of three separate deletion AS cases as compared to a non-deletion AS case (UBE3A gene mutation) (Holopainen et al., 2001). In a PET study focusing on ASD, four out of the nine ASD children examined were found to exhibit a significant reduction in whole brain [^{11}C]flumazenil binding with focal decreases in the cerebellum as compared to eight age-matched children with temporal lobe epilepsy and seven normal adults (Pfund et al., 2001). Significant reductions in binding density of the benzodiazepine radioligand [^{3}H]flunitrazepam (15–35%) has also been observed in hippocampus from four ASD cases compared to three controls (Blatt et al., 2001). Another six receptor densities (5HT1A, 5HT2, M1, NMDA, kainate, choline uptake sites) were also assessed in these brains without significant differences being observed. Lastly, a significant reduction in the expression of the β_3 subunit protein was reported in postmortem cerebral cortex in five out of nine (56%) ASD cases, two out of three (66%) Rett syndrome cases (an ASD subtype), and two out of two (100%) AS cases relative to 11 control brains (Samaco et al., 2005). Taken together the previously mentioned studies provide compelling evidence for a reduction in the density of GABA$_A$ receptors in both ASD and AS brains.

V. Electrophysiology

A. DORSAL ROOT GANGLIA

Electrophysiological recordings from sensory neurons isolated from the dorsal root ganglia (DRG) of neonatal gabrb3 null mice revealed a dramatic decrease (~80%) in the maximal amplitude of GABA-activated chloride currents in a population of these neurons compared to those taken from wildtype siblings (Homanics et al., 1997). The same study also found sensory neurons taken from heterozygous mice lacking one allele of the gabrb3 gene, likewise exhibited a statistically significant reduction (25%) in GABA-activated current amplitude, relative to neurons taken from wildtype mice. These findings are in agreement with mRNA expression studies on DRG cells that suggest that mRNA for β_3 subunit is the predominant β subunit expressed (Ma *et al.*, 1993). Pharmacological studies in sensory neurons taken from both gabrb3 null mice and control littermates further indicate there is little or no compensation for the loss of the β_3 subunit by the other two β subunits (Krasowski *et al.*, 1998). GABA$_A$ receptors on the terminals of sensory afferents in the dorsal and ventral horn of the spinal cord are expected to provide presynaptic inhibition (Eccles *et al.*, 1961). Therefore, it has been suggested that some of the motor manifestations of the hyperexcitability/hyperresponsivity seen in the gabrb3 null mice may be due to the ineffectiveness of spinal presynaptic inhibition resulting from the decline in GABA$_A$ receptor expression on the terminals of primary afferents (Homanics *et al.*, 1997). As direct assessment of DRG electrophysiology in individuals with ASD or AS is not possible, this mouse model provides a means by which to study the consequences of impaired GABAergic inhibition in the DRG. If GABAergic inhibition were compromised in the DRG of ASD and/or AS individuals one would expect these individuals to be hyperresponsive to sensory input. Consequently, there is evidence that ASD individuals are hyperresponsive to touch (Ayres and Tickle, 1980).

B. THALAMUS

Electroencephalographic oscillations that occur during sleep (partial synchronous oscillations) and those that occur during absence epilepsy (hypersynchronous oscillations) result from synchronous activity in the cerebral cortex through interactions with inhibitory circuits arising in the reticular thalamic nucleus (RTN) and the thalamocortical relay nuclei (Huguenard and Prince, 1994; Steriade *et al.*, 1993; Warren *et al.*, 1994). Inhibitory postsynaptic currents (IPSCs) in RTN neurons are mediated by GABA$_A$ receptors that likely contain the β_3 subunit, as this is one of the limited GABA$_A$ receptor subunits found in this nucleus (Pirker *et al.*, 2000; Wisden *et al.*, 1992). By examining the inhibitory function of the RTN in thalamic slices taken from gabrb3 null mice, one can assess whether intra-RTN inhibition is suppressed, thereby promoting intrathalamic synchrony. GABA mediated inhibition and spontaneous IPSCs were observed to be nearly abolished in the RTN of gabrb3 null mice, resulting in a dramatic intensification of the oscillatory synchrony (Huntsman *et al.*, 1999). The same study found GABA mediated inhibition and spontaneous IPSCs to be unaffected in the thalamic relay neurons of the ventrobasal complex, which contains other β subunits. The previous findings demonstrate how the inactivation of inhibitory postsynaptic GABA$_A$ receptors in the RTN can alter oscillatory functions with the likely consequence of disrupting sleep architecture

(McCormick and Bal, 1997) and causing epilepsy (Huguenard and Prince, 1994); both features are observed in gabrb3 null mice (DeLorey *et al.*, 1998; Homanics *et al.*, 1997; Wisor *et al.*, 2002) as well as in ASD (Limoges *et al.*, 2005; Rossi *et al.*, 1995) and AS (Bruni *et al.*, 2004; Minassian *et al.*, 1998). It is noteworthy that the RTN is also involved in attentional processing (Guillery *et al.*, 1998), a feature often reported as being impaired in ASD (Allen and Courchesne, 2001).

C. Hippocampus

In situ hybridization studies demonstrate there to be a substantial amount of mRNA from the gabrb3 gene present in CA1 neurons of the hippocampus (Laurie *et al.*, 1992). Hippocampal neurons isolated from neonatal gabrb3 null mice and cultured for 4–5 days exhibited significant reductions (50%) in the maximal amplitude of GABA-evoked whole cell currents when compared to hippocampal neurons from wildtype littermates (Krasowski *et al.*, 1998). This is in agreement with the 40% reduction in [^{3}H]Ro15-4513 binding to hippocampus from neonatal gabrb3 null mice compared to wildtype controls, discussed earlier (Homanics *et al.*, 1997). GABA$_A$ receptor function is less impaired in the hippocampal neurons of the neonatal gabrb3 null mice, compared to RTN neurons. This is likely reflective of the larger proportion of GABA$_A$ receptors in the hippocampus that contain either the β_1 or β_2 subunits than is found in the RTN (Pirker *et al.*, 2000), making the hippocampus less dependent on the β_3 subunit. A reduction of GABAergic inhibition in the hippocampus would be expected to cause cognitive deficits and contribute to a heightened seizure susceptibility, both of which are observed in the gabrb3 null mice (DeLorey *et al.*, 1998; Homanics *et al.*, 1997) as well as in ASD (Rossi *et al.*, 1995; Vig and Jedrysek, 1999) and AS (Williams *et al.*, 1995).

D. Olfactory Bulb

The gabrb3 gene is abundantly expressed in the rodent olfactory bulb (OB) with the granule cells expressing only the β_3 variant of the β subunit and the principal cells (mitral and tufted cells) expressing all three β subunits (Laurie *et al.*, 1992; Nusser *et al.*, 1999). In the olfactory bulb, odors trigger synchronous oscillatory activity that is believed to arise from coherent and rhythmic discharges of large numbers of neurons (Lagier *et al.*, 2004). The gabrb3 null mice exhibited a >93% reduction in GABA-mediated synaptic inhibition of granule cells, the local inhibitory interneurons of the OB, as well as an augmentation of inhibitory postsynaptic currents in the principal cells. These two effects lead to an increase in network oscillations in the OB resulting in complex effects on olfactory learning, representation, and discrimination (Nusser *et al.*, 2001). Consequently, the same study found the gabrb3 null mice to be better at discriminating a particular odor from closely related odors but poorer at discriminating closely related mixtures of odors than their wildtype littermates. Sensory symptoms associated with ASD, including olfaction, have been less studied than those dealing with social and cognitive functioning. However, a study employing a parental questionnaire of sensory reactivity reported that ASD children (2–4 yrs old) exhibited significantly higher scores on taste/smell sensitivity than normal children of the same age (Rogers *et al.*, 2003). It is not clear how one would best go about correlating the previous findings in mice with those found in ASD, as rodents rely more on their sense of olfaction than do humans. A more in depth study of olfaction in ASD individuals would be required before one could assess whether gabrb3 null mice model any of the olfactory abnormalities observed in ASD. To date no study of olfaction in AS has been performed.

VI. Epilepsy

A. Seizures and EEG Abnormalities

Virtually all gabrb3 null mice experience some form of recurring spontaneous seizure starting at about 10 weeks of age and becoming more frequent as they age. Seizure severity ranged from twitching of mouth muscles, face, whiskers, and ears to the more robust seizures that involved head and bilateral forelimb jerks, arching of the back, straub tail, and the mouse falling on its side. The severest observed seizures involved strong clonic shaking that progressed into a wild running/bouncing phase (DeLorey *et al.*, 1998). Seizures were usually followed by a period of behavioral quiescence. Seizures were also noted in mice heterozygous for the gabrb3 gene, although with less frequency than observed in the homozygous gabrb3 null mice (DeLorey *et al.*, 1998; Homanics *et al.*, 1997). EEG measurements performed on gabrb3 null mice revealed an evolving electrocortical phenomena in which young (<10 weeks old) gabrb3 null mice display relatively normal EEG traces that became markedly abnormal as the mice age (DeLorey *et al.*, 1998). Behavioral observations, coupled with EEG recordings, indicate that these mice are subject to an evolving epileptogenic condition from disorganized electrocortical activity with high amplitude slow and sharp waves to interictal spiking culminating in spontaneous seizures as they mature (DeLorey *et al.*, 1998; Homanics *et al.*, 1997). These observations plausibly suggest, but do not prove, that seizure activity evolves as a self-propelled process engendered by the lack of gabrb3 gene expression at an earlier critical period of development. As discussed earlier, gabrb3 null mice exhibit a reduction in sIPSC in the RTN, which is associated with the generation of absence type seizures (Huguenard and Prince, 1994).

The incidence of epilepsy in ASD has been reported to be about 33% (Tuchman *et al.*, 1991), however, it is important to note that between 43–68% of ASD individuals exhibit epileptiform EEG activity in sleep without actually manifesting clinical seizures (Chez *et al.*, 2004). Interestingly, when one views separately ASD cases attributed to chromosome 15q11–13 duplications (isodicentric chr 15), the incidence of epilepsy rises to about 71% (Mann *et al.*, 2004; Wolpert *et al.*, 2000), consistent with the involvement of a GABAergic mechanism. ASD and AS patients are frequently described as having EEG paroxysmal abnormalities, multiple seizure types, and an age-dependent evolution of the epilepsy (Kumada *et al.*, 2005; Matsumoto *et al.*, 1992; Rossi *et al.*, 1995). The prevalence rate for seizures in AS is >80% (Minassian *et al.*, 1998; Moncla *et al.*, 1999) with AS patients possessing the large chromosome 15q11–13 region deletion typically exhibiting the more severe seizure phenotype, which is usually harder to treat than the seizures associated with AS patients that just have a point mutation in the UBE3A gene (Minassian *et al.*, 1998). Interestingly, only 29% of the mice with a homozygous disruption of the ube3a gene on the same mixed hybrid background as the gabrb3 null mice exhibit seizures and require audiogenic-induction (Jiang *et al.*, 1998). This is in contrast to the 100% of gabrb3 null mice that exhibit spontaneous seizures (DeLorey *et al.*, 1998). The previous observations in both AS patients and the mouse models suggest a contributing role for the gabrb3 gene in the more robust epileptic phenotype observed in AS deletion patients. Moreover, a revealing case involved a patient that had an unusual maternal 15q deletion that eliminated the GABA$_A$ receptor subunit gene cluster but not the UBE3A and ATP10C genes as usually occurs in AS. Although this individual exhibited moderate mental retardation (IQ 50) he did not meet the full AS diagnostic criteria. However, this patient exhibited a strikingly similar electrocortical disturbance to those seen during sleep in AS patients with large deletions (Michaelis *et al.*, 1995), again implicating the genes in the GABA$_A$ receptor gene cluster (GABRB3, GABRA5, and/or GABRG3) in the electrocortical disturbances that likely contribute to the seizures often associated with AS. The high-amplitude polymorphic slow wave activity seen in EEG of both AS children and gabrb3 null mice, instead of the normally expected alpha and beta frequencies, is reminiscent of the EEG patterns associated with disturbances of cortical-thalamic and cortical-cortical physiology.

B. ANTIEPILEPTIC DRUGS

A variety of antiepileptic drugs (AED) have been administered to gabrb3 null mice, with ethosuximide being the most potent in lessening seizures and normalizing EEG abnormalities (DeLorey *et al.*, 1998). Ethosuximide is generally used for the control of absence seizures and works by inhibiting T-type calcium channels involved in synchronization of thalamocortical circuitry (Huguenard, 1999). Carbamazepine was found to worsen seizures and EEG in gabrb3 null mice (DeLorey et al., 1998).

Anticonvulsants that have been most often reported as being beneficial in treating seizures in AS include valproic acid and benzodiazepines, with ethosuximide and topiramate also receiving favorable results (Franz et al., 2000; Laan et al., 1996; Ostergaard and Balslev, 2001; Sugiura et al., 2001). Carbamazepine, oxcarbazepine, and vigabatrin have been reported to exacerbate seizures in both frequency and severity in AS (Guerrini et al., 2003; Minassian et al., 1998; Ostergaard and Balslev, 2001). Interestingly, lamotrigine, which has no direct effect on $GABA_A$ receptors but has been shown to increase the expression of the GABRB3 gene, is also considered effective in the treatment of the epilepsy associated with AS (Gibbs et al., 2002; Ruggieri and McShane, 1998; Wang et al., 2002). Few controlled clinical trials of AEDs in the treatment of ASD associated epilepsy have been conducted, likely due to the high heterogeneity observed in this disorder. However, one study reported that carbamazepine triggered new-onset epileptic seizures in an individual with ASD (Monji et al., 2004). Interestingly, lamotrigine was found to improve autistic symptoms in 8/13 ASD children, separate of whether there was an improvement in their epilepsy or not (Uvebrant and Bauziene, 1994).

VII. Behavior

A. Social Behavior

To date there have been no reports of the social characteristics of the gabrb3 null mouse, other than the noted failure of dams to display appropriate nurturing behavior toward their offspring, irrespective of the genotype of the offspring (Homanics et al., 1997). Impaired social behavior is a core feature of ASD (Gillberg et al., 1996) and has also been reported in AS (Penner et al., 1993). Efforts to establish whether social parameters are altered in gabrb3 null mouse have become of paramount importance and are currently underway.

B. Stereotypical Behavior

Gabrb3 null mice are easily excited, which elicits hyperactive behavior in the gabrb3 null mice, during which time they often display repetitive peripheral circling or tight turning in place, as if they were chasing their own tails (Homanics *et al.*, 1997). This type of behavior has likewise been observed in mice with extensive Purkinje cell loss in the cerebellum (Fransen *et al.*, 1998). Interestingly, recent studies in rat have implicated GABAergic mechanisms in the substantia nigra pars reticulata working in relationship with the dopaminergic system in the substantia nigra pars compacta as contributing to circling behavior (Velisek *et al.*, 2005).

A diagnostic criteria of both ASD and AS is stereotypical behavior, in which affected individuals exhibit motor responses that are repetitive, invariant, and seemingly without purpose or goal (Barthelemy *et al.*, 1997; Summers *et al.*, 1995). Interestingly, a circling (spinning) behavior is often reported in individuals with ASD (Bracha *et al.*, 1995).

C. MOTOR COORDINATION AND LOCOMOTOR ACTIVITY

The cerebellum is one of the first brain structures to begin to differentiate, yet it is one of the last to achieve maturity; its cellular organization continues to change well after birth. During this transitional period a strong correlation exists between the high expression of β_3 subunit message and the formation of multiple connections between the inferior olive climbing fibers (axons) and Purkinje cells (Frostholm *et al.*, 1992). This protracted developmental process creates a special susceptibility to disruptions during development; such disruptions would likely reveal themselves through changes in motor coordination and locomotor abilities. When held by the tail gabrb3 null mice tend to ball-up by clasping their paws together rather than splaying their paws outwards as is observed in wildtype mice and the progenitor strains used in creating the gabrb3 null mice; this trait is often associated with neurological impairment (Homanics *et al.*, 1997). In addition, gabrb3 null mice have difficulty swimming, walking on wire grid floors, repeatedly fall off platforms, and perform poorly on the accelerating rota-rod task (DeLorey *et al.*, 1998; Homanics *et al.*, 1997). Similarly, homozygous p^{cp} mice, which lack the GABA$_A$ receptor subunit genes, gabra5 and gabrg3 in addition to the gabrb3 gene, are described as being ataxic (Nakatsu *et al.*, 1993). Gabrb3 null mice are also hyperactive as compared to their wildtype littermates in measures of cage crossings and velocity and are easily discerned from their wildtype littermates by casual observation (DeLorey *et al.*, 1998). This study also found gabrb3 null mice to be strikingly different from wildtype mice in both duration of their active period and overall total activity when monitored over a 3-day period (DeLorey *et al.*, 1998).

AS patients exhibit strong motor disturbances including ataxia and poor motor control (Williams *et al.*, 1995). ASD individuals likewise exhibit deficits in complex or skilled motor movements and are often described as being clumsy (Minshew *et al.*, 1997) with some individuals being described as having "a bizarre gait" (Barthelemy *et al.*, 1997). In addition, there are reports that the achievement of motor milestones in the developing autistic child is delayed (Teitelbaum *et al.*, 1998). Lastly, a behavioral hallmark often reported in both ASD and AS is hyperactivity as well as easy excitability (Barthelemy *et al.*, 1997; Summers *et al.*, 1995; Williams *et al.*, 1995).

D. HYPERRESPONSIVITY

One of the earliest traits noted in gabrb3 null mice was that they were hyperresponsive to being handled or exposed to other sensory stimuli, which usually culminated in the expression of hyperactive behavior and stereotypical circling mentioned earlier (Homanics *et al.*, 1997). Gabrb3 null mice were found to display enhanced responsiveness, compared to wildtype mice, to low-intensity thermal stimuli in the tail-flick and hot-plate tests (Ugarte *et al.*, 2000). In addition, the same study reported that the gabrb3 null mice exhibited enhanced responsiveness to innocuous tactile stimuli compared to wildtype mice as assessed with von Frey filaments. They suggested that the occurrence of such post-stimulus behaviors as vocalization, fending behavior, and licking or shaking of hindpaws suggests that the decrease in mechanical thresholds is a sensory effect rather than a secondary facilitated motor response. The presence of thermal hyperalgesia and tactile allodynia in the gabrb3 null mice is consistent with a loss of inhibition of somatosensory transmission mediated by presynaptic and post-synaptic $GABA_A$ receptors in the spinal cord where the majority of $GABA_A$ receptors express mRNA for the β_3 subunit (Ma *et al.*, 1993; Persohn *et al.*, 1991; Zhang *et al.*, 1991). A wealth of evidence indicates that GABA and $GABA_A$ receptors in the spinal cord, medulla, and pons play important roles in the modulations of nociception (Hammond, 1997). The previous results are also consistent with electrophysiological investigations in which $GABA_A$ receptor antagonists were applied to the spinal cord resulting in an enhanced response to light touch, further supporting a role for GABA and $GABA_A$ receptors in somatosensation (Sivilotti and Woolf, 1994). The previous findings likewise implicate $GABA_A$ receptors in the tonic inhibition of low threshold afferent inputs to the spinal cord and also suggest high threshold thermoreceptive inputs to the spinal cord are tonically inhibited to a lesser extent.

A sensory profile of 26 children with ASD (age range 2–3.5 yrs) and 24 normal children (age range 1–3 yrs), found the ASD children to exhibit significantly heightened tactile sensitivity compared to the normal children (Rogers *et al.*, 2003). Other studies likewise reported ASD children to exhibit hyperresponsivity to touch (Ayres and Tickle, 1980) and tactile defensiveness (Baranek *et al.*, 1997). In addition, in a study of 25 ASD individuals, 44% were reported to exhibit hypo/hypersensitivity to temperature (Harrison and Hare, 2004). Neither tactile defensiveness or heat sensitivity has been investigated in AS, however, parents often anecdotally report their AS child as exhibiting tactile defensiveness and/or hypersensitivity to heat.

E. SLEEP ARCHITECTURE

Assessment of sleep states and sleep electroencephalography in gabrb3 null mice revealed little difference from wildtype mice in non-rapid eye movement sleep (NREMS) time but found substantial differences between the two genotypes in regards to the EEG spectral characteristics during NREM sleep (Wisor *et al.*, 2002). In addition, gabrb3 null mice exhibited significantly less REM sleep time compared to wildtype mice measured during the light portion of a 24-hr light-dark cycle (Wisor *et al.*, 2002). This finding is in line with studies supporting a critical role for GABAergic transmission in the regulation of REM sleep (Rye, 1997), thereby implicating the β_3 subunit of the $GABA_A$ receptor in the regulation of the cortical expression of sleep states. The observed differences seen in the gabrb3 null mice are likely the result of the disrupted RTN allowing hypersynchronous activity to occur in the thalamic circuitry, discussed earlier, which regulates cortical expression of NREMS. Significant differences in EEG spectral power during the transition from wake-NREMS were also noted between the two genotypes. In addition, the transient increase in EEG power in the 12–16 Hz range that occurs in wildtype mice during the transition from NREM to REM sleep was significantly blunted in the gabrb3 null mice. The ramifications of this finding are unclear, however sleep spindles, transient 11–16 Hz oscillations occurring in the cortical EEG and are prominent during the transition from NREMS to REMS, are generated by the RTN (Destexhe *et al.*, 1998; McCormick and Bal, 1997) and are impaired in the gabrb3 null mice as discussed earlier. Interestingly, circadian rhythms monitored over a nine-day period in the mutant mice appear to remain intact. The previous findings clearly indicate that disrupting the gabrb3 gene in these mice leads to significant changes in normal sleep architecture.

Sleep disturbances and atypical sleep architecture are frequently reported in both ASD (Limoges *et al.*, 2005) and AS (Bruni *et al.*, 2004; Clayton-Smith, 1993; Miano *et al.*, 2004). Likewise, AS patients of 10 years of age or younger exhibit similar sleep disturbances to those characterized in ASD including problems with initiating sleep with long latencies to fall asleep, excessive nocturnal awakenings, reduced hours of sleep per night, poor sleep quality, unusual and jerky movements during sleep, and daytime sleepiness (Bruni *et al.*, 2004). A reduction in REM sleep time has also been reported in AS individuals compared to controls (Miano *et al.*, 2004). ASD individuals (age range: 16–27 years) also exhibit longer latencies to fall asleep, frequent nocturnal awakenings, lower sleep efficiency and quality, more daytime sleepiness, decreased non-REM sleep, less sleep spindles during stage 2 sleep, and a lower number of rapid eye movements during REM sleep (Limoges *et al.*, 2005).

F. LEARNING DEFICITS

Gabrb3 null mice display a deficiency in Pavlovian fear conditioned contextual memory compared to wildtype littermates (DeLorey *et al.*, 1998). This type of conditioning is a rapidly acquired form of learning thought to be a model of human explicit memory that is dependent on the induction of long-term potentiation in the hippocampus (Kim and Fanselow, 1992) and the functional integrity of the cerebellar vermis (Sacchetti *et al.*, 2002). Both the hippocampus and the cerebellum have been implicated in ASD (Bauman and Kemper, 1985) and contain abundant amounts of the β_3 subunit during development and in adulthood. Gabrb3 null mice also exhibited defective operant learning as measured in the passive avoidance task (DeLorey *et al.*, 1998), which has likewise been suggested to involve the hippocampus (Lorenzini *et al.*, 1996).

A prominent feature of ASD and AS is an association with mental retardation (Edwards and Bristol, 1991; Williams *et al.*, 1995) with reports of autistic individuals exhibiting impairments in spatial working memory (Minshew *et al.*, 1999).

VIII. Concluding Remarks

The growing evidence implicating GABAergic mechanisms in the etiology of ASD and AS, mirrored by the numerous neurochemical, electrophysiological, and behavioral similarities shared between these disorders and the gabrb3 null mice (Table I), lend support for the involvement of the GABRB3 gene in the etiology of these disorders. Moreover, because ASD is generally believed to be caused by multiple mechanisms, it may be worthwhile to also consider that the co-inheritance of a combination of hypomorphic alleles, each affecting overall GABAergic tone, could likewise be the basis of some forms of ASD. As GABAergic neurons are known to be essential to information processing in almost every brain region, one would expect that a reduction in GABAergic function, of any nature, would result in these regions becoming hyper-excitable, consequently impairing neural information processing, a feature considered to be central to ASD (Belmonte *et al.*, 2004). Gabrb3 gene deficient mice, as well as other animal models of GABAergic dysfunction, remain valuable tools in the critical examination of the interconnection between development, pathology, and behavior as they relate to disorders of neurodevelopment.

References

Alcantara, S., Ruiz, M., D'Arcangelo, G., Ezan, F., de Lecea, L., Curran, T., Sotelo, C., and Soriano, E. (1998). Regional and cellular patterns of reelin mRNA expression in the forebrain of the developing and adult mouse. *J. Neurosci.* **18,** 7779–7799.

Allen, G., and Courchesne, E. (2001). Attention function and dysfunction in autism. *Front Biosci.* **6,** D105–D119.

Anderson, S. A., Qiu, M., Bulfone, A., Eisenstat, D. D., Meneses, J., Pedersen, R., and Rubenstein, J. L. (1997). Mutations of the homeobox genes Dlx-1 and Dlx-2 disrupt the striatal subventricular zone and differentiation of late born striatal neurons. *Neuron* **19,** 27–37.

Ayres, A. J., and Tickle, L. S. (1980). Hyper-responsivity to touch and vestibular stimuli as a predictor of positive response to sensory integration procedures by autistic children. *Am. J. Occup. Ther.* **34,** 375–381.

Ballaban-Gil, K., and Tuchman, R. (2000). Epilepsy and epileptiform EEG: Association with autism and language disorders. *Ment. Retard. Dev. Disabil. Res. Rev.* **6,** 300–308.

Baranek, G. T., Foster, L. G., and Berkson, G. (1997). Tactile defensiveness and stereotyped behaviors. *Am. J. Occup. Ther.* **51,** 91–95.

Barthelemy, C., Roux, S., Adrien, J. L., Hameury, L., Guerin, P., Garreau, B., Fermanian, J., and Lelord, G. (1997). Validation of the revised behavior summarized evaluation scale. *J. Autism Dev. Disord.* **27,** 139–153.

Bass, M. P., Menold, M. M., Wolpert, C. M., Donnelly, S. L., Ravan, S. A., Hauser, E. R., Maddox, L. O., Vance, J. M., Abramson, R. K., Wright, H. H., Gilbert, J. R., Cuccaro, M. L., DeLong, G. R., and Pericak-Vance, M. A. (2000). Genetic studies in autistic disorder and chromosome 15. *Neurogenetics* **2,** 219–226.

Bauman, M., and Kemper, T. L. (1985). Histoanatomic observations of the brain in early infantile autism. *Neurology* **35,** 866–874.

Belmonte, M. K., Cook, E. H., Jr., Anderson, G. M., Rubenstein, J. L., Greenough, W. T., Beckel-Mitchener, A., Courchesne, E., Boulanger, L. M., Powell, S. B., Levitt, P. R., Perry, E. K., Jiang, Y. H., DeLorey, T. M., and Tierney, E. (2004). Autism as a disorder of neural information processing: Directions for research and targets for therapy(1). *Mol. Psychiatry* **9,** 646–663.

Blatt, G. J., Fitzgerald, C. M., Guptill, J. T., Booker, A. B., Kemper, T. L., and Bauman, M. L. (2001). Density and distribution of hippocampal neurotransmitter receptors in autism: An autoradiographic study. *J. Autism Dev. Disord.* **31,** 537–543.

Bracha, H. S., Livingston, R., Dykman, K., Edwards, D. R., and Adam, B. (1995). An automated electronic method for quantifying spinning (circling) in children with autistic disorder. *J. Neuropsychiatry Clin. Neurosci.* **7,** 213–217.

Bruni, O., Ferri, R., D'Agostino, G., Miano, S., Roccella, M., and Elia, M. (2004). Sleep disturbances in Angelman syndrome: A questionnaire study. *Brain Dev.* **26,** 233–240.

Burd, L., Stenehjem, A., Franceschini, L. A., and Kerbeshian, J. (2000). A 15-year follow-up of a boy with pyridoxine (vitamin B6)-dependent seizures with autism, breath holding, and severe mental retardation. *J. Child Neurol.* **15,** 763–765.

Buxbaum, J. D., Silverman, J. M., Smith, C. J., Greenberg, D. A., Kilifarski, M., Reichert, J., Cook, E. H., Jr., Fang, Y., Song, C. Y., and Vitale, R. (2002). Association between a GABRB3 polymorphism and autism. *Mol. Psychiatry* **7,** 311–316.

Carod, F. J., Prats, J. M., Garaizar, C., and Zuazo, E. (1995). [Clinical-radiological evaluation of infantile autism and epileptic syndromes associated with autism]. *Rev. Neurol.* **23,** 1203–1207.

Cassidy, S. B., and Morris, C. A. (2002). Behavioral phenotypes in genetic syndromes: Genetic clues to human behavior. *Adv. Pediatr.* **49,** 59–86.

Chez, M. G., Buchanan, T., Aimonovitch, M., Mrazek, S., Krasne, V., Langburt, W., and Memon, S. (2004). Frequency of EEG abnormalities in age-matched siblings of autistic children with abnormal sleep EEG patterns. *Epilepsy Behav.* **5,** 159–162.

Clayton-Smith, J. (1993). Clinical research on Angelman syndrome in the United Kingdom: Observations on 82 affected individuals. *Am. J. Med. Genet.* **46,** 12–15.

Condie, B. G., Bain, G., Gottlieb, D. I., and Capecchi, M. R. (1997). Cleft palate in mice with a targeted mutation in the gamma-aminobutyric acid-producing enzyme glutamic acid decarboxylase 67. *Proc. Natl. Acad. Sci. USA* **94,** 11451–11455.

Cook, E. H., Jr., Courchesne, R. Y., Cox, N. J., Lord, C., Gonen, D., Guter, S. J., Lincoln, A., Nix, K., Haas, R., Leventhal, B. L., and Courchesne, E. (1998). Linkage-disequilibrium mapping of autistic disorder, with 15q11–13 markers. *Am. J. Hum. Genet.* **62,** 1077–1083.

Cox, R. D., and Schopler, E. (1993). Aggression and self-injurious behaviors in persons with autism–the TEACCH (Treatment and Education of Autistic and related Communications Handicapped Children) approach. *Acta Paedopsychiatr.* **56,** 85–90.

Crawley, J. N. (2004). Designing mouse behavioral tasks relevant to autistic-like behaviors. *Ment. Retard. Dev. Disabil. Res. Rev.* **10,** 248–258.

Crestani, F., Keist, R., Fritschy, J. M., Benke, D., Vogt, K., Prut, L., Bluthmann, H., Mohler, H., and Rudolph, U. (2002). Trace fear conditioning involves hippocampal alpha5 GABA(A) receptors. *Proc. Natl. Acad. Sci. USA* **99,** 8980–8985.

Culiat, C. T., Stubbs, L. J., Woychik, R. P., Russell, L. B., Johnson, D. K., and Rinchik, E. M. (1995). Deficiency of the beta 3 subunit of the type A gamma-aminobutyric acid receptor causes cleft palate in mice. *Nat. Genet.* **11,** 344–346.

De Felipe, J., Marco, P., Fairen, A., and Jones, E. G. (1997). Inhibitory synaptogenesis in mouse somatosensory cortex. *Cereb. Cortex* **7,** 619–634.

DeLorey, T. M., Handforth, A., Anagnostaras, S. G., Homanics, G. E., Minassian, B. A., Asatourian, A., Fanselow, M. S., Delgado-Escueta, A., Ellison, G. D., and Olsen, R. W. (1998). Mice lacking the beta3 subunit of the GABAA receptor have the epilepsy phenotype and many of the behavioral characteristics of Angelman syndrome. *J. Neurosci.* **18,** 8505–8514.

Destexhe, A., Contreras, D., and Steriade, M. (1998). Mechanisms underlying the synchronizing action of corticothalamic feedback through inhibition of thalamic relay cells. *J. Neurophysiol.* **79,** 999–1016.

Eccles, J. C., Hubbard, J. I., and Oscarsson, O. (1961). Intracellular recording from cells of the ventral spinocerebellar tract. *J. Physiol. (Paris)* **158,** 486–516.

Edwards, D. R., and Bristol, M. M. (1991). Autism: Early identification and management in family practice. *Am. Fam. Physician* **44,** 1755–1764.

Fombonne, E. (2003). The prevalence of autism. *Jama* **289,** 87–89.

Fransen, E., D'Hooge, R., Van Camp, G., Verhoye, M., Sijbers, J., Reyniers, E., Soriano, P., Kamiguchi, H., Willemsen, R., Koekkoek, S. K., De Zeeuw, C. I., De Deyn, P. P., Van der Linden, A., Lemmon, V., Kooy, R. F., and Willems, P. J. (1998). L1 knockout mice show dilated ventricles, vermis hypoplasia and impaired exploration patterns. *Hum. Mol. Genet.* **7,** 999–1009.

Franz, D. N., Glauser, T. A., Tudor, C., and Williams, S. (2000). Topiramate therapy of epilepsy associated with Angelman's syndrome. *Neurology* **54,** 1185–1188.

Frostholm, A., Zdilar, D., Luntz-Leybman, V., Janapati, V., and Rotter, A. (1992). Ontogeny of GABAA/benzodiazepine receptor subunit mRNAs in the murine inferior olive: Transient appearance of beta 3 subunit mRNA and [3H]muscimol binding sites. *Brain Res. Mol. Brain Res.* **16,** 246–254.

Ghaziuddin, M., and Butler, E. (1998). Clumsiness in autism and Asperger syndrome: A further report. *J. Intellect. Disabil. Res.* **42**(Pt 1), 43–48.

Gibbs, J. W., 3rd., Zhang, Y. F., Ahmed, H. S., and Coulter, D. A. (2002). Anticonvulsant actions of lamotrigine on spontaneous thalamocortical rhythms. *Epilepsia* **43,** 342–349.

Gillberg, C., Nordin, V., and Ehlers, S. (1996). Early detection of autism. Diagnostic instruments for clinicians. *Eur. Child Adolesc. Psychiatry* **5,** 67–74.

Gillott, A., Furniss, F., and Walter, A. (2001). Anxiety in high-functioning children with autism. *Autism* **5,** 277–286.

Guerrini, R., Carrozzo, R., Rinaldi, R., and Bonanni, P. (2003). Angelman syndrome: Etiology, clinical features, diagnosis, and management of symptoms. *Paediatr. Drugs* **5,** 647–661.

Guerrini, R., De Lorey, T. M., Bonanni, P., Moncla, A., Dravet, C., Suisse, G., Livet, M. O., Bureau, M., Malzac, P., Genton, P., Thomas, P., Sartucci, F., Simi, P., and Serratosa, J. M. (1996). Cortical myoclonus in Angelman syndrome. *Ann. Neurol.* **40,** 39–48.

Guillery, R. W., Feig, S. L., and Lozsadi, D. A. (1998). Paying attention to the thalamic reticular nucleus. *Trends Neurosci.* **21,** 28–32.

Hammond, D. L. (1997). Inhibitory neurotransmitters and nociception:role of GABA and glycine. *In* "The Pharmacology of Pain" (A. Dickenson and J.-M. Besson, Eds.), pp. 361–384. Springer, Berlin.

Harrison, J., and Hare, D. J. (2004). Brief report: Assessment of sensory abnormalities in people with autistic spectrum disorders. *J. Autism Dev. Disord.* **34,** 727–730.

Harvey, M. T., and Kennedy, C. H. (2002). Polysomnographic phenotypes in developmental disabilities. *Int. J. Dev. Neurosci.* **20,** 443–448.

Holopainen, I. E., Metsahonkala, E. L., Kokkonen, H., Parkkola, R. K., Manner, T. E., Nagren, K., and Korpi, E. R. (2001). Decreased binding of [11C]flumazenil in Angelman syndrome patients with GABA(A) receptor beta3 subunit deletions. *Ann. Neurol.* **49,** 110–113.

Homanics, G. E., DeLorey, T. M., Firestone, L. L., Quinlan, J. J., Handforth, A., Harrison, N. L., Krasowski, M. D., Rick, C. E., Korpi, E. R., Makela, R., Brilliant, M. H., Hagiwara, N., Ferguson, C., Snyder, K., and Olsen, R. W. (1997). Mice devoid of gamma-aminobutyrate type A receptor beta3 subunit have epilepsy, cleft palate, and hypersensitive behavior. *Proc. Natl. Acad. Sci. USA* **94,** 4143–4148.

Hornig, M., and Lipkin, W. I. (2001). Infectious and immune factors in the pathogenesis of neurodevelopmental disorders: Epidemiology, hypotheses, and animal models. *Ment. Retard. Dev. Disabil. Res. Rev.* **7,** 200–210.

Hughes, J. R., and Melyn, M. (2005). EEG and seizures in autistic children and adolescents: Further findings with therapeutic implications. *Clin. EEG Neurosci.* **36,** 15–20.

Huguenard, J. R. (1999). Neuronal circuitry of thalamocortical epilepsy and mechanisms of antiabsence drug action. *Adv. Neurol.* **79,** 991–999.

Huguenard, J. R., and Prince, D. A. (1994). Intrathalamic rhythmicity studied *in vitro*: Nominal T-current modulation causes robust antioscillatory effects. *J. Neurosci.* **14,** 5485–5502.

Huntsman, M. M., Porcello, D. M., Homanics, G. E., DeLorey, T. M., and Huguenard, J. R. (1999). Reciprocal inhibitory connections and network synchrony in the mammalian thalamus. *Science* **283,** 541–543.

Jiang, Y. H., Armstrong, D., Albrecht, U., Atkins, C. M., Noebels, J. L., Eichele, G., Sweatt, J. D., and Beaudet, A. L. (1998). Mutation of the Angelman ubiquitin ligase in mice causes increased cytoplasmic p53 and deficits of contextual learning and long-term potentiation. *Neuron* **21,** 799–811.

Jurand, A., and Martin, L. V. (1994). Cleft palate and open eyelids inducing activity of lorazepam and the effect of flumazenil, the benzodiazepine antagonist. *Pharmacol. Toxicol.* **74,** 228–235.

Kanner, L. (1943). Autistic disturbances of affective contact. *Nervous Child* **2,** 217–250.

Kim, J. J., and Fanselow, M. S. (1992). Modality-specific retrograde amnesia of fear. *Science* **256,** 675–677.

Kishino, T., Lalande, M., and Wagstaff, J. (1997). UBE3A/E6-AP mutations cause Angelman syndrome. *Nat. Genet.* **15,** 70–73.

Krasowski, M. D., Rick, C. E., Harrison, N. L., Firestone, L. L., and Homanics, G. E. (1998). A deficit of functional GABA(A) receptors in neurons of beta 3 subunit knockout mice. *Neurosci. Lett.* **240,** 81–84.

Kumada, T., Ito, M., Miyajima, T., Fujii, T., Okuno, T., Go, T., Hattori, H., Yoshioka, M., Kobayashi, K., Kanazawa, O., Tohyama, J., Akasaka, N., Kamimura, T., Sasagawa, M., Amagane, H., Mutoh, K., Yamori, Y., Kanda, T., Yoshida, N., Hirota, H., Tanaka, R., and Hamada, Y. (2005).

Multi-institutional study on the correlation between chromosomal abnormalities and epilepsy. *Brain Dev.* **27,** 127–134.

Laan, L. A., den Boer, A. T., Hennekam, R. C., Renier, W. O., and Brouwer, O. F. (1996). Angelman syndrome in adulthood. *Am. J. Med. Genet.* **66,** 356–360.

Lagier, S., Carleton, A., and Lledo, P. M. (2004). Interplay between local GABAergic interneurons and relay neurons generates gamma oscillations in the rat olfactory bulb. *J. Neurosci.* **24,** 4382–4392.

Laurie, D. J., Seeburg, P. H., and Wisden, W. (1992). The distribution of 13 GABAA receptor subunit mRNAs in the rat brain. II. Olfactory bulb and cerebellum. *J. Neurosci.* **12,** 1063–1076.

Limoges, E., Mottron, L., Bolduc, C., Berthiaume, C., and Godbout, R. (2005). Atypical sleep architecture and the autism phenotype. *Brain* **128**(Pt 5), 1049–1061.

London, E. A. (2000). The environment as an etiologic factor in autism: A new direction for research. *Environ. Health Perspect.* **108**(Suppl. 3), 401–404.

Lorenzini, C. A., Baldi, E., Bucherelli, C., Sacchetti, B., and Tassoni, G. (1996). Role of dorsal hippocampus in acquisition, consolidation and retrieval of rat's passive avoidance response: A tetrodotoxin functional inactivation study. *Brain Res.* **730,** 32–39.

Lyon, M. F., King, T. R., Gondo, Y., Gardner, J. M., Nakatsu, Y., Eicher, E. M., and Brilliant, M. H. (1992). Genetic and molecular analysis of recessive alleles at the pink-eyed dilution (p) locus of the mouse. *Proc. Natl. Acad. Sci. USA* **89,** 6968–6972.

Ma, W., Saunders, P. A., Somogyi, R., Poulter, M. O., and Barker, J. L. (1993). Ontogeny of GABAA receptor subunit mRNAs in rat spinal cord and dorsal root ganglia. *J. Comp. Neurol.* **338,** 337–359.

Mann, S. M., Wang, N. J., Liu, D. H., Wang, L., Schultz, R. A., Dorrani, N., Sigman, M., and Schanen, N. C. (2004). Supernumerary tricentric derivative chromosome 15 in two boys with intractable epilepsy: Another mechanism for partial hexasomy. *Hum. Genet.* **115,** 104–111.

Marin, O., and Rubenstein, J. L. (2001). A long, remarkable journey: Tangential migration in the telencephalon. *Nat. Rev. Neurosci.* **2,** 780–790.

Martin, E. R., Menold, M. M., Wolpert, C. M., Bass, M. P., Donnelly, S. L., Ravan, S. A., Zimmerman, A., Gilbert, J. R., Vance, J. M., Maddox, L. O., Wright, H. H., Abramson, R. K., DeLong, G. R., Cuccaro, M. L., and Pericak-Vance, M. A. (2000). Analysis of linkage disequilibrium in gamma-aminobutyric acid receptor subunit genes in autistic disorder. *Am. J. Med. Genet.* **96,** 43–48.

Martinsson, T., Johannesson, T., Vujic, M., Sjostedt, A., Steffenburg, S., Gillberg, C., and Wahlstrom, J. (1996). Maternal origin of inv dup(15) chromosomes in infantile autism. *Eur. Child Adolesc. Psychiatry* **5,** 185–192.

Matsumoto, A., Kumagai, T., Miura, K., Miyazaki, S., Hayakawa, C., and Yamanaka, T. (1992). Epilepsy in Angelman syndrome associated with chromosome 15q deletion. *Epilepsia* **33,** 1083–1090.

Matsuura, T., Sutcliffe, J. S., Fang, P., Galjaard, R. J., Jiang, Y. H., Benton, C. S., Rommens, J. M., and Beaudet, A. L. (1997). *De novo* truncating mutations in E6-AP ubiquitin-protein ligase gene (UBE3A) in Angelman syndrome. *Nat. Genet.* **15,** 74–77.

McCormick, D. A., and Bal, T. (1997). Sleep and arousal: Thalamocortical mechanisms. *Annu. Rev. Neurosci.* **20,** 185–215.

Miano, S., Bruni, O., Leuzzi, V., Elia, M., Verrillo, E., and Ferri, R. (2004). Sleep polygraphy in Angelman syndrome. *Clin. Neurophysiol.* **115,** 938–945.

Michaelis, R. C., Skinner, S. A., Lethco, B. A., Simensen, R. J., Donlon, T. A., Tarleton, J., and Phelan, M. C. (1995). Deletion involving D15S113 in a mother and son without Angelman syndrome: Refinement of the Angelman syndrome critical deletion region. *Am. J. Med. Genet.* **55,** 120–126.

Minassian, B. A., DeLorey, T. M., Olsen, R. W., Philippart, M., Bronstein, Y., Zhang, Q., Guerrini, R., Van Ness, P., Livet, M. O., and Delgado-Escueta, A. V. (1998). Angelman syndrome: Correlations between epilepsy phenotypes and genotypes. *Ann. Neurol.* **43,** 485–493.

Minshew, N. J., Goldstein, G., and Siegel, D. J. (1997). Neuropsychologic functioning in autism: Profile of a complex information processing disorder. *J. Int. Neuropsychol. Soc.* **3,** 303–316.

Minshew, N. J., Luna, B., and Sweeney, J. A. (1999). Oculomotor evidence for neocortical systems but not cerebellar dysfunction in autism. *Neurology* **52,** 917–922.

Moncla, A., Malzac, P., Voelckel, M. A., Auquier, P., Girardot, L., Mattei, M. G., Philip, N., Mattei, J. F., Lalande, M., and Livet, M. O. (1999). Phenotype-genotype correlation in 20 deletion and 20 non-deletion Angelman syndrome patients. *Eur. J. Hum. Genet.* **7,** 131–139.

Monji, A., Maekawa, T., Yanagimoto, K., Yoshida, I., and Hashioka, S. (2004). Carbamazepine may trigger new-onset epileptic seizures in an individual with autism spectrum disorders: A case report. *Eur. Psychiatry* **19,** 322–323.

Nakatsu, Y., Tyndale, R. F., DeLorey, T. M., Durham-Pierre, D., Gardner, J. M., McDaniel, H. J., Nguyen, Q., Wagstaff, J., Lalande, M., Sikela, J. M., Olsen, R. W., Allan, J., Tobin, A. J., and Brilliant, M. H. (1993). A cluster of three GABAA receptor subunit genes is deleted in a neurological mutant of the mouse p locus. *Nature* **364,** 448–450.

Nurmi, E. L., Bradford, Y., Chen, Y., Hall, J., Arnone, B., Gardiner, M. B., Hutcheson, H. B., Gilbert, J. R., Pericak-Vance, M. A., Copeland-Yates, S. A., Michaelis, R. C., Wassink, T. H., Santangelo, S. L., Sheffield, V. C., Piven, J., Folstein, S. E., Haines, J. L., and Sutcliffe, J. S. (2001). Linkage disequilibrium at the Angelman syndrome gene UBE3A in autism families. *Genomics* **77,** 105–113.

Nurmi, E. L., Dowd, M., Tadevosyan-Leyfer, O., Haines, J. L., Folstein, S. E., and Sutcliffe, J. S. J. (2003). Exploratory subsetting of autism families based on savant skills improves evidence of genetic linkage to 15q11-q13. *J. Am. Acad. Child Adolesc. Psychiatry* **42,** 856–863.

Nusser, Z., Kay, L. M., Laurent, G., Homanics, G. E., and Mody, I. (2001). Disruption of GABA(A) receptors on GABAergic interneurons leads to increased oscillatory power in the olfactory bulb network. *J. Neurophysiol.* **86,** 2823–2833.

Nusser, Z., Sieghart, W., and Mody, I. (1999). Differential regulation of synaptic GABAA receptors by cAMP-dependent protein kinase in mouse cerebellar and olfactory bulb neurones. *J. Physiol.* **521**(Pt 2), 421–435.

Odano, I., Anezaki, T., Ohkubo, M., Yonekura, Y., Onishi, Y., Inuzuka, T., Takahashi, M., and Tsuji, S. (1996). Decrease in benzodiazepine receptor binding in a patient with Angelman syndrome detected by iodine-123 iomazenil and single-photon emission tomography. *Eur. J. Nucl. Med.* **23,** 598–604.

Olsen, R. W., and DeLorey, T. M. (1999). GABA and glycine. *In* "Basic Neurochemistry" (G. J. Siegel, B. W. Agranoff, R. W. Albers, S. K. Fisher, and M. D. Uhler, Eds.), pp. 335–346. Lippincott-Raven, New York.

Ostergaard, J. R., and Balslev, T. (2001). Efficacy of different antiepileptic drugs in children with Angelman syndrome associated with 15q11-13 deletion: The Danish experience. *Dev. Med. Child Neurol.* **43,** 718–719.

Owens, D. F., Boyce, L. H., Davis, M. B., and Kriegstein, A. R. (1996). Excitatory GABA responses in embryonic and neonatal cortical slices demonstrated by gramicidin perforated-patch recordings and calcium imaging. *J. Neurosci.* **16,** 6414–6423.

Penner, K. A., Johnston, J., Faircloth, B. H., Irish, P., and Williams, C. A. (1993). Communication, cognition, and social interaction in the Angelman syndrome. *Am. J. Med. Genet.* **46,** 34–39.

Persohn, E., Malherbe, P., and Richards, J. G. (1991). *In situ* hybridization histochemistry reveals a diversity of GABAA receptor subunit mRNAs in neurons of the rat spinal cord and dorsal root ganglia. *Neuroscience* **42,** 497–507.

Petersen, M. B., Brondum-Nielsen, K., Hansen, L. K., and Wulff, K. (1995). Clinical, cytogenetic, and molecular diagnosis of Angelman syndrome: Estimated prevalence rate in a Danish county. *Am. J. Med. Genet.* **60,** 261–262.

Pfund, Z., Chugani, D., Behen, M., Juhasz, C., Muzik, O., Lee, J., and Chugani, H. (2001). Abnormalities of GABAA receptors measured with [C-11]flumazenil PET in autisitic children, vol. 31, p. 879.3. Society for Neuroscience, San Diego.

Pickles, A., Bolton, P., Macdonald, H., Bailey, A., Le Couteur, A., Sim, C. H., and Rutter, M. (1995). Latent-class analysis of recurrence risks for complex phenotypes with selection and measurement error: A twin and family history study of autism. *Am. J. Hum. Genet.* **57,** 717–726.

Pirker, S., Schwarzer, C., Wieselthaler, A., Sieghart, W., and Sperk, G. (2000). GABA(A) receptors: Immunocytochemical distribution of 13 subunits in the adult rat brain. *Neuroscience* **101,** 815–850.

Rineer, S., Finucane, B., and Simon, E. W. (1998). Autistic symptoms among children and young adults with isodicentric chromosome 15. *Am. J. Med. Genet.* **81,** 428–433.

Risch, N., Spiker, D., Lotspeich, L., Nouri, N., Hinds, D., Hallmayer, J., Kalaydjieva, L., McCague, P., Dimiceli, S., Pitts, T., Nguyen, L., Yang, J., Harper, C., Thorpe, D., Vermeer, S., Young, H., Hebert, J., Lin, A., Ferguson, J., Chiotti, C., Wiese-Slater, S., Rogers, T., Salmon, B., Nicholas, P., Petersen, P. B., Pingree, C., McMahon, W., Wong, D. L., Cavalli-Sforza, L. L., Kraemer, H. C., and Myers, R. M. (1999). A genomic screen of autism: Evidence for a multilocus etiology. *Am. J. Hum. Genet.* **65,** 493–507.

Rogers, S. J., Hepburn, S., and Wehner, E. (2003). Parent reports of sensory symptoms in toddlers with autism and those with other developmental disorders. *J. Autism Dev. Disord.* **33,** 631–642.

Rossi, P. G., Parmeggiani, A., Bach, V., Santucci, M., and Visconti, P. (1995). EEG features and epilepsy in patients with autism. *Brain Dev.* **17,** 169–174.

Ruggieri, M., and McShane, M. A. (1998). Parental view of epilepsy in Angelman syndrome: A questionnaire study. *Arch. Dis. Child* **79,** 423–426.

Rutter, M., and Schopler, E. (1987). Autism and pervasive developmental disorders: Concepts and diagnostic issues. *J. Autism Dev. Disord.* **17,** 159–186.

Rye, D. B. (1997). Contributions of the pedunculopontine region to normal and altered REM sleep. *Sleep* **20,** 757–788.

Sacchetti, B., Baldi, E., Lorenzini, C. A., and Bucherelli, C. (2002). Cerebellar role in fear-conditioning consolidation. *Proc. Natl. Acad. Sci. USA* **99,** 8406–8411.

Samaco, R. C., Hogart, A., and La Salle, J. M. (2005). Epigenetic overlap in autism-spectrum neurodevelopmental disorders: MECP2 deficiency causes reduced expression of UBE3A and GABRB3. *Hum. Mol. Genet.* **14,** 483–492.

Schroer, R. J., Phelan, M. C., Michaelis, R. C., Crawford, E. C., Skinner, S. A., Cuccaro, M., Simensen, R. J., Bishop, J., Skinner, C., Fender, D., and Stevenson, R. E. (1998). Autism and maternally derived aberrations of chromosome 15q. *Am. J. Med. Genet.* **76,** 327–336.

Shao, Y., Cuccaro, M. L., Hauser, E. R., Raiford, K. L., Menold, M. M., Wolpert, C. M., Ravan, S. A., Elston, L., Decena, K., Donnelly, S. L., Abramson, R. K., Wright, H. H., DeLong, G. R., Gilbert, J. R., and Pericak-Vance, M. A. (2003). Fine mapping of autistic disorder to chromosome 15q11–q13 by use of phenotypic subtypes. *Am. J. Hum. Genet.* **72,** 539–548.

Simon, J., Wakimoto, H., Fujita, N., Lalande, M., and Barnard, E. A. (2004). Analysis of the set of GABA(A) receptor genes in the human genome. *J. Biol. Chem.* **279,** 41422–41435.

Sinkkonen, S. T., Homanics, G. E., and Korpi, E. R. (2003). Mouse models of Angelman syndrome, a neurodevelopmental disorder, display different brain regional GABA(A) receptor alterations. *Neurosci. Lett.* **340,** 205–208.

Sivilotti, L., and Woolf, C. J. (1994). The contribution of GABAA and glycine receptors to central sensitization: Disinhibition and touch-evoked allodynia in the spinal cord. *J. Neurophysiol.* **72,** 169–179.

Steffenburg, S., Gillberg, C. L., Steffenburg, U., and Kyllerman, M. (1996). Autism in Angelman syndrome: A population-based study. *Pediatr. Neurol.* **14,** 131–136.

Steriade, M., McCormick, D. A., and Sejnowski, T. J. (1993). Thalamocortical oscillations in the sleeping and aroused brain. *Science* **262,** 679–685.

Sugiura, C., Ogura, K., Ueno, M., Toyoshima, M., and Oka, A. (2001). High-dose ethosuximide for epilepsy in Angelman syndrome: Implication of GABA(A) receptor subunit. *Neurology* **57,** 1518–1519.

Summers, J. A., Allison, D. B., Lynch, P. S., and Sandler, L. (1995). Behaviour problems in Angelman syndrome. *J. Intellect. Disabil. Res.* **39**(Pt 2), 97–106.

Teitelbaum, P., Teitelbaum, O., Nye, J., Fryman, J., and Maurer, R. G. (1998). Movement analysis in infancy may be useful for early diagnosis of autism. *Proc. Natl. Acad. Sci. USA* **95,** 13982–13987.

Tuchman, R. F., Rapin, I., and Shinnar, S. (1991). Autistic and dysphasic children. II: Epilepsy. *Pediatrics* **88,** 1219–1225.

Ugarte, S. D., Homanics, G. E., Firestone, L. L., and Hammond, D. L. (2000). Sensory thresholds and the antinociceptive effects of GABA receptor agonists in mice lacking the beta3 subunit of the GABA(A) receptor. *Neuroscience* **95,** 795–806.

Uvebrant, P., and Bauziene, R. (1994). Intractable epilepsy in children. The efficacy of lamotrigine treatment, including non-seizure-related benefits. *Neuropediatrics* **25,** 284–289.

Veenstra-VanderWeele, J., Gonen, D., Leventhal, B. L., and Cook, E. H., Jr. (1999). Mutation screening of the UBE3A/E6-AP gene in autistic disorder. *Mol. Psychiatry* **4,** 64–67.

Velisek, L., Veliskova, J., Ravizza, T., Giorgi, F. S., and Moshe, S. L. (2005). Circling behavior and [14C]2-deoxyglucose mapping in rats: Possible implications for autistic repetitive behaviors. *Neurobiol. Dis.* **18,** 346–355.

Viani, F., Romeo, A., Viri, M., Mastrangelo, M., Lalatta, F., Selicorni, A., Gobbi, G., Lanzi, G., Bettio, D., Briscioli, V., Di Segni, M., Parini, R., and Terzoli, G. (1995). Seizure and EEG patterns in Angelman's syndrome. *J. Child Neurol.* **10,** 467–471.

Vig, S., and Jedrysek, E. (1999). Autistic features in young children with significant cognitive impairment: Autism or mental retardation? *J. Autism Dev. Disord.* **29,** 235–248.

Wang, J. F., Sun, X., Chen, B., and Young, L. T. (2002). Lamotrigine increases gene expression of GABA-A receptor beta3 subunit in primary cultured rat hippocampus cells. *Neuropsychopharmacology* **26,** 415–421.

Warren, R. A., Agmon, A., and Jones, E. G. (1994). Oscillatory synaptic interactions between ventroposterior and reticular neurons in mouse thalamus *in vitro. J. Neurophysiol.* **72,** 1993–2003.

Williams, C. A., Angelman, H., Clayton-Smith, J., Driscoll, D. J., Hendrickson, J. E., Knoll, J. H., Magenis, R. E., Schinzel, A., Wagstaff, J., Whidden, E. M., and Zori, R. T. (1995). Angelman syndrome: Consensus for diagnostic criteria. Angelman Syndrome Foundation. *Am. J. Med. Genet.* **56,** 237–238.

Wisden, W., Laurie, D. J., Monyer, H., and Seeburg, P. H. (1992). The distribution of 13 GABAA receptor subunit mRNAs in the rat brain. I. Telencephalon, diencephalon, mesencephalon. *J. Neurosci.* **12,** 1040–1062.

Wisor, J. P., DeLorey, T. M., Homanics, G. E., and Edgar, D. M. (2002). Sleep states and sleep electroencephalographic spectral power in mice lacking the beta 3 subunit of the GABA(A) receptor. *Brain Res.* **955,** 221–228.

Wolpert, C., Pericak-Vance, M. A., Abramson, R. K., Wright, H. H., and Cuccaro, M. L. (2000). Autistic symptoms among children and young adults with isodicentric chromosome 15. *Am. J. Med. Genet.* **96,** 128–129.

Zhang, J. H., Sato, M., and Tohyama, M. (1991). Region-specific expression of the mRNAs encoding beta subunits (beta 1, beta 2, and beta 3) of GABAA receptor in the rat brain. *J. Comp. Neurol.* **303,** 637–657.

THE REELER MOUSE: ANATOMY OF A MUTANT

Gabriella D'Arcangelo

The Cain Foundation Laboratories, Texas Children's Hospital, Departments of Pediatrics,
Neuroscience, Programs in Developmental Biology, Translational Biology, and
Molecular Medicine, Baylor College of Medicine, Houston, Texas 77030, USA

The neurological mouse mutant *reeler* is characterized by ataxia and disruption of cellular layers in the brain. This mutant has long been studied as a model to understand how cortical structures of the vertebrate brain are formed during development. The *reeler* phenotype arises from homozygous loss-of-function mutations in *reelin*, a gene that encodes an extracellular glycoprotein secreted by distinct neuronal populations during embryonic and postnatal brain development. Reelin is a key regulator of cortical development that appears to function as a switch, causing neurons to terminate their migration phase and to begin the assembly of cortical layers. The Reelin signal is interpreted by neurons and neuronal progenitors through signal transducing molecules, such as the VLDLR and the ApoER2 receptors, and the Dab1 adapter protein. Loss of these essential transducers in mutant mice results in the appearance of a phenotype indistinguishable from *reeler*. Much has been learned during the past few years about the molecular mechanisms that mediate the Reelin signal. This large protein is

383

thought to cluster the VLDLR and ApoER2 receptors, thereby activating src-family kinases that phosphorylate Dab1 on tyrosine residues. This event in turn causes Dab1 to interact with a variety of signal transduction molecules and proteins that regulate cytoskeleton dynamics, before being degraded by the proteosome pathway. Expression of Reelin and its transduction machinery continues long after migration is complete and may affect neuronal maturation and synaptic connectivity in the postnatal brain. Despite the tremendous progress made in this past decade, the whole spectrum of Reelin activities in brain development and its relevance to human cognitive disorders has yet to be fully unraveled.

I. Introduction

Reeler is a well-characterized autosomal mutation in the mouse that affects several aspects of brain development. The first known *reeler* mutation occurred spontaneously in 1948, in an inbred mouse colony in Edinburgh, Scotland, and was thus referred to as rl^{Ed}. Mutant mice, which appeared in this strain at the frequency expected for a recessive trait according to classical Mendelian inheritance, were severely ataxic and exhibited a characteristic reeling gait that conferred them their name. Anatomical studies indicated that all major cortical structures of the brain were present, but appeared disorganized in *reeler* mutants. Strikingly, projection neurons were shown to connect properly to their ectopic targets. For example, thalamocortical projections terminated properly on neurons destined for cortical layer IV, even though these occupied abnormal positions in the *reeler* cortex (Molnar *et al.*, 1998; Steindler and Colwell, 1976). Climbing fibers also terminate properly on ectopically positioned Purkinje cells in the *reeler* cerebellum, although subtle defects in the number and position of their synaptic boutons are present (Mariani, 1982). The observation that all major neuronal types are born at the correct time in the *reeler* brain, but end up in the wrong location led to the realization that the study of the *reeler* mouse could provide exquisite insights into the mechanisms of neuronal migration and cortical layer formation (reviewed by Caviness and Rakic, 1978; D'Arcangelo and Curran, 1998; Goffinet, 1984b; Lambert de Rouvroit and Goffinet, 1998). Indeed, an entirely novel signaling pathway was discovered following the identification in 1995 of the *reeler* gene, which was called *reelin* (D'Arcangelo *et al.*, 1995). This gene is mutated not only in the original rl^{Ed} strain, now designated Reln^{rl} and commercially available from The Jackson Laboratory, but also in several other *reeler* mouse strains. All of the known *reeler* strains lack functional Reelin protein and are therefore null (Andersen *et al.*, 2002; D'Arcangelo and Curran, 1998). Most strains lack *reelin* mRNA expression altogether, due to large deletions of the gene or exon skipping that renders the transcript highly unstable. The Orleans strain (rl^{Orl}) is unique in that a small 3' deletion in the coding sequence, due to a retroviral insertion, gives origin to a Reelin protein that lacks the C terminus and cannot be secreted (D'Arcangelo *et al.*, 1997; de Bergeyck *et al.*, 1997). The phenotype among all *reeler* strains is essentially the same. Recently, two *reelin*-deficient rat strains have been described (Kikkawa *et al.*, 2003; Yokoi *et al.*, 2003). The Shaking Rat Kawasaki (SRK) and the Komeda Zucker Creeping (KZC) rats display all of the typical anatomical features of the mouse *reeler* mutation. These animals may be of particular use for future behavioral and pharmacological studies, for which the rat is the preferred species.

This chapter will describe the most salient anatomical aspects of the *reeler* mutation, I will then recapitulate the key molecular discoveries and finally attempt to integrate these findings into a framework that explains how Reelin may regulate brain development and function.

II. Neuroanatomy of *Reeler* and *Reeler*-Like Mutants

A. NEOCORTICAL DEVELOPMENT IN *REELER*

The development of the neocortex in *reeler* has been previously reviewed in detail (Caviness and Rakic, 1978; Caviness *et al.*, 1988; Goffinet, 1979, 1984b). However, because some general concepts in neurogenesis and migration have evolved over the past few years I will briefly revisit the issue here (see Fig. 1). In the early embryonic brain, the cell bodies of neuroepithelial cells (radial glia) reside near the ventricle and extend cellular processes towards both the ventricular and the pial surface. These cells were for a long time considered postmitotic glia whose function was solely to provide support for neuronal migration, but it is now clear that radial cells are dividing progenitor cells that produce both neurons and glia, depending on the developmental stage of the cortex (Malatesta *et al.*, 2000, 2003; Noctor *et al.*, 2001, 2004). The first postmitotic neurons in the cortex, the pioneer neurons, appear superficially to the germinal layer and form a transient structure named the preplate (Marin-Padilla, 1978; Meyer *et al.*, 1998). This structure contains two transient neuronal populations, the Cajal-Retzius cells (located superficially) and the subplate cells (located deeply). Cajal-Retzius cells express high levels of Reelin throughout corticogenesis and are thus essential for normal brain development. It is now believed that these early neurons originate from extracortical sites and migrate into the neocortex by tangential migration (Takiguchi-Hayashi *et al.*, 2004). Subplate cells, on the other hand, are important for targeting of thalamocortical afferents into the early neocortex (Shatz *et al.*, 1990). Cortical neurons destined to become excitatory

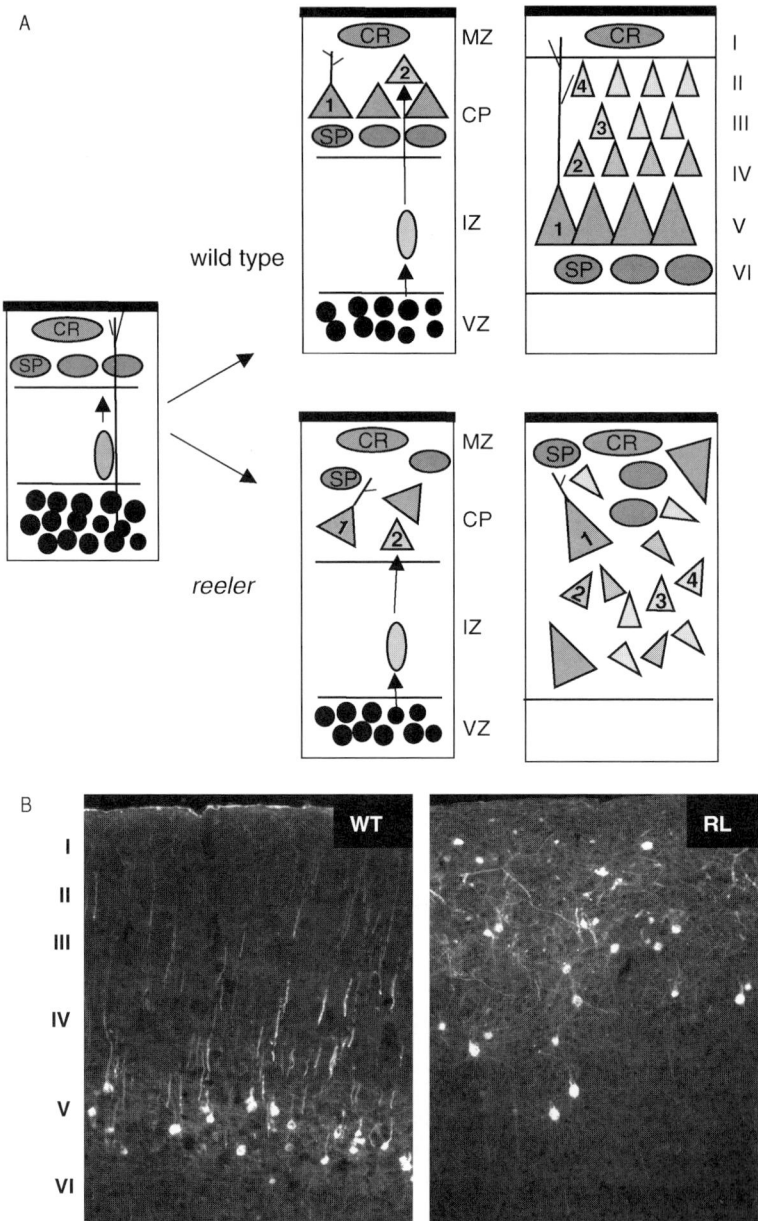

FIG. 1. A. Diagram of neocortical development in wild type and *reeler* mice. Principal neurons are born in the ventricular zone (VZ) and migrate radially toward the marginal zone (MZ). In wild type cortex, the first-born principal neurons (labeled 1) split the preplate composed of Reelin-expressing Cajal-Retzius (CR) and subplate cells (SP). Each wave of later-born neurons (labeled sequentially with

(principal neurons) are derived from radial progenitors in the germinal layer of the cortex. These neurons migrate radially toward the pial surface along the process of the mother radial cell, and split the preplate by positioning themselves between Cajal-Retzius and subplate cells (Marin-Padilla, 1978). Birthdating studies demonstrated that newborn principal neurons bypass their predecessors and terminate just beneath Reelin-rich Cajal-Retzius cells according to an inside-out patterning (Angevine and Sidman, 1961). Inhibitory cortical neurons (inter-neurons), on the other hand, are generated in subpallial areas and migrate into the developing neocortex by tangential migration, independently of radial guidance (reviewed in Marin and Rubenstein, 2001). Once they enter the neo-cortex, interneurons also move radially to terminate in specific layers, depending on their subtype, following cues provided by principal neurons or other laminar determinants.

From an examination of the anatomy of the *reeler* cortex, it is apparent that this mutation specifically alters the position of all neurons in the radial dimension of the cortex (Fig. 1). The preplate forms normally in *reeler*, but principal neurons fail to split it and instead accumulate underneath subplate cells. Each subsequent neuronal cohort fails to penetrate the preplate and to bypass older neurons. Instead of forming cellular layers, neurons either spread throughout the mutant cortex or become ectopically positioned in the wrong layer, giving rise to approx-imately inverted layers. These layers are very abnormal because they contain neurons that not only are born at a time that is inappropriate for the position they occupy, but also because they are grossly disoriented. *Reeler* neurons fail to acquire or maintain a proper radial orientation, and also maintain an abnormally tight association with the radial process of germinal cells. This observation led to the 'obstructed migration' hypothesis, which postulates that earlier principal neurons in *reeler* get stuck on radial fibers and thus prevent older neurons from moving forward (Pinto-Lord *et al.*, 1982). Interneurons also appear disoriented and ectopic in the *reeler* cortex, although it is not clear whether this is a secondary consequence of the principal neuron ectopia (Hevner *et al.*, 2004; Yabut *et al.*, in press). Another aspect of the *reeler* cortical defect is that the radial fiber scaffold is abnormal. Unlike the normal cortex where radial fibers terminate and branch in the Reelin–rich marginal zone, in *reeler* the radial fibers are slightly oblique, with end feet that often terminate below the marginal zone (Derer, 1979; Hartfuss *et al.*, 2003). The radial progenitors also disappear prematurely in *reeler*, grating neurons of their guidance (Derer, 1979; Hunter-Schaedle, 1997; Hartfuss

---

arabic numbers) bypasses their predecessors and forms cortical layers (labeled with roman numbers). In *reeler*, principal neurons do not split the preplate and fail to form layers. CP = cortical plate. IZ = intermediate zone. B. Fluorescence images of layer V neurons labeled with YFP in the adult wild type (WT) and *reeler* (RL) neocortex. (See Color Insert.)

et al., 2003). Whether the *reeler* defects in corticogenesis are due to abnormalities intrinsic to neurons and/or radial cells is still debated. *In vitro* studies indicate that Reelin directly affects the development of both neurons and radial cells and that both cell types express Reelin signaling proteins (Hartfuss *et al.*, 2003; Niu *et al.*, 2004). The contribution of each cell type to the *reeler* defect may vary in different regions of the brain and is not completely elucidated (see later discussion).

B. Hippocampal Development in *Reeler*

In the hippocampal formation the development of the hippocampus proper and that of the dentate gyrus proceed according to their own distinct morphogenetic patterns (see Fig. 2). In the hippocampus proper, principal neurons destined

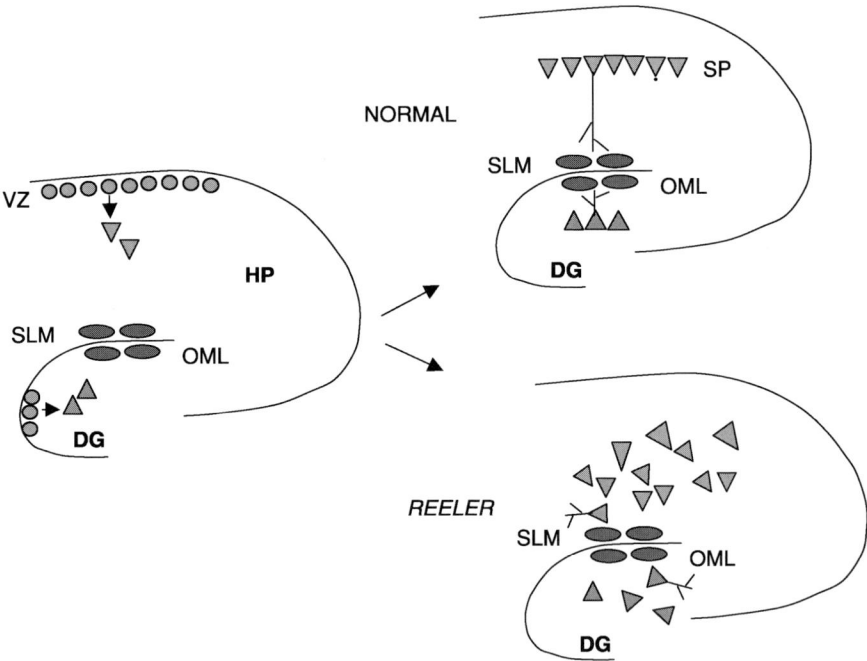

Fig. 2. Diagram of hippocampal development in normal and *reeler* mice. In the hippocampus proper (HP) pyramidal neurons are born in the ventricular zone (VZ) and migrate radially towards the stratum lacunosum moleculare (SLM), which contains Reelin-expressing Cajal-Retzius cells (large ovals). In normal mice they form a distinct pyramidal layer and project apical dendrites to the SLM. In the dentate gyrus (DG), granule neurons migrate inwardly from the outer molecular layer (OML) to form a compact layer. The OML also contains Reelin-expressing Cajal-Retzius cells. In *reeler* both pyramidal and granule cell layers fail to form. (See Color Insert.)

for the pyramidal layer migrate radially from the ventricular zone toward the stratum lacunosum moleculare (near the pial surface) where Reelin-expressing Cajal-Retzius cells reside, similarly to principal neurons in the neocortex. Unlike their neocortical counterparts, however, hippocampal neurons terminate their migration well before they reach the pial surface and accumulate according to an inside-out pattern in the pyramidal layer. In *reeler* these neurons fail to assemble in a tight layer and become disoriented and distributed throughout the hippocampus proper (Fig. 2).

In the dentate gyrus, granule neurons born mostly postnatally in a super-ficial germinal zone migrate inwardly to form a compact layer underneath Reelin-expressing Cajal-Retzius cells in the outer molecular layer. In *reeler* these neurons fail to form a distinct cellular layer and are grossly disoriented, with their dendrites often appearing oblique to the radial scaffold. In addition to malposition, a dramatic reduction in the complexity and the length of dendritic processes of hippocampal neurons has been described both *in vivo* and *in vitro* (Niu *et al.*, 2004; Stanfield and Cowan, 1979). Disorganization of the radial fiber scaffold is also prominently observed (Forster *et al.*, 2002; Weiss *et al.*, 2003; Zhao *et al.*, 2004), unlike the neocortex where this structure is only modestly affected.

An additional, unusual aspect of the *reeler* hippocampal phenotype is the delayed development of the entorhinohippocampal projections (Del Rio *et al.*, 1997). These axons terminate and branch profusely in the Reelin-rich stratum lacunosum moleculare and outer molecular layer of normal mice. In *reeler*, the axons' terminals appear less developed, at least during the first postnatal weeks. As a consequence of this delayed maturation, the formation of synaptic contacts between entorhinohippocampal neurons and their targets is impaired (Borrell *et al.*, 1999). Thus, in the hippocampus, Reelin is required not only for cell body positioning and dendrite elongation, but also for axonal branching and synaptogenesis.

C. Cerebellar Development in *Reeler*

The cerebellum is the brain structure most devastated by the *reeler* mutation and likely responsible for the ataxic gait of the mutant (see Fig. 3). The most obviously disrupted regions of the cerebellum are the hemispheres, which nor-mally consist of several folia, but in *reeler* they are severely reduced. This is a result of a remarkable reduction in cell number, which affects not only the most abundant cerebellar cell type, the granule cells, but also the Purkinje cells, although to a lesser extent (Heckroth *et al.*, 1989). This massive growth deficit is due to a cell layering defect that can be first observed during early embryogenesis (Goffinet, 1983a). In normal mice, Purkinje cells, which are the principal neurons

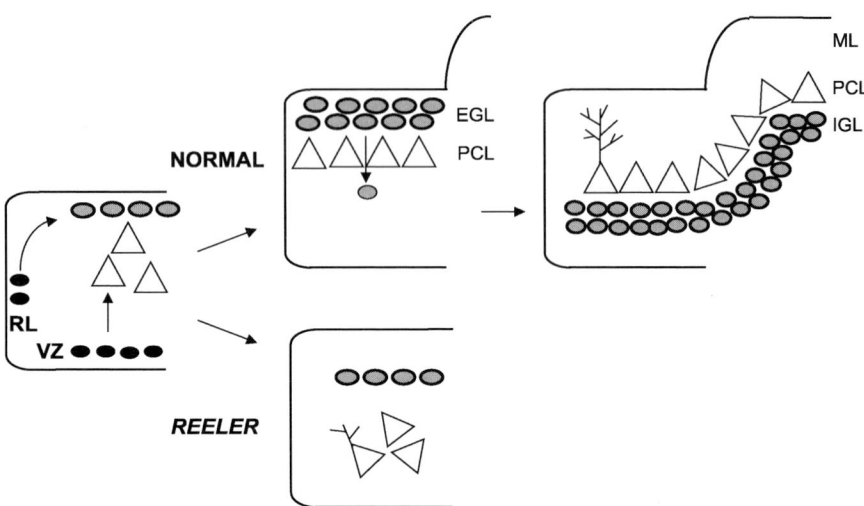

Fig. 3. Diagram of cerebellar development in normal and *reeler* mice. In the cerebellum Reelin-expressing granule cells are born in the rhombic lip (RL) and migrate tangentially to form the external granular layer (EGL). Purkinje cells (triangles) are born in ventricular zone (VZ) and migrate radially toward the EGL. In normal mice Purkinje cells form a layer (PCL). Granule cells proliferate and migrate across the PCL to form the internal granule layer (IGL). The cerebellum contains many folia. In *reeler*, Purkinje cells fail to form a layer and granule cells do not proliferate extensively, leading to reduced foliation. (See Color Insert.)

of the cerebellum, are born in the ventricular layer and migrate radially toward the pia. Along this path, they encounter Reelin-expressing cells derived from the rhombic lip, including cells destined for the granule layer and deep cerebellar nuclei. The granule cells then form a displaced germinal layer, the external granule layer (EGL), superficially to the Purkinje cells. This juxtaposition of Purkinje cells to Reelin-expressing cells during embryogenesis results in the formation of a compact Purkinje cell layer (PCL) underneath the EGL. The PCL is initially multicellular, but is later transformed into a single cell layer by the expansions of the cerebellar cortex during early postnatal ages. In *reeler* the PCL does not form and Purkinje cells remain deep in an amorphic central mass (Goffinet *et al.*, 1984; Mariani *et al.*, 1977; Mikoshiba *et al.*, 1980) (Fig. 3). Their orientation is very abnormal and the development of their dendrite trees is severely impaired, as for principal forebrain neurons. It has been proposed that the layering defect in the cerebellum, similarly to that in the neocortex, is due to obstructed migration of Purkinje cells along radial fibers (Yuasa *et al.*, 1993). An alternate possibility is that Purkinje cells fail to achieve proper radial orientation or to express adhesion molecules required for the formation of a compact cell layer. After PCL formation, the normal development of the cerebellum continues into the first 1–2 postnatal weeks with the inward migration of granule cells along glial fibers and across the PCL to form an internal granule layer (IGL). This process is obviously disrupted in *reeler*, given the absence of a PCL, and mutant granule cells remain loosely arranged in a cellular layer mostly superficial to ectopic Purkinje cells. Granule cell ectopia is likely to be secondary to the absence of a PCL and not due to a migration defect *per se*. *In vitro* studies have in fact shown that the capacity of *reeler* granule cells to migrate along glial fibers is not impaired (Nagata and Terashima, 1994). The dramatic reduction in granule cell number in *reeler* is thought to arise from the scarcity of Purkinje cell-derived growth factors in the superficial aspects of the cerebellar cortex. However, it is not clear that this is the sole cause of the abnormality, nor is it known what causes the more modest but significant reduction in Purkinje cell number. Thus, other possibilities such as a novel function of the *reelin* gene on cell proliferation or determination cannot presently be discounted.

D. Brain Stem and Spinal Cord Development in *Reeler*

In addition to the cerebellum, some laminated hindbrain structures also appear disrupted in *reeler*. In the brain stem, the facial nerve nucleus (Goffinet, 1984a; Terashima *et al.*, 1993), the mesencephalic trigeminal nucleus (Terashima, 1996), the cochlear nucleus (Martin, 1981), dopaminergic neurons of the substantia nigra (Nishikawa *et al.*, 2003), the inferior olives (Goffinet, 1983b), the nucleus ambiguous (Fujimoto *et al.*, 1998), the olivocochlear efferent neurons, and the facial visceral motor nucleus (Rossel *et al.*, 2005) all appear disorganized. In some cases the phenotype could be interpreted as a failure of recognition and adhesion among similar subtypes of neurons. In the case of the olivocochlear efferent neurons and facial visceral motor neurons, it appears that these neurons specifically fail to complete a second phase of radial-guided migration toward the Reelin-expressing region (Rossel *et al.*, 2005). Similarly, in the *reeler* spinal cord, specific defects occurring in a later migration step cause the malpositioning of autonomic neurons, and sympathetic as well as parasympathetic preganglionic neurons occupy ectopic positions (Kubasak *et al.*, 2004; Phelps *et al.*, 2002; Yip *et al.*, 2000, 2003). Sympathetic preganglionic neurons (SPNs) undergo a normal initial migration toward the ventrolateral aspect of the developing spinal cord, but fail to complete a second dorsolateral migration toward the intermediolateral column, and instead slip back toward the central canal. Since Reelin is expressed by V1 and V2 interneurons located just medially to the SPNs in the ventral spinal cord, it has been suggested that Reelin may act as a physical barrier, keeping the SPNs in place and preventing them from moving back toward the central canal (Yip *et al.*, 2004b). The recent observation that Reelin is actually transported from the cell bodies of interneurons along cellular process and released at the ventrolateral surface of the spinal cord suggests that, like in cortical structures of the brain, Reelin promotes an alignment of target neurons just beneath the pial surface (Kubasak *et al.*, 2004).

E. RETINAL DEVELOPMENT IN *REELER*

In the brain structures previously described, cells expressing Reelin during development appear to be strategically positioned to delimit radial migration and to provide a cue for proper orientation. In the retina, however, a different situation occurs. Ganglion cells express Reelin at the superficial edge of the retina during normal development, but there is no obvious disruption of cellular layer formation in the retina of *reeler* mice (Rice *et al.*, 1998). All retinal cell types in *reeler* are positioned in the appropriate layer and present a normal orientation. However, the stratification of AII amacrine cell synapses controlling the ON and OFF response to light stimuli is altered in the mutant. These cells normally establish synaptic contacts with rod and cone bipolar cells in two sublaminae of the inner plexiform layer, which correspond to ON and OFF responses. In the *reeler* retina, AII amacrine cells establish connectivity with their bipolar cell targets in ectopic positions outside the sublaminae of the inner plexiform layer, leading to the death of many rod bipolar cells and the disruption of the circuitry of the retina (Rice *et al.*, 2001). The resulting physiological defect in the *reeler* retina is the disruption of the ON-OFF organization and the attenuation of rod-driven retinal responses. These findings underscore the importance of Reelin not only in the formation of cellular layers, but also in the development of synaptic layers which in turn impacts the functional circuitry of the brain.

III. Reelin

A. THE REELIN PROTEIN

The gene responsible for the *reeler* phenotype was cloned a decade ago by taking advantage of a transgenic allele generated by insertional mutagenesis (rl^tg) (Miao *et al.*, 1994). The newly named *reelin* cDNA sequence predicted the synthesis of a large, secreted protein of 385 kDa (D'Arcangelo *et al.*, 1995). The encoded Reelin protein consists of an N terminal region containing a canonical signal peptide and a sequence of modest similarity to F-Spondin, followed by a main region containing eight consecutive repeats unique to Reelin, each featuring a cysteine pattern (EGF-like repeat) typically found in extracellular proteins, and finally a small C terminal region containing highly charged amino acids (see Fig. 4). Putative glycosylation sites were also noted throughout the sequence. Expression studies confirmed that the Reelin protein produced by cells transfected with the cloned cDNA or by wild type neuronal cultures indeed is secreted in the culture medium, and is subjected primarily to N-glycosylation (D'Arcangelo *et al.*, 1997). The Reelin protein was also immunoprecipitated by the monoclonal antibody CR-50, which selectively recognizes an epitope in the N terminal region of Reelin (D'Arcangelo *et al.*, 1997). This antibody was developed concurrently with the cloning of *reelin*, based on its ability to recognize an antigen absent in *reeler*, and to interfere with the aggregation of wild type cortical neurons *in vitro* (Ogawa *et al.*, 1995). The CR-50 epitope corresponds to a native conformation of the N terminus of Reelin and appears to mediate the formation of homodimers, which may be required for optimal activity during corticogenesis (Kubo *et al.*, 2002). Consistent with this hypothesis, the CR-50 antibody was also shown to interfere with Reelin-induced layer formation in cerebellar explants *in vitro*

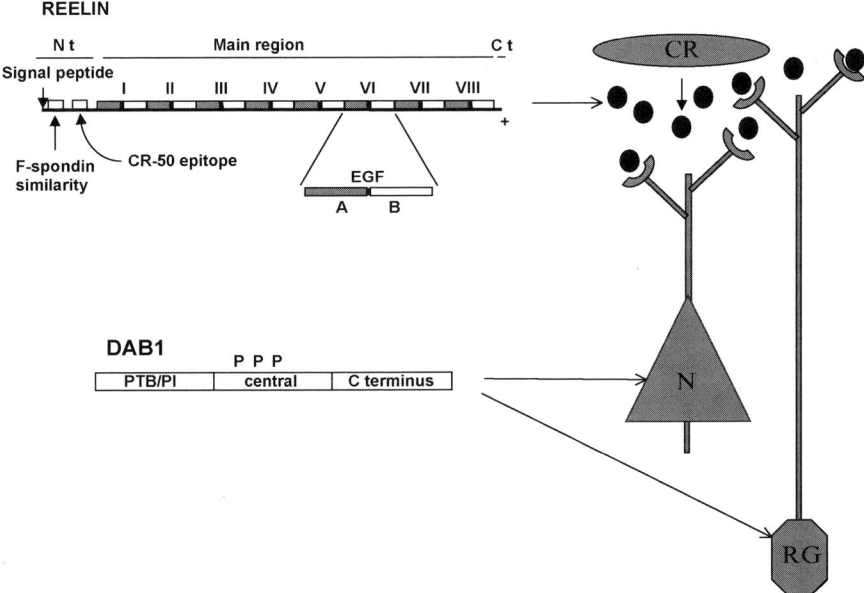

FIG. 4. Cajal-Retzius cells (CR) secrete Reelin (circles) in the extracellular environment of the marginal zone during cortical development. Migrating neurons (N) and radial glia cells (RG) express Reelin receptors and Dab1, and thus receive the Reelin signal on their superficial processes. Schematic representations of the Reelin and Dab1 proteins are shown. Roman numbers indicate Reelin repeats. Reelin-induced tyrosine phosphorylation sites on Dab1 are indicated by Ps. Nt = N terminus. Ct = C terminus.

(Miyata *et al.*, 1997), and in the hippocampus *in vivo* (Nakajima *et al.*, 1997). The C terminal region contains a domain that is necessary for Reelin secretion (D'Arcangelo *et al.*, 1997). Deletion of this domain in the rl^{Orl} strain results in the intracellular accumulation of Reelin and the appearance of a null phenotype (de Bergeyck *et al.*, 1997; Hirotsune *et al.*, 1995).

The initial biochemical studies of Reelin utilized metabolic labeling followed by CR-50 immunoprecipitation, and preferentially detected newly synthesized protein in its full-length form of approximately 400 kDa (D'Arcangelo *et al.*, 1997). Subsequent studies utilized newly developed N terminal Reelin monoclonal antibodies in Western blot assays (de Bergeyck *et al.*, 1998) and identified several proteolytic products, in addition to the full-length protein (D'Arcangelo *et al.*, 1999; de Bergeyck *et al.*, 1998). It is now known that Reelin is processed in the post-Golgi or extracellular environment to produce three major fragments, a ∼180 kDa N terminal fragment containing the CR-50 epitope and repeats 1 and 2, a central ∼120 kDa fragment containing repeats 3 to 6, and a ∼100 kDa C terminal fragment containing repeats 7 and 8 and the secretion signal (Jossin *et al.*, 2003a). Fragments derived from partial cleavage were also detected. Among these, only the central fragment has been shown to be biologically active, although it is not as effective as the full-length protein (Jossin *et al.*, 2004). Similar Reelin fragments can be detected in the cerebral spinal fluid and in the blood serum (Smalheiser *et al.*, 2000), indicating that processing is not unique to the brain environment. At the moment, the significance of this processing is not clear.

B. REELIN EXPRESSION

Reelin is abundantly and selectively expressed by Cajal-Retzius cells in the marginal zone of the embryonic cortex and hippocampus (Alcantara *et al.*, 1998; D'Arcangelo *et al.*, 1995; Ogawa *et al.*, 1995) (Fig. 4). Thus, one crucial function for these once mysterious cells is to localize Reelin to the end of the migratory path of principal neurons, providing a positional cue that signals the end of radial migration. Consistent with this role, Cajal-Retzius cells begin to disappear at the end of corticogenesis, resulting in the gradual loss of Reelin from the marginal zone of the rodent cortex and hippocampus. Interestingly, some GABAergic neurons, known as subpial granule cells, occupy the marginal zone in the late embryonic human cortex, in close apposition to Cajal-Retzius cells. These late-arriving but long-lasting Reelin-producing cells may constitute a sustained source of Reelin for prolonged corticogenesis in humans (Meyer and Goffinet, 1998). In all species examined, postnatal development is accompanied by a striking shift in the distribution of Reelin-expressing cells. Many GABAergic interneurons in the postnatal and adult brain express Reelin in all layers of the cortex and throughout the hippocampus (Alcantara *et al.*, 1998; Pesold *et al.*, 1998). The significance of this new expression pattern is not known, but it may relate to a role of Reelin in neuronal maturation and synaptic activity.

In the cerebellum, Reelin is strongly expressed at embryonic ages by granule cells forming the external granule cell layer, and by prospective deep cerebellar neurons in the nuclear transitory zone. The observation that granule cell loss in the Math-1 knock out mouse only results in a partial disruption of the PCL formation implies that other sources of Reelin significantly contribute to Purkinje cell alignment during embryonic cerebellar development (Jensen *et al.*, 2002). In the postnatal cerebellum, Reelin continues to be expressed by granule cells even after they have migrated across the PCL to form the internal granule layer. It is conceivable that this late expression relates to Purkinje cell dendrite maturation or synaptic connectivity between the axons of the granule cells (parallel fibers) and Purkinje cells dendrites, although this possibility has never been directly investigated.

Another prominent site of Reelin expression is the embryonic and postnatal olfactory bulb. Mitral cells and some periglomerular neurons in the lamina granularis externa have been shown to express Reelin (Alcantara *et al.*, 1998). One possible function of Reelin in this structure is to promote the detachment of olfactory neurons from the rostromigratory stream, thus facilitating their entry into the bulb (Hack *et al.*, 2002). However, another possibility is that Reelin may play a role in the refinement of connectivity and synaptic function in the olfactory glomeruli, which contain the apical dendrites of mitral cells and are sites of considerable synaptic plasticity even in the adult brain.

In addition to those mentioned here, many other structures of the central nervous system contain Reelin-expressing cells at some point during development or in the adult (see Alcantara *et al.*, 1998; Ikeda and Terashima, 1997; Schiffmann *et al.*, 1997, for a detailed description). In general terms, it appears that Reelin controls neuronal migration during embryonic development and modulates neuronal maturation and synaptic function postnatally. Indeed, many axonal tracts have also been shown to contain Reelin immunoreactivity, and Reelin may be secreted at the synaptic terminals (Martinez-Cerdeno *et al.*, 2003). Electron microscopy studies have also demonstrated the accumulation of Reelin in postsynaptic spine densities of the forebrain (Pappas *et al.*, 2001; Roberts *et al.*, 2005; Rodriguez *et al.*, 2000).

Reelin expression is not limited to the nervous system, but can be found in several other organs including the dentine pulp (Maurin *et al.*, 2004), liver, and blood (Ikeda and Terashima, 1997; Smalheiser *et al.*, 2000). The role of Reelin in these latter organs is completely unknown. In odontoblasts, Reelin may facilitate the innervation of the dentin-pulp complex by promoting adhesion to the dental nerve endings (Maurin *et al.*, 2004).

IV. The Reelin Signaling Pathway

A. REELIN RECEPTORS

The analysis of the anatomical defects in *reeler* cortical structures implied that radially-migrating neurons in the neocortex, pyramidal neurons of the hippocampus, and Purkinje cells in the cerebellum were all direct or indirect Reelin targets. The identification of the molecular machinery that transduces the Reelin signal in these regions enabled us to determine that these principal neurons are indeed direct targets of Reelin. Genetic and biochemical studies identified two Reelin receptors, both of which are proteins previously thought to function in lipoprotein metabolism: the very low-density lipoprotein receptor (VLDLR) and the apolipoprotein E receptor 2 (ApoER2) (D'Arcangelo *et al.*, 1999; Hiesberger *et al.*, 1999; Trommsdorff *et al.*, 1999). Double knock out mice lacking both these receptors exhibit a *reeler*-like phenotype, demonstrating that they are essential for Reelin function (Trommsdorff *et al.*, 1999). Individual receptor knock out mice do not exhibit an overt behavioral phenotype, but layering defects can be observed especially in the neocortex and hippocampus of ApoER2 mutant mice (Benhayon *et al.*, 2003; Trommsdorff *et al.*, 1999). Together with expression data showing that ApoER2 is more abundantly expressed than VLDLR, genetic data suggest that ApoER2 may be the predominant Reelin receptor, at least in the forebrain. Biochemical studies demonstrated that both ApoER2 and VLDLR bind full-length Reelin with high affinities. Binding of Reelin proteolytic fragments has also been observed, suggesting that proteolysis may not necessarily downregulate Reelin function (Benhayon *et al.*, 2003; D'Arcangelo *et al.*, 1999; Jossin *et al.*, 2004). On the intracellular side, ApoER2 and VLDLR bind the Reelin signaling protein Disabled-1 (Dab1, discussed later) through their NPxY, a motif present in many lipoprotein receptors and other transmembrane proteins that is important for ligand internalization. Similarly to other members of the lipoprotein receptor family, ApoER2 and VLDLR internalize Reelin (D'Arcangelo *et al.*, 1999; Morimura *et al.*, 2005), which thus triggers activation of a tyrosine kinase signaling pathway in the target cell. Recent studies demonstrated that Reelin promotes clustering of lipoprotein receptors and that this event is sufficient to activate the signaling cascade and to induce long-term potentiation (LTP) in hippocampal slices. Cross-linking antibodies directed toward the extracellular domain of the VLDLR and ApoER2 mimic the effect of full-length Reelin treatment on signal transduction and LTP induction (Strasser *et al.*, 2004; Weeber *et al.*, 2002), but are unable to rescue the *reeler* defect in cortical slices *in vitro* (Jossin *et al.*, 2004). This suggests that molecular events other than receptor clustering are important for the full biological activity of Reelin.

B. DISABLED-1

The transduction of the Reelin signal necessitates the presence of Disabled-1 (Dab1), an adapter protein capable of binding to the Reelin receptors VLDLR and ApoER2 (Trommsdorff *et al.*, 1999), as well as other NPxY-containing transmembrane proteins such as amyloid precursor protein (APP) (Homayouni *et al.*, 1999; Howell *et al.*, 1999b; Trommsdorff *et al.*, 1998) and integrins (Schmid *et al.*, 2005). Dab1 was originally discovered as a Src-interacting brain protein that is phosphorylated on tyrosine residues in a developmentally regulated fashion (Howell *et al.*, 1997a). Disruption of the *dab1* gene in mice by homologous recombination, or substitution of five predicted src-inducible tyrosine residues results in a *reeler*-like phenotype (Howell *et al.*, 1997b, 2000). Similarly, the disruption of the *dab1* gene by spontaneous or random mutation in the *scrambler* (scm) and *yotari* (yot) mutant mice also produces a phenotype indistinguishable from *reeler* (Sheldon *et al.*, 1997; Ware *et al.*, 1997; Yoneshima *et al.*, 1997).

Dab1 expression is mostly complementary to that of Reelin in developing brain structures, and unequivocally marks the populations of Reelin-responsive cells (Rice *et al.*, 1998). In the embryonic cortex, Dab1 is highly expressed by cortical plate neurons, but is also expressed together with ApoER2 in the ventricular and subventricular zones by progenitor cells and newborn neurons (Luque *et al.*, 2003; Meyer *et al.*, 2003; Perez-Garcia *et al.*, 2004; Rice *et al.*, 1998; Trommsdorff *et al.*, 1999) (Fig. 4). High levels of Dab1 and ApoER2 are also observed in many hippocampal cell populations including pyramidal cells and granule cells of the dentate gyrus. In contrast, expression of Dab1, VLDLR, and ApoER2 in the cerebellum is restricted to the Purkinje cell layer. A limited population of Dab1 and ApoER2/VLDLR-expressing neurons is also observed in the spinal cord, consisting of autonomic preganglionic neurons (Carroll *et al.*, 2001; Phelps *et al.*, 2002; Yip *et al.*, 2004a). Virtually all these cells are ectopic in mice lacking Reelin, ApoER2/VLDLR, or Dab1, demonstrating the essential role of these proteins in Reelin signaling.

The Dab1 protein consists of three functionally distinct domains, an N terminal phosphotyrosine binding and phosphoinositide interacting domain (PTB/PI), a central region containing Reelin-dependent tyrosine residues, and a C terminal domain (Fig. 4). Dab1 binds to the cytoplasmic tail of NPxY-containing transmembrane proteins by virtue of its PTB/PI domain. This domain also binds phosphoinositide (PI) $4,5P_2$, which strongly directs its localization to the plasma membrane. Binding to the PI occurs through a different protein interface, indicating that Dab1 most likely binds simultaneously to membrane proteins and lipids (Stolt *et al.*, 2003; Yun *et al.*, 2003). A recent study demonstrated that interaction with both the receptor tail and PI is necessary for Reelin-induced Dab1 phosphorylation (Stolt *et al.*, 2005). This event is linked to the activation of signal transduction pathways as well as the downregulation of

Dab1 expression in target cells. Addition of recombinant Reelin to cultured cortical neurons causes an acute increase in phosphorylation on tyrosine residues 198, 220, and 232 in the central domain of Dab1 (Ballif et al., 2004; Howell et al., 1999a; Keshvara et al., 2001). This induction does not occur in double ApoER2/VLDLR mutant neurons, and it is reduced in single ApoER2 mutant neurons, further indicating that Reelin signaling is mediated largely by ApoER2 in cortical neurons (Benhayon et al., 2003; Trommsdorff et al., 1999). Reelin induction of Dab1 phosphorylation is carried out by the Src family of kinases (SFK) (Arnaud et al., 2003b; Bock and Herz, 2003). Among these, Fyn appears to be the principal kinase, since Dab1 phosphorylation is significantly reduced in Fyn mutant mice (Arnaud et al., 2003b). SFK activation in turn is dependent on the presence of Dab1, suggesting a positive feedback mechanism (Bock and Herz, 2003). Dab1 phosphorylation leads to its ubiquitination and degradation by the proteosome pathway, thus resulting in the downregulation of Reelin signaling (Arnaud et al., 2003a; Suetsugu et al., 2004). These findings explain why Dab1 protein accumulates in reeler, double ApoER2/VLDLR, and Fyn mutant mice, in which Dab1 tyrosine phosphorylation is reduced (Arnaud et al., 2003a; Rice et al., 1998; Sheldon et al., 1997; Trommsdorff et al., 1999).

In addition to the N terminal and central domain, the C terminal region of Dab1 also appears to be important for Reelin signaling. A mutant mouse expressing a truncated p45 Dab1 protein lacking this terminal domain displays a hypomorphic phenotype in heterozygosity (Herrick and Cooper, 2002). Unlike full-length p80 Dab1, which causes no discernable phenotype in heterozygous mice, a single gene copy encoding p45 Dab1 results in specific lamination defects. Heterozygous p45 Dab1 mice display a split in the hippocampal pyramidal layer and a hypercellular marginal zone in the neocortex, but no cerebellar defects. These observations suggest that the C terminus of Dab1 affects the strength of the Reelin signaling, possibly by promoting interaction with downstream effectors, in a cell type-specific manner.

C. DOWNSTREAM OF DAB1

Because disruption of SFK-dependent tyrosine sites on Dab1 results in a reeler phenotype (Howell et al., 2000), it is generally believed that Dab1 phosphorylation is an event crucial for Reelin signal transduction, and not simply associated with degradation and termination of signaling. Therefore, a considerable effort has been dedicated to the identification of phospho-Dab1 interacting proteins that may transduce the downstream signal (see Fig. 5). A flurry of biochemical studies in the past few years demonstrated that phospho-Dab1 interacts with a variety of signaling proteins, all potentially important for modulating cytoskeletal dynamics, and thereby induces changes in neuronal migration and cellular

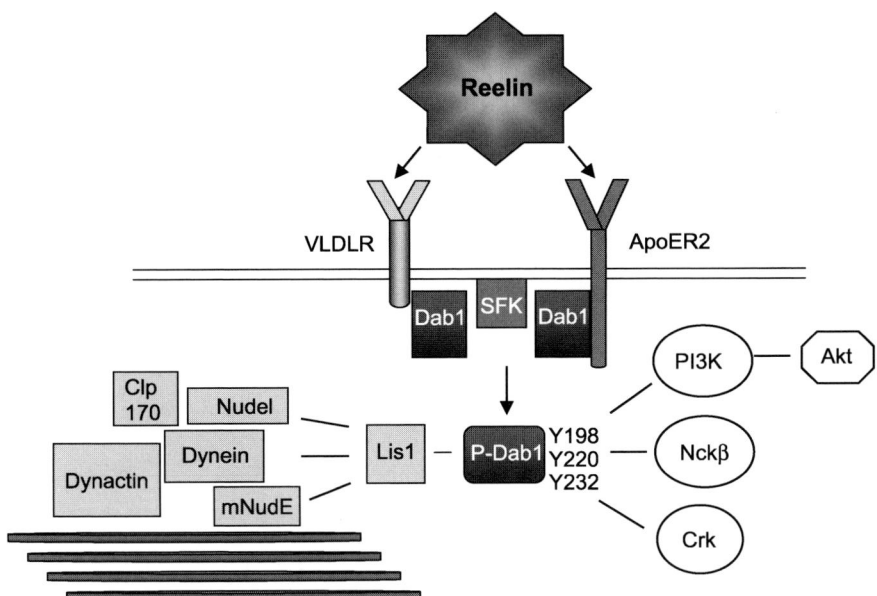

FIG. 5. The Reelin signaling pathway is activated by ligand-induced clustering of the Reelin receptors VLDLR and ApoER2, leading to SFK activation and Dab1 phosphorylation on the indicated tyrosine residues. Phospho-Dab1 interacts with a variety of intracellular proteins including Lis1, which ties the Reelin signal with the dynein complex and microtubule dynamics. Long bars indicate microtubules. (See Color Insert.)

adhesion. Reelin was first found to activate the phosphatidylinositol-3-kinase (PI3K) and Akt, while inhibiting the glycogen synthase kinase (GSK) 3β, in an ApoER2/VLDLR- and Dab1-dependent manner (Beffert *et al.*, 2002). PI3K activation results from a direct interaction between phospho-Dab1 and the PI3K subunit p85α (Bock *et al.*, 2003). Pharmacological inhibition studies suggest that PI3K is required for Reelin-induced Akt activation and cortical plate formation *in vitro*, however the specific role of this kinase in Reelin signaling cannot be fully dissected from other signaling events linked to a multiplicity of cellular processes (Bock *et al.*, 2003). Phospho-Dab1 was also shown to bind Lis1, a protein involved in neuronal migration that is mutated in human Miller-Dieker lissencephaly (Hattori *et al.*, 1994; Reiner *et al.*, 1993). Lis1 participates in two major complexes. One enzymatic complex is the platelet-activating factor acetyl-hydrolase 1b (Pafah1b), in which Lis1 functions as a non-catalytic subunit. The other includes the microtubule motor dynein and dynein-associated proteins NudE, NudEL, dynactin, and CLIP-170 (reviewed by Vallee *et al.*, 2001; Wynshaw-Boris and Gambello, 2001). These proteins form an evolutionarily conserved pathway that regulates centrosome function and mediates nucleokinesis. The interaction between phospho-Dab1 and Lis1 allows a cross-talk between Reelin signaling and Lis1-dependent molecular events that may be important for nuclear translocation and neuronal migration (Assadi et al., 2003). The adapter molecule Nck/β can also bind phospho-Dab1 and redistribute to the membrane upon Reelin stimulation (Pramatarova et al., 2003). This interaction is thought to cause remodeling of the actin cytoskeleton. Another link to the actin cytoskeleton is provided by the Dab1-binding protein N-WASP (Suetsugu et al., 2004). This interaction results in the polymerization of actin and filopodia extension in non-neuronal cells. Interestingly, Dab1 phosphorylation was shown to suppress filopodia formation in this study, suggesting that a similar function in neurons may be important Reelin-induced termination of migration. A systematic screen of phospho-Dab1 binding proteins from embryonic brain lysates revealed interaction with PLC-γ1, Shp2, Src, CrkL, and CrkII (Ballif et al., 2004). CrkL binds the Rap1 guanidine exchange factor C3G, which is phosphoryled on tyrosine residues in response to Reelin and leads to Rap1 activation. A similar screen of phospho-Dab1-binding protein resulted in the identification of RasGap, Nck-1, Nck-2, β-actin, CrkL, CrkII, and its splicing variant Crk-I (Huang et al., 2004). Crk-I and CrkII, which are both products of the Crk gene, were shown to stimulate Dab1 phosphorylation in transfected cells in a Src-dependent way. Together, these studies suggest that Crk-family proteins form complexes with phospho-Dab1 and participate in Reelin signaling, both upstream and downstream of Dab1.

Other Dab1-intaracting proteins that do not require its phosphorylation include the amyloid precursor protein (APP) and APP-like proteins (Homayouni et al., 1999; Howell et al., 1999b), and the GTPase activating protein Dab2IP (Homayouni et al., 2003). The significance of these interactions in Reelin signaling is unknown. In the case of APP, the link with Dab1 may be relevant to neurodegeneration. Mutations in genes encoding APP or APP-processing enzymes such as presenilin-1 and presenilin-2 cause increased phosphorylation of the axonal microtubule stabilizing protein tau and early onset Alzheimer's disease (Lee et al., 2001). Interestingly, it has been shown that deficiency in Reelin, ApoER2/VLDLR, or Dab1 results in tau hyperphosphorylation and that APP is a genetic modifier of tau hyperphosphorylation in Dab1 mutant mice (Brich et al., 2003; Hiesberger et al., 1999). These findings reinforce the association between Reelin signaling and neurodegeneration previously established by the identification of ApoER2 and VLDLR as Reelin receptors as well as ApoE-binding proteins; ApoE proteins compete with Reelin for binding to these receptors, and the ApoE4 allele is a risk factor for late onset Alzheimer's disease. However, a specific effect of ApoE4 compared to ApoE2 and ApoE3 on Reelin binding could not be demonstrated using recombinant proteins (D'Arcangelo et al., 1999).

Dab1 is generally thought to function downstream of Reelin receptors and to mediate signal transduction by virtue of its tyrosine phosphorylation-dependent association with many of the proteins previously mentioned. However, a recent study has raised the possibility that Dab1 may also function upstream of the receptors by regulating their trafficking to the plasma membrane (Morimura *et al.*, 2005). Cell-surface localization of Reelin was reduced in Dab1 mutant neurons, whereas cell surface expression of ApoER2 and VLDLR was augmented in Dab1-transfected cells. Phosphorylation of Dab1 appears to induce endocytosis of Reelin as well as its receptors and Dab1 itself, thus triggering a short-lived signaling cascade that terminates with Dab1 ubiquitination and degradation.

V. The Biological Activities of Reelin

A. CELLULAR LAYER FORMATION

Despite advances in understanding the molecular steps involved in Reelin signaling, it is still not exactly clear how these events result in the formation of cortical layers in the developing brain. The presence of Reelin receptors and Dab1 in neuronal populations that become ectopic in *reeler* and *reeler*-like mutants suggests that the Reelin signal is passed on directly from the extracellular environment of superficial layers where Reelin is present to migrating neurons that approach these layers. The first contact is likely to be made on the leading edge of migrating neurons. However, the presence of Reelin signaling molecules on these cellular processes has not been demonstrated. This is due to technical difficulties related to the rapid trafficking of lipoprotein receptors at the plasma membrane, and the fact that the currently available antibodies against lipoprotein receptors and Dab1 work poorly in immunohistochemical assays. An alternative hypothesis is that radial fibers are the primary targets of Reelin and that neurons become disorganized as a secondary disruption of the radial scaffold. This view is based primarily on the observation that the radial scaffold is disrupted in *reeler* and *reeler*-like mutants. This disruption is particularly prominent in the dentate gyrus (Weiss *et al.*, 2003), but is more subtle in the neocortex where radial fibers appear to be shorter than normal and less branched (Derer, 1979; Hartfuss *et al.*, 2003). Addition of recombinant Reelin to the culture medium of hippocampal slices promoted radial fiber elongation (Zhao *et al.*, 2004). Interestingly though, it did not rescue layer formation and orientation of dentate granule cells. Granule cell migration could only be rescued by the addition of Reelin in a normotopic position near the marginal zone, indicating that Reelin is a positional cue that dictates proper positioning of dentate granule neurons, possibly by acting primarily on radial fibers. In support of this hypothesis, several investigators have reported the expression of Reelin receptors and Dab1 in the ventricular zone of the neocortex where radial progenitor cells are abundant (Luque *et al.*, 2003; Rice *et al.*, 1998; Trommsdorff *et al.*, 1999) and in isolated progenitor cells (Hartfuss *et al.*, 2003). It should be noted, however, that the expression levels of Reelin transducing proteins in radial progenitor cells pail in comparison to those achieved in migrating neurons, particularly in superficial layers of the neocortex. Furthermore, in the cerebellum, ApoER2, VLDLR, and Dab1 expression has been described in Purkinje cells and not in radial cells, suggesting that Reelin acts directly on radially-migrating neurons in this structure. Nevertheless, it is possible that a defect in the radial scaffold, particularly in the dentate gyrus, contributes to layer disruption in the *reeler* brain.

To understand the molecular events that mediate Reelin function in cellular layer formation, several *in vitro* model systems have been developed. A classical model consists of cortical cell aggregates formed in rotating cultures. Cortical aggregates from *reeler* mice appear more disorganized than normal (DeLong, 1970; DeLong and Sidman, 1970), and normal aggregates can be rendered abnormal by the addition of the CR-50 interfering antibody (Ogawa *et al.*, 1995). Another assay system consists of isolated neurons lifted from embryonic cortical slices together with a radial fiber and cultured on a membrane (Dulabon *et al.*, 2000). Using this system it was reported that recombinant Reelin stops migration by causing detachment of neurons from the radial process. However, this conclusion was challenged based on the analysis of a transgenic mouse in which *reelin* is ectopically expressed in the ventricular zone from the nestin promoter (Magdaleno *et al.*, 2002). Expression of the transgene in the ventricular zone did not cause an arrest of neuronal migration in normal mice, and partially rescued the cortical phenotype in *reeler* mice. These findings suggest that Reelin acts more as an instructive molecule, enabling neurons to respond to positional cues, than a physical barrier. However, it is also possible that progenitor cells in the ventricular zone of transgenic mice transported the Reelin protein along the radial fiber and secreted it in the marginal zone, the normotopic region of the cortex, where it could function as a positional cue by signaling the end of migration. Expression studies in several regions of the central nervous system are consistent with the view that Reelin represents a stop signal to radial migration. Also supporting this is the fact that the Reelin protein is seen in marginal areas of the forebrain, cerebellum, and spinal cord where Dab1-positive neurons terminate (Kubasak *et al.*, 2004; Rice *et al.*, 1998; Yip *et al.*, 2004a).

A recently developed assay for neocortical development involves embryonic cortical slices in which two consecutive neuronal cohorts are labeled with radioactive thymidine and BrdU (Jossin *et al.*, 2003b). This model enabled investigators to examine early events in corticogenesis, such as preplate splitting and the establishment of an inside-out gradient. Addition of recombinant full-length

Reelin to the medium rescued the *reeler* defects. Fragments of recombinant Reelin similar to those generated spontaneously by proteolytic processing could be analyzed for the first time for their biological activity. This analysis led to the identification of a central domain of Reelin (consisting of repeats 3 to 6) that is necessary and sufficient for preplate splitting and early layer formation (Jossin *et al.*, 2004). N terminal fragments (up to repeat 3) or C terminal fragments (repeats 6 to 8) were not functionally active. These functional data correlated well with the ability of Reelin fragments to induce Dab1 phosphorylation in cortical slices. The central domain, even though sufficient for activity, appears to be less effective than the full-length Reelin protein. One possible explanation for this difference is that the CR50 epitope, which is absent in the central fragment but mediates the aggregation of full-length Reelin in multimers, may potentiate the biological activity of Reelin.

The cortical slice system was also exploited in a pharmacological approach to demonstrate the requirement for SFKs, non-classical protein kinase C (PKC) and PI3K in layer formation (Bock *et al.*, 2003; Jossin *et al.*, 2003b). Addition of the SFK inhibitor PP2 and the protein kinase C inhibitor bisindolylmaleimide 1 (BIM1) to wild type cultures prevented normal corticogenesis. PP2 also prevented Dab1 phosphorylation in cortical slices, as previously demonstrated in dissociated neurons (Arnaud *et al.*, 2003b; Bock and Herz, 2003), whereas BIM1 did not (Jossin *et al.*, 2003b). The role of non-classical PKCs in Reelin signal transduction has not been further elucidated. Addition of the PI3K inhibitor LY294002 also blocked the formation of a well-defined cortical plate and Akt phosphorylation (Bock *et al.*, 2003), indicating that activation of this pathway by Reelin or clustered receptors (Ballif *et al.*, 2003; Beffert *et al.*, 2002; Strasser *et al.*, 2004), is physiologically relevant to corticogenesis.

To visualize live migrating neurons in cortical slices, techniques were developed to label neurons with fluorescent dyes *in vitro* (Nadarajah *et al.*, 2001) or *in vivo* by *in utero* electroporation (Tabata and Nakajima, 2001). The migration of selected labeled neurons was then conducted by time-lapse confocal imaging. Using this approach, it was discovered that neurons migrate by three modalities, radial fiber-guided locomotion, somal translocation, and branched migration (Nadarajah *et al.*, 2001; Tabata and Nakajima, 2003). The two latter modes of migration, which had not previously been described, enable neurons to move without the support of the radial fiber scaffold. Somal translocation appears to be commonly used by migrating neurons in the beginning of corticogenesis: a long leading edge extends deep into the preplate followed by its collapse, which brings the cell body in a superficial position (Nadarajah *et al.*, 2001).

One limitation of the cortical slice system is that it is a short-term assay, allowing analysis of corticogenesis only for 3 days *in vitro*. To examine later stages of corticogenesis, *in utero* gene transfer of GFP-expressing vectors was used to label cohorts of newborn neurons in mouse embryos that were carried to term.

When GFP-labeled neurons were visualized in the postnatal *reeler* cortex, they appeared to be unable to reach the top of the cortex and developed abnormally oriented dendrites in abnormally deep positions (Tabata and Nakajima, 2002). Later born neurons tended to accumulate along dendrite-rich areas, which in normal cortex consist largely of the marginal zone. However, in *reeler,* ectopic dendrites clustered throughout the cortex and appeared to impede the progress of migrating neurons toward the surface of the cortex. These observations suggest that defective dendrite growth may also contribute to cortical layer disruption in the *reeler* cortex. A further improvement of this assay system was achieved by combining *in utero* gene transfer with time-lapse confocal imaging of cortical slices. The use of GFP-carrying retroviruses injected *in utero,* allowing for clonal analysis of dividing cells and their progeny, was instrumental in demonstrating that radial cells are precursors for both neurons and glia at different stages of corticogenesis (Malatesta *et al.*, 2000; Noctor *et al.*, 2001). When this approach was applied to Dab1-deficient *scrambler* mice (Sanada *et al.*, 2004), mutant neurons were found to be unable to extend their leading edge into the preplate. This observation suggests that Reelin may have an attractive or stabilizing function on these cellular processes, likely mediated by phospho-Dab1 and its interacting proteins through modifications of the cytoskeleton. Mutant *scrambler* neurons also appeared to be abnormally closely associated to radial fibers. Proper positioning of GFP-labeled neurons was shown to be dependent on intrinsic Dab1 expression in migrating neurons and was not inhibited by the presence of Dab1-deficient radial fibers in the *scrambler* cortical background. This provides evidence that Reelin functions by signaling to migrating neurons directly, and not indirectly through radial fibers, at least in the neocortex. Since the mutant environment did not interfere dramatically with the radial progression of Dab1-expressing neurons, this study demonstrates that obstruction due to increased adhesion to radial fibers cannot alone account for the cortical layer inversion seen in *reeler* or Dab1 mutant cortex. Dab1 phosphorylation on tyrosine residues 220 and 232 was further shown to be essential for proper detachment from radial fibers, as expression of these Dab1 proteins mutated on these sites did not rescue the *scrambler* defect (Sanada *et al.*, 2004). These mutant Dab1 proteins had a dominant negative effect, as they caused excessive adhesion to radial fibers even in the presence of normal Dab1 in the wild type cortex. One possible mechanism for increased adhesion appears to be the upregulation of $\alpha 3$ integrin. In the *scrambler* cortex $\alpha 3$ integrin levels were increased (Sanada *et al.*, 2004). It is not yet known whether a similar upregulation is also present in the *reeler* cortex. The excessive adhesion of tyrosine 220 mutant Dab1-expressing neurons could be countered by downregulation of $\alpha 3$ integrin levels with RNA interference, indicating that the detachment of migrating neurons from radial fibers is controlled by this integrin. It is interesting to note that $\alpha 3$ integrin in association with the $\beta 1$ chain has been shown to bind Reelin and Dab1 (Dulabon *et al.*, 2000;

Schmid *et al.*, 2005). However, adhesion control does not appear to be a key requirement for cortical layer formation, since lamination proceeds fairly normally in α3 (Anton *et al.*, 1999) and β1 (Graus-Porta *et al.*, 2001) knock out mice.

In summary, current data indicate that Reelin promotes cortical layer formation by a combination of distinct molecular mechanisms. By acting primarily on migrating neurons, the Reelin signal mediated by VLDLR/ApoER2 and Dab1 may promote the extension and stabilization of the leading edge into the marginal zone, which results in migration arrest and proper orientation. Reelin signal also may induce a decrease in neuronal adhesion to radial fibers through α3β1 integrin downregulation, facilitating the assembly of cellular layers. An additional effect of Reelin on the stability of the radial scaffold may also contribute to proper neuronal migration and cortical lamination.

B. NEURONAL MATURATION

In addition to its function as a key regulator of neuronal positioning, Reelin also functions as a neurotrophic factor that promotes growth and maturation of cellular processes. Earlier studies demonstrated that the axonal projections of the entorhinohippocampal pathway are defective in *reeler* mice, and that addition of the CR-50 antibody interferes with their development in normal hippocampal slices (Del Rio *et al.*, 1997). The extent of terminal branching appeared to be transiently affected in *reeler*. So far, this observation has not been extended to other fiber projection systems. The lamina-specific termination of the hippocampal projections is known to be strictly dependent on the position of the target cells, which is aberrant in *reeler* (Deller *et al.*, 1999). Thus, it appears that Reelin only indirectly affects axonal projections. Consistent with this view, it was observed that axonal targeting and fasciculation of vomeronasal projections into the accessory olfactory system are normal in *reeler* (Teillon *et al.*, 2003). Furthermore, no effect on axon outgrowth was observed using cortical explants cultured in a collagen gel, a system that enables to readily detect axon repulsion by semaphorins (Jossin and Goffinet, 2001).

Unlike axons, dendritic processes are severely affected in *reeler* brain. Earlier Golgi impregnation studies revealed the stunted and disoriented dendritic trees of cortical, hippocampal, and cerebellar neurons (Mikoshiba *et al.*, 1980; Pinto Lord and Caviness, 1979; Stanfield and Cowan, 1979). Other studies using *in utero* electroporation of GFP-expression plasmids showed that dendrite growth is delayed in the *reeler* neocortex (Tabata and Nakajima, 2002). Recent studies further demonstrated that dendrite defects in *reeler* are not only secondary to cellular ectopia, but are, at least in part directly due to the absence of trophic activity of Reelin. Dissociated hippocampal cultures from *reeler* mice, as well as from Dab1 mutant mice, exhibited reduced dendrite development (Niu *et al.*,

2004). The deficit was observed not only in homozygous mutant mice, but also in developing heterozygous mice that exhibit no positional defects. The *reeler* defect could be rescued by the addition of Reelin, whereas lipoprotein receptor antagonists and SFK inhibitors blocked normal growth (Niu *et al.*, 2004). These data indicated that the same signaling pathway that controls neuronal positioning during embryogenesis also controls dendrite growth in the postnatal brain.

Neuronal maturation during postnatal development involves not only the extension of cellular processes but also the formation of synapses. As previously mentioned, synaptic circuitry is disrupted in the *reeler* retina in the absence of cellular layer defects (Rice *et al.*, 2001). Interestingly, the density of dendritic spines appears reduced in the prefrontal cortex of heterozygous *reeler* mice, in the absence of cellular ectopia (Liu *et al.*, 2001). It is not currently known whether this synaptic deficit is widespread, nor whether it is secondary to delayed dendrite development. Reelin has been shown to induce LTP in hippocampal slice cultures, further suggesting a role in regulating synaptic function in the postnatal and adult brain (Weeber *et al.*, 2002). This Reelin-induced form of plasticity was shown to be dependent on ApoER2 and VLDLR, indicating that the Reelin signaling pathway remains active throughout brain development, even though its biological function changes from the control of neuronal migration and cellular architecture in the prenatal brain to the regulation of neuronal maturation and synaptic activity in the postnatal brain.

VI. Reelin and Human Diseases

The *reelin* gene is highly conserved among vertebrate species, including humans (DeSilva *et al.*, 1997). In the primates and human brain, *reelin* expression pattern is very similar to that of rodents (Meyer and Goffinet, 1998; Rodriguez *et al.*, 2000). Thus, it is very likely that in humans Reelin plays roles similar to those proposed above based on animal models. Homozygous loss of Reelin in humans causes a severe neurological disease known as lissencephaly with cerebellar hypoplasia (LCH) (Hong *et al.*, 2000). The phenotype of these patients is reminiscent of that of *reeler* mice, and is characterized by profound ataxia. In addition, LCH patients also display seizures, severe cognitive deficits, and lymphoedema. The reduction in the number of cortical gyri (lissencephaly) is a uniquely human feature, due to the absence of gyration in the mouse cortex. The similarity between lissencephaly due to Reelin absence and that due to Lis1 haploinsufficiency is likely explained by the cross-talk between signaling pathways controlled by these molecules (Assadi *et al.*, 2003).

Alterations of *reelin* expression not due to inherited genetic mutations have also been reported in developmental brain abnormalities affecting the cortex and the hippocampus. An excessive number of Reelin-expressing cortical Cajal-Retzius cells were observed in abnormal brain regions of patients with some forms of cortical dysplasia and polymicrogyria, conditions associated with epilepsy (Eriksson *et al.*, 2001; Garbelli *et al.*, 2001). On the other hand, decreased expression of *reelin* by hippocampal Cajal-Retzius cells correlated with granule cell dispersion in the dentate gyrus of patients with temporal lobe epilepsy (Haas *et al.*, 2002). Together, these findings suggest that altered *reelin* expression may result in neuronal migration defects associated with different forms of epilepsy.

A decrease in *reelin* mRNA expression is also consistently seen in patients with schizophrenia and bipolar disorder, leading to the suggestion that Reelin deficiency may be a risk factor for these diseases (Guidotti *et al.*, 2000; Impagnatiello *et al.*, 1998). Low levels and altered distribution of Reelin-expressing cells was also observed in the white matter of schizophrenic tissue (Eastwood and Harrison, 2003), although no differences in Reelin expression in the serum of patients with neurological disease compared to control subjects were detected (Ignatova *et al.*, 2004). No detailed information is currently available on the health status of heterozygous carriers of *reelin* mutations, however they do appear to display overt neurological dysfunctions or mental illness. Because heterozygous *reeler* mice, like patients with schizophrenia and bipolar disorders, express reduced levels of Reelin (D'Arcangelo *et al.*, 1995), they have been proposed to be good models for the diseases (Costa *et al.*, 2002). Some behavioral studies of heterozygous *reeler* mice reported deficits using tests that specifically address anxiety and fear, such as prepulse inhibition of startle, and neophobic behavior, which resemble traits of schizophrenia (Tueting *et al.*, 1999). Alterations in the mesotelencephalic dopamine pathway have also been reported in heterozygous *reeler*, which may be relevant to psychotic disorders (Ballmaier *et al.*, 2002). However, other investigators have reported no behavioral deficits in heterozygous *reeler* mice using a battery of tests for evaluating emotional traits and memory (Podhorna and Didriksen, 2004; Salinger *et al.*, 2003). Genetic studies so far have also failed to demonstrate linkage between the *reelin* locus on chromosome 7q22 and schizophrenia or bipolar disorders. This could be explained by the observation that *reelin* expression is controlled by epigenetic factors such as promoter methylation (Chen *et al.*, 2002; Tremolizzo *et al.*, 2002). DNA methyltransferase (Dnmt) 1 was shown to regulate *reelin* gene expression (Noh *et al.*, 2005). Interestingly, the expression of this gene appears to be increased in cortical GABAergic interneurons of schizophrenic brains (Veldic *et al.*, 2004). This upregulation correlates with the decrease in *reelin* mRNA expression in the same interneurons. These findings suggest that promoter hypermethylation of *reelin*, and perhaps of other genes expressed in these interneurons, is associated with schizophrenia. The recent observation that histone deacetylase inhibitors decrease *reelin* promoter methylation raises the possibility that these drugs might be used in the treatment of psychoses (Mitchell *et al.*, 2005; Tremolizzo *et al.*, 2005).

Other studies suggested a link between *reelin* polymorphisms and autism (Persico *et al.*, 2001; Skaar *et al.*, 2004; Zhang *et al.*, 2002). This link is currently being further investigated (see this volume). If a cause-and-effect relationship between *reelin* and psychiatric diseases is established, a plausible mechanism may be synaptic dysfunction due to a postnatal Reelin deficit in the developing brain of these patients, rather than defective neuronal migration in early embryogenesis.

In conclusion, studies of the *reeler* mutation have shed some light on many diverse aspects of brain development. These mutant mice will continue to serve as invaluable models for neuronal migration, maturation, and synaptic circuitry, and the Reelin signaling pathway will continue to serve as a springboard for the investigation of the molecular events underlying these processes. Given the association between Reelin and a variety of human developmental brain disorders, these studies also hold great promise for the development of new treatments for these devastating conditions.

Acknowledgments

I am grateful to Drs. Andre' Goffinet and Gretchen Wieck for critical reading of the manuscript. Supported by NIH 5 R01 NS042616.

References

Alcantara, S., Ruiz, M., D'Arcangelo, G., Ezan, F., de Lecea, L., Curran, T., Sotelo, C., and Soriano, E. (1998). Regional and cellular patterns of *reelin* mRNA expression in the forebrain of the developing and adult mouse. *J. Neurosci.* **18,** 7779–7799.

Andersen, T. E., Finsen, B., Goffinet, A. M., Issinger, O. G., and Boldyreff, B. (2002). A reeler mutant mouse with a new, spontaneous mutation in the reelin gene. *Brain Res. Mol. Brain Res.* **105,** 153–156.

Angevine, J. G., and Sidman, R. L. (1961). Autoradiographic study of cell migration during histogenesis of cerebral cortex in the mouse. *Nature* **192,** 766–768.

Anton, E. S., Kriedberg, J. A., and Rakic, P. (1999). Distinct functions of alpha3 and alpha5 integrin receptors in neuronal migration and laminar organization of the cerebral cortex. *Neuron* **22,** 277–289.

Arnaud, L., Ballif, B. A., and Cooper, J. A. (2003a). Regulation of protein tyrosine kinase signaling by substrate degradation during brain development. *Mol. Cell Biol.* **23,** 9293–9302.

Arnaud, L., Ballif, B. A., Forster, E., and Cooper, J. A. (2003b). Fyn tyrosine kinase is a critical regulator of disabled-1 during brain development. *Curr. Biol.* **13,** 9–17.

Assadi, A. H., Zhang, G., Beffert, U., McNeil, R. S., Renfro, A. L., Niu, S., Quattrocchi, C. C., Antalffy, B. A., Sheldon, M., Armstrong, D. D., Wynshaw-Boris, A., Herz, J., D'Arcangelo, G., and Clark, G. D. (2003). Interaction of reelin signaling and Lis1 in brain development. *Nat. Genet.* **35,** 270–276.

Ballif, B. A., Arnaud, L., Arthur, W. T., Guris, D., Imamoto, A., and Cooper, J. A. (2004). Activation of a Dab1/CrkL/C3G/Rap1 pathway in Reelin-stimulated neurons. *Curr. Biol.* **14,** 606–610.

Ballif, B. A., Arnaud, L., and Cooper, J. A. (2003). Tyrosine phosphorylation of Disabled-1 is essential for Reelin-stimulated activation of Akt and Src family kinases. *Brain Res. Mol. Brain Res.* **117,** 152–159.

Ballmaier, M., Zoli, M., Leo, G., Agnati, L. F., and Spano, P. (2002). Preferential alterations in the mesolimbic dopamine pathway of heterozygous reeler mice: An emerging animal-based model of schizophrenia. *Eur. J. Neurosci.* **15,** 1197–1205.

Beffert, U., Morfini, G., Bock, H. H., Reyna, H., Brady, S. T., and Herz, J. (2002). Reelin-mediated signaling locally regulates protein kinase B/Akt and glycogen synthase kinase 3beta. *J. Biol. Chem.* **277,** 49958–49964.

Benhayon, D., Magdaleno, S., and Curran, T. (2003). Binding of purified Reelin to ApoER2 and VLDLR mediates tyrosine phosphorylation of Disabled-1. *Brain Res. Mol. Brain Res.* **112,** 33–45.

Bock, H. H., and Herz, J. (2003). Reelin activates SRC family tyrosine kinases in neurons. *Curr. Biol.* **13,** 18–26.

Bock, H. H., Jossin, Y., Liu, P., Forster, E., May, P., Goffinet, A. M., and Herz, J. (2003). PI3-Kinase interacts with the adaptor protein Dab1 in response to Reelin signaling and is required for normal cortical lamination. *J. Biol. Chem.* **278,** 38772–38779.

Borrell, V., del Rio, J. A., Alcantara, S., Derer, M., Martinez, A., D'Arcangelo, G., Nakajima, K., Mikoshiba, K., Derer, P., Curran, T., and Soriano, E. (1999). Reelin regulates the development and synaptogenesis of the layer-specific entorhino-hippocampal connections. *J. Neurosci.* **19,** 1345–1358.

Brich, J., Shie, F. S., Howell, B. W., Li, R., Tus, K., Wakeland, E. K., Jin, L. W., Mumby, M., Churchill, G., Herz, J., and Cooper, J. A. (2003). Genetic modulation of tau phosphorylation in the mouse. *J. Neurosci.* **23,** 187–192.

Carroll, P., Gayet, O., Feuillet, C., Kallenbach, S., de Bovis, B., Dudley, K., and Alonso, S. (2001). Juxtaposition of CNR protocadherins and reelin expression in the developing spinal cord. *Mol. Cell Neurosci.* **17,** 611–623.

Caviness, V. S., Jr., and Rakic, P. (1978). Mechanisms of cortical development: A view from mutations in mice. *Annu. Rev. Neurosci.* **1,** 297–326.

Caviness, V. S. J., Crandall, J. E., and Edwards, M. A. (1988). "The Reeler Malformation. Implications for Neocortical Histogenesis." Plenum Press, New York.

Chen, Y., Sharma, R. P., Costa, R. H., Costa, E., and Grayson, D. R. (2002). On the epigenetic regulation of the human reelin promoter. *Nucleic Acids Res.* **30,** 2930–2939.

Costa, E., Chen, Y., Davis, J., Dong, E., Noh, J. S., Tremolizzo, L., Veldic, M., Grayson, D. R., and Guidotti, A. (2002). REELIN and schizophrenia: A disease at the interface of the genome and the epigenome. *Mol. Interv.* **2,** 47–57.

D'Arcangelo, G., and Curran, T. (1998). Reeler: New tales on an old mutant mouse. *Bioessays* **20,** 235–244.

D'Arcangelo, G., Homayouni, R., Keshvara, L., Rice, D. S., Sheldon, M., and Curran, T. (1999). Reelin is a ligand for lipoprotein receptors. *Neuron* **24,** 471–479.

D'Arcangelo, G., Miao, G. G., Chen, S. C., Soares, H. D., Morgan, J. I., and Curran, T. (1995). A protein related to extracellular matrix proteins deleted in the mouse mutant reeler. *Nature* **374,** 719–723.

D'Arcangelo, G., Nakajima, K., Miyata, T., Ogawa, M., Mikoshiba, K., and Curran, T. (1997). Reelin is a secreted glycoprotein recognized by the CR-50 monoclonal antibody. *J. Neurosci.* **17,** 23–31.

de Bergeyck, V., Naerhuyzen, B., Goffinet, A. M., and Lambert de Rouvroit, C. (1998). A panel of monoclonal antibodies against reelin, the extracellular matrix protein defective in reeler mutant mice. *J. Neurosci. Methods* **82,** 17–24.

de Bergeyck, V., Nakajima, K., Lambert de Rouvroit, C., Naerhuyzen, B., Goffinet, A. M., Miyata, T., Ogawa, M., and Mikoshiba, K. (1997). A truncated Reelin protein is produced but not secreted in the "Orleans" reeler mutation (Reln$^{rl\text{-}Orl}$). Mol. Brain Res. **50,** 85–90.

Del Rio, J. A., Heimrich, B., Borrell, V., Forster, E., Drakew, A., Alcantara, S., Nakajima, K., Miyata, T., Ogawa, M., Mikoshiba, K., Derer, P., Frotscher, M., and Soriano, E. (1997). A role for Cajal-Retzius cells and reelin in the development of hippocampal connections. Nature **385,** 70–74.

Deller, T., Drakew, A., Heimrich, B., Forster, E., Tielsch, A., and Frotscher, M. (1999). The hippocampus of the reeler mutant mouse: Fiber segregation in area CA1 depends on the position of the postsynaptic target cells. Exp. Neurology **156,** 254–267.

DeLong, G. R. (1970). Histogenesis of fetal mouse isocortex and hippocampus in reaggregating cell cultures. Dev. Biol. **563,** 563–583.

DeLong, G. R., and Sidman, R. L. (1970). Alignment defect of reaggregating cells in cultures of developing brains of reeler mutant mice. Dev. Biol. **563,** 584–600.

Derer, P. (1979). Evidence for the occurrence of early modifications in the 'glia limitans' layer of the neocortex of the reeler mouse. Neurosci. Lett. **13,** 195–202.

DeSilva, U., D'Arcangelo, G., Braden, V. V., Chen, J., Miao, G., Curran, T., and Green, E. D. (1997). The human reelin gene: Isolation, sequencing, and mapping on chromosome 7. Genome Res. **7,** 157–164.

Dulabon, L., Olson, E. C., Taglienti, M. G., Eisenhuth, S., McGrath, B., Walsh, C. A., Kreidberg, J. A., and Anton, E. S. (2000). Reelin binds alpha3beta1 integrin and inhibits neuronal migration. Neuron **27,** 33–44.

Eastwood, S. L., and Harrison, P. J. (2003). Interstitial white matter neurons express less reelin and are abnormally distributed in schizophrenia: Towards an integration of molecular and morphologic aspects of the neurodevelopmental hypothesis. Mol. Psychiatry. **8,** 821–831.

Eriksson, S. H., Thom, M., Heffernan, J., Lin, W. R., Harding, B. N., Squier, M. V., and Sisodiya, S. M. (2001). Persistent reelin-expressing Cajal-Retzius cells in polymicrogyria. Brain **124,** 1350–1361.

Forster, E., Tielsch, A., Saum, B., Weiss, K. H., Johanssen, C., Graus-Porta, D., Muller, U., and Frotscher, M. (2002). Reelin, Disabled 1, and beta 1 integrins are required for the formation of the radial glial scaffold in the hippocampus. Proc. Natl. Acad. Sci. USA **99,** 13178–13183.

Fujimoto, Y., Setsu, T., Ikeda, Y., Miwa, A., Okado, H., and Terashima, T. (1998). Ambiguos nucleus neurons innervating the abdominal esophagus are malpositioned in the reeler mouse. Brain Res. **811,** 156–160.

Garbelli, R., Frassoni, C., Ferrario, A., Tassi, L., Bramerio, M., and Spreafico, R. (2001). Cajal-Retzius cell density as marker of type of focal cortical dysplasia. Neuroreport **12,** 2767–2771.

Goffinet, A. M. (1979). An early developmental defect in the cerebral cortex of the reeler mouse. Anat. Embryol. **157,** 205–216.

Goffinet, A. M. (1983a). The embryonic development of the cerebellum in normal and reeler mutant mice. Anat. Embryol. (Berl.) **168,** 73–86.

Goffinet, A. M. (1983b). The embryonic development of the inferior olivary complex in normal and reeler (rlORL) mutant mice. J. Comp. Neurol. **219,** 10–24.

Goffinet, A. M. (1984a). Abnormal development of the facial nerve nucleus in reeler mutant mice. J. Anat. **138,** 207–215.

Goffinet, A. M. (1984b). Events governing organization of postmigratory neurons: Studies on brain development in normal and reeler mice. Brain Res. **319,** 261–296.

Goffinet, A. M., So, K. F., Yamamoto, M., Edwards, M., and Caviness, V., Jr. (1984). Architectonic and hodological organization of the cerebellum in reeler mutant mice. Brain Res. **318,** 263–276.

Graus-Porta, D., Blaess, S., Senften, M., Littlewood-Evans, A., Damsky, C., Huang, Z., Orban, P., Klein, R., Schittny, J. C., and Muller, U. (2001). Beta1-class integrins regulate the development of laminae and folia in the cerebral and cerebellar cortex. Neuron **31,** 367–379.

Guidotti, A., Auta, J., Davis, J. M., DiGiorgi Gerevini, V., Dwivedi, Y., Grayson, D. R., Impagnatiello, F., Pandey, G., Pesold, C., Sharma, R., Uzunov, D., and Costa, E. (2000). Decrease in reelin and glutamic acid decarboxylase67 (GAD67) expression in schizophrenia and bipolar disorder: A postmortem brain study. *Arch. Gen. Psychiatry* **57,** 1061–1069.

Haas, C. A., Dudeck, O., Kirsch, M., Huszka, C., Kann, G., Pollak, S., Zentner, J., and Frotscher, M. (2002). Role for reelin in the development of granule cell dispersion in temporal lobe epilepsy. *J. Neurosci.* **22,** 5797–5802.

Hack, I., Bancila, M., Loulier, K., Carroll, P., and Cremer, H. (2002). Reelin is a detachment signal in tangential chain-migration during postnatal neurogenesis. *Nat. Neurosci.* **5,** 939–945.

Hartfuss, E., Forster, E., Bock, H. H., Hack, M. A., Leprince, P., Luque, J. M., Herz, J., Frotscher, M., and Gotz, M. (2003). Reelin signaling directly affects radial glia morphology and biochemical maturation. *Development* **130,** 4597–4609.

Hattori, M., Adachi, H., Tsujimoto, M., Arai, N., and Inoue, K. (1994). Miller-Dieker lissencephaly gene encodes a subunit of brain platelet-activating factor acetylhydrolase. *Nature* **370,** 216–218.

Heckroth, J. A., Goldowitz, D., and Eisenman, L. M. (1989). Purkinje cell reduction in the reeler mutant mouse: A quantitative immunohistochemical study. *J. Comp. Neurol.* **279,** 546–555.

Herrick, T. M., and Cooper, J. A. (2002). A hypomorphic allele of dab1 reveals regional differences in reelin-Dab1 signaling during brain development. *Development* **129,** 787–796.

Hevner, R. F., Daza, R. A., Englund, C., Kohtz, J., and Fink, A. (2004). Postnatal shifts of interneuron position in the neocortex of normal and reeler mice: Evidence for inward radial migration. *Neuroscience* **124,** 605–618.

Hiesberger, T., Trommsdorff, M., Howell, B. W., Goffinet, A. M., Mumby, M. C., Cooper, J. A., and Herz, J. (1999). Direct binding of Reelin to VLDL receptor and ApoE receptor 2 induces tyrosine phosphorylation of Disabled-1 and modulates Tau phosphorylation. *Neuron* **24,** 481–489.

Hirotsune, S., Takahara, T., Sasaki, N., Hirose, K., Yoshiki, A., Ohashi, T., Kusakabe, M., Murakami, Y., Muramatsu, M., Watanabe, S., Nakao, K., Katsuki, M., and Hayashizaki, Y. (1995). The reeler gene encodes a protein with an EGF-like motif expressed by pioneer neurons. *Nat. Genet.* **10,** 77–83.

Homayouni, R., Magdaleno, S., Keshvara, L., Rice, D. S., and Curran, T. (2003). Interaction of Disabled-1 and the GTPase activating protein Dab2IP in mouse brain. *Brain Res. Mol. Brain Res.* **115,** 121–129.

Homayouni, R., Rice, D. S., Sheldon, M., and Curran, T. (1999). Dab1 binds to the cytoplasmic domain of amyloid precursor-like protein 1. *J. Neurosci.* **19,** 7507–7515.

Hong, S. E., Shugart, Y. Y., Huang, D. T., Al Shahwan, S., Grant, P. E., Hourihane, J. O. B., Martin, N. D. T., and Walsh, C. A. (2000). Autosomal recessive lissencephaly with cerebellar hypoplasia is associated with human RELN mutations. *Nature Genet.* **26,** 93–96.

Howell, B. W., Gertler, F. B., and Cooper, J. A. (1997a). Mouse disabled (mDab)1: A src binding protein implicated in neuronal development. *EMBO J.* **16,** 121–132.

Howell, B. W., Hawkes, R., Soriano, P., and Cooper, J. A. (1997b). Neuronal position in the developing brain is regulated by mouse disabled-1. *Nature* **389,** 733–736.

Howell, B. W., Herrick, T. M., and Cooper, J. A. (1999a). Reelin-induced tyrosine phosphorylation of Disabled 1 during neuronal positioning. *Genes Dev.* **13,** 643–648.

Howell, B. W., Herrick, T. M., Hildebrand, J. D., Zhang, Y., and Cooper, J. A. (2000). Dab1 tyrosine phosphorylation sites relay positional signals during mouse brain development. *Curr. Biol.* **10,** 877–885.

Howell, B. W., Lanier, L. M., Frank, R., Gertler, F. B., and Cooper, J. A. (1999b). The Disabled-1 PTB domain binds to the internalization signals of transmembrane glycoproteins and to phospholipids. *Mol. Cell. Biol.* **19,** 5179–5188.

Huang, Y., Magdaleno, S., Hopkins, R., Slaughter, C., Curran, T., and Keshvara, L. (2004). Tyrosine phosphorylated Disabled 1 recruits Crk family adapter proteins. *Biochem. Biophys. Res. Commun.* **318,** 204–212.

Hunter-Schaedle, K. E. (1997). Radial glial cell development and transformation are disturbed in reeler forebrain. *J. Neurobiol.* **33,** 459–472.

Ignatova, N., Sindic, C. J., and Goffinet, A. M. (2004). Characterization of the various forms of the Reelin protein in the cerebrospinal fluid of normal subjects and in neurological diseases. *Neurobiol. Dis.* **15,** 326–330.

Ikeda, Y., and Terashima, T. (1997). Expression of *reelin*, the gene responsible for the Reeler mutation, in embryonic development and adulthood in the mouse. *Dev. Dyn.* **210,** 157–172.

Impagnatiello, F., Guidotti, A. R., Pesold, C., Dwivedi, Y., Caruncho, H., Pisu, M. G., Uzunov, D. P., Smalheiser, N. R., Davis, J. M., Pandey, G. N., Pappas, G. D., Tueting, P., Sharma, R. P., and Costa, E. (1998). A decrease of reelin expression as a putative vulnerability factor in schizophrenia. *Proc. Natl. Acad. Sci. USA* **95,** 15718–15723.

Jensen, P., Zoghbi, H. Y., and Goldowitz, D. (2002). Dissection of the cellular and molecular events that position cerebellar Purkinje cells: A study of the math1 null-mutant mouse. *J. Neurosci.* **22,** 8110–8116.

Jossin, Y., Bar, I., Ignatova, N., Tissir, F., De Rouvroit, C. L., and Goffinet, A. M. (2003a). The reelin signaling pathway: Some recent developments. *Cereb. Cortex.* **13,** 627–633.

Jossin, Y., and Goffinet, A. M. (2001). Reelin does not directly influence axonal growth. *J. Neurosci.* **21,** RC183.

Jossin, Y., Ignatova, N., Hiesberger, T., Herz, J., Lambert de Rouvroit, C., and Goffinet, A. M. (2004). The central fragment of Reelin, generated by proteolytic processing *in vivo*, is critical to its function during cortical plate development. *J. Neurosci.* **24,** 514–521.

Jossin, Y., Ogawa, M., Metin, C., Tissir, F., and Goffinet, A. M. (2003b). Inhibition of SRC family kinases and non-classical protein kinases C induce a reeler-like malformation of cortical plate development. *J. Neurosci.* **23,** 9953–9959.

Keshvara, L., Benhayon, D., Magdaleno, S., and Curran, T. (2001). Identification of reelin-induced sites of tyrosyl phosphorylation on disabled 1. *J. Biol. Chem.* **276,** 16008–16014.

Kikkawa, S., Yamamoto, T., Misaki, K., Ikeda, Y., Okado, H., Ogawa, M., Woodhams, P. L., and Terashima, T. (2003). Missplicing resulting from a short deletion in the reelin gene causes reeler-like neuronal disorders in the mutant shaking rat Kawasaki. *J. Comp. Neurol.* **463,** 303–315.

Kubasak, M. D., Brooks, R., Chen, S., Villeda, S. A., and Phelps, P. E. (2004). Developmental distribution of reelin-positive cells and their secreted product in the rodent spinal cord. *J. Comp. Neurol.* **468,** 165–178.

Kubo, K., Mikoshiba, K., and Nakajima, K. (2002). Secreted Reelin molecules form homodimers. *Neurosci. Res.* **43,** 381–388.

Lambert de Rouvroit, C., and Goffinet, A. M. (1998). The reeler mouse as a model of brain development. *Adv. Anat. Embryol Cell Biol.* **150,** 1–108.

Lee, V. M., Goedert, M., and Trojanowski, J. Q. (2001). Neurodegenerative tauopathies. *Annu. Rev. Neurosci.* **24,** 1121–1159.

Liu, W. S., Pesold, C., Rodriguez, M. A., Carboni, G., Auta, J., Lacor, P., Larson, J., Condie, B. G., Guidotti, A., and Costa, E. (2001). Down-regulation of dendritic spine and glutamic acid decarboxylase 67 expressions in the reelin haploinsufficient heterozygous reeler mouse. *Proc. Natl. Acad. Sci. USA* **98,** 3477–3482.

Luque, J. M., Morante-Oria, J., and Fairen, A. (2003). Localization of ApoER2, VLDLR and Dab1 in radial glia: Groundwork for a new model of reelin action during cortical development. *Brain Res. Dev. Brain Res.* **140,** 195–203.

Magdaleno, S., Keshvara, L., and Curran, T. (2002). Rescue of ataxia and preplate splitting by ectopic expression of Reelin in reeler mice. *Neuron* **33,** 573–586.

Malatesta, P., Hack, M. A., Hartfuss, E., Kettenmann, H., Klinkert, W., Kirchhoff, F., and Gotz, M. (2003). Neuronal or glial progeny: Regional differences in radial glia fate. *Neuron* **37,** 751–764.

Malatesta, P., Hartfuss, E., and Gotz, M. (2000). Isolation of radial glial cells by fluorescent-activated cell sorting reveals a neuronal lineage. *Development* **127,** 5253–5263.

Mariani, J. (1982). Extent of multiple innervation of Purkinje cells by climbing fibers in the olivocerebellar system of weaver, reeler and staggerer mutant mice. *J. Neurosci.* **13,** 119–126.

Mariani, J., Crepel, F., Mikoshiba, K., Changeux, J. P., and Sotelo, C. (1977). Anatomical, physiological and biochemical studies of the cerebellum from Reeler mutant mouse. *Philos. Trans. R. Soc. Lond. Biol.* **281,** 1–28.

Marin, O., and Rubenstein, J. L. R. (2001). A long, remarkable journey: Tangential migration in the telencephalon. *Nat. Rev.* **2,** 1–11.

Marin-Padilla, M. (1978). Dual origin of the mammalian neocortex and evolution of the cortical plate. *Anat. Embryol.* **152,** 109–126.

Martin, M. R. (1981). Morphology of the cochlear nucleus of the normal and reeler mutant mouse. *J. Comp. Neurol.* **197,** 141–152.

Martinez-Cerdeno, V., Galazo, M. J., and Clasca, F. (2003). Reelin-immunoreactive neurons, axons, and neuropil in the adult ferret brain: Evidence for axonal secretion of reelin in long axonal pathways. *J. Comp. Neurol.* **463,** 92–116.

Maurin, J. C., Couble, M. L., Didier-Bazes, M., Brisson, C., Magloire, H., and Bleicher, F. (2004). Expression and localization of reelin in human odontoblasts. *Matrix Biol.* **23,** 277–285.

Meyer, G., De Rouvroit, C. L., Goffinet, A. M., and Wahle, P. (2003). Disabled-1 mRNA and protein expression in developing human cortex. *Eur. J. Neurosci.* **17,** 517–525.

Meyer, G., and Goffinet, A. M. (1998). Prenatal development of Reelin-immunoreactive neurons in the human neocortex. *J. Comp. Neurol.* **397,** 29–40.

Meyer, G., Soria, J. M., Martinez-Galan, J. R., Martin-Clemente, B., and Fairen, A. (1998). Different origins and developmental histories of transient neurons in the marginal zone of the fetal and neonatal rat cortex. *J. Comp. Neurol.* **397,** 493–518.

Miao, G. G., Smeyne, R. J., D'Arcangelo, G., Copeland, N. G., Jenkins, N. A., Morgan, J. I., and Curran, T. (1994). Isolation of an allele of reeler by insertional mutagenesis. *Proc. Natl. Acad. Sci. USA* **91,** 11050–11054.

Mikoshiba, K., Nagaike, K., Kosaka, S., Takamatsu, K., Aoki, E., and Tsukada, Y. (1980). Developmental studies on the cerebellum from reeler mutant mice *in vivo* and *in vitro*. *Dev. Biol.* **79,** 64–80.

Mitchell, C. P., Chen, Y., Kundakovic, M., Costa, E., and Grayson, D. R. (2005). Histone deacetylase inhibitors decrease reelin promoter methylation *in vitro*. *J. Neurochem.* **93,** 483–492.

Miyata, T., Nakajima, K., Mikoshiba, K., and Ogawa, M. (1997). Regulation of Purkinje cell alignment by Reelin as revealed with CR-50 antibody. *J. Neurosci.* **17,** 3599–3609.

Molnar, Z., Adams, R., Goffinet, A. M., and Blakemore, C. (1998). The role of the first postmitotic cortical cells in the development of thalamocortical innervation in the reeler mouse. *J. Neurosci.* **18,** 5746–5765.

Morimura, T., Hattori, M., Ogawa, M., and Mikoshiba, K. (2005). Disabled1 regulates the intracellular trafficking of reelin receptors. *J. Biol. Chem.* **280,** 16901–16908.

Nadarajah, B., Brunstrom, J. E., Grutzendler, J., Wong, R. O., and Pearlman, A. L. (2001). Two modes of radial migration in early development of the cerebral cortex. *Nat. Neurosci.* **4,** 143–150.

Nagata, I., and Terashima, T. (1994). Migration behavior of granule cells on laminin in cerebellar microexplant cultures from early postnatal reeler mutant mice. *Int. J. Dev. Neurosci.* **12,** 387–395.

Nakajima, K., Mikoshiba, K., Miyata, T., Kudo, C., and Ogawa, M. (1997). Disruption of hippocampal development *in vivo* by CR-50 mAb against Reelin. *Proc. Natl. Acad. Sci. USA* **94,** 8196–8201.

Nishikawa, S., Goto, S., Yamada, K., Hamasaki, T., and Ushio, Y. (2003). Lack of Reelin causes malpositioning of nigral dopaminergic neurons: Evidence from comparison of normal and Reln (rl) mutant mice. *J. Comp. Neurol.* **461,** 166–173.

Niu, S., Renfro, A., Quattrocchi, C. C., Sheldon, M., and D'Arcangelo, G. (2004). Reelin promotes hippocampal dendrite development through the VLDLR/ApoER2-Dab1 pathway. *Neuron* **41,** 71–84.

Noctor, S. C., Flint, A. C., Weissman, T. A., Dammerman, R. S., and Kriegstein, A. R. (2001). Neurons derived from radial glial cells establish radial units in neocortex. *Nature* **409,** 714–720.

Noctor, S. C., Martinez-Cerdeno, V., Ivic, L., and Kriegstein, A. R. (2004). Cortical neurons arise in symmetric and asymmetric division zones and migrate through specific phases. *Nat. Neurosci.* **7,** 136–144.

Noh, J. S., Sharma, R. P., Veldic, M., Salvacion, A. A., Jia, X., Chen, Y., Costa, E., Guidotti, A., and Grayson, D. R. (2005). DNA methyltransferase 1 regulates reelin mRNA expression in mouse primary cortical cultures. *Proc. Natl. Acad. Sci. USA* **102,** 1749–1754.

Ogawa, M., Miyata, T., Nakajima, K., Yagyu, K., Seike, M., Ikenaka, K., Yamamoto, H., and Mikoshiba, K. (1995). The reeler gene-associated antigen on Cajal-Retzius neurons is a crucial molecule for laminar organization of cortical neurons. *Neuron* **14,** 899–912.

Pappas, G. D., Kriho, V., and Pesold, C. (2001). Reelin in the extracellular matrix and dendritic spines of the cortex and hippocampus: A comparison between wild type and heterozygous reeler mice by immunoelectron microscopy. *J. Neurocytol.* **30,** 413–425.

Perez-Garcia, C. G., Tissir, F., Goffinet, A. M., and Meyer, G. (2004). Reelin receptors in developing laminated brain structures of mouse and human. *Eur. J. Neurosci.* **20,** 2827–2832.

Persico, A. M., D'Agruma, L., Maiorano, N., Totaro, A., Militerni, R., Bravaccio, C., Wassink, T. H., Schneider, C., Melmed, R., Trillo, S., Montecchi, F., Palermo, M., Pascucci, T., Puglisi-Allegra, S., Reichelt, K. L., Conciatori, M., Marino, R., Quattrocchi, C. C., Baldi, A., Zelante, L., Gasparini, P., and Keller, F. (2001). Reelin gene alleles and haplotypes as a factor predisposing to autistic disorder. *Mol. Psychiatry* **6,** 129–133.

Pesold, C., Impagnatiello, F., Pisu, M. G., Uzunov, D. P., Costa, E., Guidotti, A., and Caruncho, H. J. (1998). Reelin is preferentially expressed in neurons synthesizing γ-aminobutyric acid in cortex and hippocampus of adult rats. *Proc. Natl. Acad. Sci. USA* **95,** 3221–3226.

Phelps, P. E., Rich, R., Dupuy-Davies, S., Rios, Y., and Wong, T. (2002). Evidence for a cell-specific action of Reelin in the spinal cord. *Dev. Biol.* **244,** 180–198.

Pinto Lord, M. C., and Caviness, V. S., Jr. (1979). Determinants of cell shape and orientation: A comparative Golgi analysis of cell-axon interrelationships in the developing neocortex of normal and reeler mice. *J. Comp. Neurol.* **187,** 49–69.

Pinto-Lord, M. C., Evrard, P., and Caviness, V. S. J. (1982). Obstructed neuronal migration along radial glial fibers in the neocortex of the reeler mouse: A Golgi-EM analysis. *Dev. Brain Res.* **4,** 379–393.

Podhorna, J., and Didriksen, M. (2004). The heterozygous reeler mouse: Behavioural phenotype. *Behav. Brain Res.* **153,** 43–54.

Pramatarova, A., Ochalski, P. G., Chen, K., Gropman, A., Myers, S., Min, K. T., and Howell, B. W. (2003). Nck beta interacts with tyrosine-phosphorylated disabled 1 and redistributes in Reelin-stimulated neurons. *Mol. Cell. Biol.* **23,** 7210–7221.

Reiner, O., Carrozzo, R., Shen, Y., Wehnert, M., Faustinella, F., Dobyns, W. B., Caskey, C. T., and Ledbetter, D. H. (1993). Isolation of a Miller-Dieker lissencephaly gene containing G protein beta-subunit-like repeats. *Nature* **364,** 717–721.

Rice, D. S., Nusinowitz, S., Azimi, A. M., Martinez, A., Soriano, E., and Curran, T. (2001). The reelin pathway modulates the structure and function of retinal synaptic circuitry. *Neuron* **31,** 929–941.

Rice, D. S., Sheldon, M., D'Arcangelo, G., Nakajima, K., Goldowitz, D., and Curran, T. (1998). *Disabled-1* acts downstream of *Reelin* in a signaling pathway that controls laminar organization in the mammalian brain. *Development* **125,** 3719–3729.

Roberts, R. C., Xu, L., Roche, J. K., and Kirkpatrick, B. (2005). Ultrastructural localization of reelin in the cortex in post-mortem human brain. *J. Comp. Neurol.* **482,** 294–308.

Rodriguez, M. A., Pesold, C., Liu, W. S., Khrino, V., Guidotti, A., Pappas, G. D., and Costa, E. (2000). Colocalization of integrin receptors and reelin in dendritic spine postsynaptic densities of adult nonhuman primate cortex. *Proc. Natl. Acad. Sci. USA* **97,** 3550–3555.

Rossel, M., Loulier, K., Feuillet, C., Alonso, S., and Carroll, P. (2005). Reelin signaling is necessary for a specific step in the migration of hindbrain efferent neurons. *Development* **132,** 1175–1185.

Salinger, W. L., Ladrow, P., and Wheeler, C. (2003). Behavioral phenotype of the reeler mutant mouse: Effects of RELN gene dosage and social isolation. *Behav. Neurosci.* **117,** 1257–1275.

Sanada, K., Gupta, A., and Tsai, L. H. (2004). Disabled-1-regulated adhesion of migrating neurons to radial glial fiber contributes to neuronal positioning during early corticogenesis. *Neuron* **42,** 197–211.

Schiffmann, S. N., Bernier, B., and Goffinet, A. M. (1997). Reelin mRNA expression during mouse brain development. *Eur. J. Neurosci.* **9,** 1055–1071.

Schmid, R. S., Jo, R., Shelton, S., Kreidberg, J. A., and Anton, E. S. (2005). Reelin, integrin and Dab1 interactions during embryonic cerebral cortical development. *Cereb. Cortex.* **15,** 1632–1636.

Shatz, C. J., Ghosh, A., McConnell, S. K., Allendoerfer, K. L., Friauf, E., and Antonini, A. (1990). Pioneer neurons and target selection in cerebral cortical development. *Cold Spring Harb. Symp. Quant. Biol.* **55,** 469–480.

Sheldon, M., Rice, D. S., D'Arcangelo, G., Yoneshima, H., Nakajima, K., Mikoshiba, K., Howell, B. W., Cooper, J. A., Goldowitz, D., and Curran, T. (1997). *Scrambler* and *yotari* disrupt the *disabled* gene and produce a *reeler*-like phenotype in mice. *Nature* **389,** 730–733.

Skaar, D. A., Shao, Y., Haines, J. L., Stenger, J. E., Jaworski, J., Martin, E. R., Delong, G. R., Moore, J. H., McCauley, J. L., Sutcliffe, J. S., Ashley-Koch, A. E., Cuccaro, M. L., Folstein, S. E., Gilbert, J. R., and Pericak-Vance, M. A. (2004). Analysis of the RELN gene as a genetic risk factor for autism. *Mol Psychiatry* **10,** 563–571.

Smalheiser, N. R., Costa, E., Guidotti, A., Impagnatiello, F., Auta, J., Lacor, P., Kriho, V., and Pappas, G. D. (2000). Expression of reelin in adult mammalian blood, liver, pituitary pars intermedia, and adrenal chromaffin cells. *Proc. Natl. Acad. Sci. USA* **97,** 1281–1286.

Stanfield, B. B., and Cowan, W. M. (1979). The morphology of the hippocampus and dentate gyrus in normal and reeler mice. *J. Comp. Neurol.* **185,** 393–422.

Steindler, D. A., and Colwell, S. A. (1976). Reeler mutant mouse: Maintenance of appropriate and reciprocal connections in the cerebral cortex and thalamus. *Brain Res.* **113,** 386–393.

Stolt, P. C., Chen, Y., Liu, P., Bock, H. H., Blacklow, S. C., and Herz, J. (2005). Phosphoinositide binding by the Disabled-1 PTB domain is necessary for membrane localization and Reelin signal transduction. *J. Biol. Chem.* **280,** 9671–9677.

Stolt, P. C., Jeon, H., Song, H. K., Herz, J., Eck, M. J., and Blacklow, S. C. (2003). Origins of peptide selectivity and phosphoinositide binding revealed by structures of disabled-1 PTB domain complexes. *Structure (Camb.)* **11,** 569–579.

Strasser, V., Fasching, D., Hauser, C., Mayer, H., Bock, H. H., Hiesberger, T., Herz, J., Weeber, E. J., Sweatt, J. D., Pramatarova, A., Howell, B., Schneider, W. J., and Nimpf, J. (2004). Receptor clustering is involved in Reelin signaling. *Mol. Cell. Biol.* **24,** 1378–1386.

Suetsugu, S., Tezuka, T., Morimura, T., Hattori, M., Mikoshiba, K., Yamamoto, T., and Takenawa, T. (2004). Regulation of actin cytoskeleton by mDab1 through N-WASP and ubiquitination of mDab1. *Biochem. J.* **384,** 1–8.

Tabata, H., and Nakajima, K. (2001). Efficient *in utero* gene transfer system to the developing mouse brain using electroporation: Visualization of neuronal migration in the developing cortex. *Neuroscience* **103,** 865–872.

Tabata, H., and Nakajima, K. (2002). Neurons tend to stop migration and differentiate along the cortical internal plexiform zones in the Reelin signal-deficient mice. *J. Neurosci. Res.* **69,** 723–730.

Tabata, H., and Nakajima, K. (2003). Multipolar migration: The third mode of radial neuronal migration in the developing cerebral cortex. *J. Neurosci.* **23,** 9996–10001.

Takiguchi-Hayashi, K., Sekiguchi, M., Ashigaki, S., Takamatsu, M., Hasegawa, H., Suzuki-Migishima, R., Yokoyama, M., Nakanishi, S., and Tanabe, Y. (2004). Generation of reelin-positive marginal zone cells from the caudomedial wall of telencephalic vesicles. *J. Neurosci.* **24,** 2286–2295.

Teillon, S. M., Yiu, G., and Walsh, C. A. (2003). Reelin is expressed in the accessory olfactory system, but is not a guidance cue for vomeronasal axons. *Brain Res. Dev. Brain Res.* **140,** 303–307.

Terashima, T. (1996). Distribution of mesencephalic trigeminal nucleus neurons in the reeler mutant mouse. *Anat. Rec.* **244,** 563–571.

Terashima, T., Kishimoto, Y., and Ochiishi, T. (1993). Musculotopic organization of the facial nucleus of the reeler mutant mouse. *Brain Res.* **617,** 1–9.

Tremolizzo, L., Carboni, G., Ruzicka, W. B., Mitchell, C. P., Sugaya, I., Tueting, P., Sharma, R., Grayson, D. R., Costa, E., and Guidotti, A. (2002). An epigenetic mouse model for molecular and behavioral neuropathologies related to schizophrenia vulnerability. *Proc. Natl. Acad. Sci. USA* **99,** 17095–17100.

Tremolizzo, L., Doueiri, M. S., Dong, E., Grayson, D. R., Davis, J., Pinna, G., Tueting, P., Rodriguez-Menendez, V., Costa, E., and Guidotti, A. (2005). Valproate corrects the schizophrenia-like epigenetic behavioral modifications induced by methionine in mice. *Biol. Psychiatry* **57,** 500–509.

Trommsdorff, M., Borg, J. P., Margolis, B., and Herz, J. (1998). Interaction of cytosolic adaptor proteins with neuronal apolipoprotein E receptors and the amyloid precursor protein. *J. Biol. Chem.* **273,** 33556–33560.

Trommsdorff, M., Gotthardt, M., Hiesberger, T., Shelton, J., Stockinger, W., Nimpf, J., Hammer, R. E., Richardson, J. A., and Herz, J. (1999). Reeler/Disabled-like disruption of neuronal migration in knockout mice lacking the VLDL receptor and ApoE receptor 2. *Cell* **97,** 689–701.

Tueting, P., Costa, E., Dwivedi, Y., Guidotti, A., Impagnatiello, F., Manev, R., and Pesold, C. (1999). The phenotypic characteristics of heterozygous reeler mouse. *Neuroreport* **10,** 1329–1334.

Vallee, R. B., Tai, C., and Faulkner, N. E. (2001). LIS1: Cellular function of a disease-causing gene. *Trends Cell. Biol.* **11,** 155–160.

Veldic, M., Caruncho, H. J., Liu, W. S., Davis, J., Satta, R., Grayson, D. R., Guidotti, A., and Costa, E. (2004). DNA-methyltransferase 1 mRNA is selectively overexpressed in telencephalic GABAergic interneurons of schizophrenia brains. *Proc. Natl. Acad. Sci. USA* **101,** 348–353.

Ware, M. L., Fox, J. W., Gonzales, J. L., Davis, N. M., Lambert de Rouvroit, C., Russo, C. J., Chua, S. C. J., Goffinet, A. M., and Walsh, C. A. (1997). Aberrant splicing of a mouse *disabled* homolog, *mdab1,* in the *scrambler* mouse. *Neuron* **19,** 239–249.

Weeber, E. J., Beffert, U., Jones, C., Christian, J. M., Forster, E., Sweatt, J. D., and Herz, J. (2002). Reelin and ApoE receptors cooperate to enhance hippocampal synaptic plasticity and learning. *J. Biol Chem.* **277,** 39944–39952.

Weiss, K. H., Johanssen, C., Tielsch, A., Herz, J., Deller, T., Frotscher, M., and Forster, E. (2003). Malformation of the radial glial scaffold in the dentate gyrus of reeler mice, scrambler mice, and ApoER2/VLDLR-deficient mice. *J. Comp. Neurol.* **460,** 56–65.

Wynshaw-Boris, A., and Gambello, M. J. (2001). LIS1 and dynein motor function in neuronal migration and development. *Genes. Dev.* **15,** 639–651.

Yabut, O., Renfro, A., Swann, J. W., Marín, O., and D'Arcangelo, G. Abnormal laminar position and dendrite development of interneurons in the reeler forebrain. *Mol. Brain Res.*, in press.

Yip, J. W., Yip, Y. P., Nakajima, K., and Capriotti, C. (2000). Reelin controls position of autonomic neurons in the spinal cord. *Proc. Natl. Acad. Sci. USA* **97,** 8612–8616.

Yip, Y. P., Capriotti, C., Magdaleno, S., Benhayon, D., Curran, T., Nakajima, K., and Yip, J. W. (2004a). Components of the reelin signaling pathway are expressed in the spinal cord. *J. Comp. Neurol.* **470,** 210–219.

Yip, Y. P., Capriotti, C., and Yip, J. W. (2003). Migratory pathway of sympathetic preganglionic neurons in normal and reeler mutant mice. *J. Comp. Neurol.* **460,** 94–105.

Yip, Y. P., Zhou, G., Capriotti, C., and Yip, J. W. (2004b). Location of preganglionic neurons is independent of birthdate but is correlated to reelin-producing cells in the spinal cord. *J. Comp. Neurol.* **475,** 564–574.

Yokoi, N., Namae, M., Wang, H. Y., Kojima, K., Fuse, M., Yasuda, K., Serikawa, T., Seino, S., and Komeda, K. (2003). Rat neurological disease creeping is caused by a mutation in the reelin gene. *Brain Res. Mol. Brain Res.* **112,** 1–7.

Yoneshima, H., Nagata, E., Matsumoto, M., Yamada, M., Nakajima, K., Miyata, T., Ogawa, M., and Mikoshiba, K. (1997). A novel neurological mutation of mouse, *yotari*, which exhibits *reeler*-like phenotype but expresses *reelin*. *Neurosci. Res.* **29,** 217–223.

Yuasa, S., Kitoh, J., Oda, S., and Kawamura, K. (1993). Obstructed migration of Purkinje cells in the developing cerebellum of the reeler mutant mouse. *Anat. Embryol. (Berl.)* **188,** 317–329.

Yun, M., Keshvara, L., Park, C. G., Zhang, Y. M., Dickerson, J. B., Zheng, J., Rock, C. O., Curran, T., and Park, H. W. (2003). Crystal structures of the Dab homology domains in mouse disabled-1 and -2. *J. Biol. Chem.* **278,** 36572–36581.

Zhang, H., Liu, X., Zhang, C., Mundo, E., Macciardi, F., Grayson, D. R., Guidotti, A. R., and Holden, J. J. (2002). Reelin gene alleles and susceptibility to autism spectrum disorders. *Mol. Psychiatry* **7,** 1012–1017.

Zhao, S., Chai, X., Forster, E., and Frotscher, M. (2004). Reelin is a positional signal for the lamination of dentate granule cells. *Development* **131,** 5117–5125.

SHARED CHROMOSOMAL SUSCEPTIBILITY REGIONS BETWEEN AUTISM AND OTHER MENTAL DISORDERS

Yvon C. Chagnon

Genetic and Molecular Psychiatry Unit, Robert-Giffard Research Center, Laval University
Beauport, Québec GIJ 2G3, Canada

We have compiled significant results from 53 genome scans for five different mental disorders including autism syndrome disorder (ASD), schizophrenia (SZ) and catatonia, bipolar disorder (BP), attention-deficit/hyperactivity disorder (ADHD), and alcoholism. Eight autosomal chromosomes (1, 2, 3, 7, 9, 13, 15, and 17) showed significant linkages with ASD while five of these chromosomes (1, 7, 13, 15, and 17) shared common susceptibility loci with other mental disorders mostly SZ and BP. Chromosome 15 is particularly rich in shared regions, three to four being detected where all four other mental disorders are involved in one or the other of these regions. Chromosome 15 is a particularly unstable chromosome where numerous chromosomal rearrangement and abnormalities have been associated with ASD. Strong candidate genes such as gamma-amino-butyric acid (GABA) receptor B3, A5, and G3 have shown associations with ASD on this chromosome. Some susceptibility loci for different mental disorders have also been assigned to chromosome 15 such as schizophrenia 10, major depressive disorder 2, dyslexia 1, epilepsy juvenile myotonic 2, and spinocerebellar ataxia 11. Finally, the only significant linkage results with catatonia are also found in a region shared by ASD, SZ, and ADHD on chromosome 15.

I. Introduction

Autism spectrum disorder (ASD) is a severe neurodevelopmental disorder characterized by delayed or absent speech, impairments in social interaction and in communication, and repetitive behaviors and restricted interests. The relationships between ASD, schizophrenia (SZ), and bipolar disorder (BP) are not clear.

INTERNATIONAL REVIEW OF
NEUROBIOLOGY, VOL. 71
DOI: 10.1016/S0074-7742(05)71017-5

419

Although characteristic features of SZ and BP such as delusions, hallucinations, euphoria, or melancholia are usually not present in people with ASD, there are also a few common psychopathological dimensions. For example, catatonia has been described in ASD, SZ, and BP (Abrams and Taylor, 1976; Dhossche, 1998; Realmuto and August, 1991). Moreover, family studies have reported increased rates of schizophrenia-like and affective disorders (Larsson et al., 2005).

Parental schizophrenia-like psychosis and affective disorder were significant risk factors for autism in offspring in a nationwide Danish case-control study of 698 children diagnosed with autism between 1972–1999 (Larsson et al., 2005). Relative risks were 3.44, 95% CI 1.48–7.95 and 2.91, 95% CI 1.65–5.14, for parental schizophrenia-like disorder and affective disorder respectively. Other significant variables were breech presentation (RR = 1.63), low Apgar score at 5 minutes (RR = 1.89), and gestational age at birth less than 35 weeks (RR = 2.45). Weight for gestational age, parity, number of antenatal visits, parental age, or socioeconomic status were not significant risk factors. These findings support that perinatal factors and parental psychopathology are associated with risk of autism. It remains an open question whether perinatal adversity was due to environmental factors, factors associated with autism in the fetus, or a combination of these and possibly other (unmeasured) variables.

A parsimonious explanation is that the genetic make-up of the fetus interferes with intrauterine development leading to increased risk for perinatal complications. However, perinatal factors and parental psychiatric disorder seemed to act independently in this study. This suggests two sets of autism-related etiologies (i.e., a genetic set and an obstetric set). These findings need to be replicated in other samples. However, the Danish study will be hard to match because of its almost complete ascertainment of cases, high quality of information on all risk factors, and prospective design. In any case, further family psychiatric studies assessing psychotic disorders, including catatonic subtypes, as risk factors for autism are warranted. Previous studies have typically not separated out catatonic subtypes of schizophrenia, affective disorder, or other psychotic disorders.

A subtype of unsystematic SZ is characterized by periodic catatonia where acute psychotic episodes are followed by remission. These disorders have shown strong heritabilities with some 15 to 20 genes expected to be involved in ASD (Spence, 2004), while a major gene effect is predicted for catatonia (Stober et al., 1995). The search of the genes related to these psychiatric disorders has shown significant progress in the recent years with the identification of some strong candidates for SZ such as the dystrobevin binding protein 1 (Straub et al., 2002) and the neuregulin 1 (Stefansson et al., 2002) genes. Numerous other genes have been tested for these disorders with more or less success. However, from the literature, it is obvious that the same genes and biological pathways could be shared among the disorders. A striking example is the catechol-O-methyltransferase gene (COMT) on chromosome 22q11.2. COMT encodes a key enzyme in the elimination of dopamine in the prefrontal cortex of the human brain and this role in the degradation of catecholamine neurotransmitters may suggest a general involvement of COMT in psychiatric diseases. The COMT protein shows two forms, the membrane-bound longer form being the main form expressed in the brain, while the soluble shorter form is expressed in other tissues such as the spleen and the liver. A common COMT polymorphism (Val108/158Met) in exon 4, changing a valine for a methionine at the position 108 or 158 of the short and the long forms, respectively, affects significantly the protein abundance and the enzyme activity, but not mRNA expression (Chen et al., 2004a; Egan et al., 2001; Shield et al., 2004). Positive association of Val108/158Met with SZ have been generally reported (Chen et al., 2004b; Egan et al., 2001; Glatt et al., 2003a; Sazci et al., 2004; Wonodi et al., 2003), as weaker evidence in a recent meta-analysis (Fan et al., 2005), and in a study of Korean SZ inpatients (Park et al., 2002). COMT Val108/158Met polymorphism has also been associated with schizotypy (Avramopoulos et al., 2002) but not cognition (Stefanis et al., 2004), with prefrontal neurocognitive function in healthy (Rosa et al., 2004) but not SZ individuals (Ho et al., 2005; Rosa et al., 2004), with the 22q11.2 deletion syndrome (Bearden et al., 2004), with anxiety (Enoch et al., 2003; McGrath et al., 2004), and with anorexia nervosa (Gabrovsek et al., 2004). Two SNPs located in intron 1 and in the 3′ flanking region of COMT have also been associated with a risk for SZ (Shifman et al., 2002) and for BP (Shifman et al., 2004). It has been proposed that the effect of the intron 1 SNP in SZ could come from a linkage disequilibrium (LD) with a SNP in the P2 promoter (Palmatier et al., 2004). These SNPs lowered COMT mRNA expression in anonymous postmortem brains (Bray et al., 2003), while this lower expression was not observed in the lymphocytes of SZ patients (Chen et al., 2004a). Other examples of the communality observed between mental disorders are the serotonergic and dopaminergic family of genes. For example, the serotonin transporter gene has been associated with different mental disorders including ASD (Yirmiya et al., 2001), while the dopamine receptor D2 had showed association with attention-deficit/hyperactivity disorder (ADHD), alcoholism (ALC), and Tourette's syndrome (Comings et al., 1991).

We have compiled the significant results from published genome scans for six different mental disorders and observed shared chromosomal susceptibility regions among all of them (Chagnon et al. in preparation). From this first analysis, we observed that chromosome 15, particularly, shared common susceptibility regions for some disorders including ASD, SZ, BP, ADHD, and ALC. The only significant linkage results for catatonia in a subgroup of SZ were also observed on chromosome 15. In this chapter, we will present chromosomes with significant genome scan results for ASD putting an emphasis on chromosome 15 where chromosomal rearrangements and abnormalities, and candidate gene analyses will also be presented in relation to ASD and catatonia susceptibility loci.

A. METHODS

Relevant genome scan papers have been identified by a search in the PubMed database using the key words "linkage OR genome scan" AND "schizophrenia OR bipolar disorder OR autism OR catatonia OR alcoholism OR attention deficit hyperactivity deficiency OR Tourette" (Chagnon *et al.* in preparation). The criterion for inclusion of the results of a genome scan was a Lod score of 3.0 and greater corresponding to a P value of 0.0001 and smaller. P values of 0.05 and smaller were also included when "genome wide adjusted." From these, only chromosomes including significant susceptibility regions for ASD have been retained in the actual compilation. The chromosomal location of all the linked markers has been updated using the same and most recent version of the physical map in megabases (Mb) from the National Center for Biological Information (NCBI built 35.1). One Mb corresponds to 10^6 bases or nucleotides of DNA. When the location on the physical map was not available for a given marker, genetic maps in centimorgan (cM) units from Marshfield (Broman *et al.*, 1998) have been used to determine the relative position of the marker, where one cM corresponds roughly to 1 Mb. Additionally, the cytological locations have been updated using the predictive locations (GMAP) from the Genetic Location Database (Collins *et al.*, 1996). We have also reported for the linked markers of these chromosomes the susceptibility loci assigned by NCBI (built 35.1) where the same locus can be assigned to more then one marker. For example, four linked markers of the genome scans (D1S1631, D1S1653, D1S1679, and D1S196) are related to the Schizophrenia susceptibility 9 on chromosome 1p21.2-q24.2, while a unique linked marker (D1S484) at 1q24.1 is related to the Asperger syndrome 3. Some apparent contradictions are observed between physical and cytological locations for these susceptibility loci. For example, Autism 1 susceptibility locus at 7q31.31 is related to marker D7S486 according to NCBI ePCR result while Autism 1 was originally assigned to 15q11-q13. For chromosome 15 only, ASD and catatonia suggestive genome scan results, chromosomal rearrangements and abnormalities, and positive or negative candidate gene analyses have also been included.

B. RESULTS AND DISCUSSION

1. *Shared Chromosomes*

Table I presents the results from the 53 genome scans related to ASD (N = 18), to SZ (N = 17), to SZ catatonia (N = 2), to SZ and BP (N = 4), to BP (N = 8), to ADHD (n = 1), and to ALC (N = 3) that have been included in the table. We observed that eight autosomal chromosomes (1, 2, 3, 7, 9, 13, 15, and 17) showed significant linkages with ASD (see Table I). Chromosome 15

TABLE I

SIGNIFICANT GENOME SCAN RESULTS OBSERVED FOR ASD AND SHARED BY OTHER MENTAL DISORDERS (MD). PHYSICAL (NCBI) AND GENETIC (MARSHFIELD) ARE REPORTED

Chr	Location	Markers	NCBI	Marshfield	Statistic	MD	Loci/ Association results	References
1	1p36.32	D1S1612	8052251	16.2	Lod = 3.3	SZ		(Abecasis et al., 2004)
1	1p21.2	D1S1631	105372673	136.9	Zmax = 3.4	ASD	SCZD9	(Risch et al., 1999)
1	1q21.3	KCNN3	151655827	na	ns	SZ catatonia		(Stober et al., 2000a)
1	1q22	D1S1653	154745847	164.1	Lod = 6.1	SZ	SCZD9	(Brzustowicz et al., 2000)
1	1q24.1	D1S484	157580640	169.7	Zmax = 3.6	ASD (AS)	ASPG3	(Ylisaukko-Oja et al., 2004)
1	1q24	rs1415263	158897701	na	Lod = 6.5	SZ		(Brzustowicz et al., 2002)
1	1q24	336H14-CA1	158897701	na	p = 0.001	SZ		(Brzustowicz et al., 2004)
1	1q24.2	D1S1679	159093573	170.8	Lod = 6.1	SZ	SCZD9	(Brzustowicz et al., 2000)
1	1q24.2	D1S196	164791505	181.5	Lod = 3.2	SZ	SCZD9	(Gurling et al., 2001)
1	1q31.2	PFKFB2	203639264	na	p = 0.04*	SZ		(Stone et al., 2004)
1	1q42.12	D1S2141	211583695	233.4	Zmax = 3.8	SZ		(Hovatta et al., 1999)
1	1q42.2	D1S2709	228337174	247.2	Zmax = 3.2	SZ		(Ekelund et al., 2001)
2	2p14	D2S441	68150649	na	Lod = 3.2	BP	PARK3	(Liu et al., 2003)
2	2p11.1	D2S139	79675251	101.6	Lod = 3.0	SZ		(DeLisi et al., 2002)
2	2p11.1	D2S1790	84986947	na	Lod = 4.2	SCI/ALC		(Hesselbrock et al., 2004)
2	2q31.1	D2S335	172392096	175.9	NPL = 3.3	ASD		(Buxbaum et al., 2001)
2	2q31.3	D2S2188	175430218	180.8	MLS = 4.8	ASD		(IMGSAC, 2001a)
2	2q32.1	D2S364	182860040	186.2	NPL = 3.3	ASD	AUTS5	(Buxbaum et al., 2001)
2	2q37.3	D2S427	232031769	236.7	Zmax = 4.4	SZ		(Paunio et al., 2001)
3	3p11.1	D3S1276	85338848	111.9	NPL = 3.6	BP		(Bailer et al., 2002)
3	3q11.1	D3S1271	102217427	na	NPL = 3.2	BP		(Bailer et al., 2002)
3	3q27.1	D3S3037	na	190.4	Lod = 4.3	ASD	AUTS3 ASPG1	(Auranen et al., 2002)
3	3q29	D3S1265	197014377	222.8	NPL = 3.7–3.5	SZ/BP		(Bailer et al., 2002)
3	3q29	D3S1265	197014377	222.8	NPL = 4.0	SZ/BP		(Schosser et al., 2004)
3	3q29	D3S3550	na	227.6	NPL = 4.1–3.9	BP		(Bailer et al., 2002)
7	7p11.2	D7S1818	49166136	69.6	MLS = 3.0	ADHD		(Bakker et al., 2003)
7	7q31.1	D7S501	106034415	118.9	MLS = 3.5	SZ/BP		(Ekelund et al., 2000)

(*Continued*)

TABLE I (*Continued*)

Chr	Location	Markers	NCBI	Marshfield	Statistic	MD	Loci/ Association results	References
7	7q31.31	D7S523	111295678	123.0	MLS = 3.5	SZ/BP		(Ekelund *et al.*, 2000)
7	7q31.31	D7S486	115488713	124.1	MLS = 3.2	SZ	AUTS1	(Ekelund *et al.*, 2000)
7	7q32.1	D7S530	128216773	134.6	MLS = 3.6	ASD	AUTS1	(IMGSAC, 1998)
7	7q	D7S684	137521669	147.2	MLS = 3.6	ASD	AUTS1	(IMGSAC, 1998)
7	7q34	D7S1824	139465749	149.9	Z = 3.0	ASD		(Alarcon *et al.*, 2002)
7	7q36.2	D7S2462	152999336	169.8	Lod = 3.7	ASD		(Auranen *et al.*, 2002)
9	9p24.2	D9S288	3941795	na	Lod = 3.1	SZ		(Wilcox *et al.*, 2002)
9	9p22.3	D9S157	17618382	32.2	MLS = 3.1	ASD		(IMGSAC, 2001a)
9	9q12	D9S301	71032274	na	p = 0.0005	ALC		(Bergen *et al.*, 2003)
9	9q34.3	D9S1826	135674269	159.6	MLS = 3.6	ASD		(IMGSAC, 2001a)
9	9q34.3	D9S158	136325009	161.7	MLS = 3.2	ASD		(IMGSAC, 2001a)
13	13q13.3	D13S1491	37461442	na	Lod = 3.0	SZ		(Maziade *et al.*, 2005)
13	13q14.2	D13S1272	43983647	41.7	NPL = 4.1	BP		(Badenhop *et al.*, 2001)
13	13q14.3	D13S153	47788774	45.6	NPL = 4.1	BP		(Badenhop *et al.*, 2001)
13	13q22.32	D13S800	72772693	55.3	MML = 3.0	ASD		(IMGSAC, 2001b)
13	13q31.1	D13S317	81620060	63.9	Lod = 3.6	PBP	SCZD7	(Potash *et al.*, 2003)
13	13q32.1	D13S793	na	na	Lod = 4.4	SZ	SCZD7; PAND	(Brzustowicz *et al.*, 1999)
13	13q32.3	D13S1271	na	79.5	Lod = 3.4	BP		(Detera-Wadleigh *et al.*, 1999)
13	13q32.3	D13S779	100301956	82.9	Lod = 3.4	BP	SCZD7	(Detera-Wadleigh *et al.*, 1999)
13	13q32.3	D13S174	101752077	84.9	NPL = 4.2	SZ	SCZD7	(Blouin *et al.*, 1998)
15	15q11.2	D15S128	22681893	6.1	Z = 4.0	SZ	SCZD10	(Freedman *et al.*, 2001)
15	15q11.2	D15S128	22681893	6.1	Lod = 0.7	ASD	SCZD10	(Philippe *et al.*, 1999)
15	15q11.1	D15S122	23231137	6.1	Lod = 4.6	BP	UBE3A	(Maziade *et al.*, 2005)
15	15q12	UBE3A	23231137	6.1	p = 0.004	ASD	LD, T$_{SP}$ 5′ UTR D15S122	(Nurmi *et al.*, 2001)
15	15q12	UBE3A	23235221	na	ns	ASD	Angelman syndrome	(Veenstra-VanderWeele *et al.*, 1999)

424

Chr	Cytoband	Gene/Marker	Position		Statistic	Disease	Details	Reference
15	15q	ATP10A[#]	23659963	na	p = 0.03, 0.03	ASD	PDT exon 22 rs1047700, intron 2 rs1345098	(Nurmi et al., 2003)
15	15q11.1	GABRB3	24570020	na	Lod = 4.7	ASD		(Shao et al., 2003)
15	15q11.1	GABRB3	24570020	na	p = 0.001	ASD	MTDT 155CA-2	(Cook et al., 1998)
15	15q11.1	GABRB3	24570020	na	ns	ASD	TDT 155CA-2	(Maestrini et al., 1999)
15	15q11.1	GABRB3	24570020	na	p = 0.01, 0.04 / p = 0.03, 0.04	ASD	PDT intron 7 rs1432007, hCV2901140 PDT rs4542636, rs878960	(McCauley et al., 2004)
15	15q11.1	GABRA5	24724386	na	p = 0.03	ASD	TDT intron 6 hCV252720	(McCauley et al., 2004)
15	15q12	GABRG3	24799413	na	p = 0.02, 0.03	ASD	exon5_539T/C, intron5_687T/C	(Menold et al., 2001)
15	15q11.2	D15S217	25747918	na	Z = 1.8	ASD		(Bass et al., 2000)
15	15	APBA2	27000145	na	ns	ASD		(Sutcliffe et al., 2003)
15	15q13.3	D15S1360 (CHRNA7)	(30110018)	na	Lod = 5.3	SZ/P50		(Freedman et al., 1997)
15	15q15.3	ACTC	32870697	31.5	Lod = 3.5	BP		(Turecki et al., 2001)
15	15q15.3	D15S118	34024135	32.6	Z = 4.0	SZ	SCZD10; EJM2	(Freedman et al., 2001)
15	15q15.3	D15S118	34024135	32.6	Lod = 1.1	ASD	SCZD10; EJM2	(Philippe et al., 1999)
15	15q15.3	D15S1042	34050903	32.6	Z = 2.6, 3.9	SZ catatonia		(Stober et al., 2000b, 2002)
15	15q15.3	GATA50C03	na	34.8	MLS = 3.5	ADHD		(Bakker et al., 2003)
15	15q21.1	D15S1012	36794835	36.0	Z = 2.8	SZ catatonia		(Stober et al., 2000b)
15	15q21.1	SLC30A4	43602294	na	na	SZ catatonia	No variant detected in coding and promotor sequences	(Kury et al., 2003)
15	15q21.1	D15S659	44161300	43.5	Z = 3.9	SZ catatonia		(Stober et al., 2002)
15	15q21.2	D15S117	56266889	51.2	MLS = 2.6	ASD	Non-male affected sib-pairs	(Lamb et al., 2005)

(Continued)

TABLE I (*Continued*)

Chr	Location	Markers	NCBI	Marshfield	Statistic	MD	Loci/ Association results	References
15	15q22.31	D15S125	64880007	64.2	MLS = 2.6	ASD	Non-male affected sib-pairs	(Lamb *et al.*, 2005)
15	15q22.31	PKM2 (PK3)	70278439	na	p = 0.02*	SZ		(Stone *et al.*, 2004)
15	15q26.2	D15S652	90318371	90.0	Z = 2.3	ASD	MDD2	(Risch *et al.*, 1999)
15	15q26.3	D15S1014	95803594	107.7	Lod = 4.6	SZ/BP		(Maziade *et al.*, 2005)
15	15q26.3	D15S642	100152332	122.1	p = 0.00005	ALC		(Zinn-Justin and Abel, 1999)
17	17q12	D17S1299	36607512	62.0	MLS = 3.6	ASD	male affected sibships	(Cantor *et al.*, 2005)
17	17q21.1	D17S2180	45940225	66.9	MLS = 4.1	ASD	male affected sibships	(Cantor *et al.*, 2005)
17	17q21.32	D17S1290	55383891	82.0	p = 0.00006	ALC		(Bergen *et al.*, 2003)

#: previously named ATP10C; *: genome-wide adjusted.

Locations: cytological locations according to the GMAP estimation of the Genetic Location Database (LDB); NCBI: physical location in nucleotide number according to the National Center for Biological Information; Marshfield: genetic distance from the Marshfield map in centimorgans.

Mental disorders. SZ: schizophrenia; BP: bipolar disorder; SZ/BP: psychotic bipolar disorder; ASD: autism spectrum disorder; AS: Asperger syndrome; ADHD: attention-deficit/hyperactivity disorder; ALC: alcoholism.

Statistics. MLS: multipoint Lod score; NPL: non-parametric Lod score; MML: maximum multipoint heterogeneity LOD score; na: not available; LD: linkage disequilibrium; TDT: transmission disequilibrium test; MTDT: multiallelic transmission-disequilibrium test; PDT: pedigree disequilibrium test.

Loci: are those reported for the corresponding markers in the NCBI database. AUTS1, 3, 5: autism 1, 3, 5; ASPG1, 3: Asperger syndrome 1, 3; SCZD7, 9: schizophrenia 7, 9; SCZD10: schizophrenia 10 (periodic catatonia); EJM2: epilepsy, juvenile; MDD2: Major depressive disorder 2; PAND: Panic disorder syndrome; PARK3: Parkinson disease 3.

Genes. KCNN3: calcium-activated potassium channel; PFKFB2: 6-phosphofructo-2-kinase/fructose-2,6-bisphosphatase 2; UBE3A: E6-AP ubiquitin-protein ligase; ATP10A: ATPase, Class V, type 10A; GABRB3 GABRA5 GABRG3: gamma-aminobutyric acid (GABA(A)) receptor B3, A5, G3; APBA2: amyloid precursor protein-binding protein; CHRNA7: alpha 7-nicotinic receptor; SLC30A4: zinc transporter; PKM2 (PK3): pyruvate kinase, muscle.

showed also significant linkages with SZ catatonia (Table I), whereas a suggestive linkage (Z = 1,9) was observed on chromosome 22 (Stober *et al.*, 2000b). From these results, it can be observed that ASD shared common susceptibility loci with other mental disorders on five of these chromosomes. On chromosome 1, a susceptibility locus is shared between ASD and SZ at 1q22-q24.2 in a region located between 155 and 165 Mb. The calcium-regulated potassium channel (KCNN3) gene, located at 151,7 Mb slightly centromeric to the shared susceptibility locus, had shown a possible association with SZ but not with a subgroup of SZ showing catatonia (Stober *et al.*, 2000a). Negative association results have also been reported for KCNN3 and both SZ and BP (Glatt *et al.*, 2003b).

On chromosome 7, ASD, SZ, and BP shared a region of 30–40 Mb at 7q31.1-q36.2. The homeogene Engrailed 2 (EN2) gene is located at 7q36. EN2 is specifically involved in patterning the region that gives rise to the cerebellum and controls the plasticity of midbrain dopaminergic neurons. A PvuII polymorphism in EN2 had showed significant differences in the allele frequencies between a sample of 100 autistic children and controls (Petit *et al.*, 1995). Two EN2 intronic SNPs have demonstrated significant association with ASD which supports a neurodevelopmental defect hypothesis in the etiology of ASD (Gharani *et al.*, 2004). However, no significant difference in frequency of a CA repeat polymorphism located in the 3′ region of EN2 was observed between 165 schizophrenic subjects and 97 controls matched for age and ethnicity from a French Caucasian population (Gourion *et al.*, 2004).

On chromosome 13, a linked region of about 30 Mb between 73 and 102 Mb at 13q22.32-q32.3 is shared between ASD, SZ, BP, and psychotic bipolar disorder (PBP) (Table I). The gene D-amino acid oxidase activator (DAOA), previously named G72, is located at 13q34 at the telomeric end of this region (104,9 MB). DAOA had also been associated with SZ in different populations of Caucasians and South-Africans (Hall *et al.*, 2004), of Ashkenazi (Korostishevsky *et al.*, 2004), and of Chinese (Liu *et al.*, 2004; Wang *et al.*, 2004). DAOA has also been associated with BP (Hattori *et al.*, 2003), and with both SZ and BP (Addington *et al.*, 2004; Schumacher *et al.*, 2004). A TDT of a haplotype set including 16 SNPs covering a 157-kb region encompassing the entire complementary DNA sequences of DAOA was significant with BP in a US population (Hattori *et al.*, 2003). This result has been replicated in a second US sample (Chen *et al.*, 2004c), and in Germans (Schumacher *et al.*, 2004). Until located at the extreme end of the 13q22.32-q32.3 susceptibility region and on the opposite side of the ASD potential susceptibility locus, DAOA had also shown a possible association with scores on a premorbid phenotype measured by the Autism Screening Questionnaire (Addington *et al.*, 2004). This region also includes two QTLs: schizophrenia 7 and panic disorder syndrome (Table I). The gene encoding the propionyl Coenzyme A carboxylase alpha polypeptide (PCCA) is located at 13q32 and 99,5 Mb. PCCA, which is expressed both in brain and blood, is involved in the nicotinic acid metabolism (Van Greevenbroek et al., 2004). It is well known that patients suffering from SZ show an increased prevalence of nicotine addiction which improves, cognitive function affected in SZ patients (Cattapan-Ludewig et al., 2005).

On chromosome 17q12-q21.32 between 37–56 Mb, a susceptibility locus for ASD co-localises with a locus for ALC (Table I). Positive (Conroy et al., 2004; Klauck et al., 1997) and negative (Cook et al., 1997a; Kim et al., 2002; Maestrini et al., 1999; Persico et al., 2000; Zhong et al., 1999) associations with a length and SNP polymorphisms in the serotonin transporter gene located at 17q11.2-q12 and 25,6 Mb, 11 Mb centromeric to the shared susceptibility region, have been reported. A QTL for male serotonin levels was also reported in this region (Weiss et al., 2005a) where the beta 3 integrin gene (ITGB3) is located at 42,7 Mb. Five SNPs in ITGB3 have shown association with the serotonin level in males only, and with a cardiovascular risk factor in females (Weiss et al., 2005b).

Finally, on chromosome 16p13.2 only a close to significant linkage (MLS = 2.93) with ASD was observed for D16S3102 (IMGSAC, 2001a). But this region will be discussed since two positional candidate genes, the glutamate receptor, ionotropic, N-methyl D-aspartate 2A (GRIN2A) and the 4-aminobu-tyrate aminotransferase (ABAT), both located at 16p13.3, have showed association with ASD (Barnby et al., 2005). This region appeared to be shared with SZ and/or BP (Ekholm et al., 2003; Maziade et al., 2005), and ALC (Detera-Wadleigh, 1999). NMDA channel is involved in long-term potentiation, an activity-dependent increase in the efficiency of synaptic transmission thought to underlie certain kinds of memory and learning, while ABAT is responsible for catabolism of gamma-aminobutyric acid (GABA), an important, mostly inhibitory neurotransmitter in the central nervous system. A variable (GT) (n) repeat in the promoter region of GRIN2A has been associated with SZ (Itokawa et al., 2003b) and BP (Itokawa et al., 2003a), while a GRIN2A exon 5 polymorphism showed association with ADHD (Turic et al., 2004).

2. Chromosome 15

Table II presents some of the chromosomal rearrangements and associated syndromes related to ASD for chromosome 15. Most of the rearrangements/abnormalities are observed at 15q11-q13 where the Prader-Willi/Angelman syndrome loci are located. Two more telomeric regions have also been reported with a deletion at 15q22-q23 (Smith et al., 2000), and a duplication of the 15q25-qter region at 15p producing a true trisomy (Bonati et al., 2005). The Figure presents a compendium of all chromosomal rearrangements, linkage results, candidate gene analysis, and assigned susceptibility loci for this chromosome. From these, it could be seen that three to four different chromosomal regions are shared between ASD and other mental disorders (see Fig. 1). The first region is 15q11-q13, located between 19 Mb and 31 Mb, well known to be associated with

TABLE II

SOME OF THE SYNDROMES AND CHROMOSOMAL REARRANGEMENTS RELATED TO AUTISM SYNDROME DISORDER OBSERVED ON CHROMOSOME 15

Location	Syndrome/ rearrangements	Description	References
15p11.2	t(15;16) (p11.2;p13.3) *de novo*		(Martin *et al.*, 2003)
15p11-q15	+der(15)(pter-q15: p11-pter)		(Konstantareas and Homatidis, 1999)
15q11-q12	del(15)(q11q12)		(Schroer *et al.*, 1998)
15q11.2-q13	Prader Willi / Angelman	euchromatic variants, no duplication & clinical features	(Jalal *et al.*, 1994; Ludowese *et al.*, 1991; Mao and Jalal, 2000)
15q11.2-q13	Prader Willi / Angelman	maternal, duplication & clinical features	(Browne *et al.*, 1997; Cook *et al.*, 1997b; Repetto *et al.*, 1998)
15q11.2-q13	Prader Willi / Angelman	paternal, duplication, clinical features or not & obesity	(Cook *et al.*, 1997b; Mao and Jalal, 2000; Mohandas *et al.*, 1999)
15q11.2-q13	dup(15)(q11.2q13)		(Reddy, 2005; Wassink *et al.*, 2001)
15q11.2-q13	del(15)(q11.2q13)		(Wassink *et al.*, 2001)
15q11-q13	dup(15)(q11q13)mat		(Baker *et al.*, 1994; Cook *et al.*, 1997b; Schroer *et al.*, 1998)
15q11-q13	inv dup(15)(q11q13)		(Filipek *et al.*, 2003)
15pter-q13	invdup(15)(pter-q13::q13-pter)mat		(Flejter *et al.*, 1996)
15q22-q23	del(15)(q22-q23)		(Smith *et al.*, 2000)
15q25.2qter	dup(15)(15q25.2qter)		(Bonati *et al.*, 2005)

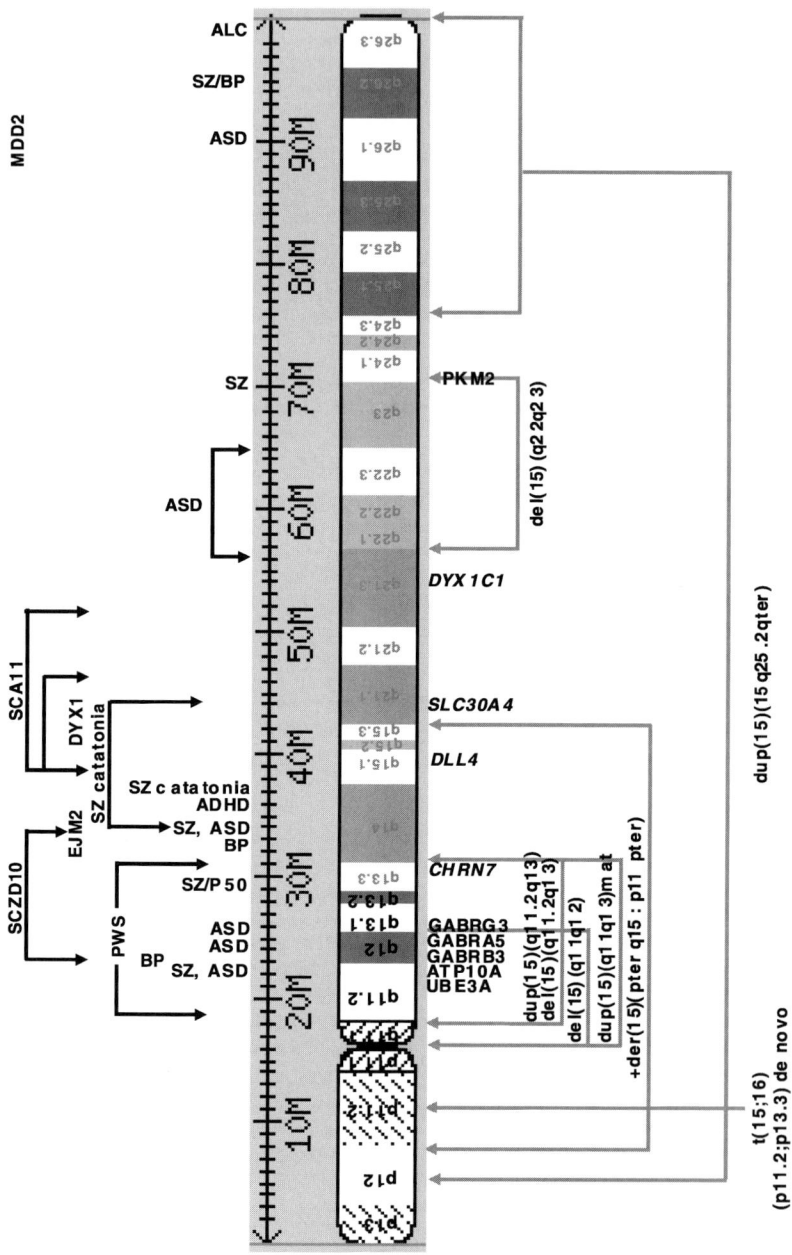

FIG. 1. Linkage results and susceptibility loci (above), and chromosomal rearrangements/abnormalities and candidate genes (below) observed on chromosome 15 in relation to different mental disorders (see text for abbreviations). Genes with negative association results are in italic.

ASD because of the numerous chromosomal abnormalities observed in this region in autistic patients (see Table II). Significant linkage and candidate gene association results for ASD have also been observed in this region (Table I). Linkages with the same two markers, both located around 23 Mb, have been observed for ASD and SZ (D15S128), and for ASD and BP (D15S122). The marker D15S122 is located in the 5'UTR region of the E6-AP ubiquitin-protein ligase (UBE3A) gene which has been shown to be associated with the Angelman syndrome (Veenstra-VanderWeele *et al.*, 1999) and ASD (Nurmi *et al.*, 2001). A significant linkage was observed between ASD and the 155CA-2 microsatellite marker located in the gamma-aminobutyric acid (GABA) receptor B3 (GABRB3) (Shao *et al.*, 2003). A positive association with the same GABRB3 155CA-2 marker (Cook *et al.*, 1998), and with two SNPs located in the intron 7 of GABRB3 (McCauley *et al.*, 2004) had also been observed with ASD. In addition to UBEA3 and GABRB3, three other genes located within 1–2 Mb in the 15q11-q13 region have showed an association with ASD: two SNPs and a haplotype block in ATP10A (Nurmi *et al.*, 2003), previously named ATP10C, six SNPs and haplotypes including these SNPs in GABRA5 (McCauley *et al.*, 2004), and two SNPs in GABRG3 (Menold *et al.*, 2001). Also, it is not clear at this time if all these genes are involved individually in ASD, or as a group or subgroups of them interacting with each other in an epistatic effect, or if markers analyzed within these genes are simply in linkage disequilibrium.

A second chromosomal region 15q14-q15.3 located between 32 and 44 Mb is very close to 15q11-q13 and could eventually be part of the same susceptibility region. However, some features could distinguish these two regions. First, most of the chromosomal abnormalities and two susceptibility loci, schizophrenia 10 and the juvenile epilepsy 2, are observed in 15q11-q13 in contrast to 15q14-q15.3 (Fig. 1). Additionally, only the region 15q14-q15.3 shared susceptibility locus with catatonia and ADHD. At 15q14-q15.3, the same marker (D15S118 at 34 Mb) is linked to ASD and SZ, while markers D15S1042, also located at 34 Mb, D15S1012 at 37 Mb, and D15S659 at 44 Mb are linked to catatonia in SZ patients. The zinc transporter gene SLC30A4, located at 15q21.1 and 44 Mb, had shown altered expression patterns in post mortem analysis of the brains of schizophrenic patients (Mirnics *et al.*, 2000). However, no genetic variants within the coding and the putative promoter region of SLC30A4 was found in affected individuals from SZ catatonia large pedigrees showing a perfect co-segregation of a chromosomal segment between marker D15S1042 and D15S659 (Kury *et al.*, 2003). Similarly, no association with catatonia has been observed with the cholinergic receptor, nicotinic, alpha polypeptide 7, and the delta-like 4 (Drosophila) both located at 15q14 and 31,0 and 39,0 Mb, respectively (McKeane *et al.*, 2005; Meyer *et al.*, 2002). Two other QTLs are located in this region: dyslexia-1 (DYX1) and spinocerebellar ataxia-11 (SCA11) loci. Developmental dyslexia is characterized by an unexpected difficulty in learning to read and write despite adequate intelligence, motivation, and education (Ylisaukko-Oja et al., 2004). The gene dyslexia susceptibility candidate 1 located at 15q21.3 and 53,6 Mb had shown no association with ASD (Ylisaukko-Oja et al., 2004). Autosomal dominant cerebellar ataxia type III is a relatively benign, late-onset, slowly progressive neurological disorder characterized by an uncomplicated cerebellar syndrome. The other spinocerebellar ataxia loci SCA1, located at 6p22.3, has been associated to SZ (Culjkovic et al., 2000; Wang et al., 1996), and it remains to be shown if it could be the same for SCA11. On the other hand, the gene encoding the huntingtin interacting protein K is located at 15q15.3 and 41,9 Mb. The Huntington disease is an autosomal dominant disease that gives rise to progressive, selective, and localized neural cell death associated with choreic movements and dementia.

The third chromosomal region is located at 15q22-q23, between 56 and 70 Mb (Fig. 1). A deletion of this region has been associated with ASD and significant linkage results have been observed for both ASD and SZ. It has to be noted that the linkage with ASD has been observed in non-male affected sib-pairs only which suggests possible parent of origin specific effects (Lamb et al., 2005). One hundred forty genes are located in this region according to NCBI. Among them, the N-methyl-D-aspartate (NMDA) receptor regulated 2 gene (NARG2) is located at 15q22.2 and 58,6 Mb. NMDA receptors play an important role in the transition from proliferation of neuronal precursors to differentiation of neurons (Sugiura et al., 2001). Postmortem brain abnormalities of the glutamate neurotransmitter system have been observed in ASD (Purcell et al., 2001). To our knowledge, NARG2 does not seem to have been studied yet in relation to ASD or SZ. The Bardet-Biedl syndrome 4 gene, characterized by a triallelism and mental retardation, is located slightly outside the shared susceptibility region at 71 Mb, close to the SZ linkage.

The fourth and last region encompassed the q terminal end of the chromosome 15 at 15q25-q26 and 90–100 Mb (Fig. 1). This relatively short region is shared by ASD, SZ, BP, and ALC. Interestingly, the Major depressive disorder 2 susceptibility locus has also been assigned to the same marker (D15S652) linked to ASD. Fifty-five genes are located in this region according to NCBI. The desmuslin gene (DMN) is located at 15q26.3 and 97.5 Mb. DMN is mainly expressed in muscular tissues where it has been involved in different myopathies, but also in brain (Mizuno et al., 2001a,b). DMN encodes an intermediate filament protein that interacts with the desmin, also an intermediate filament protein, and with the alpha form of the dystrobevin (Mizuno et al., 2001b). The beta form of the dystrobevin has been shown to be bound by the dystrobevin binding protein 1 (DTNBP1) located at 6p22.3, and which has been associated to SZ in multiple studies (Funke et al., 2004; Kohn et al., 2004; Schwab *et al.*, 2003; Straub *et al.*, 2002; VanDenBogaert *et al.*, 2003; van den Oord *et al.*, 2003) and possibly to BP in a subset of cases with psychosis (Raybould *et al.*, 2005). A QTL related to cerebellar ataxia with mental retardation (CAMOS) is also found at 15q26. In skin biopsies of children affected by CAMOS, an inversion of the usual osmiophilic pattern of the vessels is observed and it was thought that this prevented normal exchange between the blood and surrounding tissues, thus decreasing vessel permeability and modifying the production and/or migration of neuronal cells at an early stage (Delague *et al.*, 2002).

3. *Specific Chromosomes*

Finally, susceptibility regions specific to ASD are observed at 2q31.1-q32.1, and at 9p22.3 and 9q34.3. It is not clear at this point if the ASD linkage on chromosome 3 at 3q27.1 is shared or not with SZ and BP since the physical location of the linked marker (D3S3027) is not found in the NCBI database (built 35.1). However, according to the Marshfield genetic map, the ASD linked region is located at more then 30 cM centromeric to the susceptibility regions linked to SZ and BP at 3q29, and the probability that these regions shared the same susceptibility locus is weak. No significant linkage result has been reported for either of the two sexual chromosomes X and Y.

II. Conclusion

Several possible susceptibility chromosomal regions shared between different mental disorders have been identified by putting together on a same physique map scale the significant results from different genome scans. This supports the hypothesis that strong common endophenotype or intermediate phenotypes are shared between these disorders. The genes located in these shared susceptibility regions can help to shed light on metabolic pathways involved and on what could be these common intermediate phenotypes.

Chromosome 15 contains three to four regions shared by five different mental disorders and some interesting candidate genes. As such, this chromosome is a prime target for direct investigations of common intermediate phenotypes in relevant cohorts. Catatonia seems to be a shared syndrome in ASD, SZ, and BP, and may be a useful intermediate phenotype. Findings in this review suggest overlapping susceptibility regions on chromosome 15. Future studies in ASD, SZ, and BP should assess and distinguish the catatonic syndrome within study samples. The identification and validation of additional candidate genes on chromosome 15 by fine mapping and association studies should also be a priority.

References

Abecasis, G. R., Burt, R. A., Hall, D., Bochum, S., Doheny, K. F., Lundy, S. L., Torrington, M., Roos, J. L., Gogos, J. A., and Karayiorgou, M. (2004). Genomewide scan in families with schizophrenia from the founder population of Afrikaners reveals evidence for linkage and uniparental disomy on chromosome 1. *Am. J. Hum. Genet.* **74,** 403–417.

Abrams, R., and Taylor, M. A. (1976). Catatonia. A prospective clinical study. *Arch. Gen. Psychiatry* **33,** 579–581.

Addington, A. M., Gornick, M., Sporn, A. L., Gogtay, N., Greenstein, D., Lenane, M., Gochman, P., Baker, N., Balkissoon, R., Vakkalanka, R. K., Weinberger, D. R., Straub, R. E., and Rapoport, J. L. (2004). Polymorphisms in the 13q33.2 gene G72/G30 are associated with childhood-onset schizophrenia and psychosis not otherwise specified. *Biol. Psychiatry* **55,** 976–980.

Alarcon, M., Cantor, R. M., Liu, J., Gilliam, T. C., and Geschwind, D. H. (2002). Evidence for a language quantitative trait locus on chromosome 7q in multiplex autism families. *Am. J. Hum. Genet.* **70,** 60–71.

Auranen, M., Vanhala, R., Varilo, T., Ayers, K., Kempas, E., Ylisaukko-Oja, T., Sinsheimer, J. S., Peltonen, L., and Jarvela, I. (2002). A genomewide screen for autism-spectrum disorders: Evidence for a major susceptibility locus on chromosome 3q25–27. *Am. J. Hum. Genet.* **71,** 777–790.

Avramopoulos, D., Stefanis, N. C., Hantoumi, I., Smyrnis, N., Evdokimidis, I., and Stefanis, C. N. (2002). Higher scores of self reported schizotypy in healthy young males carrying the COMT high activity allele. *Mol. Psychiatry* **7,** 706–711.

Badenhop, R. F., Moses, M. J., Scimone, A., Mitchell, P. B., Ewen, K. R., Rosso, A., Donald, J. A., Adams, L. J., and Schofield, P. R. (2001). A genome screen of a large bipolar affective disorder pedigree supports evidence for a susceptibility locus on chromosome 13q. *Mol. Psychiatry* **6,** 396–403.

Bailer, U., Leisch, F., Meszaros, K., Lenzinger, E., Willinger, U., Strobl, R., Heiden, A., Gebhardt, C., Doge, E., Fuchs, K., Sieghart, W., Kasper, S., Hornik, K., and Aschauer, H. N. (2002). Genome scan for susceptibility loci for schizophrenia and bipolar disorder. *Biol. Psychiatry* **52,** 40–52.

Baker, P., Piven, J., Schwartz, S., and Patil, S. (1994). Brief report: Duplication of chromosome 15q11–13 in two individuals with autistic disorder. *J. Autism. Dev. Disord.* **24,** 529–535.

Bakker, S. C., van der Meulen, E. M., Buitelaar, J. K., Sandkuijl, L. A., Pauls, D. L., Monsuur, A. J., van 't Slot, R., Minderaa, R. B., Gunning, W. B., Pearson, P. L., and Sinke, R. J. (2003). A whole-genome scan in 164 Dutch sib pairs with attention-deficit/hyperactivity disorder: Suggestive evidence for linkage on chromosomes 7p and 15q. *Am. J. Hum. Genet.* **72,** 1251–1260.

Barnby, G., Abbott, A., Sykes, N., Morris, A., Weeks, D. E., Mott, R., Lamb, J., Bailey, A. J., and Monaco, A. P. (2005). Candidate-Gene screening and association analysis at the autism-susceptibility locus on chromosome 16p: Evidence of association at GRIN2A and ABAT. *Am. J. Hum. Genet.* **76,** 950–966.

Bass, M. P., Menold, M. M., Wolpert, C. M., Donnelly, S. L., Ravan, S. A., Hauser, E. R., Maddox, L. O., Vance, J. M., Abramson, R. K., Wright, H. H., Gilbert, J. R., Cuccaro, M. L., De Long, G. R., and Pericak-Vance, M. A. (2000). Genetic studies in autistic disorder and chromosome 15. *Neurogenetics* **2,** 219–226.

Bearden, C. E., Jawad, A. F., Lynch, D. R., Sokol, S., Kanes, S. J., McDonald-McGinn, D. M., Saitta, S. C., Harris, S. E., Moss, E., Wang, P. P., Zackai, E., Emanuel, B. S., and Simon, T. J. (2004). Effects of a functional COMT polymorphism on prefrontal cognitive function in patients with 22q11.2 deletion syndrome. *Am. J. Psychiatry* **161,** 1700–1702.

Bergen, A. W., Yang, X. R., Bai, Y., Beerman, M. B., Goldstein, A. M., and Goldin, L. R. (2003). Genomic regions linked to alcohol consumption in the Framingham Heart Study. *BMC Genet.* **4** (Suppl. 1), S101.

Blouin, J. L., Dombroski, B. A., Nath, S. K., Lasseter, V. K., Wolyniec, P. S., Nestadt, G., Thornquist, M., Ullrich, G., McGrath, J., Kasch, L., Lamacz, M., Thomas, M. G., Gehrig, C., Radhakrishna, U., Snyder, S. E., Balk, K. G., Neufeld, K., Swartz, K. L., De Marchi, N., Papadimitriou, G. N., Dikeos, D. G., Stefanis, C. N., Chakravarti, A., Childs, B., Housman, D. E., Kazazian, H. H., Antonarakis, S. E., and Pulver, A. E. (1998). Schizophrenia susceptibility loci on chromosomes 13q32 and 8p21. *Nat. Genet.* **20,** 70–73.

Bonati, M. T., Finelli, P., Giardino, D., Gottardi, G., Roberts, W., and Larizza, L. (2005). Trisomy 15q25.2-qter in an autistic child: Genotype-phenotype correlations. *Am. J. Med. Genet. A.* **133,** 184–188.

Bray, N. J., Buckland, P. R., Williams, N. M., Williams, H. J., Norton, N., Owen, M. J., and O'Donovan, M. C. (2003). A haplotype implicated in schizophrenia susceptibility is associated with reduced COMT expression in human brain. *Am. J. Hum. Genet.* **73,** 152–161.

Broman, K. W., Murray, J. C., Sheffield, V. C., White, R. L., and Weber, J. L. (1998). Comprehensive human genetic maps: Individual and sex-specific variation in recombination. *Am. J. Hum. Genet.* **63,** 861–869.

Browne, C. E., Dennis, N. R., Maher, E., Long, F. L., Nicholson, J. C., Sillibourne, J., and Barber, J. C. (1997). Inherited interstitial duplications of proximal 15q: Genotype-phenotype correlations. *Am. J. Hum. Genet.* **61,** 1342–1352.

Brzustowicz, L. M., Hayter, J. E., Hodgkinson, K. A., Chow, E. W., and Bassett, A. S. (2002). Fine mapping of the schizophrenia susceptibility locus on chromosome 1q22. *Hum. Hered.* **54,** 199–209.

Brzustowicz, L. M., Hodgkinson, K. A., Chow, E. W., Honer, W. G., and Bassett, A. S. (2000). Location of a major susceptibility locus for familial schizophrenia on chromosome 1q21-q22. *Science* **288,** 678–682.

Brzustowicz, L. M., Honer, W. G., Chow, E. W., Little, D., Hogan, J., Hodgkinson, K., and Bassett, A. S. (1999). Linkage of familial schizophrenia to chromosome 13q32. *Am. J. Hum. Genet.* **65,** 1096–1103.

Brzustowicz, L. M., Simone, J., Mohseni, P., Hayter, J. E., Hodgkinson, K. A., Chow, E. W., and Bassett, A. S. (2004). Linkage disequilibrium mapping of schizophrenia susceptibility to the CAPON region of chromosome 1q22. *Am. J. Hum. Genet.* **74,** 1057–1063.

Buxbaum, J. D., Silverman, J. M., Smith, C. J., Kilifarski, M., Reichert, J., Hollander, E., Lawlor, B. A., Fitzgerald, M., Greenberg, D. A., and Davis, K. L. (2001). Evidence for a susceptibility gene for autism on chromosome 2 and for genetic heterogeneity. *Am. J. Hum. Genet.* **68,** 1514–1520.

Cantor, R. M., Kono, N., Duvall, J. A., Alvarez-Retuerto, A., Stone, J. L., Alarcon, M., Nelson, S. F., and Geschwind, D. H. (2005). Replication of autism linkage: Fine-mapping peak at 17q21. *Am. J. Hum. Genet.* **76,** 1050–1056.

Cattapan-Ludewig, K., Ludewig, S., Jaquenoud Sirot, E., Etzensberger, M., and Hasler, F. (2005). [Why do schizophrenic patients smoke?]. *Nervenarzt* **76,** 287–294.

Chen, J., Lipska, B. K., Halim, N., Ma, Q. D., Matsumoto, M., Melhem, S., Kolachana, B. S., Hyde, T. M., Herman, M. M., Apud, J., Egan, M. F., Kleinman, J. E., and Weinberger, D. R. (2004a). Functional analysis of genetic variation in catechol-O-methyltransferase (COMT): Effects on mRNA, protein, and enzyme activity in postmortem human brain. *Am. J. Hum. Genet.* **75,** 807–821.

Chen, X., Wang, X., O'Neill, A. F., Walsh, D., and Kendler, K. S. (2004b). Variants in the catechol-o-methyltransferase (COMT) gene are associated with schizophrenia in Irish high-density families. *Mol. Psychiatry* **9,** 962–967.

Chen, Y. S., Akula, N., Detera-Wadleigh, S. D., Schulze, T. G., Thomas, J., Potash, J. B., De Paulo, J. R., McInnis, M. G., Cox, N. J., and McMahon, F. J. (2004c). Findings in an independent sample support an association between bipolar affective disorder and the G72/G30 locus on chromosome 13q33. *Mol. Psychiatry* **9,** 87–92 image 5.

Collins, A., Frezal, J., Teague, J., and Morton, N. E. (1996). A metric map of humans: 23,500 loci in 850 bands. *Proc. Natl. Acad. Sci. USA* **93,** 14771–14775.

Comings, D. E., Comings, B. G., Muhleman, D., Dietz, G., Shahbahrami, B., Tast, D., Knell, E., Kocsis, P., Baumgarten, R., Kovacs, B. W., Levy, D. L., Smith, M., Borison, R. L., Evans, D. D., Klein, D. N., MacMurray, J., Tosk, J. F., Sverd, J., Gysin, R., and Flanagan, S. D. (1991). The dopamine D2 receptor locus as a modifying gene in neuropsychiatric disorders. *JAMA.* **266,** 1793–1800.

Conroy, J., Meally, E., Kearney, G., Fitzgerald, M., Gill, M., and Gallagher, L. (2004). Serotonin transporter gene and autism: A haplotype analysis in an Irish autistic population. *Mol. Psychiatry* **9,** 587–593.

Cook, E. H., Jr., Courchesne, R., Lord, C., Cox, N. J., Yan, S., Lincoln, A., Haas, R., Courchesne, E., and Leventhal, B. L. (1997a). Evidence of linkage between the serotonin transporter and autistic disorder. *Mol. Psychiatry* **2,** 247–250.

Cook, E. H., Jr., Courchesne, R. Y., Cox, N. J., Lord, C., Gonen, D., Guter, S. J., Lincoln, A., Nix, K., Haas, R., Leventhal, B. L., and Courchesne, E. (1998). Linkage-disequilibrium mapping of autistic disorder, with 15q11–13 markers. *Am. J. Hum. Genet.* **62,** 1077–1083.

Cook, E. H., Jr., Lindgren, V., Leventhal, B. L., Courchesne, R., Lincoln, A., Shulman, C., Lord, C., and Courchesne, E. (1997b). Autism or atypical autism in maternally but not paternally derived proximal 15q duplication. *Am. J. Hum. Genet.* **60,** 928–934.

Culjkovic, B., Stojkovic, O., Savic, D., Zamurovic, N., Nesic, M., Major, T., Keckarevi, D., Romac, S., Zamurovi, B., and Vukosavic, S. (2000). Comparison of the number of triplets in SCA1, MJD/SCA3, HD, SBMA, DRPLA, MD, FRAXA and FRDA genes in schizophrenic patients and a healthy population. *Am. J. Med. Genet.* **96,** 884–887.

Delague, V., Bareil, C., Bouvagnet, P., Salem, N., Chouery, E., Loiselet, J., Megarbane, A., and Claustres, M. (2002). A new autosomal recessive non-progressive congenital cerebellar ataxia associated with mental retardation, optic atrophy, and skin abnormalities (CAMOS) maps to chromosome 15q24-q26 in a large consanguineous Lebanese Druze Family. *Neurogenetics* **4,** 23–27.

DeLisi, L. E., Shaw, S. H., Crow, T. J., Shields, G., Smith, A. B., Larach, V. W., Wellman, N., Loftus, J., Nanthakumar, B., Razi, K., Stewart, J., Comazzi, M., Vita, A., Heffner, T., and Sherrington, R. (2002). A genome-wide scan for linkage to chromosomal regions in 382 sibling pairs with schizophrenia or schizoaffective disorder. *Am. J. Psychiatry* **159,** 803–812.

Detera-Wadleigh, S. D. (1999). Chromosomes 12 and 16 workshop. *Am. J. Med. Genet.* **88,** 255–259.

Detera-Wadleigh, S. D., Badner, J. A., Berrettini, W. H., Yoshikawa, T., Goldin, L. R., Turner, G., Rollins, D. Y., Moses, T., Sanders, A. R., Karkera, J. D., Esterling, L. E., Zeng, J., Ferraro, T. N., Guroff, J. J., Kazuba, D., Maxwell, M. E., Nurnberger, J. I., Jr., and Gershon, E. S. (1999). A high-density genome scan detects evidence for a bipolar-disorder susceptibility locus on 13q32 and other potential loci on 1q32 and 18p11.2. *Proc. Natl. Acad. Sci. USA* **96,** 5604–5609.

Dhossche, D. (1998). Brief report: Catatonia in autistic disorders. *J. Autism. Dev. Disord.* **28,** 329–331.

Egan, M. F., Goldberg, T. E., Kolachana, B. S., Callicott, J. H., Mazzanti, C. M., Straub, R. E., Goldman, D., and Weinberger, D. R. (2001). Effect of COMT Val108/158 Met genotype on frontal lobe function and risk for schizophrenia. *Proc. Natl. Acad. Sci. USA* **98,** 6917–6922.

Ekelund, J., Hovatta, I., Parker, A., Paunio, T., Varilo, T., Martin, R., Suhonen, J., Ellonen, P., Chan, G., Sinsheimer, J. S., Sobel, E., Juvonen, H., Arajarvi, R., Partonen, T., Suvisaari, J., Lonnqvist, J., Meyer, J., and Peltonen, L. (2001). Chromosome 1 loci in Finnish schizophrenia families. *Hum. Mol. Genet.* **10,** 1611–1617.

Ekelund, J., Lichtermann, D., Hovatta, I., Ellonen, P., Suvisaari, J., Terwilliger, J. D., Juvonen, H., Varilo, T., Arajarvi, R., Kokko-Sahin, M. L., Lonnqvist, J., and Peltonen, L. (2000). Genome-wide scan for schizophrenia in the Finnish population: Evidence for a locus on chromosome 7q22. *Hum. Mol. Genet.* **9,** 1049–1057.

Ekholm, J. M., Kieseppä, T., Hiekkalinna, T., Partonen, T., Paunio, T., Perola, M., Ekelund, J., Lönnqvist, J., Pekkarinen-Ijäs, P., and Peltonen, L. (2003). Evidence of susceptibility loci on 4q32 and 16p12 for bipolar disorder. *Human Molecular Genetics* **12,** 1907–1915.

Enoch, M. A., Xu, K., Ferro, E., Harris, C. R., and Goldman, D. (2003). Genetic origins of anxiety in women: A role for a functional catechol-O-methyltransferase polymorphism. *Psychiatr. Genet.* **13,** 33–41.

Fan, J. B., Zhang, C. S., Gu, N. F., Li, X. W., Sun, W. W., Wang, H. Y., Feng, G. Y., St Clair, D., and He, L. (2005). Catechol-O-methyltransferase gene Val/Met functional polymorphism and risk of schizophrenia: A large-scale association study plus meta-analysis. *Biol. Psychiatry* **57,** 139–144.

Filipek, P. A., Juranek, J., Smith, M., Mays, L. Z., Ramos, E. R., Bocian, M., Masser-Frye, D., Laulhere, T. M., Modahl, C., Spence, M. A., and Gargus, J. J. (2003). Mitochondrial dysfunction in autistic patients with 15q inverted duplication. *Ann. Neurol.* **53,** 801–804.

Flejter, W. L., Bennett-Baker, P. E., Ghaziuddin, M., McDonald, M., Sheldon, S., and Gorski, J. L. (1996). Cytogenetic and molecular analysis of inv dup(15) chromosomes observed in two patients with autistic disorder and mental retardation. *Am. J. Med. Genet.* **61,** 182–187.

Freedman, R., Coon, H., Myles-Worsley, M., Orr-Urtreger, A., Olincy, A., Davis, A., Polymeropoulos, M., Holik, J., Hopkins, J., Hoff, M., Rosenthal, J., Waldo, M. C., Reimherr, F., Wender, P., Yaw, J., Young, D. A., Breese, C. R., Adams, C., Patterson, D., Adler, L. E., Kruglyak, L., Leonard, S., and Byerley, W. (1997). Linkage of a neurophysiological deficit in schizophrenia to a chromosome 15 locus. *Proc. Natl. Acad. Sci. USA* **94,** 587–592.

Freedman, R., Leonard, S., Olincy, A., Kaufmann, C. A., Malaspina, D., Cloninger, C. R., Svrakic, D., Faraone, S. V., and Tsuang, M. T. (2001). Evidence for the multigenic inheritance of schizophrenia. *Am. J. Med. Genet.* **105,** 794–800.

Funke, B., Finn, C. T., Plocik, A. M., Lake, S., De Rosse, P., Kane, J. M., Kucherlapati, R., and Malhotra, A. K. (2004). Association of the DTNBP1 locus with schizophrenia in a U.S. population. *Am. J. Hum. Genet.* **75,** 891–898.

Gabrovsek, M., Brecelj-Anderluh, M., Bellodi, L., Cellini, E., Di Bella, D., Estivill, X., Fernandez-Aranda, F., Freeman, B., Geller, F., Gratacos, M., Haigh, R., Hebebrand, J., Hinney, A., Holliday, J., Hu, X., Karwautz, A., Nacmias, B., Ribases, M., Remschmidt, H., Komel, R., Sorbi, S., Tomori, M., Treasure, J., Wagner, G., Zhao, J., and Collier, D. A. (2004). Combined family trio and case-control analysis of the COMT Val158Met polymorphism in European patients with anorexia nervosa. *Am. J. Med. Genet. B. Neuropsychiatr. Genet.* **124,** 68–72.

Gharani, N., Benayed, R., Mancuso, V., Brzustowicz, L. M., and Millonig, J. H. (2004). Association of the homeobox transcription factor, ENGRAILED 2, 3, with autism spectrum disorder. *Mol. Psychiatry* **9,** 474–484.

Glatt, S. J., Faraone, S. V., and Tsuang, M. T. (2003a). Association between a functional catechol O-methyltransferase gene polymorphism and schizophrenia: Meta-analysis of case-control and family-based studies. *Am. J. Psychiatry* **160,** 469–476.

Glatt, S. J., Faraone, S. V., and Tsuang, M. T. (2003b). CAG-repeat length in exon 1 of KCNN3 does not influence risk for schizophrenia or bipolar disorder: A meta-analysis of association studies. *Am. J. Med. Genet. B. Neuropsychiatr. Genet.* **121,** 14–20.

Gourion, D., Leroy, S., Bourdel, M. C., Goldberger, C., Poirier, M. F., Olie, J. P., and Krebs, M. O. (2004). Cerebellum development and schizophrenia: An association study of the human homeogene Engrailed 2. *Psychiatry Res.* **126,** 93–98.

Gurling, H. M., Kalsi, G., Brynjolfson, J., Sigmundsson, T., Sherrington, R., Mankoo, B. S., Read, T., Murphy, P., Blaveri, E., McQuillin, A., Petursson, H., and Curtis, D. (2001). Genomewide genetic linkage analysis confirms the presence of susceptibility loci for schizophrenia, on chromosomes 1q32.2, 5q33.2, and 8p21–22 and provides support for linkage to schizophrenia, on chromosomes 11q23.3–24 and 20q12.1–11.23. *Am. J. Hum. Genet.* **68,** 661–673.

Hall, D., Gogos, J. A., and Karayiorgou, M. (2004). The contribution of three strong candidate schizophrenia susceptibility genes in demographically distinct populations. *Genes. Brain Behav.* **3,** 240–248.

Hattori, E., Liu, C., Badner, J. A., Bonner, T. I., Christian, S. L., Maheshwari, M., Detera-Wadleigh, S. D., Gibbs, R. A., and Gershon, E. S. (2003). Polymorphisms at the G72/G30 gene locus, on 13q33, are associated with bipolar disorder in two independent pedigree series. *Am. J. Hum. Genet.* **72,** 1131–1140.

Hesselbrock, V., Dick, D., Hesselbrock, M., Foroud, T., Schuckit, M., Edenberg, H., Bucholz, K., Kramer, J., Reich, T., Goate, A., Bierut, L., Rice, J. P., and Nurnberger, J. I., Jr. (2004). The search for genetic risk factors associated with suicidal behavior. *Alcohol Clin. Exp. Res.* **28,** 70S–76S.

Ho, B. C., Wassink, T. H., O'Leary, D. S., Sheffield, V. C., and Andreasen, N. C. (2005). Catechol-O-methyl transferase Val158Met gene polymorphism in schizophrenia: Working memory, frontal lobe MRI morphology and frontal cerebral blood flow. *Mol. Psychiatry* **10,** 287–298.

Hovatta, I., Varilo, T., Suvisaari, J., Terwilliger, J. D., Ollikainen, V., Arajarvi, R., Juvonen, H., Kokko-Sahin, M. L., Vaisanen, L., Mannila, H., Lonnqvist, J., and Peltonen, L. (1999). A genomewide screen for schizophrenia genes in an isolated Finnish subpopulation, suggesting multiple susceptibility loci. *Am. J. Hum. Genet.* **65,** 1114–1124.

IMGSAC (1998). A full genome screen for autism with evidence for linkage to a region on chromosome 7q. *Hum. Mol. Genet.* **7,** 571–578.

IMGSAC (2001a). A genomewide screen for autism: Strong evidence for linkage to chromosomes 2q, 7q, and 16p. *Am. J. Hum. Genet.* **69,** 570–581.

IMGSAC (2001b). An autosomal genomic screen for autism. *Am. J. Med. Genet.* **105,** 609–615.

Itokawa, M., Yamada, K., Iwayama-Shigeno, Y., Ishitsuka, Y., Detera-Wadleigh, S., and Yoshikawa, T. (2003a). Genetic analysis of a functional GRIN2A promoter (GT)n repeat in bipolar disorder pedigrees in humans. *Neurosci. Lett.* **345,** 53–56.

Itokawa, M., Yamada, K., Yoshitsugu, K., Toyota, T., Suga, T., Ohba, H., Watanabe, A., Hattori, E., Shimizu, H., Kumakura, T., Ebihara, M., Meerabux, J. M., Toru, M., and Yoshikawa, T. (2003b). A microsatellite repeat in the promoter of the N-methyl-D-aspartate receptor 2A subunit (GRIN2A) gene suppresses transcriptional activity and correlates with chronic outcome in schizophrenia. *Pharmacogenetics* **13,** 271–278.

Jalal, S. M., Persons, D. L., Dewald, G. W., and Lindor, N. M. (1994). Form of 15q proximal duplication appears to be a normal euchromatic variant. *Am. J. Med. Genet.* **52,** 495–497.

Kim, S. J., Cox, N., Courchesne, R., Lord, C., Corsello, C., Akshoomoff, N., Guter, S., Leventhal, B. L., Courchesne, E., and Cook, E. H., Jr. (2002). Transmission disequilibrium mapping at the serotonin transporter gene (SLC6A4) region in autistic disorder. *Mol. Psychiatry* **7,** 278–288.

Klauck, S. M., Poustka, F., Benner, A., Lesch, K. P., and Poustka, A. (1997). Serotonin transporter (5-HTT) gene variants associated with autism? *Hum. Mol. Genet.* **6,** 2233–2238.

Kohn, Y., Danilovich, E., Filon, D., Oppenheim, A., Karni, O., Kanyas, K., Turetsky, N., Korner, M., and Lerer, B. (2004). Linkage disequlibrium in the DTNBP1 (dysbindin) gene region and on chromosome 1p36 among psychotic patients from a genetic isolate in Israel: Findings from identity by descent haplotype sharing analysis. *Am. J. Med. Genet. B. Neuropsychiatr. Genet.* **128,** 65–70.

Konstantareas, M. M., and Homatidis, S. (1999). Chromosomal abnormalities in a series of children with autistic disorder. *J. Autism. Dev. Disord.* **29,** 275–285.

Korostishevsky, M., Kaganovich, M., Cholostoy, A., Ashkenazi, M., Ratner, Y., Dahary, D., Bernstein, J., Bening-Abu-Shach, U., Ben-Asher, E., Lancet, D., Ritsner, M., and Navon, R. (2004). Is the G72/G30 locus associated with schizophrenia? Single nucleotide polymorphisms, haplotypes, and gene expression analysis. *Biol. Psychiatry* **56,** 169–176.

Kury, S., Rubie, C., Moisan, J. P., and Stober, G. (2003). Mutation analysis of the zinc transporter gene SLC30A4 reveals no association with periodic catatonia on chromosome 15q15. *J. Neural. Transm.* **110,** 1329–1332.

Lamb, J. A., Barnby, G., Bonora, E., Sykes, N., Bacchelli, E., Blasi, F., Maestrini, E., Broxholme, J., Tzenova, J., Weeks, D., Bailey, A. J., and Monaco, A. P. (2005). Analysis of IMGSAC autism susceptibility loci: Evidence for sex limited and parent of origin specific effects. *J. Med. Genet.* **42,** 132–137.

Larsson, H. J., Eaton, W. W., Madsen, K. M., Vestergaard, M., Olesen, A. V., Agerbo, E., Schendel, D., Thorsen, P., and Mortensen, P. B. (2005). Risk factors for autism: Perinatal factors, parental psychiatric history, and socioeconomic status. *Am. J. Epidemiol.* **161,** 916–925.

Liu, J., Juo, S. H., Dewan, A., Grunn, A., Tong, X., Brito, M., Park, N., Loth, J. E., Kanyas, K., Lerer, B., Endicott, J., Penchaszadeh, G., Knowles, J. A., Ott, J., Gilliam, T. C., and Baron, M. (2003). Evidence for a putative bipolar disorder locus on 2p13–16 and other potential loci on 4q31, 7q34, 8q13, 9q31, 10q21–24, 13q32, 14q21 and 17q11–12. *Mol. Psychiatry* **8,** 333–342.

Liu, X., He, G., Wang, X., Chen, Q., Qian, X., Lin, W., Li, D., Gu, N., Feng, G., and He, L. (2004). Association of DAAO with schizophrenia in the Chinese population. *Neurosci. Lett.* **369,** 228–233.

Ludowese, C. J., Thompson, K. J., Sekhon, G. S., and Pauli, R. M. (1991). Absence of predictable phenotypic expression in proximal 15q duplications. *Clin. Genet.* **40,** 194–201.

Maestrini, E., Lai, C., Marlow, A., Matthews, N., Wallace, S., Bailey, A., Cook, E. H., Weeks, D. E., and Monaco, A. P. (1999). Serotonin transporter (5-HTT) and gamma-aminobutyric acid receptor subunit beta3 (GABRB3) gene polymorphisms are not associated with autism in the IMGSA families. The International Molecular Genetic Study of Autism Consortium. *Am. J. Med. Genet.* **88,** 492–496.

Mao, R., and Jalal, S. M. (2000). Characteristics of two cases with dup(15)(q11.2-q12): One of maternal and one of paternal origin. *Genet. Med.* **2,** 131–135.

Martin, C. L., Ilkin, Y., Powell, C., Rao, K., Whichelio, A., and Cook, E. (2003). Breakpoint mapping of a de novo 15p:16p translocation reveals a candidate gene for autism. *Am. J. Hum. Genet.* **73,** 325.

Maziade, M., Roy, M. A., Chagnon, Y. C., Cliche, D., Fournier, J. P., Montgrain, N., Dion, C., Lavallee, J. C., Garneau, Y., Gingras, N., Nicole, L., Pires, A., Ponton, A. M., Potvin, A., Wallot, H., and Merette, C. (2005). Shared and specific susceptibility loci for schizophrenia and bipolar disorder: A dense genome scan in Eastern Quebec families. *Mol. Psychiatry* **10,** 486–499.

McCauley, J. L., Olson, L. M., Delahanty, R., Amin, T., Nurmi, E. L., Organ, E. L., Jacobs, M. M., Folstein, S. E., Haines, J. L., and Sutcliffe, J. S. (2004). A linkage disequilibrium map of the 1-Mb 15q12 GABA(A) receptor subunit cluster and association to autism. *Am. J. Med. Genet. B. Neuropsychiatr. Genet.* **131,** 51–59.

McGrath, M., Kawachi, I., Ascherio, A., Colditz, G. A., Hunter, D. J., and De Vivo, I. (2004). Association between catechol-O-methyltransferase and phobic anxiety. *Am. J. Psychiatry* **161,** 1703–1705.

McKeane, D. P., Meyer, J., Dobrin, S. E., Melmed, K. M., Ekawardhani, S., Tracy, N. A., Lesch, K. P., and Stephan, D. A. (2005). No causative DLL4 mutations in periodic catatonia patients from 15q15 linked families. *Schizophr. Res.* **75,** 1–3.

Menold, M. M., Shao, Y., Wolpert, C. M., Donnelly, S. L., Raiford, K. L., Martin, E. R., Ravan, S. A., Abramson, R. K., Wright, H. H., Delong, G. R., Cuccaro, M. L., Pericak-Vance, M. A., and Gilbert, J. R. (2001). Association analysis of chromosome 15 gabaa receptor subunit genes in autistic disorder. *J. Neurogenet.* **15,** 245–259.

Meyer, J., Ortega, G., Schraut, K., Nurnberg, G., Ruschendorf, F., Saar, K., Mossner, R., Wienker, T. F., Reis, A., Stober, G., and Lesch, K. P. (2002). Exclusion of the neuronal nicotinic acetylcholine receptor alpha7 subunit gene as a candidate for catatonic schizophrenia in a large family supporting the chromosome 15q13–22 locus. *Mol. Psychiatry* **7,** 220–223.

Mirnics, K., Middleton, F. A., Marquez, A., Lewis, D. A., and Levitt, P. (2000). Molecular characterization of schizophrenia viewed by microarray analysis of gene expression in prefrontal cortex. *Neuron* **28,** 53–67.

Mizuno, Y., Puca, A. A., O'Brien, K. F., Beggs, A. H., and Kunkel, L. M. (2001a). Genomic organization and single-nucleotide polymorphism map of desmuslin, a novel intermediate filament protein on chromosome 15q26.3. *BMC. Genet.* **2,** 8 (7 pages).

Mizuno, Y., Thompson, T. G., Guyon, J. R., Lidov, H. G., Brosius, M., Imamura, M., Ozawa, E., Watkins, S. C., and Kunkel, L. M. (2001b). Desmuslin, an intermediate filament protein that interacts with alpha-dystrobrevin and desmin. *Proc. Natl. Acad. Sci. USA* **98,** 6156–6161.

Mohandas, T. K., Park, J. P., Spellman, R. A., Filiano, J. J., Mamourian, A. C., Hawk, A. B., Belloni, D. R., Noll, W. W., and Moeschler, J. B. (1999). Paternally derived de novo interstitial duplication of proximal 15q in a patient with developmental delay. *Am. J. Med. Genet.* **82,** 294–300.

Nurmi, E. L., Amin, T., Olson, L. M., Jacobs, M. M., McCauley, J. L., Lam, A. Y., Organ, E. L., Folstein, S. E., Haines, J. L., and Sutcliffe, J. S. (2003). Dense linkage disequilibrium mapping in the 15q11-q13 maternal expression domain yields evidence for association in autism. *Mol. Psychiatry* **8,** 624–634.

Nurmi, E. L., Bradford, Y., Chen, Y., Hall, J., Arnone, B., Gardiner, M. B., Hutcheson, H. B., Gilbert, J. R., Pericak-Vance, M. A., Copeland-Yates, S. A., Michaelis, R. C., Wassink, T. H., Santangelo, S. L., Sheffield, V. C., Piven, J., Folstein, S. E., Haines, J. L., and Sutcliffe, J. S. (2001). Linkage disequilibrium at the Angelman syndrome gene UBE3A in autism families. *Genomics* **77,** 105–113.

Palmatier, M. A., Pakstis, A. J., Speed, W., Paschou, P., Goldman, D., Odunsi, A., Okonofua, F., Kajuna, S., Karoma, N., Kungulilo, S., Grigorenko, E., Zhukova, O. V., Bonne-Tamir, B., Lu, R. B., Parnas, J., Kidd, J. R., De Mille, M. M., and Kidd, K. K. (2004). COMT haplotypes suggest P2 promoter region relevance for schizophrenia. *Mol. Psychiatry* **9,** 859–870.

Park, T. W., Yoon, K. S., Kim, J. H., Park, W. Y., Hirvonen, A., and Kang, D. (2002). Functional catechol-O-methyltransferase gene polymorphism and susceptibility to schizophrenia. *Eur. Neuropsychopharmacol* **12,** 299–303.

Paunio, T., Ekelund, J., Varilo, T., Parker, A., Hovatta, I., Turunen, J. A., Rinard, K., Foti, A., Terwilliger, J. D., Juvonen, H., Suvisaari, J., Arajarvi, R., Suokas, J., Partonen, T., Lonnqvist, J., Meyer, J., and Peltonen, L. (2001). Genome-wide scan in a nationwide study sample of schizophrenia families in Finland reveals susceptibility loci on chromosomes 2q and 5q. *Hum. Mol. Genet.* **10,** 3037–3048.

Persico, A. M., Militerni, R., Bravaccio, C., Schneider, C., Melmed, R., Conciatori, M., Damiani, V., Baldi, A., and Keller, F. (2000). Lack of association between serotonin transporter gene promoter variants and autistic disorder in two ethnically distinct samples. *Am. J. Med. Genet.* **96,** 123–127.

Petit, E., Herault, J., Martineau, J., Perrot, A., Barthelemy, C., Hameury, L., Sauvage, D., Lelord, G., and Muh, J. P. (1995). Association study with two markers of a human homeogene in infantile autism. *J. Med. Genet.* **32,** 269–274.

Philippe, A., Martinez, M., Guilloud-Bataille, M., Gillberg, C., Rastam, M., Sponheim, E., Coleman, M., Zappella, M., Aschauer, H., Van Maldergem, L., Penet, C., Feingold, J., Brice, A., and Leboyer, M. (1999). Genome-wide scan for autism susceptibility genes. Paris Autism Research International Sibpair Study. *Hum. Mol. Genet.* **8,** 805–812.

Potash, J. B., Zandi, P. P., Willour, V. L., Lan, T. H., Huo, Y., Avramopoulos, D., Shugart, Y. Y., Mac Kinnon, D. F., Simpson, S. G., McMahon, F. J., De Paulo, J. R., Jr., and McInnis, M. G.

(2003). Suggestive linkage to chromosomal regions 13q31 and 22q12 in families with psychotic bipolar disorder. *Am. J. Psychiatry* **160,** 680–686.

Purcell, A. E., Jeon, O. H., Zimmerman, A. W., Blue, M. E., and Pevsner, J. (2001). Postmortem brain abnormalities of the glutamate neurotransmitter system in autism. *Neurology* **57,** 1618–1628.

Raybould, R., Green, E. K., Mac Gregor, S., Gordon-Smith, K., Heron, J., Hyde, S., Caesar, S., Nikolov, I., Williams, N., Jones, L., O'Donovan, M. C., Owen, M. J., Jones, I., Kirov, G., and Craddock, N. (2005). Bipolar disorder and polymorphisms in the dysbindin gene (DTNBP1). *Biol. Psychiatry* **57,** 696–701.

Realmuto, G. M., and August, G. J. (1991). Catatonia in autistic disorder: A sign of comorbidity or variable expression? *J. Autism. Dev. Disord.* **21,** 517–528.

Reddy, K. S. (2005). Cytogenetic abnormalities and fragile-X syndrome in Autism Spectrum Disorder. *BMC. Med. Genet.* **6,** 3 (16 pages).

Repetto, G. M., White, L. M., Bader, P. J., Johnson, D., and Knoll, J. H. (1998). Interstitial duplications of chromosome region 15q11q13: Clinical and molecular characterization. *Am. J. Med. Genet.* **79,** 82–89.

Risch, N., Spiker, D., Lotspeich, L., Nouri, N., Hinds, D., Hallmayer, J., Kalaydjieva, L., McCague, P., Dimiceli, S., Pitts, T., Nguyen, L., Yang, J., Harper, C., Thorpe, D., Vermeer, S., Young, H., Hebert, J., Lin, A., Ferguson, J., Chiotti, C., Wiese-Slater, S., Rogers, T., Salmon, B., Nicholas, P., Petersen, P. B., Pingree, C., McMahon, W., Wong, D. L., Cavalli-Sforza, L. L., Kraemer, H. C., and Myers, R. M. (1999). A genomic screen of autism: Evidence for a multilocus etiology. *Am. J. Hum. Genet.* **65,** 493–507.

Rosa, A., Peralta, V., Cuesta, M. J., Zarzuela, A., Serrano, F., Martinez-Larrea, A., and Fananas, L. (2004). New evidence of association between COMT gene and prefrontal neurocognitive function in healthy individuals from sibling pairs discordant for psychosis. *Am. J. Psychiatry* **161,** 1110–1112.

Sazci, A., Ergul, E., Kucukali, I., Kilic, G., Kaya, G., and Kara, I. (2004). Catechol-O-methyltransferase gene Val108/158Met polymorphism, and susceptibility to schizophrenia: Association is more significant in women. *Brain Res. Mol. Brain Res.* **132,** 51–56.

Schosser, A., Fuchs, K., Leisch, F., Bailer, U., Meszaros, K., Lenzinger, E., Willinger, U., Strobl, R., Heiden, A., Gebhardt, C., Kasper, S., Sieghart, W., Hornik, K., and Aschauer, H. N. (2004). Possible linkage of schizophrenia and bipolar affective disorder to chromosome 3q29; a follow-up. *J. Psychiatr. Res.* **38,** 357–364.

Schroer, R. J., Phelan, M. C., Michaelis, R. C., Crawford, E. C., Skinner, S. A., Cuccaro, M., Simensen, R. J., Bishop, J., Skinner, C., Fender, D., and Stevenson, R. E. (1998). Autism and maternally derived aberrations of chromosome 15q. *Am. J. Med. Genet.* **76,** 327–336.

Schumacher, J., Jamra, R. A., Freudenberg, J., Becker, T., Ohlraun, S., Otte, A. C., Tullius, M., Kovalenko, S., Bogaert, A. V., Maier, W., Rietschel, M., Propping, P., Nothen, M. M., and Cichon, S. (2004). Examination of G72 and D-amino-acid oxidase as genetic risk factors for schizophrenia and bipolar affective disorder. *Mol. Psychiatry* **9,** 203–207.

Schwab, S. G., Knapp, M., Mondabon, S., Hallmayer, J., Borrmann-Hassenbach, M., Albus, M., Lerer, B., Rietschel, M., Trixler, M., Maier, W., and Wildenauer, D. B. (2003). Support for association of schizophrenia with genetic variation in the 6p22.3 gene, dysbindin, in sib-pair families with linkage and in an additional sample of triad families. *Am. J. Hum. Genet.* **72,** 185–190.

Shao, Y., Cuccaro, M. L., Hauser, E. R., Raiford, K. L., Menold, M. M., Wolpert, C. M., Ravan, S. A., Elston, L., Decena, K., Donnelly, S. L., Abramson, R. K., Wright, H. H., De Long, G. R., Gilbert, J. R., and Pericak-Vance, M. A. (2003). Fine mapping of autistic disorder to chromosome 15q11-q13 by use of phenotypic subtypes. *Am. J. Hum. Genet.* **72,** 539–548.

Shield, A. J., Thomae, B. A., Eckloff, B. W., Wieben, E. D., and Weinshilboum, R. M. (2004). Human catechol O-methyltransferase genetic variation: Gene resequencing and functional characterization of variant allozymes. *Mol. Psychiatry* **9,** 151–160.

Shifman, S., Bronstein, M., Sternfeld, M., Pisante, A., Weizman, A., Reznik, I., Spivak, B., Grisaru, N., Karp, L., Schiffer, R., Kotler, M., Strous, R. D., Swartz-Vanetik, M., Knobler, H. Y., Shinar, E., Yakir, B., Zak, N. B., and Darvasi, A. (2004). COMT: A common susceptibility gene in bipolar disorder and schizophrenia. *Am. J. Med. Genet. B. Neuropsychiatr. Genet.* **128,** 61–64.

Shifman, S., Bronstein, M., Sternfeld, M., Pisante-Shalom, A., Lev-Lehman, E., Weizman, A., Reznik, I., Spivak, B., Grisaru, N., Karp, L., Schiffer, R., Kotler, M., Strous, R. D., Swartz-Vanetik, M., Knobler, H. Y., Shinar, E., Beckmann, J. S., Yakir, B., Risch, N., Zak, N. B., and Darvasi, A. (2002). A highly significant association between a COMT haplotype and schizophrenia. *Am. J. Hum. Genet.* **71,** 1296–1302.

Smith, M., Filipek, P. A., Wu, C., Bocian, M., Hakim, S., Modahl, C., and Spence, M. A. (2000). Analysis of a 1-megabase deletion in 15q22-q23 in an autistic patient: Identification of candidate genes for autism and of homologous DNA segments in 15q22-q23 and 15q11-q13. *Am. J. Med. Genet.* **96,** 765–770.

Spence, S. J. (2004). The genetics of autism. *Semin. Pediatr. Neurol.* **11,** 196–204.

Stefanis, N. C., Van Os, J., Avramopoulos, D., Smyrnis, N., Evdokimidis, I., Hantoumi, I., and Stefanis, C. N. (2004). Variation in catechol-o-methyltransferase val158 met genotype associated with schizotypy but not cognition: A population study in 543 young men. *Biol. Psychiatry* **56,** 510–515.

Stefansson, H., Sigurdsson, E., Steinthorsdottir, V., Bjornsdottir, S., Sigmundsson, T., Ghosh, S., Brynjolfsson, J., Gunnarsdottir, S., Ivarsson, O., Chou, T. T., Hjaltason, O., Birgisdottir, B., Jonsson, H., Gudnadottir, V. G., Gudmundsdottir, E., Bjornsson, A., Ingvarsson, B., Ingason, A., Sigfusson, S., Hardardottir, H., Harvey, R. P., Lai, D., Zhou, M., Brunner, D., Mutel, V., Gonzalo, A., Lemke, G., Sainz, J., Johannesson, G., Andresson, T., Gudbjartsson, D., Manolescu, A., Frigge, M. L., Gurney, M. E., Kong, A., Gulcher, J. R., Petursson, H., and Stefansson, K. (2002). Neuregulin 1 and susceptibility to schizophrenia. *Am. J. Hum. Genet.* **71,** 877–892.

Stober, G., Franzek, E., Lesch, K. P., and Beckmann, H. (1995). Periodic catatonia: A schizophrenic subtype with major gene effect and anticipation. *Eur. Arch. Psychiatry Clin. Neurosci.* **245,** 135–141.

Stober, G., Meyer, J., Nanda, I., Wienker, T. F., Saar, K., Jatzke, S., Schmid, M., Lesch, K. P., and Beckmann, H. (2000a). hKCNN3 which maps to chromosome 1q21 is not the causative gene in periodic catatonia, a familial subtype of schizophrenia. *Eur. Arch. Psychiatry Clin. Neurosci.* **250,** 163–168.

Stober, G., Saar, K., Ruschendorf, F., Meyer, J., Nurnberg, G., Jatzke, S., Franzek, E., Reis, A., Lesch, K. P., Wienker, T. F., and Beckmann, H. (2000b). Splitting schizophrenia: Periodic catatonia-susceptibility locus on chromosome 15q15. *Am. J. Hum. Genet.* **67,** 1201–1207.

Stober, G., Seelow, D., Ruschendorf, F., Ekici, A., Beckmann, H., and Reis, A. (2002). Periodic catatonia: Confirmation of linkage to chromosome 15 and further evidence for genetic heterogeneity. *Hum. Genet.* **111,** 323–330.

Stone, W. S., Faraone, S. V., Su, J., Tarbox, S. I., Van Eerdewegh, P., and Tsuang, M. T. (2004). Evidence for linkage between regulatory enzymes in glycolysis and schizophrenia in a multiplex sample. *Am. J. Med. Genet. B. Neuropsychiatr. Genet.* **127,** 5–10.

Straub, R. E., Jiang, Y., Mac Lean, C. J., Ma, Y., Webb, B. T., Myakishev, M. V., Harris-Kerr, C., Wormley, B., Sadek, H., Kadambi, B., Cesare, A. J., Gibberman, A., Wang, X., O'Neill, F. A., Walsh, D., and Kendler, K. S. (2002). Genetic variation in the 6p22.3 gene DTNBP1, the human ortholog on the mouse dysbindin gene, is associated with schizophrenia. *American Journal of Human Genetics* **71,** 337–348.

Sugiura, N., Patel, R. G., and Corriveau, R. A. (2001). N-methyl-D-aspartate receptors regulate a group of transiently expressed genes in the developing brain. *J. Biol. Chem.* **276,** 14257–14263.

Sutcliffe, J. S., Han, M. K., Amin, T., Kesterson, R. A., and Nurmi, E. L. (2003). Partial duplication of the APBA2 gene in chromosome 15q13 corresponds to duplicon structures. *BMC. Genomics.* **4,** 15.

Turecki, G., Grof, P., Grof, E., D'Souza, V., Lebuis, L., Marineau, C., Cavazzoni, P., Duffy, A., Betard, C., Zvolsky, P., Robertson, C., Brewer, C., Hudson, T. J., Rouleau, G. A., and Alda, M. (2001). Mapping susceptibility genes for bipolar disorder: A pharmacogenetic approach based on excellent response to lithium. *Molecular Psychiatry* **6,** 570–578.

Turic, D., Langley, K., Mills, S., Stephens, M., Lawson, D., Govan, C., Williams, N., Van Den, Bree, M., Craddock, N., Kent, L., Owen, M., O'Donovan, M., and Thapar, A. (2004). Follow-up of genetic linkage findings on chromosome 16p13: Evidence of association of N-methyl-D aspartate glutamate receptor 2A gene polymorphism with ADHD. *Mol. Psychiatry* **9,** 169–173.

Van Den Bogaert, A., Schumacher, J., Schulze, T. G., Otte, A. C., Ohlraun, S., Kovalenko, S., Becker, T., Freudenberg, J., Jonsson, E. G., Mattila-Evenden, M., Sedvall, G. C., Czerski, P. M., Kapelski, P., Hauser, J., Maier, W., Rietschel, M., Propping, P., Nothen, M. M., and Cichon, S. (2003). The DTNBP1 (dysbindin) gene contributes to schizophrenia, depending on family history of the disease. *Am. J. Hum. Genet.* **73,** 1438–1443.

van den Oord, E. J., Sullivan, P. F., Jiang, Y., Walsh, D., O'Neill, F. A., Kendler, K. S., and Riley, B. P. (2003). Identification of a high-risk haplotype for the dystrobrevin binding protein 1 (DTNBP1) gene in the Irish study of high-density schizophrenia families. *Mol. Psychiatry* **8,** 499–510.

Van Greevenbroek, M. M., Vermeulen, V. M., and De Bruin, T. W. (2004). Identification of novel molecular candidates for fatty liver in the hyperlipidemic mouse model, HcB19. *J. Lipid Res.* **45,** 1148–1154.

Veenstra-Vander Weele, J., Gonen, D., Leventhal, B. L., and Cook, E. H., Jr. (1999). Mutation screening of the UBE3A/E6-AP gene in autistic disorder. *Mol. Psychiatry* **4,** 64–67.

Wang, S., Detera-Wadleigh, S. D., Coon, H., Sun, C. E., Goldin, L. R., Duffy, D. L., Byerley, W. F., Gershon, E. S., and Diehl, S. R. (1996). Evidence of linkage disequilibrium between schizophrenia and the SCa1 CAG repeat on chromosome 6p23. *Am. J. Hum. Genet.* **59,** 731–736.

Wang, X., He, G., Gu, N., Yang, J., Tang, J., Chen, Q., Liu, X., Shen, Y., Qian, X., Lin, W., Duan, Y., Feng, G., and He, L. (2004). Association of G72/G30 with schizophrenia in the Chinese population. *Biochem. Biophys. Res. Commun.* **319,** 1281–1286.

Wassink, T. H., Piven, J., and Patil, S. R. (2001). Chromosomal abnormalities in a clinic sample of individuals with autistic disorder. *Psychiatr. Genet.* **11,** 57–63.

Weiss, L. A., Abney, M., Cook, E. H., Jr., and Ober, C. (2005a). Sex-specific genetic architecture of whole blood serotonin levels. *Am. J. Hum. Genet.* **76,** 33–41.

Weiss, L. A., Abney, M., Parry, R., Scanu, A. M., Cook, E. H., Jr., and Ober, C. (2005b). Variation in ITGB3 has sex-specific associations with plasma lipoprotein(a) and whole blood serotonin levels in a population-based sample. *Hum. Genet.* **117,** 81–87.

Wilcox, M. A., Faraone, S. V., Su, J., Van Eerdewegh, P., and Tsuang, M. T. (2002). Genome scan of three quantitative traits in schizophrenia pedigrees. *Biol. Psychiatry* **52,** 847–854.

Wonodi, I., Stine, O. C., Mitchell, B. D., Buchanan, R. W., and Thaker, G. K. (2003). Association between Val108/158 Met polymorphism of the COMT gene and schizophrenia. *Am. J. Med. Genet. B. Neuropsychiatr. Genet.* **120,** 47–50.

Yirmiya, N., Pilowsky, T., Nemanov, L., Arbelle, S., Feinsilver, T., Fried, I., and Ebstein, R. P. (2001). Evidence for an association with the serotonin transporter promoter region polymorphism and autism. *Am. J. Med. Genet.* **105,** 381–386.

Ylisaukko-Oja, T., Nieminen-von Wendt, T., Kempas, E., Sarenius, S., Varilo, T., von Wendt, L., Peltonen, L., and Jarvela, I. (2004). Genome-wide scan for loci of Asperger syndrome. *Mol. Psychiatry* **9,** 161–168.

Zhong, N., Ye, L., Ju, W., Brown, W. T., Tsiouris, J., and Cohen, I. (1999). 5-HTTLPR variants not associated with autistic spectrum disorders. *Neurogenetics* **2,** 129–131.

Zinn-Justin, A., and Abel, L. (1999). Genome search for alcohol dependence using the weighted pairwise correlation linkage method: Interesting findings on chromosome 4. *Genet. Epidemiol* **17** (Suppl 1), S421–S426.

INDEX

CONTENTS OF RECENT VOLUMES

Volume 52

Volume 53

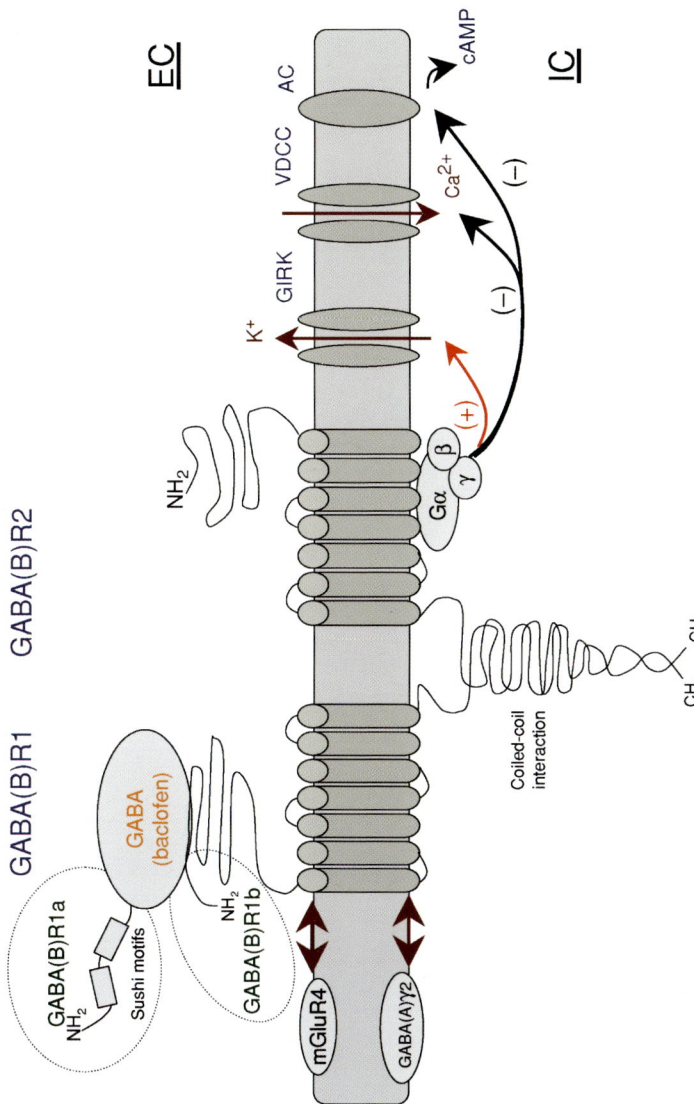

JELITAI AND MADARASZ, FIG. 1. The GABA_B receptor dimer is composed of GABA_BR1 and GABA_BR2 subunits. GABA_BR1 contains the ligand binding site, while the G-protein complex associates to GABA_BR2. Through G-protein-mediated coupling, the activated receptor can regulate inward rectifying K-channels (GIRK), voltage dependent Ca-channels (VDDC), and adenylate cyclase (AC) enzymes. The N-terminal (extracellular) domain of GABA_BR1a contain, Sushi-sequences, while GABA_BR1b does not. Several GABA_BR1 subunits can associate with different receptors, among them ae GABA_Aγ2 and mGluR4.

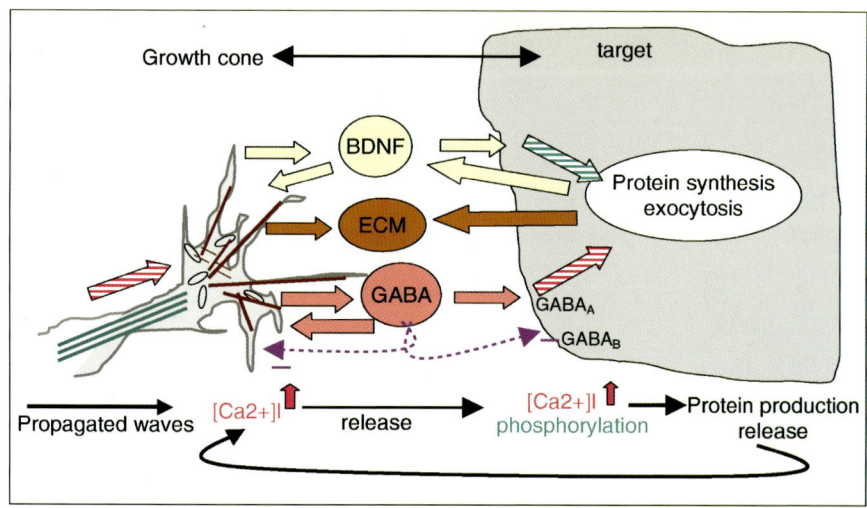

JELITAI AND MADARASZ, FIG. 2. Depolarization increases the intracellular Ca-level at the growth cone and elicits the release of growth factors (in most cases BDNF), extracellular matrix molecules (ECM), and GABA. BDNF through Trk receptors will stimulate local protein synthesis in both partners resulting in the production of further factors to be released. GABA also will depolarize both partners and trigger the release of various factors, among them BDNF, ECM molecules, and neurotransmitter substances, mainly GABA. The over-excitation by increasing concentrations of GABA can be prevented by the activation of GABA$_B$ receptors, which will reduce or even stop depolarization.

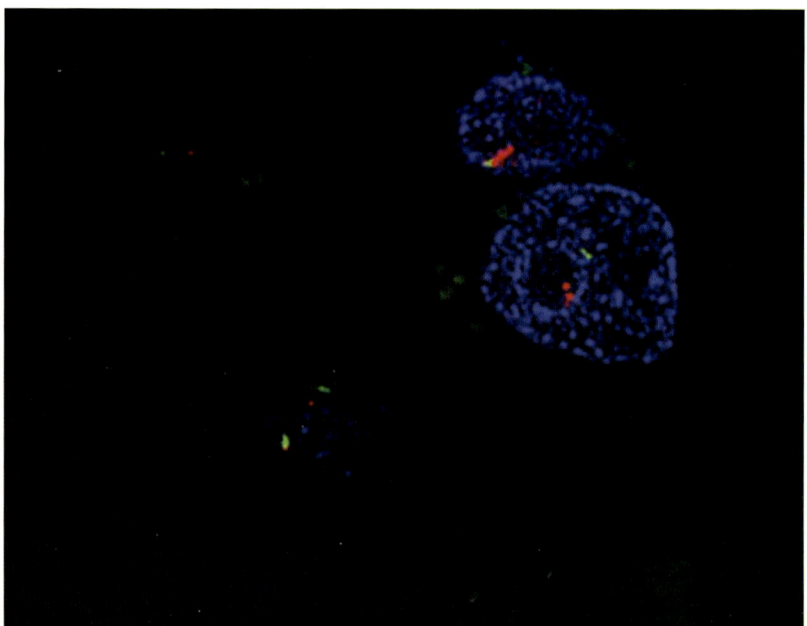

LASALLE *ET AL.*, FIG. 2. Homologous pairing of 15q11-13 loci within MeCP2 high expressing neurons in human brain. A normal human cerebral cortical section stained for MeCP2 (blue), chromosome 15 centromere (green), and 15q11-13 (*SNRPN*, red). Two nuclei with high levels of MeCP2 expression (top right) show homologous association of 15q11-13 domains around the nucleolus. In contrast, a neuronal nucleus with low MeCP2 expression (bottom left) shows no homologous pairing.

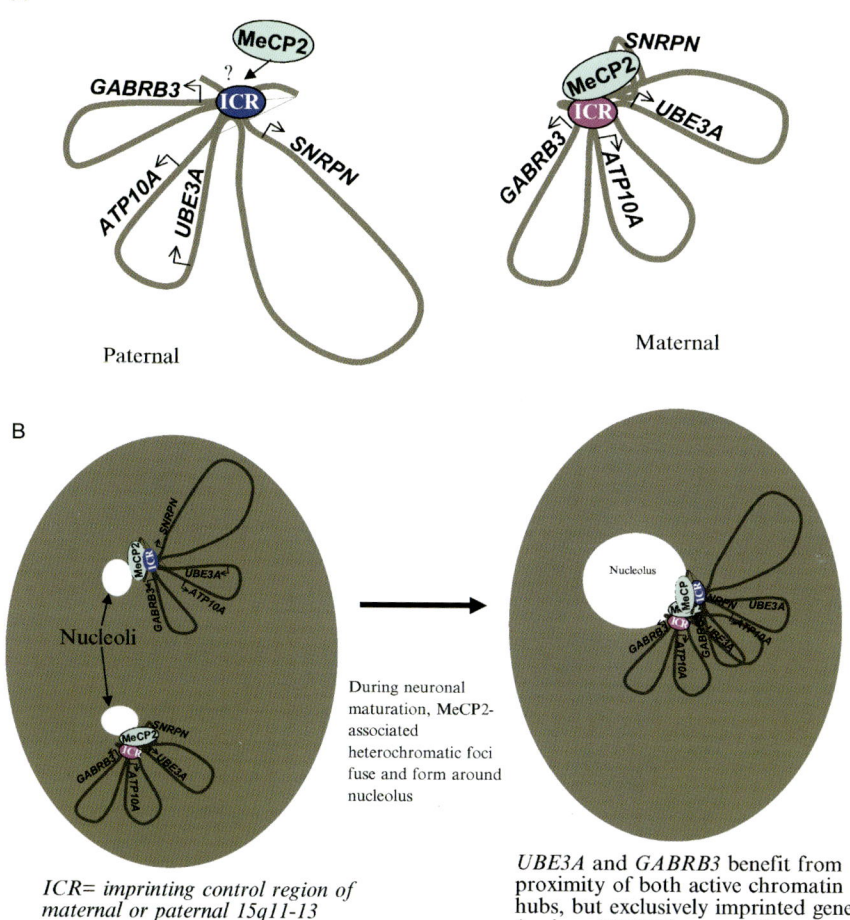

A

B

Nucleoli

During neuronal
maturation, MeCP2-
associated
heterochromatic foci
fuse and form around
nucleolus

*ICR= imprinting control region of
maternal or paternal 15q11-13*

UBE3A and *GABRB3* benefit from
proximity of both active chromatin
hubs, but exclusively imprinted genes
(such as *SNRPN*) are unaffected

LaSalle *et al.*, Fig. 3. A higher order model for MeCP2 in regulating gene expression within 15q11-13. (A) Chromatin loop structures are proposed that would change the gene expression patterns on the paternal and maternal 15q11-13 regions. Differential MeCP2 binding sites would be determined by differentially methylated regions (DMRs). One known binding site for MeCP2 is present in the maternal methylated allele of the *SNRPN* promoter, but presumably MeCP2 could also bind to sites of paternal-specific methylation. The chromatin loop structures could position the imprinting control regions (ICR) for either the paternal (PWS-ICR, blue) or maternal (AS-ICR, pink) ICR near the promoters of *Gabrb3* and *Ube3a*, thus explaining how MeCP2 deficiency can alter expression of these genes without necessarily binding at the promoters. (B) In the context of a neuronal nucleus undergoing maturational differentiation, the 15q11-13 chromatin loop structures are brought into close proximity by the formation of a large nucleolus (as in Fig. 2). MeCP2 colocalizes with the peri-nucleolar heterochromatin, bringing the associated chromatin loops together. *UBE3A* and *GABRB3* expression would benefit by being brought into the active chromatin hub of the opposite parental chromosome.

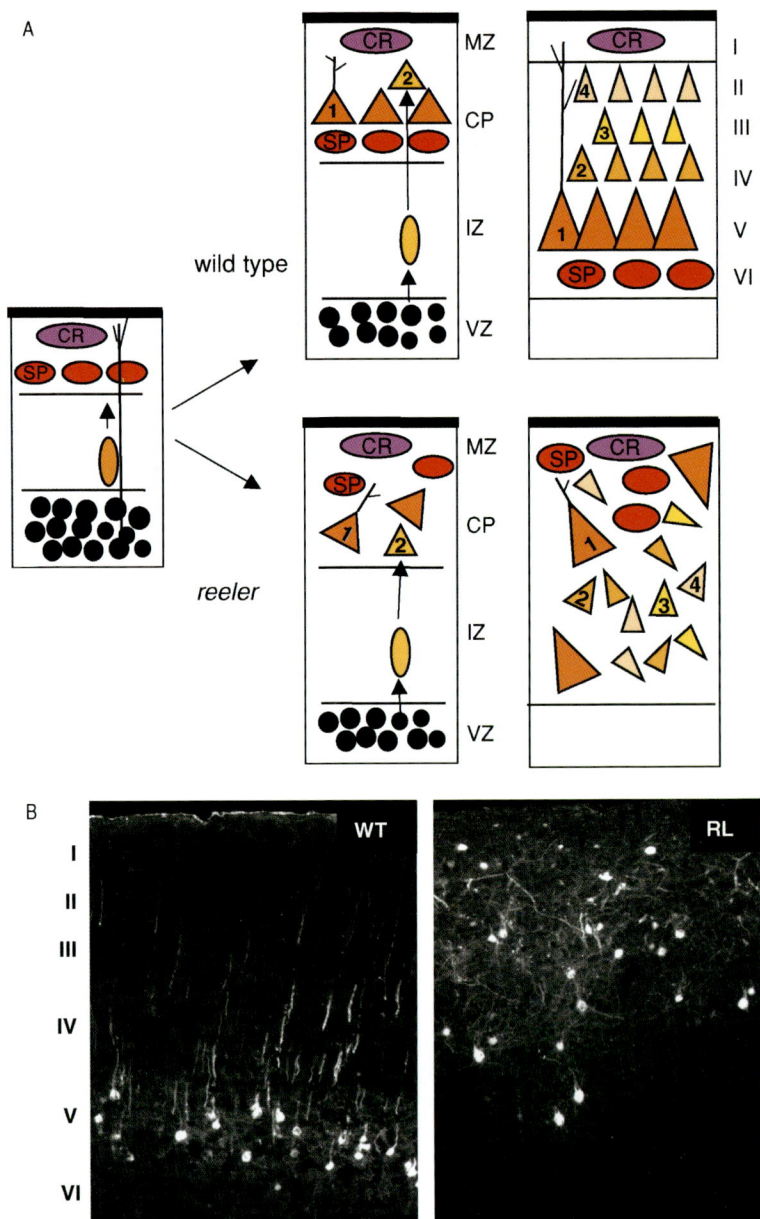

A

wild type

reeler

B

D'ARCANGELO, FIG. 1. A. Diagram of neocortical development in wild type and *reeler* mice. Principal neurons are born in the ventricular zone (VZ) and migrate radially toward the marginal zone (MZ). In wild type cortex, the first-born principal neurons (labeled 1) split the preplate composed of Reelin-expressing Cajal-Retzius (CR) and subplate cells (SP). Each wave of later-born neurons (labeled sequentially with arabic numbers) bypasses their predecessors and form cortical layers (labeled with roman numbers). In reeler, principal neurons do not split the preplate and fail to form layers. CP = cortical plate. IZ = intermediate zone. B. Fluorescence images of layer V neurons labeled with YFP in the adult wild type (WT) and *reeler* (RL) neocortex.

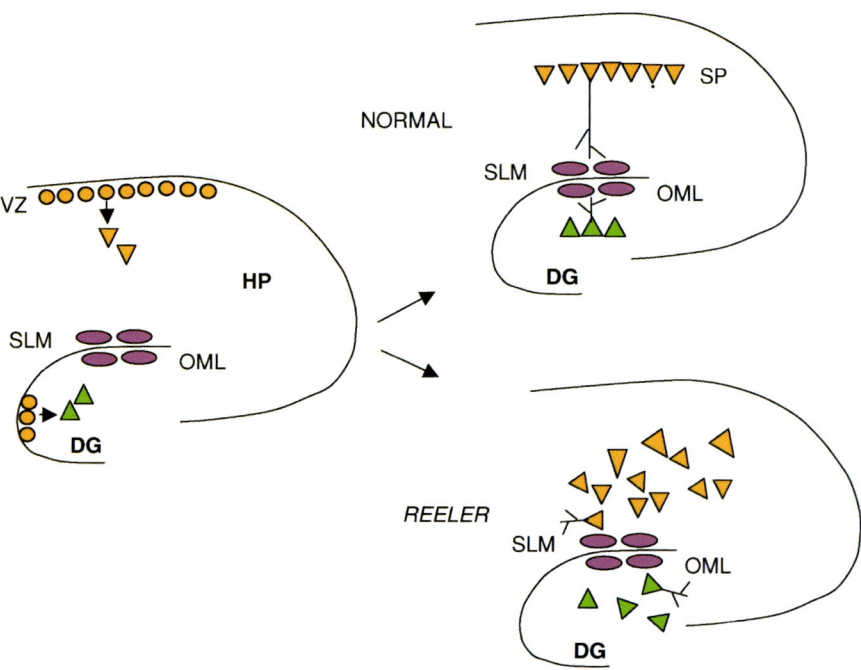

D'Arcangelo, Fig. 2. Diagram of hippocampal development in normal and *reeler* mice. In the hippocampus proper (HP) pyramidal neurons are born in the ventricular zone (VZ) and migrate radially towards the stratum lacunosum moleculare (SLM), which contains Reelin-expressing Cajal-Retzius cells (large ovals). In normal mice they form a distinct pyramidal layer and project apical dendrites to the SLM. In the dentate gyrus (DG), granule neurons migrate inwardly from the outer molecular layer (OML) to form a compact layer. The OML also contains Reelin-expressing Cajal-Retzius cells. In *reeler* both pyramidal and granule cell layers fail to form.

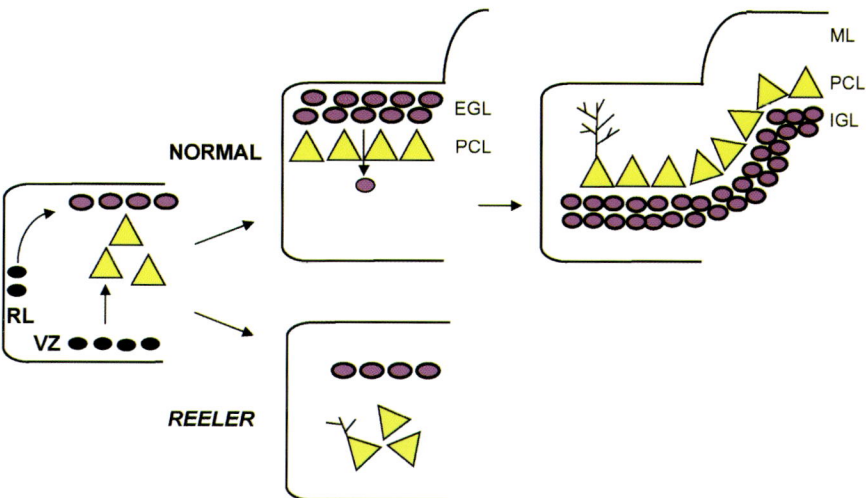

D'Arcangelo, Fig. 3. Diagram of cerebellar development in normal and *reeler* mice. In the cerebellum Reelin-expressing granule cells are born in the rhombic lip (RL) and migrate tangentially to from the external granular layer (EGL). Purkinje cells (triangles) are born in ventricular zone (VZ) and migrate radially toward the EGL. In normal mice Purkinje cells form a layer (PCL). Granule cells proliferate and migrate across the PCL to form the internal granule layer (IGL). The cerebellum contains many folia. In *reeler*, Purkinje cells fail to form a layer and granule cells do not proliferate extensively, leading to reduced foliation.

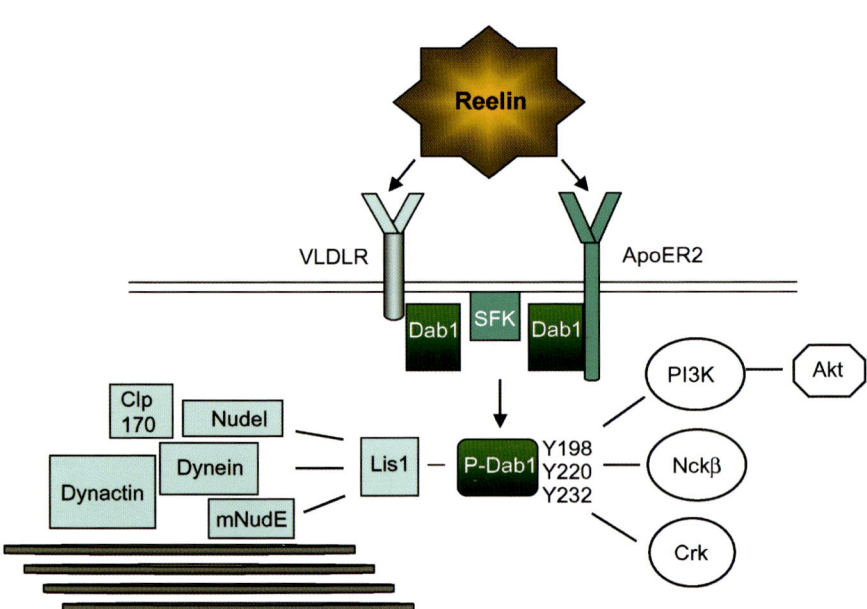

D'ARCANGELO, FIG. 5. The Reelin signaling pathway is activated by ligand-induced clustering of the Reelin receptors VLDLR and ApoER2, leading to SFK activation and Dab1 phosphorylation on the indicated tyrosine residues. Phospo-Dab1 interacts with a variety of intracellular proteins including Lis1, which ties the Reelin signal with the dynein complex and microtubule dynamics. Long bars indicate microtubules.